WATER SUPPLY SYSTEMS SECURITY

WATER SUPPLY SYSTEMS SECURITY

Larry W. Mays, Ph.D., P.E., P.H.
Editor-in-Chief

Department of Civil and Environmental Engineering
Arizona State University
Tempe, Arizona

McGRAW-HILL
New York Chicago San Francisco Lisbon London Madrid
Mexico City Milan New Delhi San Juan Seoul
Singapore Sydney Toronto

The McGraw·Hill Companies

Library of Congress Cataloging-in-Publication Data

Water supply systems security/Larry W. Mays, editor-in-chief.
    p.   cm.
  Includes index.
  ISBN 0-07-142531-4
  1. Waterworks—Security measures. 2. Water-supply—Security measures.
3. Terrorism—Prevention. I. Mays, Larry W.

TD485.W377   2004
363.6'1'0289—dc22   2003070169

Copyright © 2004 by The McGraw-Hill Companies, Inc. All rights reserved. Printed in the United States of America. Except as permitted under the United States Copyright Act of 1976, no part of this publication may be reproduced or distributed in any form or by any means, or stored in a data base or retrieval system, without the prior written permission of the publisher.

2 3 4 5 6 7 8 9 0   DOC/DOC   0 1 0 9 8 7 6 5 4

ISBN 0-07-142531-4

The sponsoring editor for this book was Larry S. Hager and the production supervisor was Sherri Souffrance. It was set in Times Roman by International Typesetting and Composition. The art director for the cover was Margaret Webster-Shapiro.

Printed and bound by RR Donnelley.

McGraw-Hill books are available at special quantity discounts to use as premiums and sales promotions, or for use in corporate training programs. For more information, please write to the Director of Special Sales, Professional Publishing, McGraw-Hill, Two Penn Plaza, New York, NY 10121-2298. Or contact your local bookstore.

        This book was printed on recycled, acid-free paper containing a minimum of 50% recycled, de-inked fiber.

Information contained in this work has been obtained by The McGraw-Hill Companies, Inc. ("McGraw-Hill") from sources believed to be reliable. However, neither McGraw-Hill nor its authors guarantee the accuracy or completeness of any information published herein, and neither McGraw-Hill nor its authors shall be responsible for any errors, omissions, or damages arising out of use of this information. This work is published with the understanding that McGraw-Hill and its authors are supplying information but are not attempting to render engineering or other professional services. If such services are required, the assistance of an appropriate professional should be sought.

CONTENTS

Contributors ix
Preface xi
Acknowledgments xiii

Chapter 1. Water Supply Security: An Introduction   1.1

1.1 History   *1.1*
1.2 The Water Supply System: A Brief Description   *1.2*
1.3 Why Water Supply Systems?   *1.5*
1.4 The Threats   *1.6*
1.5 Prior to September 11, 2001   *1.7*
1.6 Response to September 11, 2001   *1.8*
     References   *1.11*

Chapter 2. Microbiological Contaminants and Threats of Concern   2.1

2.1 Introduction   *2.1*
2.2 Etiological Groups   *2.2*
     References   *2.12*

Chapter 3. Vulnerability Assessment, Emergency Response Planning: Summary of What's Available   3.1

3.1 Introduction   *3.1*
3.2 Vulnerability Assessment   *3.1*
3.3 Emergency Response Planning: Response, Recovery, and Remediation Guidance   *3.6*
3.4 Information Sharing (www.waterisac.org)   *3.7*
     Appendix 3a: What are Some Points to Consider in a Vulnerability Assessment?   *3.11*
     Appendix 3b: Security Vulnerability Self-Assessment for Small Water Systems   *3.15*
     Appendix 3c: Water Utility Response, Recovery, and Remediation Guidelines   *3.28*
     Appendix 3d: Water System Emergency Response Plan Outline   *3.36*
     References   *3.42*

Chapter 4. Drinking Water Distribution Systems: An Overview   4.1

4.1 Introduction   *4.1*
4.2 Modeling Contaminant Transport   *4.5*
4.3 Simulation of Residential Water Demands   *4.21*
4.4 Conducting a Tracer Study   *4.35*
4.5 Summary and Conclusions   *4.45*
     References   *4.46*

Chapter 15. Hydraulic and Water Quality Modeling for Contamination Response   15.1

15.1 Operational Needs for Responding to Contamination Events   15.1
15.2 Role of Hydraulic and Water Quality Models   15.3
15.3 Conventional Modeling   15.4
15.4 Definitions   15.6
15.5 The Connectivity Matrix   15.7
15.6 Enumeration of Operating Modes   15.8
15.7 Delineation of Maximum Spread Potential   15.8
References   15.9

Chapter 16. Optimal Monitoring Stations Allocations for Water Distribution Systems Security   16.1

16.1 Introduction   16.1
16.2 Scientific Background   16.2
16.3 optiMQ   16.4
16.4 Concluding Remarks   16.12
Notation   16.14
References   16.14

Chapter 17. Contingency Planning for Emergency Water Supply in Non-Conventional Times   17.1

17.1 Introduction   17.1
17.2 The Birth of the Atomic Age and World War II—Hiroshima and Nagasaki   17.2
17.3 Three Mile Island and Chernobyl   17.3
17.4 Recent Trends and Future Risks   17.5
17.5 Possible Contamination of Water Resources in Nuclear Contingencies   17.6
17.6 Contingency Planning for Water Supply Systems in Nonconventional Times   17.6
17.7 Conclusions   17.11
References   17.11

Chapter 18. Development of the Next Generation Microbiological and Chemical Detection Capabilities for Water Supplies   18.1

18.1 Introduction   18.1
18.2 Microbiological Detection Capabilities   18.2
18.3 Concluding Remarks for Microbiological Systems of the Future   18.12
18.4 Chemical Detection Systems of the Future   18.12
References   18.17

Index   I.1

CONTRIBUTORS

Morteza Abbaszadegan *Department of Civil and Environmental Engineering, Arizona State University, Tempe, Arizona.* (CHAP. 2)

Nabil R. Adam *The Center for Information Management, Integration, and Connectivity. Director, The Meadowlands Environment Research Institute, Rutgers University, Newark, New Jersey.* (CHAP. 8)

Absar Alum *Department of Civil and Environmental Engineering, Arizona State University, Tempe, Arizona.* (CHAP. 2)

Jeff Aramini *Division of Enteric, Foodborne and Waterborne Diseases, Health Canada, Guelph, Ontario.* (CHAP. 10)

Vijaylakeshmi Atluri *Center for Information Management, Integration, and Connectivity, Rutgers University, Newark, New Jersey.* (CHAP. 8)

Ron L. Booth *CH2M-HILL, Atlanta, Georgia.* (CHAP. 12)

Francois J.-C. Bouchart *Department of Civil Engineering, University of Calgary, Calgary, Alberta, Canada.* (CHAP. 15)

Andy Bowman *SiteSecure, Inc., Sanford, Florida.* (CHAP. 12)

Cynthia Bruckner-Lea *Pacific N.W. National Lab., Richland, Washington.* (CHAP. 18)

Hendrik Bruins *Department Man in the Desert, Jacob Blaustein Institute for Desert Research, Ben-Gurion University of the Negev, Sede Boker Campus, Israel.* (CHAP. 17)

Steven Buchberger *Department of Civil and Environmental Engineering, University of Cincinnati, Cincinnati, Ohio.* (CHAPS. 4, 9)

Robert M. Clark *Environmental Engineering and Public Health Consultant, Cincinnati, Ohio.* (CHAPS. 4, 5, 9, 8, 10, 14)

Laura Cummings *Passaic Valley Water Commission, Little Falls, New Jersey.* (CHAP. 8)

Rolf Deininger *University of Michigan, Ann Arbor, Michigan.* (CHAP. 11)

James W. Davidson *Department of Civil Engineering, University of Calgary, Calgary, Alberta, Canada.* (CHAP. 15)

Malcolm S. Field *National Center for Environmental Assessment, Office of Research and Development, U. S. Environmental Protection Agency, Pennsylvania, Washington, D.C.* (CHAP. 6)

Forrest Gist *CH2M Hill, Portland, Oregon.* (CHAP. 12)

Walter M. Grayman *W. M. Grayman Consulting Engineer, Cincinnati, Ohio.* (CHAPS. 4, 10, 11)

Richard W. Gullick *American Water, Voorhees, New Jersey.* (CHAP. 11)

Milton Halem *Center for Information Management, Integration, and Connectivity, Rutgers University, Newark, New Jersey.* (CHAP. 8)

Benjamin Harding *Hydrosphere Resource Consultants, Boulder, Colorado.* (CHAP. 10)

David Hartman *Greater Cincinnati Water Works, Cincinnati, Ohio.* (CHAP. 4)

Roy C. Haught *National Risk Management Research Laboratory, U.S. Environmental Protection Agency, Cincinnati, Ohio.* (CHAP. 14)

Scott Harvey *Pacific N.W. National Lab., Richland, Washington.* (CHAP. 18)

Eva Ibrahim *Water Quality and Operations, American Water, Voorhees, New Jersey.* (CHAP. 8)

Yeongho Lee *Greater Cincinnati Water Works, Cincinnati, Ohio.* (CHAP. 4)

Richard M. Males *RMM Technical Services, Cincinnati, Ohio.* (CHAP. 11)

Morris Maslia *Agency for Toxic Substances and Disease Registry, Atlanta, Georgia.* (CHAP. 10)

Larry W. Mays *Department of Civil and Environmental Engineering, Arizona State University, Tempe, Arizona.* (CHAPS. 1, 3, 7, 13)

Avi Ostfeld *Faculty of Civil and Environmental Engineering, Technion—Israel Institute of Technology, Haifa, Israel.* (CHAP. 16)

Sukru Ozger *Department of Civil and Environmental Engineering, Arizona State University, Tempe, Arizona.* (CHAP 13)

Srinivas Panguluri *Shaw Environmental, Inc., Cincinnati, Ohio.* (CHAPS. 5, 14)

William R. Phillips *CH2M-Hill, Gainesville, Florida.* (CHAP. 5)

James R. Ringold *Protection Group, Inc., Dunedin, Florida.* (CHAP. 12)

Richard Skaggs *Pacific N.W. National Lab., Richland, Washington.* (CHAP. 18)

Tim Straub *Pacific N.W. National Lab., Richland, Washington.* (CHAP. 18)

Pen C. Tao *North District Water Supply Commission, Wanague, New Jersey.* (CHAP. 8)

Yeou-Koung Tung *Department of Civil Engineering, Hong Kong University of Science and Technology, Clearwater Bay, Kowloon, Hong Kong, China.* (CHAP. 7)

Eric F. Vowinkel *U. S. Geological Survey, Edison, New Jersey.* (CHAP. 8)

Bob Wright *Pacific N.W. National Lab., Richland, Washington.* (CHAP. 18)

PREFACE

In the mid-1980s I chaired a task committee for the American Society of Civil Engineers (ASCE), which culminated in the book, *Reliability Analysis of Water Distribution Systems*, published by ASCE in 1989. As the editor and a major contributor to this book I was extremely excited about this effort, which emphasized various methodologies for the reliability assessment of water distribution systems and their components. The risk/reliability methodologies we presented were state-of-the-art methodologies, many of which, are used in the nuclear power industry, the chemical processing industry, and the electrical power industry. Unfortunately to this date these methodologies are still not used by the water utility industry. We defined failures in the framework of mechanical failures and the resulting hydraulic performance failures. At that time we never fathomed the idea of terrorist threats to our water infrastructure. Even eleven years later in 2000, as editor-in-chief of the *Water Distribution Systems Handbook*, published by McGraw-Hill, the topic of security from terrorist threats never crossed my mind as important for that handbook. The two succeeding McGraw-Hill books on urban water supply that I developed as editor-in-chief, *Urban Water Supply Handbook* in 2002 and *Urban Water Supply Management Tools* in 2003, each have a chapter on the topic of security, but nothing to the extent of *Water Supply Systems Security*.

The events of September 11, 2001 have forced a new focus on our water utility infrastructure in the United States. Prior to these events the consideration of terrorist threats to drinking water supply systems was minimal. Now our approach to the management of these systems has changed significantly with the passage of two very important acts in 2002, that focus on the water infrastructure and give directives for the future. The Public Health, Security, and Bioterrorism Preparedness and Response Act (PL 107-188) (June 2002), requires community water systems serving populations greater than 3300 to conduct vulnerability assessments and submit these assessments to the U. S. Environmental Protection Agency. The Homeland Security Act (PL 107-296), November 25, 2002, directed the greatest reorganization of the federal government in decades by consolidating a host of security-related agencies into a single cabinet-level department.

This book was developed as a response to the critical needs of engineers and utility managers to have a resource on the security of water supply systems. The future of water supply security analysis will hopefully include some of those risk/reliability methodologies that exist for our future design methodologies, vulnerability assessments, and emergency response planning.

Water Supply Systems Security presents state-of-the-art methodologies for the various aspects of water supply infrastructure. The topics have been chosen to represent what we feel are the most important for engineers, utility managers, and others working on the security of water supply infrastructure. The wide set of topics range from the various types of threats to various levels of vulnerability assessments, risk assessments, reliability assessments, to surveillance hardware, the next generation of contaminant detection devices, to many other topics. All chapter authors are leading experts and were chosen because of their proven knowledge in the specific area of their contribution.

Each book that I have worked on has been a part of my lifelong journey in water resources, and *Water Supply Systems Security* certainly is no exception. I have gained more from my experiences in developing books than can ever be measured in words. This book certainly will never have the impact upon history that the treatises of Vitruvius and Frontinus of the Roman Empire have had, but hopefully it will provide a little insight for the present generation of water distribution systems engineers and managers. Vitruvius and Frontinus of the Roman Empire were also faced with security issues of their water supply systems, but of a far different nature than what we face today.

This book has been a part of my personal journey in life to learn as much as possible about water and to use this knowledge in my teaching, research, and writing. I hope that you will be able to use this book in your own journey of learning about water. As I continue my efforts in the study and photography of ancient water structures, especially those built by the Romans, I am placing many of my photographs on my web site (www.public.asu.edu/~lwmays/). These may be of interest to some of you readers.

Larry W. Mays
Scottsdale, Arizona

ACKNOWLEDGMENTS

I must first acknowledge the authors who made *Water Supply Systems Security* possible. It has been a sincere privilege to work with such an excellent group of dedicated people. I would especially like to acknowledge that Dr. Robert Clark was involved in developing six of the chapters in this book and was lead author on four of these chapters. His distinguished career with the U.S. EPA along with his extensive research and many publications on water distribution systems, without a doubt, has made him one of the leading experts in the world on the security of water distribution systems. Dr. Clark's contributions to this book are very much appreciated by myself and I am certain the entire profession will appreciate his efforts for this book. I would also like to give a special thanks to Dr. Walter Grayman who was involved in the development of three chapters and was lead author on three of these. All the authors are experienced professionals who are among the leading experts in their fields and I would like to express my sincere appreciation to them for their efforts. Any references to material in this handbook should be attributed to the respective chapter authors.

This has been the eighth book that I have developed with Larry Hager of McGraw-Hill. I sincerely appreciate his efforts and he is always a joy to talk to, as he is one of the few willing to listen to my adventures hiking, fly fishing, and snow skiing in Colorado. I would also like to acknowledge Arizona State University, especially for the time afforded me to pursue this book.

During the 28 years of my academic career as a professor I have received help and encouragement from so many people that it is not possible to name all of them. These people represent a wide range of universities, research institutions, government agencies, and professions. They also represent all of my former students and particularly my former Ph.D. students. To all of you I express deepest thanks.

I must also acknowledge my three children, Travis, Elyssa, and Tyler for their love and willingness to enjoy the water-related sports with me, especially at our second home in Pagosa Springs, Colorado. They may never read this book, but they do represent the most valuable part of my life.

My hope in life has always been to get the most out of what the world has to offer and the most out of the talent with which God has blessed me. I've realized that once I shed the burden of the real or imagined values and expectations of others, it has been much easier to tell where my real passions lie. Certainly one of my real passions in the journey of life has been in the development of the books such as this one. I am having a great journey in life and hope the same for all of you.

I dedicate this book to humanity and human welfare.

Larry W. Mays
Scottsdale, Arizona

WATER SUPPLY SYSTEMS SECURITY

CHAPTER 1
WATER SUPPLY SECURITY: AN INTRODUCTION

Larry W. Mays
Department of Civil and Environmental Engineering
Arizona State University
Tempe, Arizona

1.1 HISTORY

A long history, in fact since the dawn of history, of threats to drinking water systems during conflicts has plagued humans. Water has been a strategic objective in armed conflicts throughout history. Gleick (1994, 1998, 2000) has developed a water conflict chronology in which he categorizes the conflicts as the following: control of water resources, military tool, political tool, terrorism, military target, and development disputes. Terrorism is defined as, "water resources, or water systems, are either targets or tools of violence or coercion by nonstate actors." There are many historical conflicts that caused flooding by diversion or eliminated water supplies by building dams or other structures, whereas in the following only a few examples of some of the water conflicts that included water supply systems are summarized.

During the time of King Hezekiah—the period of the First Temple (the latter part of the eighth century B.C.), Jerusalem was under military threat from Assyria (2 Kings 20:20; Isaiah 22:11; 2 Chronicles 32:2-4,30). The Gihon spring, located just outside the city walls, was the main water source for the ancient city of Jerusalem (Bruins, 2002), requiring strategic planning on King Hezekiah's part. He had a water tunnel (533 m) dug to channel the water underground into the city, with the outlet at a reservoir known as the Pool of Siloam. Two crews of miners dug through solid limestone from both ends of the tunnel, meeting at the same spot (Bruins, 2002).

During the second Samnite War, ca. 310 B.C., the Romans realized the need for alternate water sources for Rome due to the insufficient and unreliable local supplies. The Roman Senate procured and distributed water rights from estates surrounding Rome in order to develop the supply and security needed for Rome.

In 1503, Leonardo da Vinci and Machiavelli planned to divert the Arno River away from Pisa during the conflict between Pisa and Florence.

During the Civil War (1863) in the United States, General U.S. Grant cut levees in the battle against the Confederates during the campaign against Vicksburg.

In 1948, during the first Arab-Israeli War, Arab forces cut off the West Jerusalem water supply.

In 1982, Israel cut off the water supply of Beirut during the siege.

In 1990 in South Africa, the pro-apartheid council cut off water to the Wesselton township of 50,000 blacks following protests over miserable sanitation and living conditions.

During the 1991 Gulf War, the Allied coalition targeted Baghdad's water supply and sanitation system. Discussions were held about using the Attaturk Dam to cut off flows to the Euphrates to Iraq. Also during the Gulf War, Iraq destroyed much of Kuwait's desalination capacity during retreat. In 1993, Saddam Hussein reportedly poisoned and drained water supplies of southern Shiite Muslims.

In Kosovo (1999), water supplies/wells were contaminated by Serbs who disposed of the bodies of Kosovar Albanians in local wells. Serbian engineers shut down the water system in Pristina prior to occupation by NATO. Also during that same year in Yugoslavia, NATO targeted utilities and shut down water supplies in Belgrade.

Gleick (2000) developed a water conflict chronology (1503 to 2000) that can be found at the following site: http://www.worldwater.org/conflict.htm.

1.2 THE WATER SUPPLY SYSTEM: A BRIEF DESCRIPTION

The events of September 11, 2001 have significantly changed the approach to management of water utilities. Previously, the consideration of the terrorist threat to the U.S. drinking water supply was minimal. Now we have an intensified approach to the consideration of terrorist threat. The objective of this chapter is to provide an introduction to the very costly process of developing water security measures for U.S. water utilities.

Figure 1.1 illustrates a typical municipal water utility showing the water distribution system as a part of this overall water utility. In some locations, where excellent quality groundwater is available, water treatment may include only chlorination. Other handbooks on the subject of water supply/water distribution systems include Mays (1989, 2000, 2002, 2003).

Water distribution systems are composed of three major components: pumping stations, distribution storage, and distribution piping. These components may be further divided into subcomponents, which in turn can be divided into sub-subcomponents. For example, the pumping station component consists of structural, electrical, piping, and pumping unit subcomponents. The pumping unit can be divided further into sub-subcomponents: pump, driver, controls, power transmission. The exact definition of components, subcomponents, and sub-subcomponents depends on the level of detail of the required analysis and, to a somewhat greater extent, the level of detail of available data. In fact, the concept of component-subcomponent-sub-subcomponent merely defines a hierarchy of building blocks used to construct the water distribution system. Figure 1.2 shows the hierarchical relationship of system, components, subcomponents, and sub-subcomponents for a water distribution system.

A water distribution system operates as a system of independent components. The hydraulics of each component is relatively straightforward; however, these components depend directly upon each other and as a result effect the performance of one another. The purpose of design and analysis is to determine how the systems perform hydraulically under various demands and operation conditions. These analyses are used for the following situations:

- Design of a new distribution system
- Modification and expansion of an existing system

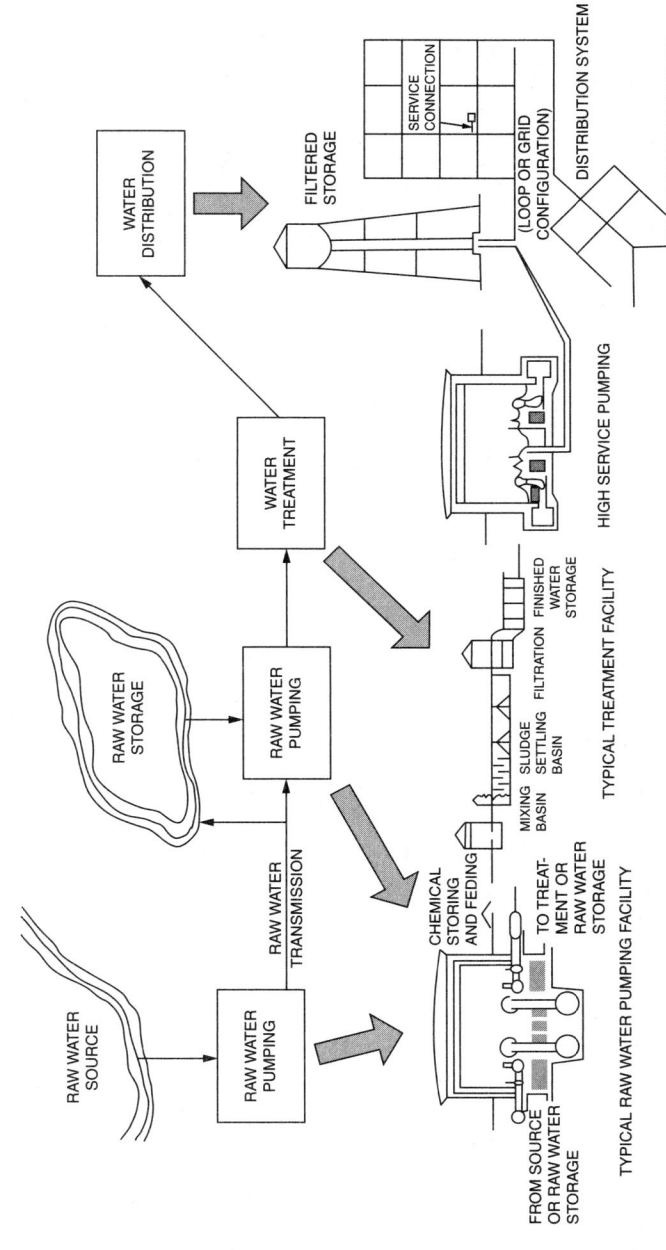

FIGURE 1.1 A typical water distribution system.

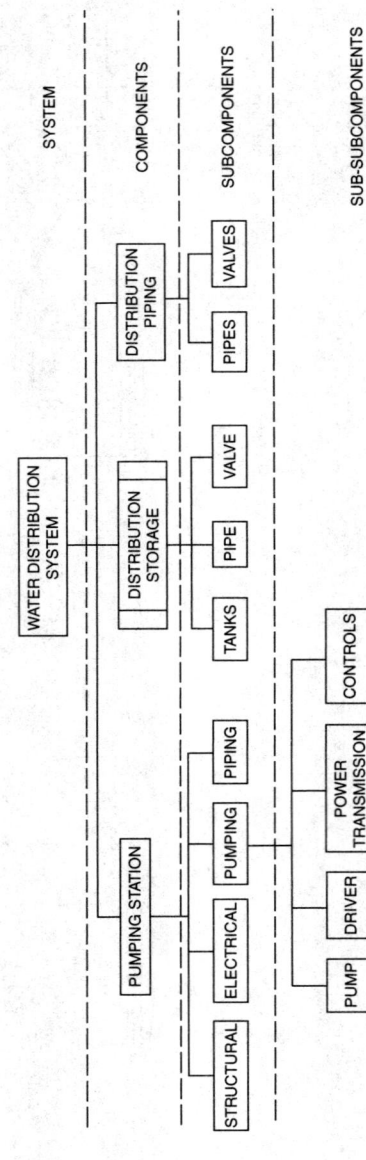

FIGURE 1.2 Hierarchy of building blocks in water distribution systems.

- Analysis of system malfunction such as pipe breaks, leakage, valve failure, pump failure
- Evaluation of system reliability
- Preparation for maintenance
- System performance and operation optimization

1.3 WHY WATER SUPPLY SYSTEMS?

A distribution system of pipelines, pipes, pumps, storage tanks, and the appurtenances such as various types of valves, meters, etc. offers the greatest opportunity for terrorism because it is extensive, relatively unprotected and accessible, and often isolated. The physical destruction of a water distribution system's assets or the disruption of water supply could be more likely than contamination. A likely avenue for such an act of terrorism is a bomb, carried by car or truck, similar to the recent events listed in Table 1.1. Truck or car bombs require less preparation, skill, or manpower than complex attacks such as those of September 11, 2001. However, we must consider all the possible threats no matter how remote we may think that they could be.

TABLE 1.1 Recent Terrorist Attacks Against American Targets Using Car-Bomb Technologies

Date	Target/location	Delivery/ material	TNT equivalent (lb)	Reference
Apr. 1983	U.S. Embassy Beirut, Lebanon	Van	2,000	www.beirut-memorial.org
Oct. 1983	U.S. Marine Barracks Beirut, Lebanon	Truck, TNT with gas enhancement	12,000	www.usmc.mil
Feb. 1993	World Trade Center New York, U.S.A.	Van, urea nitrate and hydrogen gas	2,000	www.interpol.int
Apr. 1995	Murrah Federal Bldg Oklahoma City, U.S.A.	Truck, ammonium nitrate fuel oil	5,000	U.S. Senate documents
June 1996	Khobar Towers Dhahran, Saudi Arabia	Tanker truck, plastic explosive	20,000	www.fbi.gov
Aug. 1998	U.S. Embassy Nairobi, Kenya	Truck, TNT, possibly Semtex	1,000	news reports, U.S. Senate documents
Aug. 1998	U.S. Embassy Dar es Salaam, Tanzania	Truck	1,000	U.S. Senate documents
Oct. 2000	Destroyer USS Cole Aden Harbor, Yemen	Small watercraft, possibly C-4	440	www.al-bab.com news.bbc.co.uk

Source: Peplow et al. (2003).

1.4 THE THREATS

The probability of a terrorist threat to drinking water is probably very low; however, the consequences could be extremely severe for exposed populations. Various types of threats may have higher probabilities than others. The following four major types of threats are discussed further throughout this book. The term *weaponized* when referring to chemical and biological agents means that it can be produced and disseminated in large enough quantities to cause the desired effect (Hickman, 1999).

1.4.1 Cyber Threats

- Physical disruption of a supervisory control and data acquisition (SCADA) network
- Attacks on central control system to create simultaneous failures
- Electronic attacks using worms/viruses
- Network flooding
- Jamming
- Disguise data to neutralize chlorine or add no disinfectant allowing addition of microbes

1.4.2 Physical Threats

- Physical destruction of system's assets or disruption of water supply could be more likely than contamination. A single terrorist or a small group of terrorists could easily cripple an entire city by destroying the right equipment.
- Loss of water pressure compromises firefighting capabilities and could lead to possible bacterial buildup in the system.
- Potential for creating a water hammer effect by opening and closing major control valves and turning pumps on and off too quickly that could result in simultaneous main breaks.

1.4.3 Chemical Threats

Table 1.2 lists some chemicals that are effective in drinking water. The list includes both chemical warfare agents and industrial chemical poisons. There are five types of chemical warfare (CW) agents: nerve agents, blister agents, choking agents, blood agents, and hallucinogens. The list includes some of the chemical warfare agents and some of the industrial chemical poisons along with their acute concentrations.

1.4.4 Biological Threats

Several pathogens and biotoxins (see Chap. 2) exist that have been weaponized, are potentially resistant to disinfection by chlorination, and are stable for relatively long periods in water (Burrows and Renner, 1998, 1999). The pathogens include, *Clostridium perfringens*, plague and others, and biotoxins that include botulinum, aflatoxin, ricin, and others. Even though water provides dilution potential, a neutrally buoyant particle of any size could be used to disperse pathogens into drinking water systems. Other more sophisticated systems such as microcapsules also could be used to disperse pathogens in drinking water systems.

TABLE 1.2 Summary of Chemicals Effective in Drinking Water

Chemical agents ((mg/L) unless otherwise noted)	Acute concentration* (0.5 L)	Recommended guidelines[†]	
		5 L/day	15 L/day
Chemical warfare agents			
Hydrogen cyanide	25	6.0	2.0
Tabun (GA, µg/L)	50	70.0	22.5
Sarin (GB, µg/L)	50	13.8	4.6
Soman (GD, µg/L)	50	6.0	2.0
VX (µg/L)	50	7.5	2.5
Lewisite (arsenic fraction)	100–130	80.0	27.0
Sulfur mustard (µg/L)		140.0	47.0
3-Quinucli dinyl benzilate (BZ, µg/L)		7.0	2.3
Lysergic acid diethylamide (LSD)	0.050		
Industrial chemical poisons			
Cyanides	25	6.0	2.0
Arsenic	100–130	80.0	27.0
Fluoride	3000		
Cadmium	15		
Mercury	75–300		
Dieldrin	5000		
Sodium fluoroacetate[‡]		Not provided	
Parathion[‡]		Not provided	

*Major John Garland, *Water Vulnerability Assessments*, (Armstrong Laboratory, AL-TR-1991-0049), April 8–9, 1991. The author assumes acute effects (death or debilitation) after consumption of 0.5 L.

[†]National Research Council, Committee on Toxicology, *Guidelines for Chemical Warfare Agents in Military Field Drinking Water*, 1995, 10. Listed doses are *safe*.

[‡]W. Dickinson Burrows, J. A. Valcik, and Alan Seitzinger, "Natural and Terrorist Threats to Drinking Water Systems," presented at the American Defense Preparedness Association 23rd Environmental Symposium and Exhibition, 7–10 April 1997, New Orleans, LA, 2. The authors consider the organophospate nerve agent VX, the two hallucinogens BZ and LSD, sodium cyanide, fluoroacetate and parathion as potential threat agents. They do not provide acute concentrations or lethal doses.

Hydrogen cyanide (blood agent), the nerve agents Tabun, Sarin, Soman, and VX, the blistering agents Lewisite and sulfur mustard, and the hallucinogen BZ are potential drinking water poisons. Garland focuses on LSD (a hallucinogen), nerve agents (VX is listed as most toxic), arsenic (Lewisite) and cyanide (hydrogen cyanide). Burrows, et al., list BZ, LSD, and VX. These agents, however, are not the only chemicals a saboteur might use in drinking water.

Source: As presented in Hickman (1999).

Because of dilution effects, the effectiveness of a bioattack would be enhanced by introduction of the bioweapon near the tap.

Water storage and distribution systems can facilitate the delivery of an effective dose of toxicant to a potentially very large population. These systems also can facilitate a lower-level of chronic dose (for chemicals) with longer-term effects and lower-detection thresholds (Foran and Brosnan, 2000).

1.5 PRIOR TO SEPTEMBER 11, 2001

Prior to September 11, 2001 the literature contained numerous articles concerning the threat of terrorist attacks to our water supply infrastructure. A few of these included: Burrows and Renner (1998, 1999), DeNileon (2001), Dickey (2000), Foran and Brosnan (2000),

Grayman et al. (2001), Haimes et al. (1998), Hickman (1999), and many others. The topic was receiving a little attention but basically the water utility industry was not implementing mitigation measures to such threats.

The following news article summarizes this:

> Washington, MSNBC, Jan 14. 2002—The vulnerability of the nation's water supply isn't in the headlines, it's in the details of the country's 54,065 public and private water systems. For years, experts have warned about the need to upgrade, repair and thoroughly assess the risk of terrorists targeting the nation's water supply and distribution channels. Yet most of those warnings have been ignored, under-funded or relegated to the back burner as policy-makers addressed "more important" projects. (By Brock N. Meeks, Washington, D.C. correspondent, MSNBC).

Before the events of September 11, 2001, there was a growing concern, by some, with the potential for terrorist use of biological weapons (bioweapons) to cause civilian harm (Lederberg, 1997; Simon, 1997; Burrows and Renner, 1998; Ableson, 1999; Waeckerle, 2000; Foran and Brosnan, 2000; and many others). These assumptions were focused around two assumptions (Foran and Brosnan, 2000): that a terrorist is most likely to effectively disperse bioweapons through air (Simon, 1997), and that we must be prepared to address terrorists use of bioweapons through treatment of affected individuals, with emphasis on strengthening the response of the health-care community (Simon, 1997; Waeckerle, 2000; Macintyre et al., 2000). For the most part, concern was not focused on the use of bioweapons in drinking systems (Ableson, 1999; Burrows and Renner, 1999), and much less attention was given to preattack detection than to postattack treatment (Foran and Brosnan, 2000).

Conferences such as the one *Early Warning Monitoring to Detect Hazardous Events in Water Supplies* (held May 1999 in Reston, Virginia) concluded that terrorist use of bioweapons poses a significant threat to drinking water. Other experts have agreed that introducing a toxin into a raw water reservoir would have little impact considering the dilution effect that several millions of gallons of water would have on a biohazard. However, the effectiveness of an attack could be enhanced by introducing the bioweapon near the tap, such as in the distribution system after postdisinfection (Foran and Brosnan, 2000).

The President's Commission on Critical Infrastructure Protection (PCCIP) was established by President Clinton in 1996. The PCCIP determined that the water infrastructure is highly vulnerable to a range of potential attacks and convened a public-private partnership called the Water Sector Critical Infrastructure Advisory Group. According to the PCCIP (1997), three attributes (which are obvious) are crucial to water supply users:

- There must be adequate quantities of water on demand.
- It must be delivered at sufficient pressure.
- It must be safe to use.

The first two are influenced by physical damage and the third attribute (water quality) is susceptible to physical damage as well as the introduction of microorganisms, toxins, chemicals, or radioactive materials. Actions (terrorist activities) that affect any one of these three attributes can be debilitating for the water supply system.

1.6 RESPONSE TO SEPTEMBER 11, 2001

Within a very short time after September 11, 2001 we began to see a concerted effort at all levels of government to begin addressing issues related to the threat of terrorist activities to U.S. water supply. Articles began appearing including: Bailey (2001), Blomgren (2002),

Copeland and Cody (2002), Haas (2002), and many others. We saw a number of Acts passed such as the *Security and Bioterrorism Preparedness and Response Act* and the *Homeland Security Act* that addressed the U.S. water supply. These acts resulted in agencies such as the U.S. Environmental Protection Agency (EPA) developing new protocols to address their new responsibilities under these acts.

1.6.1 Public Health, Security and Bioterrorism Preparedness and Response Act ("Bioterrorism Act") (PL 107-188), June, 2002

This act requires every community water system that serves a population of more than 3,300 persons to

- Conduct a vulnerability assessment,
- Certify and submit a copy of the assessment to the EPA Administer,
- Prepare or revise an emergency response plan that incorporates the results of the vulnerability assessment, and
- Certify to the EPA Administer, within 6 months of completing the vulnerability assessment, that the system has completed or updated their emergency response plan.

Table 1.3 lists the key provisions of the security-related amendments.

TABLE 1.3 Security-Related Amendments to Bioterrorism Act*

1. Requires community water systems serving populations more than 3300 to conduct vulnerability assessments and submit them to U.S. EPA.
2. Requires specific elements to be included in a vulnerability assessment.
3. Requires each system that completes a vulnerability assessment to revise an emergency response plan and coordinate (to the extent possible) with local emergency planning committees.
4. Identifies specific completion dates for both vulnerability assessments and emergency response plans.
5. U.S. EPA is to develop security protocols as may be necessary to protect the copies of vulnerability assessments in its possession.
6. U.S. EPA is to provide guidance to community water systems serving populations of 3300 or less on how to conduct vulnerability assessments, prepare emergency response plans, and address threats.
7. U.S. EPA is to provide baseline information to community water systems regarding types of probable terrorist or other intentional threats.
8. U.S. EPA is to review current and future methods to prevent, detect, and respond to the intentional introduction of chemical, biological, or radiological contaminants into community water systems and their respective source waters.
9. U.S. EPA is to review methods and means by which terrorists or other individuals or groups could disrupt the supply of safe drinking water.
10. Authorizes funds to support these activities.

*In June 2002, the President signed PL 107-108, the Public Health, Security, and Bioterrorism Preparedness and Response Act (Bioterrorism Act) that includes provisions to help safeguard the nation's public drinking water systems against terrorist and other intentional acts. Key provisions of the new security-related amendments are summarized in this Table.

1.6.2 Homeland Security Act (PL 107-296), November 25, 2002

This act directs the greatest reorganization of the federal government in decades by consolidating a host of security-related agencies into a single cabinet-level department to be headed by Tom Ridge, head of the White House Office of Homeland Security. It creates four major directorates to be led by White House-appointed undersecretaries. These are the Directorates of Information Analysis and Infrastructure Protection (IAIP)—most directly affects USEPA and the drinking water community; Science of Technology; Border and Transportation Security; and Emergency Preparedness and Response.

The law grants IAIP access to all pertinent information, including infrastructure vulnerabilities, and directs all federal agencies to *promptly provide* IAIP with all information they have on terrorism threats and infrastructure vulnerabilities. IAIP is responsible for overseeing transferred functions of the NIPC, the CIAO, the Energy Department's National Infrastructure Simulation and Analysis Center and Energy Assurance Office, and the General Services Administration's Federal Computer Incident Response Center. IAIP will administer the Homeland Security Advisory System, which is the government's voice for public advisories about homeland threats as well as specific warnings and counterterrorism advice to state and local governments, the private sector and the public.

1.6.3 USEPA's Protocol

On October 2, 2002 the U.S. EPA announced their Strategic Plan for Homeland Security (www.epa.gov/epahome/headline_100202.htm). The goals of the plan are separated into four distinct mission areas: critical infrastructure protection; preparedness, response, and recovery; communication and information; and protection of EPA personnel and infrastructure. EPA's strategic plan lays out goals, tactics, and results for each of these areas.

The U.S. EPA has developed a compilation of water infrastructure security website links and tools located at www.epa.gov/safewater/security/index.html. Table 1.4 lists the U.S. EPA's strategic objectives to address drinking water system and wastewater utility security needs to meet the requirements of the Bioterrorism Act for public drinking water security.

TABLE 1.4 U.S. EPA Objectives to Ensure Safe Drinking Water*

1. Providing tools and guidance to drinking water systems and wastewater utilities.
2. Providing training and technical assistance including "Train-the-Trainer" programs.
3. Providing financial assistance to undertake vulnerability assessments and emergency response plans as funds are made available.
4. Build and maintain reliable communication processes.
5. Build and maintain reliable information systems.
6. Improve knowledge of potential threats, methods to detect attacks, and effectiveness of security enhancements in the water sector.
7. Improve networking among groups involved in security-related matters—water, emergency response, laboratory, environmental, intelligence, and law enforcement communities.

*U.S. EPA has developed several strategic objectives to address drinking water system and wastewater utility security needs and also meet requirements set forth in the Bioterrorism Act for public drinking water security. These strategic objectives are as summarized in this table.

REFERENCES

Ableson, P. H., "Biological Warfare," *Science* 286: 1677, 1999.

Bailey, K. C., *The Biological and Toxin Weapons Threat to the United States,* National Institute for Public Policy, Fairfax, VA, October 2001.

Blomgren, P., "Utility Managers Need to Protect Water Systems from Cyberterrorism," *U.S. News,* 19: 10, October 2002.

Bruins, H. J., "Israel: Urban Water Infrastructure in the Desert," in L. W. Mays (ed.), *Urban Water Supply Handbook,* McGraw-Hill, New York, 2002.

Burns, N. L., C. A. Cooper, D. A. Dobbins, J. C. Edwards, and L. K. Lampe, "Security Analysis and Response for Water Utilities," in L. W. Mays (ed.), *Urban Water Supply Handbook,* McGraw-Hill, New York, 2002.

Burrows, W. D., and S. E. Renner, "Biological Warfare Agents as Potable Water Threats," U.S. Army Combined Arms Support Command, Fort Lee, VA, 1998.

Burrows, W. D. and S. E. Renner, "Biological Warfare Agents as Threats to Potable Water," *Environmental Health Perspectives.* 107(12): 975–984, December 1999.

Cheng, S.-T., B. C. Yen, and W. H. Tang, "Stochastic Risk Modeling of Dam Overtopping," in B. C. Yen and Y.-K. Tung, (eds.), *Reliability and Uncertainty Analyses in Hydraulic Design,* American Society of Civil Engineers, New York, pp. 123–132, 1993.

Clark, R. M., and R. A. Deininger, "Protecting the Nations Critical Infrastructure: The Vulnerability of U.S. Water Supply Systems," in L. W. Mays (ed.), *Urban Water Supply Handbook,* McGraw-Hill, New York, 2002.

Copeland, C., and B. Cody, "Terrorism and Security Issues Facing the Water Infrastructure Sector," Order Code RS21026, CRS Report for Congress, *Congressional Research Service,* The Library of Congress, Washington, DC, June 18, 2002.

DeNileon, G. P., "The Who, Why, and How of Counterterrorism Issues," *J. Am. Water Works Assoc.,* 93(5): 78–85, May 2001.

Dickey, M. E., "Biocruise: A Contemporary Threat," Counterproliferation Paper No. 7, Future Warfare Series No. 7 available at: www.au.af.mil/au/awc/awcgate/cpc-pubs/dickey.htm, USAF Counterproliferation Center, Air War College, Air University, Maxwell Air Force Base, Alabama, September 2000.

Foran, J. A., and T. M. Brosnan, "Early Warning Systems for Hazardous Biological Agents in Potable Water," *Environ. Health Perspect.,* 108(10): 993–996, October 2000.

Gleick, P. H., "Water, War, and Peace in the Middle East," *Environment,* vol. 36, no. 3, Heldref Publishers, Washington, DC, p. 6, 1994.

Gleick, P. H., "Water and Conflict," in: P. H. Gleick (ed.), *The World's Water 1998–1999,* Island Press, Washington, DC, pp. 105–135, 1998.

Gleick, P. H., "Water Conflict Chronology," available at: http://www.worldwater.org/conflict.htm, 2000.

Grayman, W. M., R. A. Deininger, and R. M. Males, "Design of Early Warning and Predictive Sourcewater Monitoring Systems," *AWWA Research Foundation and AWWA,* 2001.

Haas, C. N., "The Role of Risk Analysis in Understanding Bioterrorism," *Risk Anal.,* 22(2): 671–677, 2002.

Haimes, Y. Y., et al., "Reducing Vulnerability of Water Supply Systems to Attack," *J. Infrastruc. Syst., ASCE* 4(4): December 1998.

Hickman, D. C., "A Chemical and Biological Warfare Threat: USAF Water Systems at Risk," Counterproliferation Paper No. 3, Future Warfare Series No. 3, available at: www.au.af.mil/au/awc/awcgate/cpc-pubs/hickman.htm, USAF Counterproliferation Center, Air War College, Air University, Maxwell Air Force Base, Alabama, September 1999.

Lederberg, J., "Infectious Disease and Biological Weapons: Prophylaxis and Mitigation," *JAMA* 278: 435–438, 1997.

Macintyre, A. J., G. W. Christopher, E. Eitzen, R. Gum, S. Weir, C. DeAtley, K. Tonat, and J. A. Barbera, "Weapons of Mass Destruction Events with Contaminated Casualties," *JAMA* 283(2): 242–249, 2000.

Mays, L. W. (ed.), *Reliability Analysis of Water Distribution Systems*, American Society of Civil Engineers, New York, 1989.

Mays, L. W. (ed.), *Water Distribution Systems Handbook*, McGraw-Hill, New York, 2000.

Mays, L. W. (ed.), *Urban Water Supply Handbook*, McGraw-Hill, New York, 2002.

Mays, L. W. (ed.), *Urban Water Supply Management Tools*, McGraw-Hill, New York, 2003.

Simon, J. D., "Biological Terrorism: Preparing to Meet the Threat," *JAMA* 278: 428–430, 1997.

President's Commission on Critical Infrastructure Protection, Appendix A, Sector Summary Reports, *Critical Foundations: Protecting America's Infrastructure*: A-45, available at: http://www.ciao.gov/PCCIP/PCCIP_Report.pdf.

U.S. Army Medical Research Institute of Infectious Disease, *USAMRID's Medical Management of Biological Causalities Handbook*, available at: www.usamriid.army.mil/education/bluebook.html, 2001.

U.S. EPA, Guidance for Water Utility Response, Recovery, and Remediation Actions for Man-Made and/or Technological Emergencies, available at: http://www.epa.gov/safewater/security/er-guidance.pdf.

U.S. EPA, Guidance for Water Utility Response, Recovery & Remediation Actions for Man-Made and/or Technological Emergencies, EPA 810-R-02-001, Office of Water (4601), available at: www.epa.gov/safewater April 2002.

U.S. EPA, "Water Security Strategy for Systems Serving Populations Less than 100,000/15 MGD or Less," July 9, 2002.

U.S. EPA, "Instructions to Assist Community Water Systems in Complying with the Public Health Security and Bioterrorism Preparedness and Response Act of 2002," EPA 810-R-02-001, Office of Water, available at: www.epa.gov/safewater/security, January 2003.

U.S. EPA, "Vulnerability Assessment Fact Sheet 12-19," EPA 816-F-02-025, Office of Water, available at: www.epa.gov/safewater/security/va fact sheet 12-19.pdf, also at www.epa.gov/ogwdw/index.html, November 2002.

U.S. EPA, available at: http://www.epa.gov/swercepp/cntr-ter.html.

Waeckerle, J. F., "Domestic Preparedness for Events Involving Weapons of Mass Destruction," *JAMA* 283(2): 252–254, 2000.

CHAPTER 2
MICROBIOLOGICAL CONTAMINANTS AND THREATS OF CONCERN

Morteza Abbaszadegan and Absar Alum
Department of Civil and Environmental Engineering
Arizona State University
Tempe, Arizona

2.1 INTRODUCTION

Waterborne pathogens had been a threat to human societies since time immemorial but during the early part of the twentieth century, evolution of drinking-water treatment processes and wastewater collection and discharge systems and has led to a remarkable decrease in the level of threat posed by waterborne infectious diseases. While the modern concepts of resource protection and drinking water treatment and control have virtually eradicated waterborne diseases from developed countries (except for sporadic cases), the municipal water acquisition, processing, and distribution systems have emerged as vulnerable points in the emerging national security scenario.

Water-related microbial pathogens of public significance could be categorized into two broad groups; water-based pathogens and waterborne pathogens. Water-based pathogens spend part of their life cycle in water and need a vector to reach and infect their host. Some of the best-known examples of water-based pathogens are the West Nile virus and malarial parasite, which use mosquitoes as their vector. Since such microorganisms are not transmitted solely through water, therefore, they are not potential agents of bioterrorism. The waterborne pathogens are those transmitted through ingestion of contaminated water and generally are orally transmitted fecal microorganisms. In such cases water acts as a passive carrier of the infectious agents. Some of waterborne pathogens, which can potentially cause problems in drinking-water production and distribution, include newly recognized pathogens from fecal sources like *Campylobacter jejuni*, pathogenic *Escherichia coli*, *Yersinia enterocolitica*, new enteric viruses like rotavirus, calicivirus, astrovirus, and the parasites *Giardia lamblia*, *Cryptosporidium parvum*, and Microsporidia (Table 2.1). Besides these, some species of environmental bacteria that are able to grow in water distribution systems have recently been recognized as human pathogens. The examples of such bacteria include *Legionella* spp., *Aeromonas* spp., *Mycobacterium* spp., and *Pseudomonas aeruginosa*.

TABLE 2.1 Pathogens of Public Health Concern

Pathogen	Disease	Incubation period	Immunization
Salmonella enteritidis	Salmonellosis	8–10 h up to 48 h	None
Salmonella enteritidis var. *paratyphi A*	Paratyphoid fever	1–10 days	Heat killed vaccine
Salmonella typhi	Typhoid fever	1–2 weeks some times 3 weeks	Heat killed vaccine
Salmonella choleraesuis	Salmonella Septicemia	1–10 days	Heat killed vaccine
Shigella dysenteriae *S. flexneri* *S. boydii* *S. sonnei*	Shigellosis Bacillary dysentery	1–4 days Not more than 7 days	None
Vibrio cholerae	Cholera	Few hours to 2–3 days	Killed vaccine
Vibrio parahaemolyticus	Gastroenteritis	8–48 h	—
Yersinia enterocolitica	Gastroenteritis	8–48 h	—
Clostridium perfringens	Gastroenteritis	8–48 h	—
Bacillus cereus	Gastroenteritis	8–48 h	—
Escherichia coli enteropathogenic	Endemic diarrhea	2–4 days Max. 3 weeks	Heat killed vaccine
Adenovirus	Gastroenteritis	1–3 days	
Coxsackei virus	Gastroenteritis	3–5 days	None
Hepatitis A virus	Hepatitis	15–50 days	Passive
Norovirus	Gastroenteritis	48–72 h	Passive
Polio virus	Poliomyelitis	Usually 1–2 weeks, Range 3 days to 4–5 weeks	Salk vaccine (killed); Sabin vaccine (live)
Cryptosporidium	Cryptosporidiosis	~7 days	
Entamoeba histolytica	Amoebiasis	3–4 weeks	
Naegleria fowleri	Primary amebic meningoencephalitis	3–14 days	

2.2 ETIOLOGICAL GROUPS

In the literature a wide variety of bacterial, viral, and parasitic pathogens have been reported with characteristics that make them potential agents of concern in municipal water systems. The factors include latency (period between pathogen excretion and acquisition of infectious power), survival, disinfection kinetics, infectious dose, virulence, and disease episode.

2.2.1 Bacterial Pathogens

Bacteria are unicellular organisms which reproduce by binary fission. They vary rather widely in size with diameter ranging from 0.5 to 1.5 µm and length ranging from 1 to 6 µm. The three basic shapes displayed by bacterial cells include spherical cells called "*cocci*" (*coccus* a Greek word for berry), rod shaped cells called "*bacilli*" (*bacillus* a Greek word for staff or rod), and spiral shaped cells called "*spirilla*" (*spirillum* a Greek word for coil).

Infectious gastroenteritis may be caused by a variety of bacterial pathogens. The most common clinical symptom of bacterial gastroenteritis includes cramps, abdominal distress, diarrhea, nausea, and vomiting with occasional chills, headache, and mild fever. Bacterial pathogens are relatively less resistant in the environment, and chlorine has been shown to be an effective disinfectant to inactivate pathogenic bacteria in drinking water.

Salmonella. The genus *Salmonella* is a very diverse group within the family Enterobacteriaceae. Salmonellae are 0.8 to 1.5 μm wide and 2 to 5 μm long. *Salmonella typhi* (the cause of typhoid fever), *Salmonella choleraesuis* (occasionally causes septicemia in humans) and *Salmonella enteritidis* (common cause of diarrheal infections) are some important human pathogenic species of the genus *Salmonella*.

Gastroenteritis. *Salmonella enteritidis*, the most common cause of water- and food-borne gastroenteritis, contains more than 2000 serotypes. Some of the most common examples of *S. enteritidis* serotypes are *paratyphi* and *typhimuruim*. The symptoms of *Salmonella* infection start to appear approximately 6 to 24 h after ingestion of contaminated water and can last for up to a week. Early symptoms include nausea and vomiting followed by (few hours) abdominal cramps and diarrhea. Fever may also be experienced in occasional cases. The severity of abdominal cramps and diarrhea varies greatly among subjects. The infected persons continue to shed bacteria for up to 3 months (in few cases up to 1 year) even after subsidence of typical symptoms. Such chronic carriers act as sources of secondary infections in communities.

Typhoid Fever. Enteric fever is caused by *Salmonella typhi*, which is specifically a human pathogen causing typhoid fever. *Salmonella enteritidis* var. *paratyphi* A, B, and C are also known to cause a relatively milder type of typhoid fever. The incubation period for *Salmonella enteritidis* var. *paratyphi* can vary from 1 to 10 days, whereas, *Salmonella typhi* have a longer incubation period ranging from 7 to 30 days. Bacteria enter the body through M cells in the intestinal tract and multiply in the spleen and liver. Thereafter large numbers of bacteria are released into the bloodstream resulting in high fever. These symptoms can persist for 2 to 3 weeks. Finally bacteria move to the gallbladder and in a few cases can persist there for years. Such chronic carriers, which can shed bacteria for years are a major public health concern. A classic case of chronic carrier is the infamous Typhoid Mary (Mary Mallon), a professional cook in New York City, who was responsible for several outbreaks in the eastern United States during early years of the twentieth century.

The infection with *S. typhimurium* is usually self-limiting but may become systemic in infants, toddlers and immunocompromised subjects. In case of *S. typhimurium* infection antibiotics administration is recommended only when the infection becomes systemic, whereas, *Salmonella typhi* infection generally needs to be treated with antibiotics. People with cancer and AIDS when infected with *S. choleraesuis*, *S. enteritidis* var. *paratyphi*, and *S. typhimurium* have been reported to develop typhoid like disease.

The survival of *Salmonella* in an aquatic environment is affected by a variety of physical-chemical factors such as UV light, temperature, natural organic matter, nutrients, and antibiotics. Conventional water treatment processes (coagulation, sedimentation and filtration) along with chlorination have been proved effective against *Salmonella*. A chlorine residual level of 0.2 mg/L, in water distribution systems is needed to ensure the safety of water at the consumer end.

Shigella. The genus *Shigella* is a member of the family Enterobacteriaceae. Shigellae are 0.3- to 1μm-wide and 1- to 6-μm-long gram-negative bacilli. The genus consists of four species: *S. dysenteriae*, *S. flexneri*, *S. boydii*, and *S. sonnei*. Shigellae are not part of normal flora of human digestive tract, and they have well-developed virulence factors. There has been a gradual increase in the number of reported Shigellosis outbreaks in the United States during the last 50 years as against Salmonellosis outbreaks, which have been relatively stable during the same time period.

Shigellae cause an acute intestinal disease called *bacillary dysentery*. Clinical symptoms include diarrhea, abdominal cramps, vomiting, and fever. The colonic mucosa is damaged with the progression of disease resulting in ulceration. The liquid stools often contain blood and mucus.

Shigella species are highly infectious with an ID_{50} for humans of 10 to 100 bacteria. This low ID_{50} is partly due to the fact that Shigella species are resistant to an acid environment. *S. dysenteriae*, the most virulent species causing the severest form of dysentery, is uncommon in the United States and is prevalent in the eastern hemisphere. *S. sonnei* and *S. flexneri* have been the major cause of shigellosis in the United States. The incubation period for shigellosis is usually 24 to 72 h. The disease is normally self-limiting in adults and can last from 4 to 7 days. Antibiotics are known to reduce the severity and duration of the disease episode and also help reduce the risk of secondary complications but appear to prolong the carrier state of subjects. Organisms are shed over a period of 1 to 2 weeks.

Shigellae are easy to control as they are sensitive to the chlorine concentrations normally used in water treatment processes and maintained in distribution system (Table 2.2). They are not good competitors in the environment and are reported to survive for up to 4 days in river water.

TABLE 2.2 Inactivation of Microbes using Chlorine

Bacteria	Cl_2 (mg/L)	Temp. (°C)	pH	Contact time (min)	Reduction (%)
Campylobacter jejuni	0.1	25	8.0	5	99.99
E. coli	0.2	25	7.0	15	99.99
Legionella pneumophila	0.25	21	7.6–8.0	60–90	99
Mycobacterium chelonei	0.7	25	7.0	60	99.95
Mycobacterium fortuitum	1.0	—	7.0	30	99.4
Mycobacterium intracellulare	0.15	—	7.0	60	70
Salmonella typhi	0.5	20	—	6	99
Shigella dysentriae	0.05	20–29	7.0	10	99.6–100
Vibrio cholerae S Strain	1.0	20	7.0	<1	100
Vibrio cholerae R Strain	2.0	20	7.0	30	>5 logs
Yersinia enterocolitica	1.0	20	7.0	30	92
Adenovirus	0.2	25	8.8–9.0	40–50 s	99.8
Hepatitis A	0.42	25	6	1	99.99
Norovirus	0.5–1.0	25	7.4	30	Not completely inactivated
Rotavirus	0.5–1.0	25	7.4	30	100
Cryptosporidium parvum	80	25	7.0	90	90
Entamoeba histolytica	1.0	22–25	7.0	50	100
Giardia lamblia	1.5	25	6.0–8.0	10	100
Naegleria fowleri	0.5–1.0	25	7.3–7.4	60	99.99

Source: Centers for Diseases Control and Prevention, *Effect of Chlorination on Inactivating Selected Microorganisms*, Safe Water System, CDC.

***Escherichia coli* O157:H7.** The genus *Escherichia* is a member of the family Enterobacteriaceae. *Escherichia* are 0.5 to 2 μm in size. There are hundreds of strains of the bacterium *Escherichia coli*, and most strains are harmless and live in the intestines of healthy humans and animals.

The serological classification of *E. coli* is based on the two types of surface components: O antigen of LPS (O), which identifies the serogroup of the strain and H antigen of flagella (H), which identifies its serotype. Pathogenic strains have been categorized into five virotypes—enteropathogenic *E. coli* (EPEC), enterotoxigenic *E. coli* (ETEC), enteroinvasive *E. coli* (EIEC), enterohemorrhagic *E. coli* (EHEC), and enteroaggregative *E. coli* (EAggEC). The diarrhea due to EPEC and ETEC strains is more commonly seen in developing countries, whereas, EHEC is more prevalent in developed countries such as the United States and Canada. This could be due to lack of EHEC virotypes identification capabilities in developing countries. The infectious dose for all the virotypes is fairly high (in the range of 10^8 to 10^{10}) except for enterohemorrhagic *E. coli* (EHEC) in which case it is quite low like *Shigella*.

The disease caused by EHEC strains is more similar to *Shigella* dysentery than to the diarrhea caused by ETEC and EPEC strains. *E. coli* O157:H7 is the predominant serogroup among EHEC strains. *E. coli* O157:H7 produces a powerful toxin (Verotoxin), which is similar to Shiga toxin and can cause intense inflammatory response. Infection with this virotype can be fatal due to acute kidney failure (hemolytic-uremic syndrome) caused by toxins. *E. coli* O157:H7 has become a source of concern as the dose required to trigger infection can be as low 200 cells or less.

E. coli O157:H7 was first recognized as a human pathogen in 1982. It can cause extremely bloody to nonbloody diarrhea and renal failure in humans. The disease is characterized by severe abdominal pain and cramps followed by sudden onset of diarrhea (bloody or nonbloody), rare vomiting and lack of fever. The incubation period ranges from 72 to 96 h and the illness generally lasts for 1 week but it may persist longer under certain conditions. *E. coli* has been reported to survive fairly well in source and finished waters. Water temperature, nutrient levels and UV light are some of the factors affecting its survival in natural waters. In source waters similar survival times have been reported for both the pathogenic and nonpathogenic *E. coli*. Chlorine and ozone are effective for rapid inactivation of *E. coli* strains. The $C \times T_{99}$ value for *E. coli* is reported at 0.2 mg min/L (Table 2.2). The inactivation kinetics of various *E coli* strains does not vary widely.

Yersinia. The genus *Yersinia* is a member of the family Enterobacteriaceae. Yersiniae are 0.5 to 0.8 μm in diameter and 1 to 3 μm in length. There are eleven species of this bacterium, however *Yersinia enterocolitica and Yersinia pseudotuberculosisis* are the two species, which can cause disease through ingestion of contaminated water. One of the unique features of both these strains is that they can grow at 4°C. The most common illness caused by *Yersinia enterocolitica* is acute enterocolitis, which can occur in all age groups but is common in children. The most prevalent serotypes of *Y. enterocolitica*, found in the United States and Canada, are O:3, O:5, O:8, O:9, and O:27.

This bacterium has a relatively long incubation period, which can range from 4 to 10 days and symptoms may last for 2 weeks. The typical symptoms include diarrhea, abdominal pain, and sometimes fever. In some cases bacteria may infect the mesenteric lymph nodes, resulting in symptoms suggestive of acute appendicitis. In some cases (subjects with histocompatibility antigen HLA-B27) arthritis of peripheral joints may develop 2 to 6 weeks after gastrointestinal infections with *Y. enterocolitica* has cleared. Immunocompromised people have a greater risk of developing bacteriemia (disseminated Yersiniosis), which can be life threatening. The infection with *Yersinia pseudotuberculosisis* results in symptoms similar to those caused by *Y. enterocolitica* but is more likely to become systemic.

In general, all strains of *Yersinia* are sensitive to chlorine, but strain-to-strain variation in sensitivity has been reported. However, this variation in chlorine sensitivity has not shown to present significant risk. A residual chlorine level of 0.2 mg/L in distribution systems is enough to kill Yersinia strains. The routine water treatment process, with adequate chlorination practices, is deemed to prevent the public health risk from *Yersinia* infection.

Vibrio. The genus *Vibrio* is a member of the family Vibrioaceae. Vibrios are gram-negative, aerobic, nonsporulating curved rods with a single polar flagellum. Cholera is caused by toxigenic *Vibrio cholerae*, which is divided into two major serogroups—the O1 and non-O1 (now referred as O139 Bengal). Although the serogroup O1 has different variants it is divided into two biotypes—El Tor biotype and classical biotype. Historically cholera is considered an old world disease. The most recent epidemics have occurred in the Indian subcontinent and in central and south America.

This disease is characterized by a sudden onset of diarrhea accompanied by nausea, vomiting, abdominal pain, and severe dehydration resulting in the loss of skin plasticity and visibly sunken eyes of cholera victims. A person with full-blown cholera can lose 20 L of water daily. In a typical case stools are termed "rice water stool" due to the fact that stools become so dilute that it is almost clear and contains flecks of mucus. A prompt response with continuous rehydration is required to avoid collapse, shock, and death. The incubation period may range from few hours to 3 days. The organisms persist in the intestinal tract for several weeks after recovery and convalescents may shed Vibrios in feces for 2 to 3 weeks. Cholera is strictly an intestinal infection and never reaches bloodstream or results in systemic infection and is ordinarily self-limiting.

Vibrio cholerae is an autochthonous member of aquatic microflora of rivers and estuaries. The close association of *Vibrio cholerae* with surface waters plays an essential role in its spread. It survives better in saline water than freshwater. It persists in the environment because it can grow in saltwater or in freshwater. Chlorine is effective for inactivation of *Vibrio cholerae* and variation in inactivation kinetics of various *Vibrio cholerae* strains has been observed (Table 2.2).

Campylobacter. The genus *Campylobacter* belongs to the family Vibrioaceae, which was described in 1919 as an animal pathogen. Campylobacters are gram-negative curved rods with polar flagellum and are 0.2 to 0.5 μm wide and 0.5 to 5.0 μm long. There are 14 species of this genus but *C. jejuni, C. upsaliensis,* and *C. coli* are of the greatest concern to humans. These species have been known as animal pathogens since the early part of the last century, but their pathogenic potential for humans was not discovered until the 1970s. Campylobacters have become increasingly important because their infective dose is low.

The disease symptoms are similar to those of Salmonellosis. The incubation period is 2 to 10 days. The bacterium invades the epithelium of the small intestine, causing inflammation. Symptoms include a sudden onset of diarrhea preceded by abdominal cramps or severe pain, high fever and severe inflammation of intestine along with ulceration. The diarrhea may be profuse, watery, or in some cases bloody.

Campylobacters have been shown to survive for a few hours at high temperatures (37°C) but their survival potential increases with decreasing temperatures, and they can survive for several days at 4°C. Survival was enhanced by the presence of other microorganisms and especially in biofilms (Buswell et al., 1998). In cold groundwater, campylobacters have been shown to survive for several weeks (Gondrosen, 1986) and are able to exist as viable but nonculturable (VBNC) cells (Cappelier and Federighi, 1998; Höller, 1998), the virulence of which is not clear (Koenraad et al., 1997; van der Giessen et al., 1996). The survival potential and the role of VBNC stages of campylobacters in drinking-water distribution systems is not clear. Normal disinfection procedures in conventional water treatment plants are effective for inactivation of campylobacters as they are more susceptible to chlorine than *E. coli*.

Legionella. The genus *Legionella* belongs to the family Legionellaceae. They are gram-negative nonspore forming rods, 0.2 to 0.8 μm wide and 2 to 20 μm long. This genus includes about 40 species and approximately 60 serotypes. *Legionella pneumophila* serotype 1 is of the greatest concern for human health. It was discovered in 1976 when an outbreak occurred among veterans in a hotel in Philadelphia. *Legionella longbeachae, L. micdadei*, and *L. bozemanii* are among some of the other species which have been reported as human pathogens.

Legionella species can cause two different types of diseases, Legionnaires disease, and Pontiac fever. While ingestion as a cause of infection is very rare, both diseases are caused by inhaling aerosols containing *Legionella* organisms. Legionnaires disease is a severe respiratory illness. The incubation period of this disease ranges from 2 to 10 days, with an attack rate of 1 to 6 percent. Bacterial infection results in severe lung inflammation characterized by pneumonia. Individuals with weak immune system and underlying disease conditions are at greater risk of contracting the disease. Pontiac fever is a nonpneumonic, influenzalike self-limiting disease. The incubation of Pontiac fever is much shorter than Legionnaires disease, ranging from 24 to 48 h.

Legionella species are commonly found in freshwater and soil and thrive over a wide temperature range (5 to 60°C). *Legionella* are known to be able to evade a water purification system by internalizing in protozoan parasites such as Amoeba. *Legionella* species, introduced into drinking water from the environment, are able to grow under favorable conditions in cold- and hot-water distribution systems, heaters, pools, and spas. Growth occurs also in cooling towers. The amount of nutrients and the temperature are the major determining factors in the growth of this *Legionella* in distribution systems. The highest numbers of *Legionella* organisms are found in water samples with temperatures of 30 to 40°C and tend to decrease at temperatures greater than 50°C. Temperatures from 60 to 70°C cause a rapid die off (within minutes or seconds) of *Legionella* (Hoge and Breimen, 1991).

2.2.2 Viral Pathogens

More than 140 different types of enteric viruses belonging to approximately 15 different groups have been reported to cause waterborne outbreaks in the human population. The magnitude of the waterborne epidemics with a viral etiology can be quite large, like the one that occurred in Delhi, India, in 1955, during which the number of infected persons was estimated at 1 million (Dennis, 1959). Enteric viruses have been detected in many groundwater sources in the United States (Abbaszadegan et al., 2003). Based on the epidemiological significance, waterborne enteric viruses can be divided into two major categories. The first category, gastroenteritis viruses, comprises the viruses implicated as agents of gastroenteritis such as human calicivirus, rotavirus, and astroviruses. The second category, pathogenic viruses, encompasses viruses that cause viruses unrelated to human gut epithelium such as hepatitis A, poliovirus, and hepatitis E virus. Volunteer studies have revealed that some enteric viruses are highly infective (i.e, one or a few tissue culture infectious units suffice to initiate an infection) with the risk of infection being 10 to 10,000-fold higher than that for bacteria at the same level of exposure (Rose and Gerba, 1991).

Adenovirus. Adenoviruses belong to the family Adenoviridae, which includes human and animal serotypes. About 50 different serotypes of Adenovirus have been reported, which can cause respiratory, ocular, genitourinary and enteric infections in humans, but only the serotypes 31, 40, 41, 43–47 have been reported to cause gastroenteritis. The icosahedron virions are about 80 nm in diameter and have a double stranded DNA genome.

Although Adenovirus infection is more common in children, it can infect all age groups. The incubation period can range from 1 to 3 days. Typical gastroenteritis symptoms include vomiting and watery diarrhea, which can last for several days. The infected individuals are known to shed viruses in their feces for weeks or months after recovery.

Adenovirus serotype 40 and 41 are the two most common causes of gastroenteritis. In aquatic and terrestrial environments these adenoviruses serotypes have shown better survival rate than other enteric viruses such as Hepatitis A and enteroviruses. The UV doses of 30 and 23.6 $mW \cdot s/cm^2$ are required to achieve 90 percent inactivation of serotypes 40 and 41, respectively. Whereas in the case of poliovirus type 1 the same level of inactivation can be achieved by a UV dose of 4.1 $mW \cdot s/cm^2$.

Astroviruses. Astroviruses were first described in 1975 and belong to the family Astroviridae. They are named astro because they have a characteristic star-shaped surface when observed by an electron microscope. Astroviruses have a nonenveloped virion 28 to 30 nm in diameter. They have a positive strand single stranded RNA genome with a poly-A tract at 3′ terminus. Seven serotypes of human Astrovirus have been reported.

Astrovirus gastroenteritis is principally seen in young children, and it is estimated that 70 percent of children will have at least one incident of gastroenteritis with this virus. It can cause serious infection in immunocompromised and geriatric patients. The incubation period can last up to 4 days. Symptoms include abdominal discomfort, vomiting, and watery diarrhea, which can last for 1 to 4 days. Infected individuals are known to shed high number of viruses ($\sim 10^8$ per gram of feces) even after recovery.

Astroviruses have been isolated from river water. Astrovirus virions are stable at 60°C for 5 minutes, but the information on disinfection of Astroviruses using chemical and other physical agents is lacking. However recent data on astrovirus survival in chlorinated drinking water show that inactivation requires at least 0.5 to 1 mg/L of free chlorine (Gofti-Laroche, 2003).

Hepatitis A. Hepatitis A belongs to the family Picornaviridae. The virion is a nonenveloped icosahedral particle 27 nm in diameter. Hepatitis A has been grouped into several different genotypes. Genotypes III and I are primarily responsible for human illness. The symptoms of Hepatitis A infection include, malaise, loss of appetite, nausea, vomiting, dark urine, scleral icterus (yellowing of eyes), jaundice (yellowing of skin), and tender liver. The incubation period can be from 15 to 50 days. Fecal shedding of viruses starts even two weeks before the infected individuals start to show clear symptoms, and the shedding declines with the onset of acute phase of illness.

Hepatitis A is one of the most resistant waterborne viruses, which can survive high temperatures, drying desiccation, and extreme pH levels for hours. At lower temperature it can survive for months to a year in environmental water (Alum, 2001). Compared to other waterborne viruses Hepatitis A has shown a greater degree of resistance to chemical oxidants or disinfectants, and during conventional water treatment processes a 2-log removal of Hepatitis A has been reported. Chemical or physical disinfectants when used under optimal conditions can effectively achieve 3 to 4 logs inactivation.

Hepatitis E. Hepatitis E belongs to the family Caliciviridae. The virion is a nonenveloped icosahedron 32 nm in diameter, with a single stranded RNA genome. This virus is prevalent in Asia (especially the Indian subcontinent), North Africa, and part of Eastern Europe but is rarely seen in North America. Hepatitis E can cause subclinical symptoms in children but in adult population the symptoms are indistinguishable form Hepatitis A infection. The symptoms (jaundice) of Hepatitis E infection are generally more severe than Hepatitis A infection and last longer with greater bilirubin levels. In the general population, the fatality rate for Hepatitis E infection ranges from 0.1 to 4 percent, but in pregnant women the death rate can be as high as 20 percent. The incubation period of Hepatitis E infection in humans can range between 2 to 8 weeks with an average of 5 to 6 weeks. The infected individuals start to shed viruses with feces even before the appearance of symptoms and continue to shed viruses during the acute phase of illness. The virus particle is extremely labile and loses its outer layer during storage at 4 C. The standard chlorination practice has been shown to be effective in controlling the waterborne spread of this virus.

Norovirus. Human calicivirus was first discovered in 1972. Since then many different strains have been identified from around the world. These viruses belong to the family Caliciviridae, which recently has been reclassified with four genera: Vesivirus, Lagovirus, Norovirus, and Sapovirus. The genus Norovirus, which includes Norovirus and the genus Sapovirus, which includes Sapporo virus, are of public health concern. Personnel on aircraft carriers and cruise ships have traditionally been crippled due to one of the Norovirus strains (Cruise ship strain). Viral gastroenteritis due to a Norovirus was the single most common cause of disability of the American troops deployed in the Persian Gulf during Operation Desert Shield. The huge economic losses suffered by the cruise ship industry in the United States during the summer of 2002, points to the magnitude of public health threat from Noroviruses.

The virus particles are nonenveloped and icosahedral with diameters ranging from 27 to 40 nm, and then have a single stranded RNA genome. Based on volunteer studies it is assumed that the incubation period can range from 48 to 72 h. The principal symptom in children is diarrhea with vomiting, whereas adults tend to have diarrhea with no vomiting. Occasionally additional symptoms such as abdominal pain, cramping, nausea, and low-grade fever, are also seen. Depending on the strain, the symptoms can persist from 1 to 11 days. Due to the lack of in-vitro infectivity assay, no information on the stability of human caliciviruses in the environment is available at this time. The data based on molecular techniques suggest that these viruses are more resistant than most enteroviruses.

Rotaviruses. First discovered in 1973, rotaviruses belong to the family Rotaviridae. Based on the capsid antigen VP6, rotaviruses are divided into six serogroups, A-F. Groups A, B, and C are found in humans and animals, whereas groups D, E, and F are exclusively found in animals. Most human rotaviruses fall into group A, which is the most studied serogroup. Serogroup A has been further divided into various serotypes using two different classification systems. Fourteen serotypes (G1 to G14) have been defined on the bases of variation in antigen VP7, whereas 8 serotypes (P1 to P8) have been defined on bases of VP4 antigen variation. Human group A rotaviruses fall into 5 P serotypes (P1A, P1B, P2A P3A and P4), and into 10 G serotypes (G1 to G6, G8, G9, G10, and G12). The G serotypes 1, 2, 3, and 4 are commonly found in humans and 6, 8, 9, 10, and 12 are rare. The serotypes G1 and P3A are the major cause of rotavirus gastroenteritis worldwide. Rotavirus is a major cause of diarrhea in infants and children and also a primary cause of traveler's diarrhea. By age 5 nearly every child has had an episode of rotavirus gastroenteritis. The incubation period can be 1 to 3 days. A typical symptom is vomiting, which generally precedes diarrhea that lasts for 4 to 5 days and can result in severe dehydration. The infected person continues to shed large number of viruses in feces for 3 to 7 days (Polanco-Marin et al., 2003). The chlorination practices normally used by water utilities in the United States have been shown to be effective in controlling the waterborne spread of this virus.

2.2.3 Parasitic Pathogens

Waterborne parasites have played a major role in shaping the history of mankind and they continue to challenge human civilization until today. Because of their larger size and visibility they have been known since ancient times. *Dracunculus medinensis* the "fiery serpent of Moses" was mentioned in biblical writings. *Giardia lamblia* was discovered by

Leewonhoek in 1681, and still continues to be a significant public health threat. Of nearly 20,000 protozoan parasites, about 20 genera are known to cause diseases in humans. Water utilities will continue to face the challenges posed by centuries old and newly emerging parasites and their resting stages.

Acanthamoeba. Members of the genus *Acanthamoeba* are commonly found in freshwater, brackish water, sewage, and soil. Approximately 20 different species of *Acanthamoeba* have been reported of which six species (*A. castellanii, A. culbertsoni, A. divionensis, A. healyi, A. rhysodes, and A. polyphaga*) are reported to cause chronic and fatal infection of the human brain, skin, and eye, and in the environment they feed on bacteria. *Acanthamoeba* has two stages in its life cycle—the trophozoite and the cyst. The trophozoites, which are the infectious stage of Acanthamoeba, are uninucleate and 15 to 45 μm in size. Under adverse conditions *Acanthamoebas* differentiate into cysts, which are double-walled.

Acanthamoebas cause a fatal disease of the central nervous system called granulomatous amebic encephalitis (GAE), and immunocompromised individuals are more likely to contract this disease. There is no clearly defined incubation of GAE, and the disease progresses slowly. Some of the characteristic symptoms include headache, irritability, dizziness, drowsiness, confusion, and seizures. Hemorrhagic necrosis of CNS parenchyma is visible under microscopic examination. Trophozoites and cysts can be found in lymph nodes, skin, liver, lungs, and kidney. *Acanthamoebas* are also known to cause a disease called *Acanthamoeba keratitis*, which can result in chronic ulceration of cornea and eventual blindness.

Acanthamoeba cysts are able to withstand adverse conditions and can survive for a long time in environment. *Acanthamoeba* cysts are very resistant to chlorine.

Cryptosporidium Parvum. *Cryptosporidium* is a protozoan parasite that has been known for almost 100 years, but it was not until 1955 that this parasite was recognized as an animal pathogen and in 1976 as a human pathogen. It has several species, which can infect animals such as chicken (*C. baileyi*)), turkey (*C. meleagridis*), cat (*C. felis*), mouse (*C. muris*), guinea pig (*C. ucarairi*), reptiles (*C. serpentis*), and fish (*C. nasorum*). *Cryptosporidium parvum* is the principal species that can infect humans and livestock. *Cryptosporidium* is responsible for large outbreaks in the United States, including the 1993 epidemic in Milwaukee during which an estimated 403,000 people became ill with more than 100 deaths.

C. parvum is an obligate parasite that can multiply only within its host. *C. parvum* has a complex life cycle that results in the production of a hardy stage, the oocyst (5 to 7 μm in diameter), which is shed with the feces. These oocysts are able to survive for weeks to months in the environment (DeRegnier et al., 1989; Rogers et al., 1994).

The average incubation period varies widely but is usually about 7 days (Ungar, 1990; Dupont et al., 1995). The most prominent symptom of intestinal *C. parvum* infection is watery diarrhea, which can result in dehydration and weight loss (Arrowood, 1997; Fayer & Ungar, 1986; Ungar, 1990). Other symptoms include nausea, abdominal cramps, vomiting, and mild fever.

In healthy persons, *C. parvum* causes subclinical infections and self-limiting diarrhea. The infection is limited by the immune response that eventually clears the parasite. Infections in immunocompromised persons, or those with underlying illnesses, are persistent and heavy, and can be fatal. In immunocompetent individuals the duration of the infection can range from 7 to 14 days, but in immunocompromised individuals the duration can be 23 to 32 days (van Asperen et al., 1996). Volunteer studies have demonstrated that the infectious dose of this parasite can be very low ranging from 10 to 100 oocysts (Smith et al., 1993; Teunis et al., 1996). Infected individuals are known to excrete high numbers of oocysts (1.44×10^{10} cysts/day/patient) with the feces.

Cryptosporidium oocysts are resistant to environmental stresses and can survive for months under cold and moist conditions, but are known to lose infectivity under excessive heat and dry conditions. *Cryptosporidium* oocysts are very resistant to chlorine (Table 2.2).

Entamoeba Histolytica. Although amoebic protozoans are known to parasitize a wide range of animals, only members of the genus *Entamoeba* are known to be pathogenic to their hosts. The genus *Entamoeba* belongs to the family Entamoebidae in the Superclass Rhizopoda. The members of the genus *Entamoeba* commonly found in man are *E. histolytica, E. hartmani, E. coli,* and *E. gingivalis* (Stanley, 2003; Kitchen, 1999).

Entamoeba historlytica is potentially the most pathogenic parasite and the third leading cause of sickness and death worldwide. It is estimated to infect about half a billion people and causes about 100,000 deaths per year. *Entamoeba histolytica* has two stages in its life cycle—the trophozoites (the feeding stage of the parasite) that live in the host's large intestine and cysts that are passed in the host's feces. The trophozoites have finely granular cytoplasm with single pseudopodia and are 10 to 35 μm in size. Cysts are rounded or cigar-shaped with condensed chromatoidal material and are 10 to 25 μm in size.

In the case of mild intestinal infection the symptoms of "amoebiasis" are intermittent and mild (various gastrointestinal upsets, including colitis and diarrhea). The moderate intestinal infection results in diarrhea, constipation, and malaise. In more severe cases the trophozoites destroy the mucosal lining of the host's large intestine resulting in dysentery, low-grade fever, and deep flask-shaped ulcers. The trophozoites can enter the circulatory system and infect other organs, such as liver, lungs, and brain, which are often fatal (Noble et al., 1989).

The *Entamoeba histolytica* cysts are the infective stage and are acid tolerant (Lyles, 1969; Kulkarni et al., 1993; Stanley; 2003) as against the trophozoites that rarely survive in the stomach and are commonly found in the lower small intestine, cecum, and large intestine (Lyles, 1969; Read, 1972; Noble et al., 1989). *Entamoeba histolytica* can survive for a week in soil, whereas the cysts can survive for months in moist conditions and are able to withstand a range of temperatures and drying out (Read, 1969; Moat, 1979; Noble et al, 1989). In water cysts can remain viable for up to 30 days. Chlorine has been reported to be effective for inactivation of *Entamoeba histolytica* (Table 2.2).

Microsporidia. Microsporidia is the common name for a group of very small (0.5 × 1.2 μm), obligate intracellular parasites belonging to the phylum *Microspora* (Shadduck and Greeley 1989). It had been known for some time that they can cause diseases in animals but recently they have also been recognized as human pathogens, primarily in immunocompromised patients. The genera that infect humans include *Encephalitozoon, Enterocytozoon,* and *Nosema* (Despommier et al., 1995). The principal species, which cause diarrhea and cholangiopathy in AIDS patients, are *Entercytozoon bieneusi* and *Encephalitozoon intestinalis* (Pol et al., 1993). A recent survey showed that almost 40 percent of AIDS patients with diarrhea excrete these pathogens with feces (Kotler, 1995).

The variations in the survival capabilities of Microsporidia species under different environmental conditions have been reported. Very little is known about the occurrence and distribution of human-pathogenic microsporidia in the environment and most of the information available is based on animal Microsporidia. *E. cuniculi* spores can remain viable for 4 months in environment.

Naegleria. The genus *Naegleria* is a member of the super-class *Rhizopodea*, which also includes the free-living amoebae *Acanthamoeba, Entamoeba histolytica, Hartmannella,* and *Balamuthia*. There are several species of *Naegleria* but *N. fowleri* is the only species in this genus known to produce human disease. *N. fowleri* is a free-living amoeba, which can cause primary amebic meningoencephalitis (PAM) in humans. *N. fowleri* infections are more common in children and young adults but they can infect individuals in all age groups. Although swimming or diving in warm water contaminated with *N. fowleri* is the most common way of contracting the disease, other modes of transmission cannot be ruled out. *Naegleria fowleri* contaminating a community drinking water supply in the United States was responsible for the deaths of two 5-year-old boys in October 2002. The difficulty in early diagnosis of this disease along with low infective dose and rapid fatality make it a significant waterborne public health concern.

The incubation period can range from 3 to 14 days. The symptoms include abrupt onset of fever, headache, nausea, and vomiting. In some cases a prodromal stage of altered taste (ageusia) and smell (parosmia) has also been associated with the disease. About 75 percent patients show changed mental status, which is followed by rapid deterioration to coma and death. The trophozoites of *N. fowleri* enter the nose and invade the olfactory mucosa. They cross the cribriform plate after penetrating the submucosal nervous system, and eventually gain access to the subarachnoid space.

N. fowleri is thermophilic in nature and survives temperatures up to 45°C (113°F). It is found in surface and ground water, water distribution systems, heated swimming pools, and even hot springs. When water temperatures fall, *N. fowleri* encyst and enter a dormant stage, which allows them to survive until the next summer. Chlorine has also been reported to be effective against *N. fowleri* in water distribution systems (Table 2.2).

REFERENCES

Abbaszadegan, M., M. LeChevallier, and C. P. Gerba, "Occurrence of viruses in US groundwaters," *J. Am. Water Works Assoc.* 95(9): 107–120, 2003.

Alum, A., "Control of Viral Contamination of Reclaimed Irrigated Vegetable Using Drip Irrigation," Ph.D. dissertation, The University of Arizona, Tucson, Arizona, 2001.

Arrowood, M. J. "Diagnosis," in R. Fayer (ed.) *Cryptosporidium and Cryptosporidiosis*. CRC Press, Boca Raton, FL, pp. 43–64, 1997.

Buswell, C. M., Y. M. Herlihy, L. M. Lawrence, J. T. M. McGuiggan, and P. D. Marsh, "Extended Survival and Persistence of *Campylobacter* spp. in Water and Aquatic Biofilms and Their Detection by Immunofluorescent-Antibody and -rRNA Staining," *Appl. Environ. Microbiol.* 64: 733–741, 1998.

Cappelier, J. M., and M. Federighi "Demonstration of Viable but Nonculturable State," *Campylobacter jejuni. Rev. Med. Vet.* 149: 319–326, 1998.

Dennis, J. M., "Infectious hepatitis epidemic in Delhi, India," *J. Am. Water Works Assoc.* 51: 1288–1298, 1959.

DeRegnier, D. P, L. Cole, D. G. Schupp, and S. L. Erlandsen, "Viability of *Giardia* Cysts Suspended in Lake, River and Tap Water," *Appl. Environ. Microbiol.* 55: 1223–1229, 1989.

Despommier, D. D., R. W. Gwadz and P. J. Hotez *Parasitic Diseases*, 3d ed., Springer Verlag, New York, 333 pp., 1995.

Dupont, H. L., "The Infectivity of *Cryptosporidium parvum* in Healthy Volunteers," *N. Engl. J. Med.* 332: 855–859, 1995.

Fayer, R., B. L. P. Ungar, "*Cryptosporidium* spp. and Cryptosporidiosis," *Microbiol. Rev.* 50: 458–483, 1986.

Gofti-Laroche, L., B. Gratacap-Cavallier, D. Demanse, O. Genoulaz, J.-M. Seigneurin, and D. Zmirou, "Are Waterborne Astrovirus Implicated in Acute Digestive Morbidity (E.MI.R.A. Study)?" *J. Clin. Virol.* 27(1): 74–82, 2003.

Gondrosen, B., "Survival of Thermotolerant Campylobacters in Water," *Acta Vet. Scand.* 79: 1–47, 1986.

Hoge, C. W., and R. F. Breiman, "Advances in the Epidemiology and Control of Legionella Infections," *Epedemiol. Rev.* 13: 329–340, 1991.

Höller, C., D. Witthuhn and B. Janzen-Blunck "Effect of Low Temperatures on Growth, Structure, and Metabolism of *Campylobacter coli* SP10," *Appl. Environ. Microbiol.* 64: 581–587, 1998.

Koenraad, P. M. F. J., F. M. Rombouts, S. H. W. Notermans, "Epidemiological Aspects of Thermophilic Campylobacter in Water-Related Environments: A Review," *Water Environ. Res.* 69: 52–63, 1997.

Kotler, D. P., "Gastrointestinal Manifestations of Immunodeficiency Infection," *Adv. Intern. Med.* 40: 197–241, 1995.

Pol, S., C. A. Romana, S. Richard, P. Amouyal, I. Desportes-Livage, et al., "Microsporidia Infection in Patients with the Human Immunodeficiency Virus and Unexplained Cholangitis," *N. Engl. J. Med.* 328: 95–99, 1993.

Polanco-Marin, G., M. R. González-Losa, E. Rodriguez-Angulo, L. Manzano-Cabrera, J Cámara-Mejiá, and M. Puerto-Solis, "Clinical Manifestation of the Rotavirus Infection and His Relation with the Electropherotypes and Serotypes Detected During 1998 and 1999 in Merida, Yucatan Mexico," *J. Clin. Viro.* 27: 242–246, 2003.

Ridgway, H. F., and B. H. Olson "Scanning Electron Microscope Evidence for Bacterial Colonization of a Drinking-Water Distribution System," *Appl. Environ. Microbiol.* 41: 274–287, 1981.

Rose, J. B., and C. P. Gerba, "Use of Risk Assessment for Development of Microbial Standards," *Water Sci. Technol.* 24: 29–34, 1991.

Shadduck, J. A., and E. Greeley, "Microsporidia and Human Infections," *Clin. Microbiol. Rev.* 2: 158–165, 1989.

Smith, H. V., J. F. W. Parker, Z. Bukhari, D. M. Campbell, C. Benton, et al., "Significance of Small Numbers of *Cryptosporidium* sp. oocysts in Water," *Lancet* 342: 312–313, 1993.

Teunis, P. F. M., O. G. van der Heijden, J. W. B. van der Giessen, and A. H. Havelaar, *The Dose-Response Relation in Human Volunteers for Gastro-intestinal Pathogens, Rep. 284550002*, Nat. Inst. Public Health Environ., Bilthoven, The Netherlands, 87 pp., 1996.

Ungar, B. L. P., "Cryptosporidiosis in Humans (*Homo sapiens*)," in J. P. Dubey, C. A Speer, R. Fayer (eds.), *Cryptosporidiosis of Man and Animals*, CRC Press, Boca Raton, FL, pp. 59–82, 1990.

van Asperen, A., et al., "An Outbreak of Cryptosporidiosis in the Netherlands," *Euro. Communic. Diseases Bull.* 1: 11–12, 1996.

van der Giessen, A. W., C. J. Heuvelman, T. Abee, and W. C. Hazeleger, "Experimental Studies on the Infectivity of Non-culturable Forms of Campylobacter spp. in Chicks and Mice," *Epidemiol. Infect.* 117: 463–470, 1996.

CHAPTER 3
VULNERABILITY ASSESSMENT, EMERGENCY RESPONSE PLANNING: SUMMARY OF WHAT'S AVAILABLE

Larry W. Mays
Department of Civil and Environmental Engineering
Arizona State University
Tempe, Arizona

3.1 INTRODUCTION

Title IV of PL 107–188 (Public Health, Security, and Bioterrorism Preparedness and Response Act, "Bioterrorism Act of 2002") requires community water systems (CWSs) serving populations greater than 3300 to conduct vulnerability assessments and submit them to the USEPA. In addition the community water systems must prepare an emergency response plan that incorporates the results of the vulnerability assessment. The dates of compliance are listed in Table 3.1.

3.2 VULNERABILITY ASSESSMENT

3.2.1 What is Vulnerability Assessment?

The common elements of vulnerability assessments are as follows:

- Characterization of the water system, including its mission and objectives.
- Identification and prioritization of adverse consequences to avoid.
- Determination of critical assets that might be subject to malevolent acts that could result in undesired consequences.
- Assessment of the likelihood (qualitative probability) of such malevolent acts from adversaries.

TABLE 3.1 Provides the Dates by Which CWSs Must Comply with the Above Requirements

Column A Systems serving population of	Column B Submit VA and VA Certification[†] prior to	Column C Certify ERP within 6 months of VA but no later than[‡]
100,000 persons or greater	March 31, 2003	September 30, 2003
50,000 to 99,999 persons	December 31, 2003	June 30, 2004
3301 to 49,999 persons	June 30, 2004	December 31, 2004

[†]Compliance with these deadlines is determined by the date of the postmark or the date the courier places on the mailing label of the submission.
[‡]VA certifications submitted to EPA earlier than the dates shown in Column B means that the CWS must submit an ERP certification *earlier* than the dates shown in Column C.

- Evaluation of existing countermeasures.
- Analysis of current risk and development of a prioritized plan for risk reduction.

Obviously the complexity of vulnerability assessments will range based upon the design and operation of the water system. The relative points to consider for each of the above basic elements are in Appendix 3A.

The Association of State Drinking Water Administrators and the National Rural Water Association (www.asdwa.org and www.nwra.org) developed the utility guide for security decision-making shown in Fig. 3.1. The Washington State Department of Health (2003) (http://www.doh.wa.gov/ehp/dw/Security/Tools.htm) presented Table 3.2, which shows a simple way to consider a system.

3.2.2 Security Tools

Several methods are available to perform vulnerability assessments including the following:

- Risk Assessment Methodology for Water Utilities (RAM-W^{SM}) was developed in cooperation with the Energy Department's Sandia National Laboratories with funding from U.S. EPA. RAM-W^{SM} compares system components against each other to determine which components are most critical.
- The Vulnerability Self-Assessment Tool (VSAT) was developed by the Association of Metropolitan Sewerage Agencies (AMSA) in collaboration with two consulting firms with U.S. EPA funding. This is also a software-based system, which can be ordered online at www.vsatusers.net/index.html.
- The National Rural Water Association (NRWA) and the Association of State Drinking Water Administrators (ASDWA) with U.S. EPA assistance developed the *Security Self-Assessment Guide for Small Systems Serving Between 3300 and 10,000*. This self-assessment can be downloaded at www.vulnerabilityassessment.org and is provided in Appendix 3B.
- ASSET was developed by NEWWA in conjunction with the U.S. EPA and other private firms. This tool is geared toward systems that serve between 3300 and 50,000 people (small and medium systems). This is a software-based tool, which was mailed to all New England public water suppliers in June 2003. This tool allows water systems to organize system information into an organized database format.

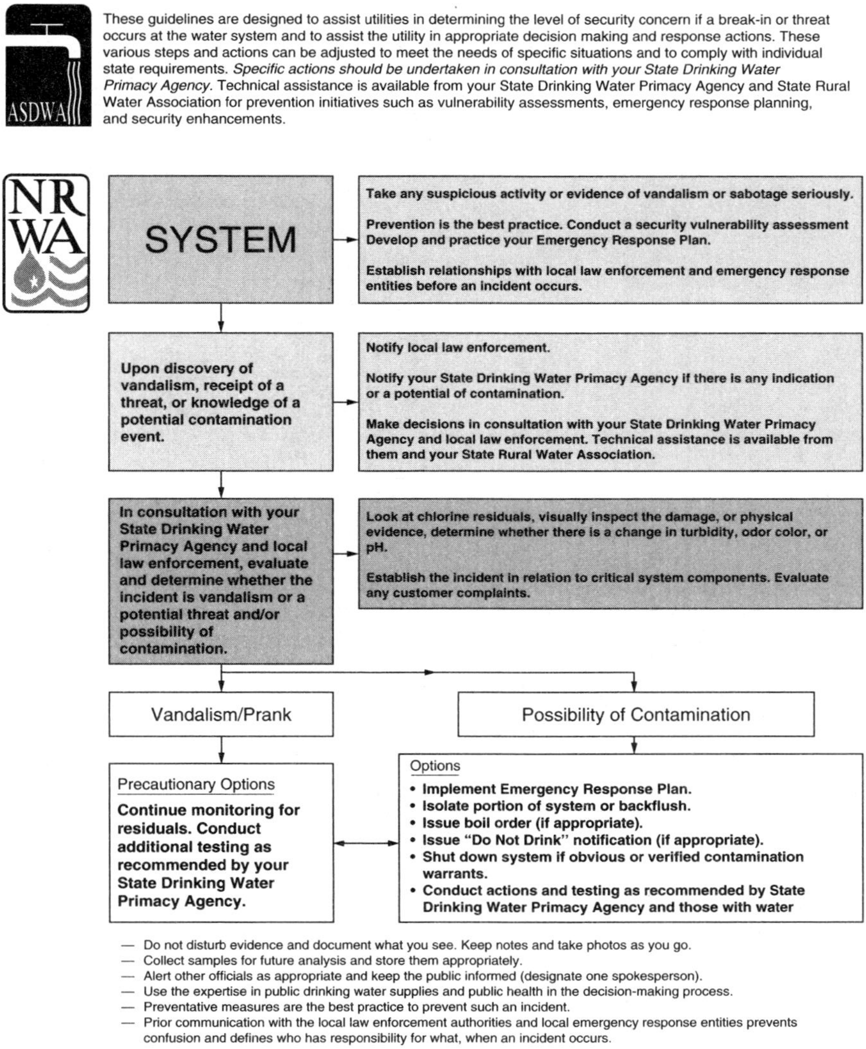

VULNERABILITY ASSESSMENT, EMERGENCY RESPONSE

 These guidelines are designed to assist utilities in determining the level of security concern if a break-in or threat occurs at the water system and to assist the utility in appropriate decision making and response actions. These various steps and actions can be adjusted to meet the needs of specific situations and to comply with individual state requirements. *Specific actions should be undertaken in consultation with your State Drinking Water Primacy Agency.* Technical assistance is available from your State Drinking Water Primacy Agency and State Rural Water Association for prevention initiatives such as vulnerability assessments, emergency response planning, and security enhancements.

SYSTEM

- Take any suspicious activity or evidence of vandalism or sabotage seriously.
- Prevention is the best practice. Conduct a security vulnerability assessment Develop and practice your Emergency Response Plan.
- Establish relationships with local law enforcement and emergency response entities before an incident occurs.

Upon discovery of vandalism, receipt of a threat, or knowledge of a potential contamination event.

- Notify local law enforcement.
- Notify your State Drinking Water Primacy Agency if there is any indication or a potential of contamination.
- Make decisions in consultation with your State Drinking Water Primacy Agency and local law enforcement. Technical assistance is available from them and your State Rural Water Association.

In consultation with your State Drinking Water Primacy Agency and local law enforcement, evaluate and determine whether the incident is vandalism or a potential threat and/or possibility of contamination.

- Look at chlorine residuals, visually inspect the damage, or physical evidence, determine whether there is a change in turbidity, odor color, or pH.
- Establish the incident in relation to critical system components. Evaluate any customer complaints.

Vandalism/Prank

Precautionary Options
Continue monitoring for residuals. Conduct additional testing as recommended by your State Drinking Water Primacy Agency.

Possibility of Contamination

Options
- Implement Emergency Response Plan.
- Isolate portion of system or backflush.
- Issue boil order (if appropriate).
- Issue "Do Not Drink" notification (if appropriate).
- Shut down system if obvious or verified contamination warrants.
- Conduct actions and testing as recommended by State Drinking Water Primacy Agency and those with water

— Do not disturb evidence and document what you see. Keep notes and take photos as you go.
— Collect samples for future analysis and store them appropriately.
— Alert other officials as appropriate and keep the public informed (designate one spokesperson).
— Use the expertise in public drinking water supplies and public health in the decision-making process.
— Preventative measures are the best practice to prevent such an incident.
— Prior communication with the local law enforcement authorities and local emergency response entities prevents confusion and defines who has responsibility for what, when an incident occurs.

FIGURE 3.1 A utility guide for security decision making. (Association of State Drinking Water Administrators (www.asdwa.org), and National Rural Water Associations (www.nrwa.org)

3.2.3 Security Vulnerability Assessment Using VSAT

The Vulnerability Self-Assessment Methodology (www.vsatusers.net) as developed in the VSAT software is depicted in Fig. 3.2. The methodology is based upon a qualitative risk assessment approach shown in the figure, adopting a broadbased approach that assesses vulnerability, prepares for extreme events, responds should extreme events occur, and restores to normal business conditions thereafter. The self-assessment framework can be

TABLE 3.2 Example Facility Vulnerability Assessment and Improvements Identification

System component	Description and condition	Vulnerability	Improvements or mitigating actions	Security improvements
Source	Two 150-ft deep groundwater wells supply the system. They are located within a few hundred feet of town and its developed areas. The sources are in excellent condition.	The wells are most vulnerable to contamination from above ground activities because they are only 150ft deep. The well houses are not highly secure so they could be vulnerable to acts of vandalism.	Implement wellhead protection program.	Upgrade wellhouses: Install fencing, and deadbolts. Secure wellhouses to foundation and install lighting around wellhouse.
Storage	Storage reservoirs are in sound condition, but reservoir hatches could be accessed and locks could be broken.	Vandals could access reservoir hatches. Also, the reservoir could be prone to shaking and settling resulting from an earthquake.	Provide earthquake strapping to secure reservoir to the foundation.	Install fencing, lighting, and signage to protect against unauthorized entry and access to reservoir hatches.
Treatment	There is a chlorination system in each well/pumphouse. Both are in sound operating condition.	Chlorination systems are subject to power outages and vandalism if a pumphouse is vandalized. Tanks are not secured and may tip over during an earthquake.	Purchase a backup generator and have it wired in or have system wired with a jack where a backup generator could be rented and plugged in. Secure tanks with earthquake straps.	Install fencing, lighting, and signage to protect against unauthorized entry.
Pumphouse and pumping facilities	The pumphouse and pumping facilities are in good condition.	Pumphouse does not have security fencing or lighting and is prone to vandalism.		Install fencing, lighting, and signage to protect against unauthorized entry.
Computer and telemetry system	Computer and telemetry systems are located in the water systems main office. All systems are in good operating condition.	Main office does not have adequate security measures. Also, computers should be better protected against cyber attack or hacking.		Install lighting and security system to guard against theft and vandalism. Hire consultant to secure computers and telemetry.

Source: Washington State Department of Health (2003) (www.doh.wa.gov/ehp/dw/Security/Tools.htm)

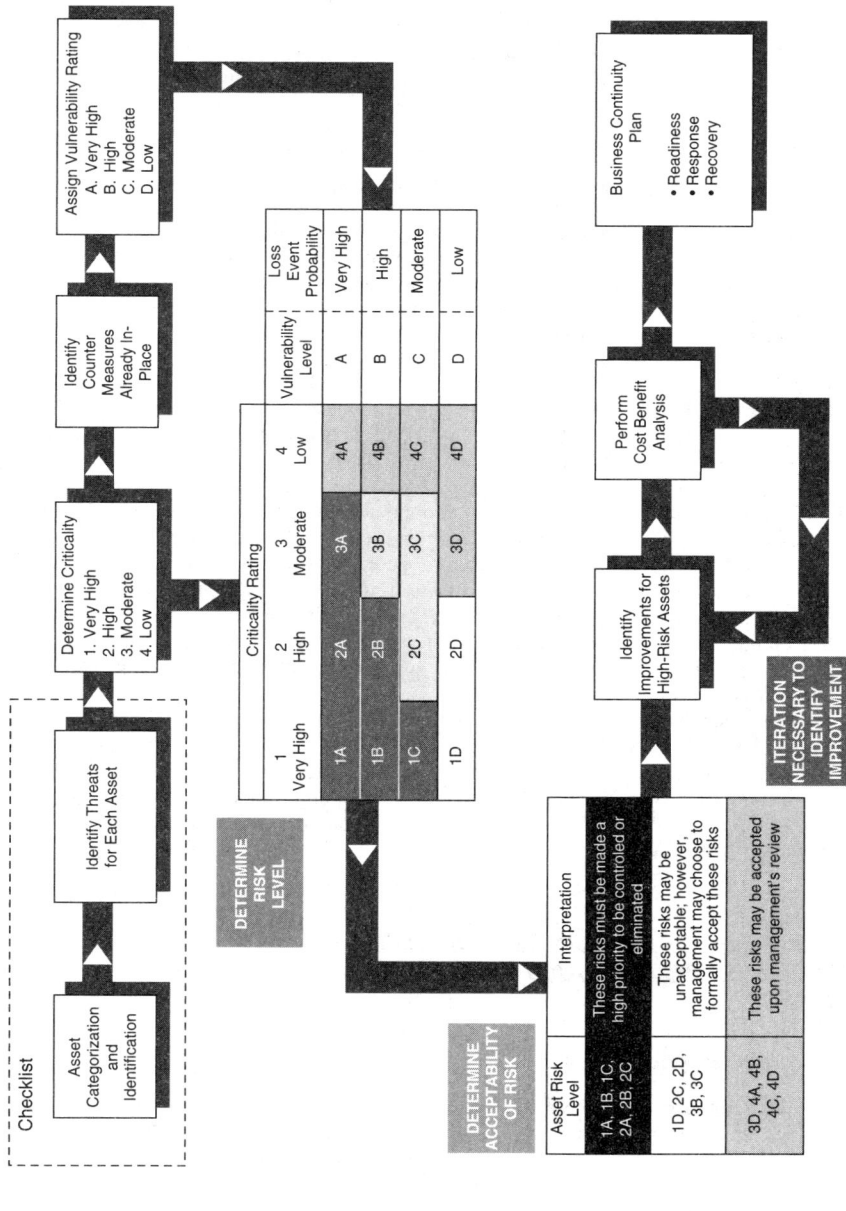

FIGURE 3.2 AMSA asset-based vulnerability analysis and response planning approach. *(Association of Metropolitan Sewage Agencies, 2002)*

divided into dimensions, with the first examining the utility assets including the physical plant, people, knowledge base, information technology (IT platform), and the customers. The second dimension of the framework recognizes that there is a process over time beginning with early assessment and planning activities, followed by later response actions as a result of an extreme event and eventually business recovery activities that occur postevent. VSAT is a tool that can help utilities identify vulnerabilities and evaluate the potential vulnerabilities, along with documentation of the decision process, rationale employed, and relative ranking of risks.

3.2.4 Security Vulnerability Assessment for Small Drinking Water Systems

The Association of State Drinking Water Administrators and the National Rural Water Association (2002) published the Security Vulnerability Self-Assessment Guide for small drinking water systems serving between 3300 and 10,000 people. This self-assessment guide (Appendix 3B) is meant to encourage smaller water system administrators, local officials, and water systems owners to review their vulnerabilities. The intent, however, is not to take the place of a comprehensive review by security experts. Completion of the documents does meet the requirement for conducting a vulnerability assessment as directed under Public Health Security and Bioterrorism Preparedness and Response Act of 2002. The goal of the vulnerability assessment is to develop a system-specific list of priorities intended to reduce risks to threats of attack.

The self-assessment guide is designed for use by water system personnel to perform the vulnerability assessment. The assessment "should include, but not be limited to a review of pipes and constructed conveyances, physical barriers, water collection, pretreatment, treatment, storage and distribution facilities, electronic, computer or other automated systems which are utilized by the public water system, the use, storage, or handling of various chemicals, and the operation and maintenance of such system." The self-assessment should be conducted on all components of the water system (wellhead or surface water intake, treatment, plant, storage tank(s), pumps, distribution system, and other important components of the water system.

The self-assessment has a simple design, with the final product being the list of priority actions based on the most likely threats to the water system. The inventory of small water system critical components and the security vulnerability self-assessment general questions are in Appendix 3B. Once the questions have been answered, then the prioritization of actions is developed. A certificate of completion must be completed and sent to the state drinking water primacy agency.

3.3 EMERGENCY RESPONSE PLANNING: RESPONSE, RECOVERY, AND REMEDIATION GUIDANCE

3.3.1 What's Needed?

Water utilities need to have an emergency operations/response plan that is coordinated with state and local emergency response organizations, regulatory authorities and local government officials. An emergency response plan is a "living" document requiring periodic updates. The updates should occur as frequent as a major change is made or at least annually (U.S. EPA, 2003).

Emergency response planning primarily needs to be a local responsibility. The U.S. EPA (2002) developed a water utility response, recovery, and remediation guidance for man-made and/or technological emergencies, as a result of their responsibilities under Presidential Decision Directive (PDD) 63. This guidance (presented in Appendix 3C) was developed for five different incident types:

- Contamination event: articulated threat with unspecified material
- Contamination threat at a major event
- Notification from health officials of potential water contamination
- Intrusion through supervisory control and data acquisition (SCADA)
- Significant structural damage resulting from a intentional act

Appendix 3C presents the response, recovery (recovery notifications and appropriate utility elements), and remediation actions in a table form for each incident type. Each category contains a section on notifications and utility actions. This guidance can be used to supplement existing water utility emergency operations plans developed to prepare and respond to natural disasters and emergencies. Even though the guidance is oriented toward the response, recovery, and remediation actions for the five incident types listed above, it could be utilized for other threatened or actual intentional acts.

Outlines for water system emergency response plans can be quite helpful. An example of one by the Association of State Drinking Water Administrators (ASDWA) has an outline on their website www.asdwa.org, which is presented in Table 3.3.

3.3.2 Rural and Small Community Water Systems

The National Rural Water Association (www.nrwa.org) developed an emergency response plan template for rural and small water and wastewater system emergency response plans.

3.3.3 Large Water System Emergency Plan Outline

The U.S. EPA developed Large Water System Emergency Response Plan Outline: Guidance to Assist Community Water systems in complying with the Public Health Security and Bioterrorism Preparedness and Response Act of 2002, dated July 2003. This plan provides guidance and recommendations to aid facilities in the preparation of emergency response plans under the PL 107–188. The outline is provided in Appendix 3D.

3.4 INFORMATION SHARING
(www.waterisac.org)

WaterISAC is an information service developed to provide America's drinking and wastewater systems with a secure web-based environment for early warning of potential threats and a source of knowledge about water system security. The WaterISAC is open to all U.S. drinking water and wastewater systems. A list of website locations provided by WaterISAC is in Table 3.4.

TABLE 3.3 ASDWA Water System Emergency Response Plan Outline

I. **Introduction, Goals, and Importance**
II. **Emergency Planning Process**
   a. Planning Partnerships: The planning process should include those parties who will need to help the utility in an emergency situation (e.g., first responders, law enforcement, public health officials, etc.)
   b. Scenarios: Incorporate VA findings to develop scenarios (events that could cause emergencies/severity of emergencies) in order to flesh out response needed
III. **Emergency Response Plan**
   1. **System Specific Information**
      a. PWS ID, owner, contact person
      b. Population served and service connections
      c. System components
         - Source water
         - Storage
         - Treatment plant
         - Distribution system
   2. **Alternative Water Sources**
   3. **Chain of Command Chart in Coordination with Local Emergency Planning Committee** (Internal and External Emergency Responders)
      - Contact name
      - Organization and responsibility
      - Telephone number
   4. **Communication Procedures**
      Who, what, when (using Chain of Command chart and following notification lists)
      a. Internal notification lists
      b. External notification lists
         - First responders (local police and emergency squad)
         - State personnel
         - Health department
         - Customers
         - Service/mutual aid
         - Others?
      c. Public/media notification (How to Communicate)
   5. **Emergency Response Protocols** (To implement in the event of a terrorist attack or intentional act in order to lessen the impact)
      a. Protocols must include:
         - Plans/actions
         - Procedures
         - Equipment identified
      b. Protocols should provide for the following activities:
         i. Assess the problem
         ii. Isolate and fix the problem
         iii. Monitoring
         iv. Recovery
         v. Return to safety
         vi. Report of findings
IV. **Next Steps**
   a. Plan approval
   b. Practice and plan to update (as necessary; once every year recommended)
V. **Appendix of Resources/Links**

Source: www.asdwa.org

TABLE 3.4 A List of Website Locations Provided by WaterISAC

Water Sector Links
American Water Works Association	http://www.awwa.org
Association of Metropolitan Sewerage Agencies	http://www.amsa-cleanwater.org
Association of Metropolitan Water Agencies	http://www.amwa.net
Association of State Drinking Water Administrators	http://www.asdwa.org
AWWA Research Foundation	http://www.awwarf.com
National Association of Water Companies	http://www.nawc.org
National Rural Water Association	http://www.nrwa.org
Water Environment Federation	http://www.wef.org
Water Environment Research Foundation	http://www.werf.org
Presidential Decision Directive 63: Critical Infrastructure Protection	http://www.ciao.gov/resource/directive.html
Executive Order 13231: Critical Infrastructure Protection	http://www.whitehouse.gov/news/releases/2001/10/20011016-12.htm

Contaminant Resources
Centers for Disease Control Biological Agents	http://www.bt.cdc.gov/agent/agentlist.asp
Centers for Disease Control Chemical Agents	http://www.bt.cdc.gov/agent/agentlistchem.asp
Physician Preparedness for Acts of Water Terrorism	http://www. WaterHealthConnection.org
Public Health Laboratory Service (UK) Biological Agents	http://www.phls.co.uk/topics_az/deliberate_release/menu.htm
Public Health Laboratory Service (UK) Chemical Agents	http://www.phls.co.uk/topics_az/deliberate_release/chemicalhomepag

Federal Links
Department of Homeland Security	http://www.dhs.gov
Environmental Protection Agency	http://www.epa.gov/safewater/security
National Infrastructure Protection Center (NIPC)	http://www.nipc.gov
Federal Bureau of Investigation (FBI)	http://www.fbi.gov
Critical Infrastructure Assurance Office (CIAO)	http://www.ciao.gov
Partnership for Critical Infrastructure Security (PCIS)	http://www.pcis-forum.org
Federal Emergency Management Agency (FEMA)	http://www.fema.gov
Office for State and Local Domestic Preparedness Support (OSLDPS)	http://www.ojp.usdoj.gov/osldps
Office of Emergency Preparedness (OEP)	http://ndms.dhhs.gov
US Army Soldier and Biological Chemical Command (SBCCOM)	http://www.sbccom.army.mil
CDC, "Bioterrorism Alleging Use of Anthrax and Interim Guidelines for Management-United States, 1998"	http://www.cdc.gov/epo/mmwr/preview/mmwrhtml/00056353.htm
EPA National Homeland Security Research Center (EPA ORD)	http://www.epa.gov/ordnhsrc

Other Links
International Association of Emergency Managers (IAEM)	http://www.iaem.com
National Drinking Water Clearinghouse	http://www.ndwc.wvu.edu
Chemical and Biological Defense Information Analysis Center	http://www.cbiac.apgea.army.mil/resources/direc_home.html

(Continued)

TABLE 3.4 A List of Website Locations Provided by Water ISAC (*Continued*)

National Emergency Management Association (NEMA)	http://www.nemaweb.org/index.cfm
National Volunteer Organizations Active in Disasters (NVOAD)	http://www.nvoad.org
Terrorism Research Center	http://www.terrorism.com
Emergency.com: Crisis, Conflict, and Emergency Service News, analysis, and Reference Information	http://www.emergency.com
Monterey Institute of International Studies' Center for Nonproliferation Studies (CNS): Chemical and Biological Weapons Resource Page	http://cns.miis.edu/research/cbw/index.htm
NSF	http://www.nsf.org/consumer/consumer_biolinks.html
The Henry L. Stimson Center's Chemical and Biological Weapons Nonproliferation Project	http://www.stimson.org/cwc/index.html
National Memorial Institute for the Prevention of Terrorism (MIPT)	http://www.mipt.org
EPA Security Product Guide	http://www.epa.gov/safewater/security/guide/index.html
Books, Periodicals, News	
FEMA's Terrorism Consequence Management Courses for Local Jurisdictions	http://www.fema.gov/emi/termng.htm
Washingtonpost.com, "Terror Strikes: A Special Report-Twenty Years of Violence"	http://www.washingtonpost.com/wp-srv/world/terror/intro.htm
Other Important Links	
Baseline Threat Information for Vulnerability Assessment of Community Water Systems	Request document here

Source: http://www.waterisac.org/resourceshtml

APPENDIX 3A WHAT ARE SOME POINTS TO CONSIDER IN A VULNERABILITY ASSESSMENT?

Some points to consider, related to the six basic elements, are included in the following tables. The manner in which the vulnerability assessment is performed is determined by each individual water utility. It will be helpful to remember throughout the assessment process that the ultimate goal is twofold: *to safeguard public health and safety and to reduce the potential for disruption of a reliable supply of pressurized water.*

Basic element	Points to consider

1. Characterization of the water system, including its mission and objectives*	• What are the important missions of the system to be assessed? Define the highest priority services provided by the utility. Identify the utility's customers:
	  • General public
	  • Government
	  • Military
	  • Industrial
	  • Critical care
	  • Retail operations
	  • Firefighting
	• What are the most important facilities, processes, and assets of the system for achieving the mission objectives and avoiding undesired consequences? Describe the
	  • Utility facilities
	  • Operating procedures
	  • Management practices that are necessary to achieve the mission objectives
	  • How the utility operates (e.g., water source including ground and surface water)
	  • Treatment processes
	  • Storage methods and capacity
	  • Chemical use and storage
	  • Distribution system
	In assessing those assets that are critical, consider critical customers, dependence on other infrastructures (e.g., electricity, transportation, other water utilities), contractual obligations, single points of failure (e.g., critical aqueducts, transmission systems, aquifers, etc.), chemical hazards and other aspects of the utility's operations, or availability of other utility capabilities that may increase or decrease the criticality of specific facilities, processes, and assets.

(Continued)

Vulnerability Assessment (*Continued*)

Basic element	Points to consider

2. Identification and prioritization of adverse consequences to avoid	• Take into account the impacts that could substantially disrupt the ability of the system to provide a safe and reliable supply of drinking water or otherwise present significant public health concerns to the surrounding community. Water systems should use the vulnerability assessment process to determine how to reduce risks associated with the consequences of significant concern.
	• Ranges of consequences or impacts for each of these events should be identified and defined. Factors to be considered in assessing the consequences may include
	  • Magnitude of service disruption
	  • Economic impact (such as replacement and installation costs for damaged critical assets or loss of revenue due to service outage)
	  • Number of illnesses or deaths resulting from an event
	  • Impact on public confidence in the water supply
	  • Chronic problems arising from specific events
	  • Other indicators of the impact of each event as determined by the water utility
	Risk reduction recommendations at the conclusion of the vulnerability assessment should strive to prevent or reduce each of these consequences.

3. Determination of critical assets that might be subject to malevolent acts that could result in undesired consequences.	• What are the malevolent acts that could reasonably cause undesired consequences? Consider the operation of critical facilities, assets, and/or processes and assess what an adversary could do to disrupt these operations. Such acts may include physical damage to or destruction of critical assets, contamination of water, intentional release of stored chemicals, interruption of electricity, or other infrastructure interdependencies.
	• The "Public Health Security and Bioterrorism Preparedness and Response Act of 2002" (PL 107-188) states that a community water system which serves a population of greater than 3300 people must review the vulnerability of its system to a terrorist attack or other intentional acts intended to substantially disrupt the ability of the system to provide a safe and reliable supply of drinking water. The vulnerability assessment shall include, but not be limited to, a review of
	  • Pipes and constructed conveyances
	  • Physical barriers
	  • Water collection, pretreatment, and treatment facilities
	  • Storage and distribution facilities
	  • Electronic, computer, or other automated systems which are utilized by the public water system (e.g., Supervisory Control and Data Acquisition (SCADA))

- The use, storage, or handling of various chemicals
- The operation and maintenance of such systems

- Determine the possible modes of attack that might result in consequences of significant concern based on the critical assets of the water system. The objective of this step of the assessment is to move beyond what is merely possible and determine the likelihood of a particular attack scenario. This is a very difficult task as there is often insufficient information to determine the likelihood of a particular event with any degree of certainty.
- The threats (the kind of adversary and the mode of attack) selected for consideration during a vulnerability assessment will dictate, to a great extent, the risk reduction measures that should be designed to counter the threat(s). Some vulnerability assessment methodologies refer to this as a "Design Basis Threat" (DBT) where the threat serves as the basis for the design of countermeasures, as well as the benchmark against which vulnerabilities are assessed. It should be noted that there is no single DBT or threat profile for all water systems in the United States. Differences in geographic location, size of the utility, previous attacks in the local area and many other factors will influence the threat(s) that water systems should consider in their assessments. Water systems should consult with the local FBI and/or other law enforcement agencies, public officials, and others to determine the threats upon which their risk reduction measures should be based. Water systems should also refer to EPA's "Baseline Threat Information for Vulnerability Assessments of Community Water Systems" to help assess the most likely threats to their system. This document is available to community water systems serving populations greater than 3300 people. If your system has not yet received instructions on how to receive a copy of this document, then contact your regional EPA office immediately. You will be sent instructions on how to securely access the document via the Water Information Sharing and Analysis Center (ISAC) website or obtain a hardcopy that can be mailed directly to you. Water systems may also want to review their incident reports to better understand past breaches of security.

4. Assessment of the likelihood (qualitative probability) of such malevolent acts from adversaries (e.g., terrorists, vandals).

5. Evaluation of existing countermeasures. (Depending on countermeasures already in place, some critical assets may already be sufficiently protected. This step will aid in identification of the areas of greatest concern, and help to focus priorities for risk reduction.)

- What capabilities does the system currently employ for detection, delay, and response?†
  - Identify and evaluate current detection capabilities such as intrusion detection systems, water quality monitoring, operational alarms, guard post orders, and employee security awareness programs.
  - Identify current delay mechanisms such as locks and key control, fencing, structure integrity of critical assets and vehicle access checkpoints.
  - Identify existing policies and procedures for evaluation and response to intrusion and system malfunction alarms, adverse water quality indicators, and cyber system intrusions.
- What cyber protection system features does the utility have in place? Assess what protective measures are in place for the SCADA and business-related computer information systems such as‡
  - Firewalls
  - Modem access
  - Internet and other external connections, including wireless data and voice communications
  - Security policies and protocols

(Continued)

3.13

Vulnerability Assessment (*Continued*)

Basic element	Points to consider

	• What security policies and procedures exit, and what is the compliance record for them? Identify existing policies and procedures concerning
	  • Personnel security
	  • Physical security
	  • Key and access badge control
	  • Control of system configuration and operational data
	  • Chemical and other vendor deliveries
	  • Security training and exercise records

6. Analysis of current risk and development of a prioritized plan for risk reduction	• Information gathered on threat, critical assets, water utility operations, consequences, and existing countermeasures should be analyzed to determine the current level of risk. The utility should then determine whether current risks are acceptable or risk reduction measures should be pursued.
	• Recommended actions should measurably reduce risks by reducing vulnerabilities and/or consequences through improved deterrence, delay, detection, and/or response capabilities or by improving operational policies or procedures. Selection of specific risk reduction actions should be completed prior to considering the cost of the recommended action(s). Utilities should carefully consider both short- and long-term solutions. An analysis of the cost of short- and long-term risk reduction actions may impact which actions the utility chooses to achieve its security goals.
	• Utilities may also want to consider security improvements in light of other planned or needed improvements. Security and general infrastructure may provide significant multiple benefits. For example, improved treatment processes or system redundancies can both reduce vulnerabilities and enhance day-to-day operation.
	• General, strategies for reducing vulnerabilities fall into three broad categories:
	  Sound business practices affect policies, procedures, and training to improve the overall security-related culture at the drinking water facility. For example, it is important to ensure rapid communication capabilities exist between public health authorities and local law enforcement and emergency responders.
	  System upgrades include changes in operations, equipment, processes, or infrastructure itself that make the system fundamentally safer.
	  Security upgrades improve capabilities for detection, delay, or response.

*Answers to system-specific questions may be helpful in characterizing the water system.

†It is important to determine the performance characteristics. Poorly operated and maintained security technologies provide little or no protection.

‡It is important to identify whether vendors have access rights and/or "backdoors" to conduct system diagnostics remotely.

Source: Vulnerability Assessment Fact Sheet, U.S. EPA (2002).

APPENDIX 3B SECURITY VULNERABILITY SELF-ASSESSMENT FOR SMALL WATER SYSTEMS

General Questions for the Entire Water System

The first 15 questions in this vulnerability self-assessment are general questions designed to apply to all components of your system [wellhead or surface water intake, treatment plant, storage tank(s), pumps, distribution system, and offices]. These are followed by more specific questions that look at individual system components in greater detail.

Question	Answer	Comment	Action needed/taken
1. Do you have a written emergency response plan (ERP)?	Yes ☐ No ☐	• Under the provisions of the Public Health Security and Bioterrorism Preparedness and Response Act of 2002 you are required to develop and/or update an ERP within 6 months after completing this assessment. If you do not have an ERP, you can obtain a sample from your state drinking water primacy agency. As a first step in developing your ERP, you should develop your emergency contact list. • A plan is vital in case there is an incident that requires immediate response. Your plan should be reviewed at least annually (or more frequently if necessary) to ensure that it is up-to-date and addresses security emergencies including ready access to laboratories capable of analyzing water samples. You should coordinate with your Local Emergency Planning Committee (LEPC). • You should designate someone to be contacted in case of emergency regardless of the day of the week or time of day. This contact information should be kept up-to-date and made available to all water system personnel and local officials (if applicable). • Share this ERP with police, emergency personnel, and your state primacy agency. Posting contact information is a good idea only if authorized personnel are the only ones seeing the information. These signs could pose a security risk if posted for public viewing since it gives people information that could be used against the system.	

(Continued)

3.15

General Questions for the Entire Water System (*Continued*)

Question	Answer	Comment	Action needed/taken
2. Have you reviewed U.S. EPA's Baseline Threat Information Document?	Yes ☐ No ☐	• The U.S. EPA baseline threat document is available through the Water Information Sharing and Analysis Center at www.waterisac.org. It is important you use this document to determine potential threats to your system and to obtain additional security related information. U.S. EPA should have provided a certified letter to your system that provided instructions on obtaining the threat document.	
3. Is access to the critical components of the water system (i.e., a part of the physical infrastructure of the system that is essential for water flow and/or water quality) restricted to authorized personnel only?	Yes ☐ No ☐	• You should restrict or limit access to the critical components of your water system to authorized personnel only. This is the first step in security enhancement for your water system. Consider the following: • Issue water system photo identification cards for employees, and require them to be displayed within the restricted area at all times. • Post signs restricting entry to authorized personnel and ensure that assigned staff escort people without proper ID.	
4. Are all critical facilities fenced, including wellhouses and pump pits, and are gates locked where appropriate?	Yes ☐ No ☐	• Ideally, all facilities should have a security fence around the perimeter. • The fence perimeter should be walked periodically to check for breaches and maintenance needs. All gates should be locked with chains and a tamper-proof padlock that at a minimum protects the shank. Other barriers such as concrete "jersey" barriers should be considered to guard certain critical components from accidental or intentional vehicle intrusion.	
5. Are all critical doors, windows, and other points of entry such as tank and roof hatches and vents kept closed and locked?	Yes ☐ No ☐	• Lock all building doors and windows, hatches and vents, gates, and other points of entry to prevent access by unauthorized personnel. Check locks regularly. Dead bolt locks and lock guards provide a high level of security for the cost. • A daily check of critical system components enhances security and ensures that an unauthorized entry has not taken place. • Doors and hinges to critical facilities should be constructed of heavy-duty reinforced material. Hinges on all outside doors should be located on the inside.	

6. Is there external lighting around all critical components of your water system?	Yes ☐ No ☐	• To limit access to water systems, all windows should be locked and reinforced with wire mesh or iron bars, and bolted on the inside. Systems should ensure that this type of security meets with the requirements of any fire codes. Alarms can also be installed on windows, doors, and other points of entry • Adequate lighting of the exterior of water systems' critical components is a good deterrent to unauthorized access and may result in the detection or deterrence of trespassers. Motion detectors that activate switches that turn lights on or trigger alarms also enhance security.
7. Are warning signs (tampering, unauthorized access, etc.) posted on all critical components of your water system (e.g., well houses and storage tanks)?	Yes ☐ No ☐	• Warning signs are an effective means to deter unauthorized access. • "Warning—tampering with this facility is a federal offense" should be posted on all water facilities. These are available from your state rural water association. • "Authorized personnel only," "Unauthorized access prohibited," and "Employees only" are examples of other signs that may be useful.
8. Do you patrol and inspect all source intake, buildings, storage tanks, equipment, and other critical components?	Yes ☐ No ☐	• Frequent and random patrolling of the water system by utility staff may discourage potential tampering. It may also help identify problems that may have arisen since the previous patrol. • All systems are encouraged to initiate personal contact with the local law enforcement to show them the drinking water facility. The tour should include the identification of all critical components with an explanation of why they are important. Systems are encouraged to review, with local law enforcement, the *NRWA/ASDWA Guide for Security Decisions* or similar state document to clarify respective roles and responsibilities in the event of an incident. Also consider asking the local law enforcement to conduct periodic patrols of your water system.
9. Is the area around all the critical components of your water system free of objects that may be used for breaking and entering?	Yes ☐ No ☐	• When assessing the area around your water system's critical components, look for objects that could be used to gain entry (e.g., large rocks, cement blocks, pieces of wood, ladders, valve keys, and other tools).

(Continued)

General Questions for the Entire Water System (*Continued*)

Question	Answer	Comment	Action needed/taken
10. Are the entry points to all of your water system easily seen?	Yes ☐ No ☐	• You should clear fence lines of all vegetation. Overhanging or nearby trees may also provide easy access. Avoid landscaping that will permit trespassers to hide or conduct unnoticed suspicious activities. • Trim trees and shrubs to enhance the visibility of your water system's critical components. • If possible, park vehicles and equipment in places where they do not block the view of your water system's critical components.	
11. Do you have an alarm system that will detect unauthorized entry or attempted entry at all critical components?	Yes ☐ No ☐	• Consider installing an alarm system that notifies the proper authorities or your water system's designated contact for emergencies when there has been a breach of security. Inexpensive systems are available. An alarm system should be considered whenever possible for tanks, pump houses, and treatment facilities. • You should also have an audible alarm at the site as a deterrent and to notify neighbors of a potential threat.	
12. Do you have a key control and accountability policy?	Yes ☐ No ☐	• Keep a record of locks and associated keys, and to whom the keys have been assigned. This record will facilitate lock replacement and key management (e.g., after employee turnover or loss of keys). Vehicle and building keys should be kept in a lockbox when not in use. • You should have all keys stamped (engraved) "DO NOT DUPLICATE."	
13. Are entry codes and keys limited to water system personnel only?	Yes ☐ No ☐	• Suppliers and personnel from co-located organizations (e.g., organizations using your facility for telecommunications) should be denied access to codes and/or keys. Codes should be changed frequently if possible. Entry into any building should always be under the direct control of water system personnel.	
14. Do you have an updated operations and maintenance manual that includes evaluations of security systems?	Yes ☐ No ☐	• Operation and maintenance plans are critical in assuring the ongoing provision of safe and reliable water service. These plans should be updated to incorporate security considerations and the ongoing reliability of security provisions, including security procedures and security-related equipment.	
15. Do you have a neighborhood watch program for your water system?	Yes ☐ No ☐	• Watchful neighbors can be very helpful to a security program. Make sure they know whom to call in the event of an emergency or suspicious activity.	

Water Sources

In addition to the preceding general checklist for your entire water system (questions 1 to 15), you should give special attention to the following issues, presented in separate tables, related to various water system components. Your water sources (surface water intakes or wells) should be secured. Surface water supplies present the greatest challenge. Typically they encompass large land areas. Where areas cannot be secured, steps should be taken to initiate or increase law enforcement patrols. Pay particular attention to surface water intakes. Ask the public to be vigilant and report suspicious activity.

Question	Answer	Comment	Action needed/taken
16. Are your wellheads sealed properly?	Yes ☐ No ☐	• A properly sealed wellhead decreases the opportunity for the introduction of contaminants. If you are not sure whether your wellhead is properly sealed, contact your well drilling/maintenance company, your state drinking water primacy agency, your state rural water association, or other technical assistance providers.	
17. Are well vents and caps screened and securely attached?	Yes ☐ No ☐	• Properly installed vents and caps can help prevent the introduction of a contaminant into the water supply. • Ensure that vents and caps serve their purpose, and cannot be easily breached or removed.	
18. Are observation/test and abandoned wells properly secured to prevent tampering?	Yes ☐ No ☐	• All observation/test and abandoned wells should be properly capped or secured to prevent the introduction of contaminants into the aquifer or water supply. Abandoned wells should be either removed or filled with concrete.	
19. Is your surface water source secured with fences or gates? Do water system personnel visit the source?	Yes ☐ No ☐	• Surface water supplies present the greatest challenge to secure. Often, they encompass large land areas. Where areas cannot be secured, steps should be taken to initiate or increase patrols by water utility personnel and law enforcement agents.	

Treatment Plant and Suppliers

Some small systems provide easy access to their water system for suppliers of equipment, chemicals, and other materials for the convenience of both parties. This practice should be discontinued.

Question	Answer	Comment	Action needed/taken
20. Are deliveries of chemicals and other supplies made in the presence of water system personnel?	Yes ☐ No ☐	• Establish a policy that an authorized person, designated by the water system, must accompany all deliveries. Verify the credentials of all drivers. This prevents unauthorized personnel from having access to the water system.	
21. Have you discussed with your supplier(s) procedures to ensure the security of their products?	Yes ☐ No ☐	• Verify that your suppliers take precautions to ensure that their products are not contaminated. Chain of custody procedures for delivery of chemicals should be reviewed. You should inspect chemicals and other supplies at the time of delivery to verify that they are sealed in and in unopened containers. Match all delivered goods with purchase orders to ensure that they were, in fact, ordered by your water system. • You should keep a log or journal of deliveries. It should include the driver's name (taken from the driver's photo ID), date, time, material delivered, and the supplier's name.	
22. Are chemicals, particularly those that are potentially hazardous (e.g., chlorine gas) or flammable, properly stored in a secure area?	Yes ☐ No ☐	• All chemicals should be stored in an area designated for their storage only, and the area should be secure and access to the area restricted. Access to chemical storage should be available only to authorized employees. Pay special attention to the storage, handling, and security of chlorine gas because of its potential hazard. • You should have tools and equipment on site (such as a fire extinguisher, drysweep, etc.) to take immediate actions when responding to an emergency.	
23. Do you monitor raw and treated water so that you can detect changes in water quality?	Yes ☐ No ☐	• Monitoring of raw and treated water can establish a baseline that may allow you to know if there has been a contamination incident. • Some parameters for raw water include pH, turbidity, total and fecal coliform, total organic carbon, specific conductivity, ultraviolet adsorption, color, and odor.	

		• Routine parameters for finished water and distribution systems include free and total chlorine residual, heterotrophic plate count (HPC), total and fecal coliform, pH, specific conductivity, color, taste, odor, and system pressure. • Chlorine demand patterns can help you identify potential problems with your water. A sudden change in demand may be a good indicator of contamination in your system. • For those systems that use chlorine, absence of chlorine residual may indicate possible contamination. Chlorine residuals provide protection against bacterial and viral contamination that may enter the water supply.
24. Are tank ladders, access hatches, and entry points secured?	Yes ☐ No ☐	• The use of tamper-proof padlocks at entry points (hatches, vents, and ladder enclosures) will reduce the potential for of unauthorized entry. • If you have towers, consider putting physical barriers on the legs to prevent unauthorized climbing.
25. Are vents and overflow pipes properly protected with screens and/or grates?	Yes ☐ No ☐	• Air vents and overflow pipes are direct conduits to the finished water in storage facilities. Secure all vents and overflow pipes with heavy-duty screens and/or grates.
26. Can you isolate the storage tank from the rest of the system?	Yes ☐ No ☐	• A water system should be able to take its storage tank(s) out of operation or drain its storage tank(s) if there is a contamination problem or structural damage. Install shutoff or bypass valves to allow you to isolate the storage tank in the case of a contamination problem or structural damage. • Consider installing a sampling tap on the storage tank outlet to test water in the tank for possible contamination.

Distribution

Hydrants are highly visible and convenient entry points into the distribution system. Maintaining and monitoring positive pressure in your system is important to provide fire protection and prevent introduction of contaminants.

Question	Answer	Comment	Action needed/taken
27. Do you control the use of hydrants and valves?	Yes ☐ No ☐	• Your water system should have a policy that regulates the authorized use of hydrants for purposes other than fire protection. Require authorization and backflow devices if a hydrant is used for any purpose other than fire fighting. • Consider designating specific hydrants for use as filling station(s) with proper backflow prevention (e.g., to meet the needs of construction firms). Then, notify local law enforcement officials and the public that these are the only sites designated for this use. • Flush hydrants should be kept locked to prevent contaminants from being introduced into the distribution system, and to prevent improper use.	
28. Does your system monitor for, and maintain, positive pressure?	Yes ☐ No ☐	• Positive pressure is essential for fire fighting and for preventing backsiphonage that may contaminate finished water in the distribution system. Refer to your state primacy agency for minimum drinking water pressure requirements.	
29. Has your system implemented a backflow prevention program?	Yes ☐ No ☐	• In addition to maintaining positive pressure, backflow prevention programs provide an added margin of safety by helping to prevent the intentional introduction of contaminants. If you need information on backflow prevention programs, contact your state drinking water primacy agency.	

Personnel

You should add security procedures to your personnel policies.

Question	Answer	Comment	Action needed/taken
30. When hiring personnel, do you request that local police perform a criminal background check, and do you verify employment eligibility (as required by the Immigration and Naturalization Service, Form I-9)?	Yes ☐ No ☐	• It is good practice to have all job candidates fill out an employment application. You should verify professional references. Background checks conducted during the hiring process may prevent potential employee-related security issues. • If you use contract personnel, check on the personnel practices of all providers to ensure that their hiring practices are consistent with good security practices.	
31. Are your personnel issued photo-identification cards?	Yes ☐ No ☐	• For positive identification, all personnel should be issued water system photoidentification cards and be required to display them at all times. • Photoidentification will also facilitate identification of authorized water system personnel in the event of an emergency.	
32. When terminating employment, do you require employees to turn in photo IDs, keys, access codes, and other security-related items?	Yes ☐ No ☐	• Former or disgruntled employees have knowledge about the operation of your water system, and could have both the intent and the physical capability to harm your system. Requiring employees who will no longer be working at your water system to turn in their IDs, keys, and access codes helps limit these types of security breaches.	
33. Do you use uniforms and vehicles with your water system name prominently displayed?	Yes ☐ No ☐	• Requiring personnel to wear uniforms, and requiring that all vehicles prominently display the water system name, helps inform the public when water system staff is working on the system. Any observed activity by personnel without uniforms should be regarded as suspicious. The public should be encouraged to report suspicious activity to law enforcement authorities.	

(Continued)

Personnel (*Continued*)

Question	Answer	Comment	Action needed/taken
34. Have water system personnel been advised to report security vulnerability concerns and to report suspicious activity?	Yes ☐ No ☐	• Your personnel should be trained and knowledgeable about security issues at your facility, what to look for, and how to report any suspicious events or activity. • Periodic meetings of authorized personnel should be held to discuss security issues.	
35. Do your personnel have a checklist to use for threats or suspicious calls or to report suspicious activity?	Yes ☐ No ☐	• To properly document suspicious or threatening phone calls or reports of suspicious activity, a simple checklist can be used to record and report all pertinent information. Calls should be reported immediately to appropriate law enforcement officials. Checklists should be available at every telephone. • Also consider installing caller ID on your telephone system to keep a record of incoming calls.	

Information Storage, Computers, Controls, and Maps

Security of the system, including computerized controls such as a Supervisory Control and Data Acquisition (SCADA) system, goes beyond the physical aspects of operation. It also includes records and critical information that could be used by someone planning to disrupt or contaminate your water system.

Question	Answer	Comment	Action needed/taken
36. Is computer access "password-protected?" Is virus protection installed and software upgraded regularly and are your virus definitions updated at least daily? Do you have Internet firewall software installed on your computer? Do you have a plan to back up your computers?	Yes ☐ No ☐	• All computer access should be password-protected. Passwords should be changed every 90 days and (as needed) following employee turnover. When possible, each individual should have a unique password that is not shared with others. If you have Internet access, a firewall protection program should be installed on your side of the computer and reviewed and updated periodically. • Also consider contacting a virus protection company and subscribing to a virus update program to protect your records. • Backing up computers regularly will help prevent the loss of data in the event that your computer is damaged or breaks. Backup copies of computer data should be made routinely and stored at a secure off-site location.	

37. Is there information on the Web that can be used to disrupt your system or contaminate your water?	Yes ☐ No ☐	• Posting detailed information about your water system on a Web site may make the system more vulnerable to attack. Web sites should be examined to determine whether they contain critical information that should be removed. • You should do a Web search (using a search engine such as Google, Yahoo!, or Lycos) using key words related to your water supply to find any published data on the Web that is easily accessible by someone who may want to damage your water supply.
38. Are maps, records, and other information stored in a secure location?	Yes ☐ No ☐	• Records, maps, and other information should be stored in a secure location when not in use. Access should be limited to authorized personnel only. • You should make backup copies of all data and sensitive documents. These should be stored in a secure off-site location on a regular basis.
39. Are copies of records, maps, and other sensitive information labeled confidential, and are all copies controlled and returned to the water system?	Yes ☐ No ☐	• Sensitive documents (e.g., schematics, maps, and plans and specifications) distributed for construction projects or other uses should be recorded and recovered after use. You should discuss measures to safeguard your documents with bidders for new projects.
40. Are vehicles locked and secured at all times?	Yes ☐ No ☐	• Vehicles are essential to any water system. They typically contain maps and other information about the operation of the water system. Water system personnel should exercise caution to ensure that this information is secure. • Water system vehicles should be locked when they are not in use or left unattended. • Remove any critical information about the system before parking vehicles for the night. • Vehicles also usually contain tools (e.g., valve wrenches) and keys that could be used to access critical components of your water system. These should be secured and accounted for daily.

Public Relations

You should educate your customers about your system. You should encourage them to be alert and to report any suspicious activity to law enforcement authorities.

Question	Answer	Comment	Action needed/taken
41. Do you have a program to educate and encourage the public to be vigilant and report suspicious activity to assist in the security protection of your water system?	Yes ☐ No ☐	• Advise your customers and the public that your system has increased preventive security measures to protect the water supply from vandalism. Ask for their help. Provide customers with your telephone number and the telephone number of the local law enforcement authority so that they can report suspicious activities. The telephone number can be made available through direct mail, billing inserts, notices on community bulletin boards, flyers, and consumer confidence reports.	
42. Does your water system have a procedure to deal with public information requests, and to restrict distribution of sensitive information?	Yes ☐ No ☐	• You should have a procedure for personnel to follow when you receive an inquiry about the water system or its operation from the press, customers, or the general public. • Your personnel should be advised not to speak to the media on behalf of the water system. Only one person should be designated as the spokesperson for the water system. Only that person should respond to media inquiries. You should establish a process for responding to inquiries from your customers and the general public.	
43. Do you have a procedure in place to receive notification of a suspected outbreak of a disease immediately after discovery by local health agencies?	Yes ☐ No ☐	• It is critical to be able to receive information about suspected problems with the water at any time and respond to them quickly. Written procedures should be developed in advance with your state drinking water primacy agency, local health agencies, and your local emergency planning committee and reviewed periodically.	

44. Do you have a procedure in place to advise the community of contamination immediately after discovery?	Yes ☐ No ☐	• As soon as possible after a disease outbreak, you should notify testing personnel and your laboratory of the incident. In outbreaks caused by microbial contaminants, it is critical to discover the type of contaminant and its method of transport (water, food, etc.). Active testing of your water supply will enable your laboratory, working in conjunction with public health officials, to determine if there are any unique (and possibly lethal) disease organisms in your water supply. • It is critical to be able to get the word out to your customers as soon as possible after discovering a health hazard in your water supply. In addition to your responsibility to protect public health, you must also comply with the requirements of the Public Notification Rule. Some simple methods include announcements via radio or television, door-to-door notification, a phone tree, and posting notices in public places. The announcement should include accepted uses for the water and advice on where to obtain safe drinking water. Call large facilities that have large populations of people who might be particularly threatened by the outbreak such as hospitals, nursing homes, the school district, jails, large public buildings, and large companies. Enlist the support of local emergency response personnel to assist in the effort.
45. Do you have a procedure in place to respond immediately to a customer complaint about a new taste, odor, color, or other physical change (oily, filmy, burns on contact with skin)?	Yes ☐ No ☐	• It is critical to be able to respond to and quickly identify potential water quality problems reported by customers. Procedures should be developed in advance to investigate and identify the cause of the problem, as well as to alert local health agencies, your state drinking water primacy agency, and your local emergency planning committee if you discover a problem.

Note: Now that you have completed the *Security Vulnerability Self-Assessment Guide for Small Water Systems Serving Populations Between 3300 and 10,000*, review your needed actions and then prioritize them on the basis of the most likely threats.

Source: www.asdwa.org

APPENDIX 3C WATER UTILITY RESPONSE, RECOVERY, AND REMEDIATION GUIDELINES

Water Utility Response, Recovery & Remediation Guidance for Man-Made and/or Technological Emergencies (April 15, 2002)

I. Contamination Event: Articulated Threat with Unspecified Material

Event Description. This event is based on the threat of intentional introduction of a contaminant into the water system (at any point within the system) without specification of the contaminant by the perpetrator.

Initial notifications
• Notify local law enforcement • Notify local FBI field office • Notify National Response Center • Notify local/state emergency management organization • Notify ISAC • Notify other associated system authorities (wastewater, water) • Notify local government official • Notify local/state health and/or environmental department • Notify critical care facilities • Notify employees • Consider when to notify customers and what notification to issue • Notify state governor
Response actions

1. Source water
   - Increase sampling at or near system intakes
   - Consider whether to isolate the water source if possible

2. Drinking water treatment facility
   - Preserve latest full-battery background test as baseline
   - Increase sampling efforts
   - Consider whether to continue normal operations (if determination is made to reduce or stop water treatment, provide notification to customers and issue alerts)
   - Coordinate alternative water supply

3. Water distribution/storage
   - Consider whether to isolate the water in the affected area if possible

4. Wastewater collection system
   - Assess what to do with potentially contaminated water within the system based on contaminant, contaminant concentration, potential for system contamination, and ability to bypass treatment plant.
   - If bypassed, notify local and appropriate state authorities, and downstream users; increase monitoring of receiving stream.

5. Wastewater treatment facility
   - Preserve latest full-battery background test as baseline
   - Increase sampling efforts
   - Consider whether to continue normal operations (if determination is made to reduce or stop water treatment, provide notification to customers and issue alerts)

VULNERABILITY ASSESSMENT, EMERGENCY RESPONSE

Recovery actions
Recovery actions should begin once the contaminant is through the system.

Recovery notifications
• Notify customers • Notify media • Notify ISAC

Appropriate utility elements
• Sample appropriate system elements (storage tanks, filters, sediment basins, solids handling) to determine if residual contamination exists • Flush system on the basis of results of sampling • Monitor health of employees • Plan for appropriate disposition of personal protection equipment (PPE) and other equipment

Remediation actions
• On the basis of sampling results, assess need to remediate storage tanks, filters, sediment basins, solids handling • Plan for appropriate disposition of PPE and other equipment • If wastewater treatment plant was bypassed, sample and establish monitoring regime for receiving stream and potential remediation based on sampling results

Note: Response, recovery and remediation actions may be tailored to a specified (identified) material if the physical properties for the material are known.

II. Contamination Threat at a Major Event

Event Description. This event is based on the threat of, or actual, intentional introduction of a contaminant into the water system at a sports arena, convention center, or similar facility.

Initial notifications
• Notify local law enforcement • Notify local FBI field office • Notify National Response Center • Notify ISAC • Notify local/state emergency management organization • Notify wastewater facility • Notify state governor • Notify other associated system authorities (wastewater, water) • Notify local government official • Notify local/state health and/or environmental department • Notify critical care facilities • Notify employees • Consider when to notify customers and what notification to issue

Response actions

1. Source water
   • No recommended action to take

2. Drinking water treatment facility
   - No recommended action to take
3. Water distribution storage
   - Coordinate isolation of water
   - Assist in plan for draining the contained water
   - Assist in developing a plan for sampling water for potential contamination based on threat notification
   - Provide alternate water source
4. Wastewater collection system
   - Coordinate acceptance of isolated water
   - Monitor accepted water
   - Assist in plan for draining the contained water
   - Assist in developing a plan for sampling water for potential contamination based on threat notification
5. Wastewater treatment facility
   - Coordinate acceptance of isolated water
   - Monitor accepted water
   - Assist in plan for draining the contained water
   - Assist in developing a plan for sampling water for potential contamination based on threat notification

Recovery actions

Recovery actions should begin once the contaminant is through the system.

Recovery notifications

- Notify customers in the area of the facility of actions to take
- Notify customers in affected area once contaminant-free clean water is reestablished
- Notify downstream users such as water suppliers, irrigators, and electric generating plants

Water distribution storage

- Consider flushing system via hydrants in distribution systems

Remediation actions

1. Water distribution/storage
   - Assess need to decontaminate or replace distribution system components
2. Wastewater treatment plant
   - On the basis of sampling results, assess need to remediate storage tanks, filters, sediment basins, solids handling
   - Plan for appropriate disposition of PPE and other equipment
   - If wastewater treatment plant was bypassed, sample and establish monitoring regime for receiving stream and potential remediation based on sampling results

VULNERABILITY ASSESSMENT, EMERGENCY RESPONSE   **3.31**

III. Notification from Health Officials of Potential Water Contamination

Event Description. This event is based on the water utility being notified by public health officials of potential contamination based on symptoms of patients.

Initial notifications
• Ask notifying official who else has been notified and request information on symptoms, potential contaminants, and potential area affected • Notify local law enforcement • Notify local FBI field office • Notify National Response Center • Notify local/state emergency management organization • Notify other associated system authorities (wastewater, water) • Notify local government official • Notify state governor • Notify local/state health and/or environmental department • Notify critical care facilities • Notify employees • Consider when to notify customers and what notification to issue • Notify ISAC
Response actions

1. Source water
   - Increase sampling at or near system intakes
   - Consider whether to isolate
2. Drinking water treatment facility
   - Preserve latest full-battery background test result as baseline
   - Increase sampling efforts
   - Consider whether to continue normal operations (if determination is to reduce or stop water treatment, provide notification to customers and issue alerts)
   - Coordinate alternative water supply (if needed)
3. Water distribution/storage
   - Increase sampling in the area potentially affected and at locations where the contaminant could have migrated to; it is important to consider the time between exposure and onset of symptoms to select sampling sites
   - Consider whether to isolate
   - Consider whether to increase residual disinfectant levels
4. Wastewater collection system
   - Increase sampling at pump stations, specifically in the area potentially affected
   - Assess what to do with potentially contaminated water within the system according to contaminant, contaminant concentration, potential for system contamination, and ability to bypass treatment plant
   - If bypassed, notify local and appropriate state authorities, downstream users (especially drinking water treatment facilities), and increase monitoring of receiving stream
5. Wastewater treatment facility
   - Increase sampling at pump stations, specifically in the area potentially affected
   - Assess what to do with potentially contaminated water within the system according to contaminant, contaminant concentration, potential for system contamination, and ability to bypass treatment plant
   - If bypassed, notify local and appropriate state authorities, downstream users (especially drinking water treatment facilities), and increase monitoring of receiving stream

Recovery actions

Recovery actions should begin once the contaminant is through the system.

Recovery notifications

- Assist health department with notifications to customers, media, downstream users, and other organizations

Appropriate utility elements

- Sample appropriate system elements (storage tanks, filters, sediment basins, solids handling) to determine if residual contamination exists
- Flush system according to results of sampling
- Monitor health of employees
- Plan for appropriate disposition of personal protection equipment (PPE) and other equipment

Remediation actions

- On the basis of sampling results, assess need to remediate storage tanks, filters, sediment basins, solids handling, and drinking water distribution system
- Plan for appropriate disposition of PPE and other equipment
- If wastewater treatment plant was bypassed, sample and establish monitoring regime for receiving stream and potential remediation based on sampling results

Note: Patient symptoms should be used to narrow the list of potential contaminants.

IV. Intrusion through Supervisory Control and Data Acquisition (SCADA)

Event Description: This event is based on internal or external intrusion of the SCADA system to disrupt normal water system operations.

Initial notifications

- Notify local law enforcement
- Notify local FBI field office
- Notify National Infrastructure Protection Center (NIPC) at 1-888-585-9078 (or 202-323-3204/5/6)
- Notify other associated system authorities (wastewater, water)
- Notify employees
- If the water is assessed to be unfit for consumption, consider when to notify customers and what notification to issue

Response actions

1. Source water
    - Increase sampling at or near system intakes
    - Consider whether to isolate
2. Drinking water treatment facility
    - Preserve latest full-battery background test as baseline
    - Increase sampling efforts
    - Temporarily shut down SCADA system and go to manual operation using established protocol
    - Consider whether to shut down system and provide alternate water
3. Water distribution/storage
    - Monitor unmanned components (storage tanks and pumping stations)
    - Consider whether to isolate

VULNERABILITY ASSESSMENT, EMERGENCY RESPONSE

4. Wastewater collection system
   - Temporarily shut down SCADA system and go to manual operation using established protocol
   - Monitor unmanned components (pumping stations)—required only if wastewater SCADA system is compromised
   - If SCADA intrusion caused release of improperly treated water, consider whether to continue normal operations (if determination is made to reduce or stop water treatment, provide notification to customers/issue alerts)
5. Wastewater treatment facility
   - Temporarily shut down SCADA system and go to manual operation using established protocol
   - Monitor unmanned components (pumping stations)—required only if wastewater SCADA system is compromised
   - If SCADA intrusion caused release of improperly treated water, consider whether to continue normal operations (if determination is made to reduce or stop water treatment, provide notification to customers/issue alerts)

Recovery actions

Recovery actions should begin once the intrusion has been eliminated and the contaminant/unsafe water (if this occurs) is through the system.

Recovery notifications

- Employees
- Local law enforcement
- Notify customers and media if the event resulted in contamination and the full range (see scenario I) of standard notifications were made

---
Appropriate utility elements

- With FBI assistance, make an image copy of all system logs to preserve evidence
- With FBI assistance, check for implanted backdoors and other malicious code and eliminate them before restarting SCADA system
- Install safeguards before restarting SCADA
- Bring SCADA system up and monitor system

Remediation actions

- Assess and/or implement additional protections for SCADA system
- Check for an NIPC water sector warning, based on the intrusion that may contain additional protective actions to be considered; NIPC warnings can be found at www.NIPC.gov or www.infragard.org for secure access Infragard members

V. Significant Structural Damage Resulting from an Intentional Act

Event Description. This event is based on intentional structural damage to water system components to disrupt normal system operations.

Initial notifications

- Notify local law enforcement
- Notify local FBI field office
- Notify National Response Center
- Notify local/state emergency management organization
- Notify state governor
- Notify ISAC

- Notify other associated system authorities (wastewater, water)
- Notify local government officials
- Notify local/state health and/or environmental department
- Notify critical care facilities
- Notify employees
- Consider when to notify customers and what notification to issue

Response actions

1. Source water
   - Deploy damage assessment teams; if damage appears to be intentional, then treat as crime scene—consult local/state law enforcement and FBI on evidence preservation
   - Inform law enforcement and FBI of potential hazardous materials
   - Coordinate alternative water supply, as needed
   - Consider increasing security measures
   - Based on extent of damage, consider alternate (interim) treatment schemes to maintain at least some level of treatment
2. Drinking water treatment system
   - Deploy damage assessment teams; if damage appears to be intentional, then treat as crime scene—consult local/state law enforcement and FBI on evidence preservation
   - Inform law enforcement and FBI of potential hazardous materials
   - Coordinate alternative water supply, as needed
   - Consider increasing security measures
   - Based on extent of damage, consider alternate (interim) treatment schemes to maintain at least some level of treatment
3. Water distribution/storage
   - Deploy damage assessment teams; if damage appears to be intentional, then treat as crime scene—consult local/state law enforcement and FBI on evidence preservation
   - Inform law enforcement and FBI of potential hazardous materials
   - Coordinate alternative water supply, as needed
   - Consider increasing security measures
   - Based on extent of damage, consider alternate (interim) treatment schemes to maintain at least some level of treatment
4. Wastewater collection system
   - Deploy damage assessment teams; if damage appears to be intentional, then treat as crime scene—consult local/state law enforcement and FBI on evidence preservation
   - Inform law enforcement and FBI of potential hazardous materials
   - Coordinate alternative water supply, as needed
   - Consider increasing security measures
   - Based on extent of damage, consider alternate (interim) treatment schemes to maintain at least some level of treatment
5. Wastewater treatment facility
   - Deploy damage assessment teams; if damage appears to be intentional, then treat as crime scene—consult local/state law enforcement and FBI on evidence preservation
   - Inform law enforcement and FBI of potential hazardous materials
   - Coordinate alternative water supply, as needed
   - Consider increasing security measures
   - Based on extent of damage, consider alternate (interim) treatment schemes to maintain at least some level of treatment

Recovery actions

- Recovery actions should begin as soon as practical after damaged facility is isolated from the rest of the utility facilities

Recovery notifications

- Employees
- Local law enforcement
- Notify local FBI office

Appropriate utility elements

- Dependent on the feedback from damage assessment teams
- Implement damage recovery plan

Remediation actions

- Repair damage
- Assess need for additional protection/security measures for damaged facility and other critical facilities within the utility

APPENDIX 3D WATER SYSTEM EMERGENCY RESPONSE PLAN OUTLINE*

I. Introduction

Safe and reliable drinking water is vital to every community. Emergency response planning is an essential part of managing a drinking water system. The introduction should identify the requirement to have a documented emergency response plan (ERP), the goal(s) of the plan (e.g., be able to quickly identify an emergency and initiate timely and effective response action, be able to quickly respond and repair damages to minimize system downtime), and how access to the plan is limited. Plans should be numbered for control. Recipients should sign and date a statement that includes their (1) ERP number, (2) agreement not to reproduce the ERP, and (3) they have read the ERP.

ERPs do not necessarily need to be one document. They may consist of an overview document, individual Emergency Action Procedures, check lists, additions to existing operations manuals, appendices, etc. There may be separate, more detailed plans for specific incidents. There may be plans that do not include particularly sensitive information and those that do. Existing applicable documents should be referenced in the ERP (e.g., chlorine Risk Management Program, contamination response).

II. Emergency Planning Process

A. Planning Partnerships. The planning process should include those parties who will need to help the utility in an emergency situation (e.g., first responders, law enforcement, public health officials, nearby utilities, local emergency planning committees, testing laboratories, etc.). Partnerships should track from the Water Utility Department up through local, state, regional, and federal agencies, as applicable and appropriate, and could also document compliance with governmental requirements.

B. General Emergency Response Policies, Procedures, Actions, Documents. A short synopsis of the overall emergency management structure, how other utility emergency response, contingency, and risk management plans fit into the ERP for water emergencies, and applicable polices, procedures, actions plans, and reference documents should be cited. Policies should include interconnect agreements with adjacent communities and just how the ERP may affect them. Policies should also address how to handle services to other public utility providers such as gas and electric.

C. Scenarios. Use your vulnerability assessment (VA) findings to identify specific emergency action steps required for response, recovery, and remediation for each of the five (5) incident types (if applicable) outlined in *The Guidance for Water Utility Response, Recovery & Remediation Actions for Man-Made and/or Technological Emergencies*, Office of Water (4610M) EPA 810-R-02-001, April 2002 available at *www.epa.gov/safewater*. In this section, a short paragraph referencing the VA and findings should be provided. Specific details identifying vulnerabilities should not be included. In Section V of this plan, specific emergency actions procedures addressing each of the incident types should be addressed.

---

*Source: U.S. EPA, Large Water System Emergency Response Plan Outline: Guidance to Assist Community Water System in Complying with the Public Health Security and Bioterrorism Preparedness and Response Act of 2002.

III. Emergency Response Plan—Policies

A. System Specific Information. In an emergency, a water system needs to have basic information for system personnel and external parties such as law enforcement, emergency responders, repair contractors/vendors, the media, and others. The information needs to be clearly formatted and readily accessible so system staff can find and distribute it quickly to those who may be involved in responding to the emergency. Basic information that may be presented in the emergency response plan are the system's ID number, system name, system address or location, directions to the system, population served, number of service connections, system owner, and information about the person in charge of managing the emergency. Distribution maps, detailed plan drawings, site plans, source water locations, and operations manuals may be attached to this plan as appendices or referenced.

1. PWS ID, owner, contact person
2. Population served and service connections
3. System components

   (a) Pipes and constructed conveyances
   (b) Physical barriers
   (c) Isolation valves
   (d) Water collection, pretreatment, treatment, storage, and distribution facilities
   (e) Electronic, computer, or other automated systems which are utilized by the public water system
   (f) Emergency power generators (onsite & portable)
   (g) The use, storage, or handling of various chemicals
   (h) The operation and maintenance of such system components

B. Identification of Alternative Water Sources

1. Amount of water needed for various durations
2. Emergency water shipments
3. Emergency water supply sources
4. Identification of alternate storage and treatment sources
5. Regional aid agreements (interconnections)

Also consider in this section, a discussion of backup wells, adjacent water systems, certified bulk water haulers, etc.

C. Chain-of-Command Chart Developed in Coordination with Local Emergency Planning Committee (Internal and/or External Emergency Responders, or both)

1. Contact name
2. Organization and emergency response responsibility
3. Telephone number(s) (hardwire, cell phones, faxes, e-mail)
4. State 24-h Emergency Communications Center telephone

D. Communication Procedures: Who, What, When. During most emergencies, it will be necessary to quickly notify a variety of parties both internal and external to the water utility. Using the Chain-of-Command Chart and all appropriate personnel from the lists below, indicate who activates the plan, the order in which notification occurs, and the members of the Emergency Response Team. All contact information should be available for routine updating and readily available. The following lists are not intended to be all inclusive—they should be adapted to your specific needs.

1. Internal notification lists

    (a) Utilities dispatch
    (b) Water source manager
    (c) Water treatment manager
    (d) Water distribution manager
    (e) Facility managers
    (f) Chief water utility engineer
    (g) Director of water utility
    (h) Data (IT) manager
    (i) Wastewater treatment plant
    (j) Other

2. Local notification

    (a) Head of local government (i.e., mayor, city manager, chairman of board, etc.)
    (b) Public safety officials—fire, local law enforcement (LLE), police, EMS, safety if a malevolent act is suspected, LLE should be immediately notified and in turn will notify the FBI, if required. The FBI is the primary agency for investigating sabotage to water systems or terrorist incidents.
    (c) Other government entities: health, schools, parks, finance, electric, etc.

3. External notification lists
    (a) State PWSS regulatory agency (or agencies)
    (b) Regional water authority (where one exists)
    (c) EPA
    (d) State police
    (e) State health department (lab)
    (f) Critical customers (Special considerations for hospitals, Federal, State and County government centers, etc.)
    (g) Service/mutual aid
    (h) Water Information Sharing and Analysis Center (ISAC)
    (i) Residential and commercial customers not previously notified

4. Public/media notification: When and how to communicate. Effective communications is a key element of emergency response, and a media or communications plan is essential to good communications. Be prepared by organizing basic facts about the crisis and your water system. Develop key messages to use with the media that are clear, brief, and accurate. Make sure your messages are carefully planned and have been coordinated with local and state officials. Considerations should be given to establishing protocols for both field and office staff to respectfully defer questions to the utility spokesperson.

    Be prepared to list geographic boundaries of the affected area (e.g., west of highway a, east of highway b, north of highway c, and south of highway d to ensure the public clearly understands the system boundaries.)

E. Personnel Safety. This should provide direction as to how operations staff, emergency responders, and the public should respond to a potential toxic release (e.g., chlorine plume release from a water treatment plant or other chemical agents), including facility evacuation, personnel accountability, proper Personnel Protective Equipment as dictated by the Risk Management Program and Process Safety Management Plan, and whether the nearby public should be "in-place sheltered" or evacuated.

F. Equipment. The ERP should identify equipment that can obviate or significantly lessen the impact of terrorist attacks or other intentional actions on the public health and protect the safety and supply of drinking water provided to communities and individuals. The water utility should maintain an updated inventory of current equipment and repair parts for normal maintenance work.

Because of the potential for extensive or catastrophic damage that could result from a malevolent act, additional equipment sources should be identified for the acquisition and installation of equipment and repair parts in excess of normal usage. This should be based on the results of the specific scenarios and critical assets identified in the vulnerability assessment that could be destroyed. For example, numerous high-pressure pumps, specifically designed for the water utility, could potentially be destroyed. A certain number of "long-lead" procurement equipment should be inventoried and the vendor information for such unique and critical equipment maintained. In addition, mutual aid agreements with other utilities, and the equipment available under the agreement, should be addressed. Inventories of current equipment, repair parts, and associated vendors should be indicated under Item 29 "Equipment Needs/Maintenance of Equipment" of Section IV "Emergency Action Procedures."

G. Property Protection. A determination should be made as to what water system facilities should be immediately "locked down," specific access control procedures implemented, initial security perimeter established, a possible secondary malevolent event considered. The initial act may be a divisionary act.

H. Training, Exercises, and Drills. Emergency response training is essential. The purpose of the training program is to inform employees of what is expected of them during an emergency situation. The level of training on an ERP directly affects how well a utility's employees can respond to an emergency. This may take the form of orientation scenarios, table-top workshops, functional exercises, etc.

I. Assessment. To evaluate the overall ERP's effectiveness and to ensure that procedures and practices developed under the ERP are adequate and are being implemented, the water utility staff should audit the program on a periodic basis.

IV. Emergency Action Procedures (EAPs)

These are detailed procedures used in the event of an operational emergency or malevolent act. EAPs may be applicable across many different emergencies and are typically common core elements of the overall municipality ERP (e.g., responsibilities, notifications lists, security procedures, etc.) and can be referenced.

    A. Event classification/severity of emergency
    B. Responsibilities of emergency director
    C. Responsibilities of incident commander
    D. Emergency Operations Center (EOC) activation
    E. Division internal communications and reporting
    F. External communications and notifications
    G. Emergency telephone list (division internal contacts)
    H. Emergency telephone list (off-site responders, agencies, state 24-h emergency phone number, and others to be notified)
    I. Mutual aid agreements
    J. Contact list of available emergency contractor services/equipment
    K. Emergency equipment list (including inventory for each facility)
    L. Security and access control during emergencies
    M. Facility evacuation and lockdown and personnel accountability
    N. Treatment and transport of injured personnel (including chemical/biological exposure)

O. Chemical records—to compare against historical results for base line
P. List of available laboratories for emergency use
Q. Emergency sampling and analysis (chemical/biological/radiological)
R. Water use restrictions during emergencies
S. Alternate temporary water supplies during emergencies
T. Isolation plans for supply, treatment, storage, and distribution systems
U. Mitigation plans for neutralizing, flushing, disinfecting tanks, pump stations, or distribution systems, including shock chlorination
V. Protection of vital records during emergencies
W. Record keeping and reporting (FEMA, OSHA, EPA, and other requirements) (It is important to maintain accurate financial records of expenses associated with the emergency event for possible federal reimbursement.)
X. Emergency program training, drills/and tabletop exercises
Y. Assessment of emergency management plan and procedures
Z. Crime scene preservation training and plans
AA. Communication plans:
   1. Police
   2. Fire
   3. Local government
   4. Media
   5. Etc.
BB. Administration and logistics, including EOC, when established
CC. Equipment needs/maintenance of equipment
DD. Recovery and restoration of operations
EE. Emergency event closeout and recovery

V. Incident-Specific Emergency Action Procedures (EAPs)

Incident-specific EAPs are action procedures that identify specific steps in responding to an operational emergency or malevolent act. The Guidance for Water Utility Response, Recovery & Remediation Actions for Man-Made and/or Technological Emergencies, Office of Water (4610M) EPA 810-R-02-001, April 2002, identifies three major steps in developing procedures—response, recovery, and remediation with a list of initial and recovery notifications required. "Response" refers to actions immediately following awareness of the incident, "recovery" refers to actions to bring the system back into operations, and "remediation" refers to long-term restoration actions. When developing an EAP for those incidents identified in Section V.2, the EAP must consider the impact of the incident on system elements and the potential impacts on upstream and downstream components of the incident location. *If during the VA process, a specific incident type was judged as not credible, then it should be noted as to why it is not applicable to the ERP. If additional incident types were identified, then these should be included in the ERP.* For those that use the Sandia National Laboratory methodology (RAM-W) the adversary sequence diagrams provide incident-specific malevolent acts, which may fit under Section V.2.

A. General Response to Terrorist Threats (Other than Bomb Threat and Incident-Specific Threats)
B. Incident-Specific Response to Man-Made or Technological Emergencies
   1. Contamination event (articulated threat with unspecified materials)
   2. Contamination threat at a major event
   3. Notification from health officials of potential water contamination
   4. Intrusion through supervisory control and data acquisition (SCADA)

VULNERABILITY ASSESSMENT, EMERGENCY RESPONSE

C. Significant structural damage resulting from intentional act
D. Customer complaints
E. Severe weather response (snow, ice, temperature, lightning)
F. Flood response
G. Hurricane and/or tornado response
H. Fire response
I. Explosion response
J. Major vehicle accident response
K. Electrical power outage response
L. Water supply interruption response
M. Transportation accident response—barge, plane, train, semitrailer/tanker
N. Contaminated/tampered with water treatment chemicals
O. Earthquakes response
P. Disgruntled employees response (i.e., workplace violence)
Q. Vandals response
R. Bomb threat response
S. Civil disturbance/riot/strike
T. Armed intruder response
U. Suspicious mail handling and reporting
V. Hazardous chemical spill/release response (including Material Safety Data Sheets)
W. Cyber-security/Supervisory Control and Data Acquisition (SCADA) system attack response (other than incident-specific, e.g., hacker)

VI. Next Steps

A. Plan Review and Approval
B. Practice and Plan to Update (as necessary; once every year recommended)

　　1. Training requirements
　　2. Who is responsible for conducting training, exercises, and emergency drills
　　3. Update and assessment requirements
　　4. Incident-specific exercises/drills

VII. Annexes:

A. Facility and Location Information

　　1. Facility maps
　　2. Facility drawings
　　3. Facility descriptions/layout
　　4. Etc.

VIII. References and Links

A. Department of Homeland Security—http://www.dhs/gov/dhspublic
B. Environmental Protection Agency—http://www.epa.gov
C. The American Water Works Association (AWWA)—http://www.awwa.org
D. The Center for Disease Control and Prevention—http://www.bt.cdc.gov
E. Federal Emergency Management Agency—http://www.fema.gov
F. Local Emergency Planning Committees—http://www.epa.gov/ceppo/lepclist.htm

REFERENCES

American Water Works Association, *Emergency Planning for Water Utilities (M19)*, Denver, CO, 2001.

American Water Works Association, *Water System Security: A Field Guide*, Denver, CO, 2002.

Association of State Drinking Water Administrators and the National Rural Water Association, Security Vulnerability Self-Assessment Guide for Small Drinking Water Systems Serving Between 3300 and 10,000, 2002.

Fullwood, R. R., and R. E. Hall, *Probabilistic Risk Assessment in Nuclear Power Industry*, Pergamon Press, Oxford, England, 1988.

Mays, L. W. (ed.), *Reliability Analysis of Water Distribution Systems*, American Society of Civil Engineers, New York, 1989.

Mays, L. W. (ed.), *Water Distribution Systems Handbook*, McGraw-Hill, New York, 2000.

Mays, L. W. (ed.), *Urban Water Supply Handbook*, McGraw-Hill, New York, 2002.

National Rural Water Association, Rural and Small Water and Wastewater System Emergency Response Plan Template, www.nrwa.org, Duncan, OK, no date.

Peplow, D. E., C. D. Sulfredge, R. L. Saunders, R. H. Morris, and T. A. Hann, "Calculating Nuclear Power Plant Vulnerability Using Integrated Geometry and Event/Fault Tree Models," Oak Ridge National Laboratory, Oak Ridge, TN, 2003.

President's Commission on Critical Infrastructure Protection, Appendix A, Sector Summary Reports, *Critical Foundations: Protecting America's Infrastructure*: A-45, available at: http://www.ciao.gov/PCCIP/PCCIP_Report.pdf.

Sulfredge, C. D., R. L. Saunders, D. E. Peplow, and R. H. Morris, "Graphical Expert System for Analyzing Nuclear Facility Vulnerability," Oak Ridge National Laboratory, Oak Ridge, TN, 2003.

U.S. EPA, "Guidance for Water Utility Response, Recovery, and Remediation Actions for Man-Made and/or Technological Emergencies," available at: http://www.epa.gov/safewater/security/er-guidance.pdf

U.S. EPA, "Guidance for Water Utility Response, Recovery, and Remediation Actions for Man-Made and/or Technological Emergencies," EPA 810-R-02-001, Office of Water (4601), available at: www.epa.gov/safewater, April 2002.

U.S. EPA, "Water Security Strategy for Systems Serving Populations Less than 100,000/15 MGD or Less," July 9, 2002.

U.S. EPA, "Vulnerability Assessment Fact Sheet 12-19," EPA 816-F-02-025, Office of Water, available at: www.epa.gov/safewater/security/va fact sheet 12-19.pdf, also at www.epa.gov/ogwdw/index.html, November 2002.

U.S. EPA, "Instructions to Assist Community Water Systems in Complying with the Public Health Security and Bioterrorism Preparedness and Response Act of 2002," EPA 810-R-02-001, Office of Water, available at: www.epa.gov/safewater/security, January 2003.

U.S. EPA, Large Water System Emergency Response Plan Outline: Guidance to Assist Community Water Systems in Complying with the Public Health Security and Bioterrorism Preparedness and Response Act of 2002, available at: www.epa.gov/safewater/, June 2003.

U.S. EPA, information available at: http://www.epa.gov/swercepp/cntr-ter.html

Washington State Department of Health, "Emergency Response Planning Guide for Public Drinking Water Systems," DOH PUB. #331-211, Olympia, Washington, May 2003.

CHAPTER 4
DRINKING WATER DISTRIBUTION SYSTEMS: AN OVERVIEW

Robert M. Clark,[*] Walter M. Grayman,[†] Steven G. Buchberger,[‡] Yeongho Lee[§] and David J. Hartman[§]

4.1 INTRODUCTION

Prior to the passage of the U.S. Safe Drinking Water Act of 1974, the focus of most of the water utilities was on treating water, even though it has long been recognized that water quality can deteriorate in the distribution system. However, after the SDWA was amended in 1986, a number of rules and regulations were promulgated which had direct impact on water quality in distribution systems. To decrease what was considered an unreasonably high risk of waterborne illness, the U.S. EPA promulgated the Total Coliform Rule (TCR), and Surface Water Treatment Rule (SWTR) in 1989 (U.S. EPA, 1989a & b). More recently there has been an increased focus on distribution systems and their importance in maintaining water quality in drinking water distribution systems. There has also been general agreement that the most vulnerable part of a water supply system is the distribution network.

This chapter discusses the general features associated with water supply distribution systems, including design considerations, points of vulnerability to accidental or deliberate contamination, and the potential for using network modeling to assess system vulnerability. The role of tanks and storage reservoirs in operating and managing water systems is discussed, including their effect on water quality. Water quality models and their potential for use in tracking and predicting water quality movement and changes in water quality are examined. Because a key aspect of assessing the performance of drinking water systems is the response of the network and the resulting network model to the pattern of water demands, new research for characterizing water systems is presented. If water quality models are to be used for predicting water quality in networks, they should be verified. Therefore the use of

---

[*]Environment Engineering and Public Health Consultant, Cincinnati, Ohio.
[†]W.M. Grayman Consulting Engineer, Cincinnati, Ohio.
[‡]University of Cincinnati, Cincinnati, Ohio.
[§]Greater Cincinnati Water Works, Cincinnati, Ohio.

tracers for verifying water quality models including planning, conducting tracer studies in the field, and the analysis of results is discussed.

4.1.1 Features and Functionality

Drinking water transmission and distribution systems are designed to deliver water from a source (possibly a treatment facility) in the required quantity and at satisfactory pressure to individual consumers in a utilities service area. Moving water between the source and the customer requires a network of pipes, pumps, valves, and other appurtenances (Clark and Tippen, 1990). The system of pipes that provides this service are generally categorized as transmission and distribution mains. Transmission mains usually convey large amounts of water over long distances such as from a treatment facility to a storage tank within the system. Distribution mains are typically smaller in diameter than transmission mains and generally follow the city streets. They are the intermediate step in delivering water to the customer. The most commonly used pipes for water mains are ductile iron, prestressed concrete cylinder, polyvinyl chloride (PVC), reinforced plastic, and steel. Service lines are pipes, including accessories, which carry water from the main to the building or property being served. Service lines can be of any size depending on how much water is required to serve a particular customer and are sized so that the utilities design pressure is maintained at the customer's property for the desired flows. Valves are used in the distribution system to isolate sections for maintenance and repair and should be located in the system so that the areas isolated will cause a minimum of inconvenience to the customer. Care should be used to ensure that only the number of valves necessary are installed. Storage of water in tanks and reservoirs is required to accommodate fluctuations in demand for fire protection and to accommodate the varying rates of usage. This entire infrastructure is typically referred to as the water distribution system.

The distribution system is generally the major investment by a municipal water works, although most of the assets are either buried or located inconspicuously. Distribution reservoirs are used to provide storage to meet fluctuations in use, to provide storage for firefighting use, and to stabilize pressures in the distribution system. It is desirable to locate reservoirs as close to the center of use as possible. Broken or leaking water mains should be repaired as soon as possible to minimize property damage and loss of water. In the past it has been standard practice to maintain the carrying capacity of the pipe in the distribution system as high as possible to provide the design flow and keep pumping costs as low as possible. However, recently there has been concern that this practice can lead to excessively long residence times and thus contribute to a deterioration in water quality.

Customers and the nature in which they use water drive the behavior of a water distribution system. Water use varies spatially and temporally. A detailed understanding of how water is used is critical to adequate water distribution system design. A major function of most distribution systems is to provide adequate fire flow. A key factor in providing fire flow is the use of fire hydrants which should be installed in areas which are easily accessible by the fire hydrant and are not obstacles to pedestrians and vehicles.

When possible, mains should be placed in areas along the public right of way, which provide for ease of installation, repair, and maintenance. The branch and loop are the two basic configurations for most water distribution systems. A branch system is similar to that of a tree branch with smaller pipes branching off larger pipes throughout the service area. This type of system is most frequently used in rural areas and the water has only one possible pathway from the source to the consumer. The grid or looped system is the most widely used configuration in municipal systems and consists of connected pipe loops throughout the area to be served. In a looped system there may be several pathways that the water can follow from the source to the consumer. A typical design would be to space larger transmission mains, 24 in (61 cm) in diameter or larger, 1.5 to 2 mi (2400 to 3200 m) apart. Feeder mains are normally 16 to 20 in (40.6 to 50.8 cm) in diameter and are spaced 3000 to 4000 ft (900 to 1200 m apart). The remaining grid is usually served by 6 to 12 in (15 to 30 cm) diameter mains in every street. Looped systems provide a high degree of reliability should a line break occur, because the break can be isolated with little impact on consumers outside the immediate area (Clark and Tippen, 1990)

4.1.2 Points of Vulnerability

The events of September 11, 2001 have raised concerns over the security of the U.S. critical infrastructure including water and waste water systems. Security of water systems is not a new issue and the potential for natural, accidental, and purposeful contamination has been the subject of many studies. For example, the American Water Works Association publishes a manual on emergency planning (M19) entitled "Emergency Planning for Water Utilities" (American Water Works Association, 2001). In May 1998, President Clinton issued Presidential Decision Directive (PDD) 63 that outlined a policy on critical infrastructure protection including the U.S. water supplies (President, 1998). However, it wasn't until after September 11, 2001 that the water industry truly focused on the vulnerability of the U.S. water supplies to security threats. Recently requirements have been established for conducting drinking water vulnerability studies (PL 107-188).

There are nearly 60,000 community water supplies in the United States serving over 226 million people (U.S. EPA, 1999). Over 63 percent of these systems supply water to less then 2.4 percent of the population and 5.4 percent supply water to 78.5 percent of the population. Most of these systems provide water to less then 500 people. In addition there are 140,000 non-community water systems that serve schools, recreational areas, trailer parks, etc.

Some of the common elements associated with water supply systems in the United States are as follows:

- A water source which may be a surface impoundment such as a lake, reservoir, river or groundwater from an aquifer
- Surface supplies that generally have conventional treatment facilities including filtration, which removes particulates and potentially pathogenic microorganisms, followed by disinfection
- Transmission systems which include tunnels, reservoirs and/or pumping facilities, and storage facilities
- A distribution system carrying finished water through a system of water mains and subsidiary pipes to consumers

Community water supplies are designed to deliver water under pressure and generally supply most of the water for firefighting purposes. Loss of water or a substantial loss of pressure could disable firefighting capability, interrupt service and disrupt public confidence. This loss might result from sabotaging pumps that maintain flow and pressure, or disabling electric power sources could cause long-term disruption. Many of the major pumps and power sources in water systems have custom designed equipment and could take months or longer to replace (Clark and Deininger, 2000).

Vulnerability of Water Systems. Water systems are spatially diverse and many of the system components such as tanks and pumps are located in isolated locations. Water distribution networks, therefore, have an inherent potential to be vulnerable to a variety of threats—physical, chemical, and biological—that may compromise the system's ability to reliably deliver safe water. These areas of vulnerability include (1) the raw water source (surface or groundwater), (2) raw water channels and pipelines, (3) raw water reservoirs, (4) treatment facilities, (5) connections to the distribution system, (6) pump stations and valves, and (7) finished water tanks and reservoirs. Each of these system elements presents a unique challenge to the water utility in safeguarding the water supply (Clark and Deininger, 2000).

Physical Disruption. The ability of a water supply system to provide water to its customers can be compromised by destroying or disrupting key physical elements of the water system. Key elements include raw water facilities (dams, reservoirs, pipes, and channels), treatment facilities, and distribution system elements (transmission lines and pump stations). Physical disruption may result in significant economic cost, inconvenience, and loss of confidence by customers, but has a limited direct threat to human health. Exceptions to this generalization include (1) destruction of a dam that causes loss of life and property in the accompanying flood wave and (2) an explosive release of chlorine gas at a treatment plant (Clark and Deininger, 2001).

Contamination. Contamination is generally viewed as the most serious potential terrorist threat to water systems. Chemical or biological agents could spread throughout a distribution system and result in sickness or death among the consumers and for some agents, the presence of the contaminant might not be known until emergency rooms reported an increase in patients with a particular set of symptoms (Clark, 2002). Even without serious health impacts, just the knowledge that a group had breached a water system could seriously undermine customers' confidence in the water supply (Grayman et al., 2002). Accidental contamination of water systems has resulted in many fatalities as well. Examples of such outbreaks include cholera contamination in Peru (Clark et al., 1995), *Cryptosporidium* contamination in Milwaukee, Wisconsin (Clark et al., 1995), and *Salmonella* contamination in Gideon, Missouri (Clark et al., 1996). In Gideon, the likely culprit was identified as pigeons infected with *Salmonella* that had entered a tank's corroded vents and hatches.

The U.S. Army has conducted extensive testing and research on potential biological agents (Burrows and Renner, 1998) and produced a list of biological agents most likely to have an impact on water systems. Though much is known about these agents, there is still research needed to fully characterize the impacts, stability, and tolerance of many of these agents to chlorine. Other agencies such as CDC and EPA and many water utilities have produced other lists of most likely contaminants that include dozens or hundreds of potential agents.

However, it would be infeasible to identify all possible contaminants and impossible to assess the impacts of all potential contaminants. As an alternative, the U.S. EPA's Baseline Threat Report (U.S. EPA, 2002) has specified characteristics of nine "model contaminants" that can be used as part of the design basis threat assessment in vulnerability assessments of water systems. These model contaminants represent the following categories:

- Radionuclides
- Biological weapons
- Chemical weapons
- Biotoxins
- Viruses
- Parasites
- Bacterial spores

- Pesticides
- Toxic chemicals

Many locations within the overall water supply system are vulnerable to the introduction of chemical or biological agents. In many cases, the most accessible location is in the raw surface water source. However, an agent introduced in a surface water source is subject to dilution, exposure to sunlight, and treatment. Therefore it follows that the most serious threats may be posed by an agent introduced into the finished water at a treatment facility or within the distribution system. Possible points of entry include the treatment plant clear well, distribution system storage tanks and reservoirs, pump stations, and direct connections to distribution system mains.

4.2 MODELING CONTAMINANT TRANSPORT

4.2.1 Network Demand Modeling

Water consumption or water demand is the driving force behind the operation of a water distribution system. Anywhere water is used or leaves the system can be characterized as a demand on the system. It is critical to be able to characterize those uses or demands in order to develop a hydraulic or water quality model. It is important to be able to determine the amount of water being used, where it is being used, and how this usage varies with time (Walski et al., 2003).

We might categorize these demands as follows:

Baseline Demands. Baseline demands usually include consumer demands and unaccounted-for water and can often be acquired from a utility's existing records such as customer's meters and billing records. The spatial assignment of these demands is extremely important and should include the assignment of customer classes such as industrial, residential, and commercial use. Special types of uses such as water uses for schools must also be determined. It might be possible to use typical demands which have been developed for various types of uses. These values can be found in many text books and handbooks.

Demand Multipliers. Water use varies over time and varies with activities over the course of a day. When developing a steady-state model the baseline demand can be modified by multipliers in order to reflect some of the variations that occur in water systems. These include average day demand, the average rate of demand for an average day; maximum day demand, the average rate of use on the maximum usage day; per-hour demand, the average rate of use during the maximum hour of usage; maximum day of record, the highest average rate of demand for the historical record.

Time Varying Demands. In all water systems these are unsteady due to continuously varying demands. It is important to account for these variations to have an adequate hydraulic model. Diurnal varying demand curves should be developed for each major customer class. For example diurnal demand curves might be developed for industrial and commercial establishments and residential use.

Fire Demands. Water provided for fire services can be the most important consideration in developing design standards for water systems.

Projecting Future Demands. Future changes in demand should be considered when considering the development of network hydraulic models. A possible approach might be to use historical trends based on population estimates or land use (Opitz, 2002).

The most commonly used technique for developing water-use models is a simple regression model that can relate water uses to customer classes of the form:

$$y = \alpha + \beta_1 X_1 + \beta_2 X_2 + \beta_3 X_3 + \cdots + \beta_n X_n + \varepsilon \tag{4.1}$$

where y is a dependent variable which might be the water use in a given time period; X_1, $X_2, \ldots, X_n$ are independent variables that might represent water use for a specific time period in a given customer class; α, β_1, β_2, and β_n are estimated coefficients, and ε is an error term. There are a number of considerations in using these types of models. For example, it is important to be able to divide water consumption data into specific classes that have similar use such as single family residences, multiunit residential housing, small commercial and large commercial, and large industrial establishments. This type of model can be used to separate the seasonal and weather effects in the time series data. However, it is important to be able to quantify the types of error associated with these data. Section 4.3 presents recent research that characterizes water demand as a stochastic variable.

4.2.2 Network Hydraulic Modeling

In order to understand the behavior of pressurized pipeline systems it is essential to understand their hydraulic behavior. Free water surfaces are almost never found in a pressurized system with the exception of reservoirs and tanks. However, for very short periods during unsteady or transient events free surfaces may be found in the pipe itself. Pressures within a pressurized pipeline system are usually well above atmospheric.

Three basic relations describe the flow of fluid in pipes. These relationships are as follows:

- Conservation of mass
- Newton's second law of motion
- Development of principles that govern transient flows

Conservation of Mass. The conservation of mass principle requires that the sum of the mass flow in all pipes entering a junction must equal the sum of all mass flows leaving the junction.

Newton's Second Law. Newton's second law states that all forces acting on a system are equal to the change of momentum of the system with respect to time. Mathematically this is

$$\sum F_{\text{ext}} = \frac{d(mv)}{dt} \tag{4.2}$$

where t is time and F_{ext} represents the external force acting on a body of mass m moving with velocity v. If the mass of the body is constant Eq. (4.2) becomes

$$\sum F_{\text{ext}} = m \frac{dv}{dt} = ma \tag{4.3}$$

where a is acceleration of the system. This equation can be applied to water as follows:

$$\sum F_{\text{ext}} = rQ(n_{\text{out}} - n_{\text{in}}) \tag{4.4}$$

where ρ is the water density, v_{out} and v_{in} are respectively the velocity out of and the velocity into a control section, and Q is the volumetric rate of flow.

Transient Flows. Because water has a high density and because pipe lines are generally long, therefore they carry extremely large amounts of mass, momentum, and kinetic energy. Water is only slightly compressible and therefore large head changes occur if small amounts of fluid are forced into a pipeline. The consequence of these two facts may lead to shock waves or transient flow. Transient fluid flow may result in water hammer which in turn may be caused by the opening or closing of valves or starting a pump and has major consequences for system design.

Friction in Pipelines. A key factor in evaluating the flow through pipe networks is the ability to calculate friction head loss (Jeppson, 1976). Three equations commonly used are the Darcy-Weisbach, the Hazen-Williams, and the Manning equations (Larock et al., 2000).

The Darcy-Weisbach equation is

$$h_f = f(L/D)(V^2/2g) \tag{4.5}$$

where h_f = head loss, in ft/ft (m/m)
f = dimensionless friction factor
D = pipe diameter, in ft (m)
L = length of pipe, in ft (m)
V = average velocity of flows, in ft/s (m/s)
g = the acceleration of gravity, in ft/s² (m/s²)

A fundamental relationship that is important for hydraulic analysis is the Reynolds number, as follows:

$$R_e = VD/\nu \tag{4.6}$$

where R_e = Reynolds number (dimensionless)
ν = kinematic viscosity, in ft²/s (m²/s)
V and D = as defined previously

There are various equations that can be used for calculating f in the Darcy-Weisbach equation (Jeppson, 1976). Although the Darcy-Weisbach equation is fundamentally sound, the most widely used equation is the Hazen-Williams equation

$$Q = 1.318 \, CAR^{0.63}S^{0.54} \tag{4.7}$$

where Q = flow, in ft³/s (m³/s)
C = Hazen Williams roughness coefficient
A = cross-sectional area, in ft² (m²)
R = hydraulic radius (D/4) in ft (m)
S = slope of the energy grade line (h_f L[ft/ft])(m/m)

If the head loss is desired and Q is known, the Hazen-Williams equation for a pipe can be written as

$$h_f = (4.73L/C^{1.852}D^{4.87})Q^{1.815} \tag{4.8}$$

where the variables are defined previously.

Another empirical equation is the Manning equation for pipes, which has been solved for h

$$h_f = (4.637n^2 L/D^{5.33})Q^2 \qquad (4.9)$$

where n is an empirical constant and the order variables are as defined previously.

Methods of Analysis. Analyzing for the flow in pipe networks, particularly if a large number of pipes are involved, is a complex process. Deciding which pipes should be included in the analysis can be a matter of judgment. It may not be possible or practical to include all the pipes that deliver water to the consumer. Analysis is frequently conducted only on the major transmission lines in the network or on the pipes that carry water between separate sections of the network. This process is called skeletonization. As mentioned previously, there are two types of analyses usually conducted on drinking water distribution systems—steady state and dynamic.

A steady-state analysis is needed for each demand or consumption pattern. The addition of new service areas, pumps, or storage tanks changes the system and requires a new steady-state analysis. Dynamic analysis or EPS is often simply a series of steady state analyses linked by specified conditions. The oldest method of solving steady state flow in pipes is the Hardy-Cross method. It was originally developed for solution by hand but has been programmed for solution using computers. However, when applied to large networks or for certain conditions it might be slow or even fail to converge. More recently the Newton-Raphson method and the linear-theory method, both described later in this chapter, have been applied to network solutions. The Newton-Raphson method requires approximately the same computer storage requirements as the Hardy Cross method and also requires an initial solution.

Dynamic analysis or extended period simulation (EPS) deals with unsteady flows or transient problems. Steady-state analysis, which will be discussed in this section, is considered solved when the flow rate in each pipe is calculated based on a specific usage or demand pattern and consumption. The supply from reservoirs, storage tanks, and/or pumps is generally the inflow or outflow from some point in the network. If flow rates are known, then pressures or head losses can be computed throughout the system. If the heads or pressures are known at each pipe junction or network node, the flow rates can be computed in each pipe.

Other issues to be considered include the calibration and verification of the hydraulic assumptions.

Reducing Network Complexity. Pipe networks may include pipes in series, parallel, or branches (like branches of a tree). The network may also use elbows, valves, motors, and other devices that cause local disturbances and head losses. These factors can frequently be combined with or converted into equivalent pipes, which is very useful in simplifying networks. The major methods of simplification follow.

Pipes in Series. For pipes in series, the same flow must pass through both pipes, therefore an equivalent head loss is the sum of the head losses for all the pipes being considered as part of the equivalent pipes.

Pipes in Parallel. Two or more pipes in parallel can also be replaced by an equivalent pipe. In this case, the head loss between junctions where the pipes part and then join again must be equal. Although using equivalent pipes may be effective from a hydraulic viewpoint, this practice can be misleading when modeling water quality. Water quality modeling accuracy depends on accurate measurement of velocity.

Branching System. In a branching system, a number of pipes are connected to a larger pipe in the form of a tree. If the flow is from the larger pipe to the smaller laterals, the flow rate can be calculated in any pipe as the sum of the downstream consumption or demand. If the laterals supply water to the main, then the same approach can be taken. When a system is analyzed, frequently only the larger pipes are used in the network analysis.

Minor Losses. An equivalent pipe can be used in a network to approximate the minor losses associated with valves, meters, elbows, or other devices. Equivalent pipes are formed by adding a length to the actual pipe length that will result in the same head loss as in the component. The Darcy-Weisbach or Hazen-Williams equations can be used to compute the head loss.

Equations Describing Flow. Flow analysis in pipe networks is based on basic continuity and energy laws. The mass weight or volumetric flow rate into a junction must equal the mass weight or volumetric flow rate out of a junction, including demand. In addition to continuity equations that must be satisfied, the head loss or energy equations must be satisfied. If one sums the head loss around a loop, the net head loss must equal zero.

Network Characterization. Engineering analysis of water distribution systems is frequently limited to the solution of the hydraulic network problem, i.e., given the physical characteristics of a distribution system modeled as a node-link network and the demands at nodes (junctions where network components connect to one another), the flows in links (a connection of any two nodes) and head at all nodes of the network are determined. This problem is formulated as a set of simultaneous nonlinear equations, and a number of well-known solution methods exist, many of which have been coded as computer programs known as hydraulic network models (Cesario, 1995).

Mathematical methods for analyzing the flow and pressure in networks have been in use for more than 50 years. They are generally based on well-accepted hydrodynamic equations. Computer-based models for performing this type of analysis were first developed in the 1950s and 1960s and greatly expanded and made more available in the 1970s and 1980s. Currently, dozens of such models are readily available on systems ranging from personal computers to supercomputers.

Hydraulic models were developed to simulate flow and pressures in a distribution system either under steady state conditions or under time varying demand (extended period simulation) and operational conditions. Hydraulic models may also incorporate optimization components that aid the user in selecting system parameters that result in the best match between observed system performance and model results.

Application of Models. The following steps should be followed in applying network models (Clark et al., 1988):

- Model selection
- Define model requirements and select a model (hydraulic and/or water quality) that fits your requirements, style, budget, etc.
- Network representation
- Accurately represent the distribution system components in the model
- Calibration
- Adjust model parameters so that predicted results adequately reflect observed field data
- Verification
- Independently compare model and field results to verify the adequacy of the model representation

- Problem definition
- Define the specific design or operational problem to be studied and incorporated (i.e., demands and system operation) into the model
- Model application
- Use the model to study the specific problem or situation
- Display and analysis of results
- Following the model application, display and analyze the results to determine how reasonable the results are
- Finally, translate the results into a solution to the problem

All water distribution system models represent the network as a series of connected links and nodes. Links are defined by the two end nodes and generally serve as conveyance devices (e.g., pipes), while nodes represent point components (e.g., junctions, tanks, and treatment plants). In most models, pumps and valves are represented as links. Water demands are aggregated and assigned to nodes; an obvious simplification of real-world situations in which individual house taps are distributed along a pipe rather than at junction nodes.

Network representation is more of an art than a science. Most networks are skeletonized for analysis, which means that they contain only a representation of selected pipes in a service area.

A minimal skeletonization should include all pipes and features of major concern. Nodes are usually placed at pipe junctions, where pipe characteristics may change in diameter, C-value (roughness), or material of construction. Nodes may also be placed at locations of known pressure or where pressure valves are desired.

Network components include pipes, reservoirs, pumps, pressure-relief valves, control valves, altitude valves, check valves, and pressure-reducing valves. For steady state modeling, a key input is information on water consumption. Usage is assigned to nodes in most models and may be estimated several different ways. Demand may be estimated by a count of structures of different types using a representative consumption per structure, meter readings and the assignment of each meter to a node, and to general land use. A universal adjustment factor should be used so that total usage in the model corresponds to total production.

Modeling Temporal Variations. We know that most phenomena and behavior vary over time in the real world. Models that assume no variation over time are referred to as steady state models, which provide a snapshot at a single point in time. In distribution system models, steady state models assume that demands are constant and operations are constant (e.g., tank water-level elevations remain constant and pumps are either on or off at a constant speed). Though such assumptions may not be valid over long periods of time, steady state models can provide some useful information concerning the behavior of a network under various representative conditions, including fire demand, nighttime low demands, etc.

Models that allow for variations over time are referred to as temporally dynamic models. Most network distribution hydraulic models incorporate temporal variation by stringing together a series of steady state solutions and refer to this method as extended period simulation (EPS). For example, in the EPS mode, demand at a node may be assumed to be 100 gallons per minute (gpm) (378 L/min) from 2:00 p.m. to 3:00 p.m. and then 150 gpm (567 L/min) from 3:00 p.m. to 4:00 p.m. Thus, one steady-state solution may be assumed from 2:00 p.m. until 3:00 p.m. and a different solution may be calculated from 3:00 p.m. to 4:00 p.m. Further, temporal dynamics may be incorporated by checking the water level in a tank and if the water level reaches a maximum allowable level at 3:20 p.m., then another steady-state solution is started at 3:20 p.m. with the tank discharging instead of filling. Though the

EPS solution does introduce some approximations and totally ignores the transient phenomena resulting from sudden changes, such as a pump being turned on, these more refined assumptions are generally not considered significant for most distribution system studies.

Some of the characteristics of the EPS mode are as follows:

- Requires information on temporal variations in water usage over the period being modeled
- Permits temporal patterns to be defined for groups of nodes (spatially different patterns can be applied to a given node)
- Uses the best available information on temporal patterns that can be estimated (for example, some users have continuous meters)
- Can sometimes use literature values for a first guess at residential patterns (which may vary with climate)
- Can use analysis of information from Supervisory Control and Data Acquisition (SCADA) to estimate system wide temporal pattern

Model Calibration. Calibration is an important part of the "art" of modeling water distribution systems. Model calibration is the process of adjusting model input data (or, in some cases, model structure) so that the simulated hydraulic and water quality output sufficiently mirrors observed field data. As mentioned earlier, one way of viewing calibration is to think of a TV screen showing observed and predicted values. Calibration is the process of adjusting the picture so that the predicted value provides the closest estimate for the actual values.

Depending on the degree of accuracy, calibration can be difficult, costly, and time-consuming. The extent and difficulty of calibration is minimized by developing an accurate representation of the network and its components. A traditional technique for calibration is to use fire flow pressure measurements. Pressures and flow in isolated pipe sections are measured in the field and the roughness factors C are adjusted to reflect the data.

Another method is to use water quality tracers. Naturally occurring or added chemical tracers may be measured in the field and the results used to calibrate hydraulic and water quality models. The most common tracer is fluoride. It is conservative (does not degrade), safe, and can usually be added (or normal feed can be curtailed) and the movement can be traced in the system using hand-held analyzers. For conservative tracers, adjustments may be made primarily in the hydraulic model to adequately match the predicted and observed concentrations.

Another calibration technique is to measure predicted tank levels derived from computer simulations against actual tank level during a given period of record. For example, using data from SCADA systems or from online pressure and tank-level recorders, flows can be adjusted in the simulation model until they match the actual tank-level information.

4.2.3 Water Quality Models

Modeling the movement of a contaminant within the distribution systems, as it moves through the system from various points of entry (e.g., treatment plants) to water users is based on three principles:

- Conservation of mass within differential lengths of pipe
- Complete and instantaneous mixing of the water entering pipe junctions
- Appropriate kinetic expressions for the growth or decay of the substance as it flows through pipes and storage facilities

This change in concentration can be expressed by the following differential equation:

$$\frac{dC_{ij}}{dt} = -v_{ij}\frac{\partial C_{ij}}{\partial x} + k_{ij}C_{ij} \tag{4.10}$$

where C_{ij} = substance concentration (g/m³) at position x and time t is the link between nodes i and j
v_{ij} = flow velocity in the link (equal to the link's flow divided by its cross-sectional area) (m/s)
k_{ij} = rate at which the substance reacts within the link (s⁻¹)

According to Eq. (4.10) the rate at which the mass of material changes within a small section of pipe equals the difference in mass flow into and out of the section plus the rate of reaction within the section. It is assumed that the velocities in the links are known beforehand from the solution to a hydraulic model of the network. In order to solve Eq. (4.10) we need to know C_{ij} at $x = 0$ for all times (a boundary condition) and a value for k_{ij}.

Equation (4.11) represents the concentration of material leaving the junction and entering a pipe

$$C_{ij} = \frac{\sum_k Q_{ki}C_{kj}}{\sum_k Q_{kj}} \tag{4.11}$$

where C_{ij} = concentration at the start of the link connecting node i to node j, in mg/L (i.e., where $x = 0$)
C_{kj} = conncentration at the end of a link, in mg/L
Q_{kj} = flow from k to i

Equation (4.11) implies that the concentration leaving a junction equals the total mass of a substance mass flowing into the junction divided by the total flow into the junction.

Storage tanks can be modeled as completely mixed, variable volume reactors in which the change in volume and concentration over time are as follows:

$$\frac{dV_s}{dt} = \sum_k Q_{ks} - \sum_i Q_{sj} \tag{4.12}$$

$$\frac{dV_s C_s}{dt} = \sum_k Q_{ks}C_{ks} - \sum_i Q_{sj}C_s + k_{ij}(C_s) \tag{4.13}$$

where C_s = concentration for tank s, in mg/L
dt = change in time, in seconds
Q_{ks} = flow from node k to s, in ft³/s (m³/s)
Q_{sj} = flow from node s to j, in ft³/s (m³/s)
dV_s = change in volume of tank at nodes, in ft³ (m³)
V = volume of tank at nodes, in ft³ (m³)
C_{ks} = concentration of contaminant at end of links, in mg/ft³ (mg/m³)
k_{il} = decay coefficient between node i and j, in s⁻¹

There are currently several models available for modeling both the hydraulics and water quality in drinking water distribution system. However, most of this discussion will be focused on a U.S. Environmental Protection Agency (U.S. EPA) developed hydraulic/contaminant propagation model called EPANET (Rossman et al., 1994), which is based on mass transfer concepts (transfer of a substance through another on a molecular scale). Another approach to water quality contaminant propagation to be discussed is the approach developed by Biswas et al. (1993). This model uses a steady state transport equation that takes into account the simultaneous corrective transport of chlorine in the axial direction, diffusion in the radial direction, and consumption by first- order reaction in the bulk liquid phase. Islam (1995) developed a model called QUALNET, which predicts the temporal and spatial distribution of chlorine in a pipe network under slowly varying unsteady flow conditions. Boulos et al. (1995) proposed a technique called the Event Driven Method (EDM), which is based on a "next-event" scheduling approach, which can significantly reduce computing times.

Solution Methods. There are several different numerical methods that can be used to solve contaminant propagation equations. Four commonly used techniques are Eulerian finite-difference method (FDM), Eulerian discrete-volume method (DVM), Lagrangian time-driven method (TDM), and Lagrangian event-driven method (EDM).

The FDM approximates derivatives with finite-difference equivalents along a fixed grid of points in time and space. Islam used this technique to model chlorine decay in distribution systems. This technique is discussed in more detail in Islam (1995).

The DVM divides each pipe into a series of equally sized, completely mixed volume segments. At the end of each successive water quality time step, the concentration within each volume segment is first reacted and then transferred to the adjacent downstream segment. This approach was used in the models that were the basis for early U.S. EPA studies.

The TDM tracks the concentration and size of a nonoverlapping segment of water that fills each link of a network. As time progresses, the size of the most upstream segment in a link increases as water enters the link. An equal loss in size of the most downstream segment occurs as water leaves the link. The size of these segments remains unchanged.

The EDM is similar to TDM, except that rather than update an entire network at fixed time steps, individual link-node conditions are updated only at times when the leading segment in a link completely disappears through this downstream node.

EPANET. As mentioned previously, the EPANET hydraulic model has been a key component in providing the basis for water quality modeling. There are many commercially available hydraulic models that incorporate water quality models as well. EPANET is a computer program based on the EPS approach to solving hydraulic behavior of a network. In addition, it is designed to be a research tool for modeling the movement and fate of drinking water constituents within distribution systems. EPANET calculates all flows in cubic feet per second (ft^3/s) and has an option for accepting flow units in gallons per minute (gpm), million gallons per day (mgd), or litres per second (L/s). The model is available to be downloaded from the U.S. EPA website.

Hydraulic Simulation. EPANET uses the Hazen-Williams formula, the Darcy-Weisbach formula, or the Chezy-Manning formula for calculating the head loss in pipes. Pumps, valves, and minor loss calculations in EPANET are also consistent with the convention established in the previous chapters. All nodes have their elevations above sea level specified and tanks and reservoirs are assumed to have a free water surface. The hydraulic head is simply the elevation of the surface above sea level. The surface of tanks is assumed to change in accordance with the following equation:

$$\Delta y = (q/A) \, \Delta t \qquad (4.14)$$

where Δy = change in water level, in ft (m)
  q = flow rate into (+) or out of (−) tank, in ft^3/s (m^3/s)
  A = cross-sectional area of the tank, in ft (m)
  Δt = time interval, in seconds

It is assumed that water usage rates, external water supply rates, and source concentrations at nodes remain constant over a fixed period of time, although these quantities can change from one period to another. Nodes are junctions where network components connect to one another (links), as well as to tanks and reservoirs. The default period interval is one hour but can be set to any desired value. Various consumption or water usage patterns can be assigned to individual nodes or groups of nodes.

EPANET solves a series of equations for each link using the gradient algorithm. Gradient algorithms provide an interactive mechanism for approaching an optimal solution by calculating a series of slopes that lead to better and better solutions. Flow continuity is maintained at all nodes after the first iteration. The method easily handles pumps and valves.

The set of equations solved for each link (between nodes i and j) and each node k is as follows:

$$h_i - h_j = f(q_{ij}) \tag{4.15}$$

$$\sum_i q_{ik} - \sum_j q_{k_j} - Q_k = 0 \tag{4.16}$$

where q_{ij} = flow in link connecting nodes i and j, in ft³/s (m³/s)
h_j = hydraulic grade line elevation at node i (equal to elevation head plus pressure head), in ft (m)
Q_k = flow consumed (+) or supplied (−) at node k, in ft³/s (m³/s)
$f(\cdot)$ = functional relation between head loss and flow in a link

The set of equations for each storage node (tank or reservoir) in the system is as follows:

$$\frac{\partial y_z}{\partial t} = \frac{q_z}{A_s} \tag{4.17}$$

$$q_s = \sum_i q_z - \sum q_{sj} \tag{4.18}$$

$$h_s = E_s + y_s \tag{4.19}$$

where y_z = height of water stored at node s, in ft (m)
A_s = cross-sectional area of storage node s (infinite for reservoirs), in ft² (m²)
E_s = elevation of node s, in ft (m)
q_s = flow into storage node s, in ft³/s (m³/s)
t = time t, in seconds
h_s = height of water in storage tank s, in ft (m)

Equation (4.17) expresses conservation of water volume at a storage node. Equations (4.18) and (4.19) express the same relationship for pipe junctions and Eq. (4.15) represents energy loss or gain due to flow within a link. For known initial storage node levels y_s, at time zero, Eqs. (4.15) and (4.17) are solved for all flows q_{ij} and heads h_i using Eq. (4.19) as a boundary condition. This is called hydraulically balancing the network and is accomplished by using an iterative technique to solve the resulting nonlinear equations.

After a network hydraulic solution is obtained, flow into (or out of) each storage node q_s is found using Eq. (4.18) and used in Eq. (4.17) to find new storage elevations after a time step dt. This process is then repeated for all the following time steps for the remainder of the simulation period.

Water Quality Simulation. EPANET uses the flows from the hydraulic simulation to track the propagation of contaminants through a distribution system. A conservation of mass equation is solved for the substance within each link between nodes i and j as follows:

$$\frac{\partial C_{ij}}{\partial t} = \frac{q_{ij}}{A_{ij}}\left[\frac{\partial c_{ij}}{\partial x_{ij}}\right] + q(C_{ij}) \qquad (4.20)$$

where C_{ij} = concentration of substance in link i, j as a function of distance and time (i.e., $C_{ij} = C_{ij}[X_{ij}, t]$, in mass/ft³ (mass/m³)
x_{ij} = distance along link i, j, in ft (m)
q_{ij} = flow rate in link i, j at time t, in ft³/s (m³/s)
a_{ij} = cross-sectional area of link i, j, in ft² (m²)
$q(C_{ij})$ = rate of reaction of constituent within link i, j, in mass/ft³/d (mass/m³/d)

Equation (4.20) must be solved with known initial conditions at time zero. The following boundary condition at the beginning of the link, i.e., at node i where $x_{ij} = 0$ must hold

$$C_{ij}(0, t) = \frac{\Sigma_k q_{ki} c_{ki}(L_{ki}, t) + M_i}{\Sigma_k q_{ki} + Q_{si}} \qquad (4.21)$$

The summations are made over all links k, i that flow into the head node i of the link i, j,

where L_{ki} = length of link k, i
M_i = substance mass introduced by any external source at node i
Q_{si} = source's flow rate

The boundary condition for link k, i depends on the concentrations at the head of the nodes of all links k, i that flow into link i, j. Equations (4.20) and (4.21) form a coupled set of differential/algebraic equations over all links in the network. These equations are solved within EPANET by using DVM, which was described earlier.

Water quality time steps are chosen to be as large as possible without causing the flow volume of any pipe to exceed its physical volume. Therefore the water quality time step dt_{wq} source cannot be larger than the shortest time of travel through any pipe in the network, i.e.,

$$dt_{wq} = \text{Min}\left\{\frac{V_{ij}}{q_{ij}}\right\} \text{ for all pipes } i, j \qquad (4.22)$$

where V_{ij} is the volume of pipe i, j and q_{ij} is flow rate of pipe i, j.

Pumps and valves are not part of this determination because transport through them is assumed to occur instantaneously. Based on this water quality time step, the number of volume segments in each pipe (n_{ij}) is

$$n_{ij} = \text{INT}\left\{\frac{V_{ij}}{q_{ij}\, dt_{wq}}\right\} \qquad (4.23)$$

where $\text{INT}(x)$ is the largest integer less than or equal to x. There is both a default limit of 100 for pipe segements or the user can set dt_{wq} to be no smaller than a user-adjustable time tolerance.

Reaction Rate Model. Equation (4.24) provides a mechanism for evaluating the reaction of a substance as it travels through a distribution system (Rossman et al., 1994). Reactions can occur in the bulk phase or with the pipe wall. EPANET models both types of reactions using first-order kinetics. In general, within any given pipe, material in the bulk phase and at the pipe well will decrease according to the following equation:

$$q(C) = -k_b C - \left[\frac{k_f}{R_h}\right](C - C_w) \tag{4.24}$$

where $q(C)$ = total reaction rate
k_b = first-order bulk reaction rate, in s^{-1}
C = substance concentration in the bulk flow, in mass/ft^3 (mass/m^3)
k_f = mass transfer coefficient between the bulk flow and the pipe wall, in ft/s (m/s)
R_H = hydraulic radius of pipe (pipe radius/2), in ft (m)
C_w = substance concentration at the wall, in mass/ft^3 (mass/m^3)

Assuming a mass balance for the substance at the pipe wall yields:

$$k_f(C - C_w) = k_w^c C_w \tag{4.25}$$

where k_w is the rate of chlorine demand at wall (wall demand), in ft/s (m/s).

Equation (4.25) pertains to the growth or decay of a substance, with mass transfer to or from the pipe wall depending on whether the sign of the equation is positive or negative. A negative sign means decay and a positive sign means growth.

There are three coefficients used by EPANET to describe reactions within a pipe. These are the bulk rate constant k_b and the wall rate constant k_w, which must be determined empirically and supplied as input to the model. The mass transfer coefficient is calculated internally by EPANET.

Other Features. EPANET can also model the changes in age of water and travel time to a node. The percentage of water reaching any node from any other node can also be calculated. Source tracing is a useful tool for computing the percentage of water from a given source at a given node in the network over time.

The following input steps should be taken before using EPANET:

- Identify all network components and their connections. These components include pipes, pumps, valves, storage tanks, and reservoirs.
- Assign unique ID numbers to all nodes. These numbers must be between 1 and 2,147,483,647, but do not have to be in any specific order.
- Assign ID numbers to each link (pipe, pump, or valve). Both a link and a node can have the same ID number.
  - Collect information on the following system parameters: diameter, length, roughness, and minor loss coefficient for each pipe.
  - Characteristic operating curve for each pump.
  - Diameter, minor loss coefficient, and pressure or flow setting for each control valve.
  - Diameter and lower and upper water levels for each tank.
  - Control rules that determine how pump, valve, and pipe settings change with time, tank water levels, or nodal pressures.
  - Changes in water demands for each node over the time period being simulated.
  - Initial water quality at all nodes and changes in water quality over time at source nodes.

Output from EPANET includes the following:

- Color-coded network maps
- Time series plots
- Tabular reports
- Concentration plots
- Pressure plots
- Flow plots

Convective Transport Model. Biswas et al. (1993) developed a steady-state transport equation that takes into account the simultaneous convective transport of chlorine in the axial direction, diffusion in the radial direction, and consumption by first order in the bulk liquid phase. Different wall conditions are considered in the model, including a perfect sink, no wall consumption, and partial wall consumption.

QUALNET. Islam (1995) developed a model called QUALNET, which predicts the spatial and temporal distribution of chlorine residuals in pipe networks under slowly varying unsteady flow conditions. Unlike other available models, which use steady state or extended-period simulation (EPS) of steady flow conditions, QUALNET uses a lumped-system approach to compute unsteady flow conditions and includes dispersion and decay of chlorine during travel in a pipe. The pipe network is first analyzed to determine the initial steady state conditions. The slowly varying conditions are then computed by numerically integrating the governing equations by an implicit finite difference, a scheme subject to the appropriate boundary conditions. The one dimensional dispersion equation is used to calculate the concentration of chlorine over time during travel in a pipe, assuming a first-order decay rate.

Numerical techniques are used to solve the dispersion, diffusion, and decay equation. Complete mixing is assumed at the pipe junctions. The model has been verified by comparing the results with those of EPANET for two typical networks. The results are in good agreement at the beginning of the simulation model for unsteady flow; however, chlorine concentrations at different nodes vary when the flow becomes unsteady and when reverse flows occur. The model may be used to analyze the propagation and decay of any other substance for which a first-order reaction rate is valid (Chaudhry, 1987): The water quality simulation process used in the previous models is based on a one-dimensional transport model, in conjunction with the assumption that complete mixing of material occurs at the junction of pipes. These models consist of moving the substance concentrations forward in time at the mean flow velocity while undergoing a concentration change based on kinetic assumptions. The simulation proceeds by considering all the changes to the state of the system as the changes occur in chronological order. Based on this approach, the advective movement of substance defines the dynamic simulation model. Most water quality simulation models are interval oriented, which in some cases, can lead to solutions that are either prohibitively expensive or contain excessive errors.

Boulos et al. (1995) proposed a technique mentioned earlier called the event-driven method (EDM). This is extremely simple in concept and is based on a next-event scheduling approach. In this method, the simulation clock time is advanced to the time of the next event to take place in the system. The simulation scheduled is executed by carrying out all the changes to a system associated with an event, as events occur in chronological order. Since the only factors affecting the concentration at any node are the concentrations and flows at the pipes immediately upstream of the given node, the only information that must be available during the simulation are the different segment concentrations. The technique makes the water quality simulation process very efficient.

The advective transport process is dictated by the distribution system demand. The model follows a front tracking approach and explicitly determines the optimal pipe segmentation scheme with the smallest number of segments necessary to carry out the simulation process. To each pipe, pointers (concentration fronts), whose function is to delineate volumes of water with different concentrations, are dynamically assigned. Particles representing substance injections are processed in chronological order as they encounter the nodes. All concentration fronts are advanced within their respective pipes based on their velocities. As the injected constituent moves through the system, the position of the concentration fronts defines the spatial location behind which constituent concentrations exist at any given time. The concentration at each affected node is then given in the form of a time-concentration histogram.

The primary advantage of this model is that it allows for dynamic water quality modeling that is less sensitive to the structure of the network and to the length of the simulation process. In addition, numerical dispersion of the concentration front profile resolution is nearly eliminated. The method can be readily applied to all types of network configurations and dynamic hydraulic conditions and has been shown to exhibit excellent convergence characteristics.

4.2.4 Effects of Tanks and Storage

A frequently overlooked aspect of water quality and contaminant propagation in drinking water distribution systems is the effect of system storage. Although direct pumping could maximize water quality by shortening the transport time between source and consumer, it is rarely used today in systems in the United States (AWWA, 1989). Most utilities use some type of ground or elevated system storage to process water at times when treatment facilities would otherwise be idle. It is then possible to distribute and store water at one or more locations in the service area closest to the user.

The principal advantages of distribution storage are that it equalizes demands on supply sources, production works, and transmission and distribution mains. As a result, the sizes or capacities of these elements may be minimized. Additionally, system flows and pressures are improved and stabilized to better serve the customers throughout the service area. Finally, reserve supplies are provided in the distribution system for emergencies, such as firefighting and power outages.

In most municipal water systems, less than 25 percent of the volume of the storage in tanks is actively used under routine conditions. As the water level drops, tank controls, call for high-service pumps to start in order to satisfy demand and refill the tanks. The remaining water in the tanks (70 to 75 percent) is normally held in reserve as dedicated fire storage.

Storage tanks and reservoirs are the most visible components of a water distribution system but are often the least understood in terms of their effect on water quality. Although these facilities can play a major role in providing hydraulic reliability for firefighting needs and in providing reliable service, they may also serve as vessels for complex chemical and biological changes that may result in the deterioration of water quality. These storage tanks and reservoirs also contribute to increased residence time in drinking water systems.

Previous Research. P.V. Danckwerts, one of the first investigators to discuss the concept of a distribution function for residence times, explained how this concept can be defined and measured in actual systems (Danckwerts, 1958). When a fluid flows through a vessel at a constant rate, either "plug flow" (no mixing) or perfect mixing is usually assumed. In practice many systems do not achieve either. Thus calculations based on these assumptions may be inaccurate. Danckwerts illustrated the use of distribution functions by showing how they can be used to calculate the efficiencies of reactors and blenders and how models may be used to predict the distribution of residence times in large systems.

A.E. Germeles developed a model based on the concept of forced plumes and mixing of liquids in tanks (Germeles, 1975). He considered the mixing between two miscible liquids of slightly different density when one of them is injected into a tank partially filled with the other. A mathematical model for the mixing of the two liquids was developed, from which one can compute the tank stratification. The model also was verified experimentally.

Empirical Studies. Several investigators have conducted field studies and attempted to apply relatively simple models to distribution storage tanks. Kennedy et al. (1993) attempted to assess the effects of storage tank design and operation on mixing regimes and effluent water quality. The influent and effluent flows of three tanks with diameter-to-height ratios ranging from 3.5:1 to 0.4:1 were monitored for chlorine residual. Chlorine levels were also measured within the water columns of each tank. Although chlorine profiles revealed some stratification in tanks with large height-to-diameter ratios, completely mixed models were more accurate than plug-flow models in representing the mixing behavior of all three tanks. These investigators further indicated that the quality of the effluent from completely mixed tanks deteriorated with decreasing volumetric change. The authors found that standpipes were the least desirable tank design with respect to effluent water quality.

Studies conducted by Grayman and Clark (1993) indicated that water quality degrades as a result of long residence times in storage tanks. These studies highlight the importance of tank design, location, and operation on water quality. Computer models, developed to explain some of the mixing and distribution issues associated with tank operation, were used to predict the effect of tank design and operation on various water quality parameters. Because of the diversity of the effects and the wide range of design and environmental conditions, the authors concluded that general design specifications for tanks are unlikely. They also concluded that models will most likely be refined and developed to facilitate site-specific analysis.

One of the first studies to document the impact of storage tanks on water quality was conducted by the U.S. Environmental Protection Agency (U.S. EPA) in conjunction with the South Central Connecticut Regional Water Authority (SCCRWA) (Clark et al., 1991). As part of the overall U.S. EPA-sponsored project, an extensive field sampling program was performed in the Cheshire service area of the SCCRWA during November and December 1989. During this period, the fluoride feed into the water drawn from the two well fields was stopped for a seven-day period during which extensive sampling occurred throughout the system, including the water entering and leaving the Prospect tanks. The fluoride feed was later restarted and sampling was performed for the next seven-day period. Complete details on the sampling program, described in Clark et al. (1991) and Skov et al. (1991). The field study showed that storage facilities could have a significant impact on water quality in a distribution system. In order to further investigate the effects of tank location and operation on the water quality in the system, a series of simulations were performed using a hydraulic and water quality model. In the simulations, the effects of the location and operation of the tank were studied (Grayman et al., 1991). It was found that water age is a key factor in the deterioration of water quality. With age, chlorine residuals will decrease and disinfection by-product (DBP) concentrations will increase. Simulating water age, an initial age of zero was assumed at all nodes and at the tanks.

Development of a Storage Tank Model. As indicated earlier, most water quality simulations assume complete mixing of the storage tanks examined. However, sampling data from the Prospect tank and other studies (Kennedy et al., 1991; Kennedy et al., 1993) indicated that is not necessarily the case. Grayman and Clark (1993) explored the use of compartment modeling to describe tank mixing and to deal with the nonuniform mixing in tanks.

Grayman, Rossman and Arnold et al (2000) developed a suite of models called CompTank that included various types of mixing models for tanks including complete mix, plug flow and compartment models. In the same study, computational fluid dynamics (CFD) models were assessed and applied to represent the dynamics of mixing in tanks.

4.2.5 Water Quality Monitoring

Monitoring. Monitoring in distribution systems provides the means for identifying variations in water quality spatially and temporally (Grayman, Rossman and Geldreich, 2000) The resulting data can be used to track transformations that are taking place in water quality and can also be used to calibrate water quality models. Monitoring can be classified as routine or for special studies. Routine monitoring is usually conducted in order to satisfy regulatory compliance requirements. Special studies are usually conducted to provide information concerning water quality problems.

Routine monitoring in the United States is one of the requirements specified by the U.S. Safe Drinking Water Act. Although from a historical viewpoint most attention has been focused on the performance of the treatment plant, a number of the SDWA maximum contaminant levels (MCL) must be met at the tap. Samples are collected by the use of continuous monitors or as grab samples. Continuous monitoring is generally performed by sensors or remote monitoring stations. Although continuous monitoring is frequently conducted at the treatment plant (turbidity and chlorine residuals) most distribution system monitoring is based on grab samples. Continuous monitoring is capital intensive, requires maintenance and calibration. Grab samples are labor intensive and provide data based on the time of collection.

Special studies might include the following:

- Measurement of disinfectant residuals in a distribution system
- Tracer studies to assist in calibrating water quality models

Some of the issues that should be considered in preparation of a special study are as follows:

- Sampling locations
- Sampling frequency
- System operation
- Preparation of sampling sites
- Sampling collection procedures
- Analysis procedures
- Personnel organization and schedule
- Safety issues
- Data recording
- Equipment and supply needs
- Training requirement
- Contingency plans
- Communications
- Calibration and review of analytical instruments

4.3 SIMULATION OF RESIDENTIAL WATER DEMANDS

4.3.1 Background Information

In the past decade sophisticated network models have evolved to assist engineers and planners with design and operation of municipal water distribution systems. Whether the intent is to simulate pipe hydraulics or to predict water quality, the response of the network model is dictated largely by the assumed pattern of water demands. An accurate picture of network demands is essential for good model performance. An inaccurate portrayal of water demands almost surely leads to poor predictions.

Urban water users can be grouped into several consumer categories, namely, commercial, industrial, institutional, and residential. While each of these end uses are important, the focus of this chapter is on water demand in the residential sector. In the context of network security, residential water use is especially relevant because domestic consumption implies direct human contact with water and inevitable exposure to water-borne constituents.

Residential demand is the largest component of municipal water use (Flack, 1982) consuming, on average, 65 percent of the treated public supply (Solley et al., 1998). Owing to different demand characteristics, residential water use is usually split into domestic (indoor) and landscape (outdoor) demands. Indoor water use corresponds closely to winter residential demands and includes water for drinking, cooking, bathing, washing, cleaning, and waste disposal (Linaweaver et al., 1966).

A breakdown of average residential water use according to various household fixtures and appliances is given in Table 4.1. Nearly 60 percent of the indoor water use occurs in the bathroom (Maddaus, 1987). Neglecting leaks, water use at a typical single-family residence is intermittent with inflow from the mainline occurring for only a small percentage of the time. Outdoor water use, driven primarily by landscape irrigation, exhibits a strong seasonal pattern with highly variable peak demands (Litke and Kaufmann, 1993).

TABLE 4.1 Typical Residential Indoor Water Use

Fixture or server	Percent of total	Daily vol (gal)	Demand rate (gpm)	Count (pulses/day)	Duration (min)
Toilet	28	65	2–5 gpm	20	24
Laundry	22	50	20 gpm	2.5	21
Shower	22	50	2–4 gpm	3	20
Faucets	11	25	0.1–6 gpm	92	25
Baths	9	20	20 gpm	1	5
Dishwasher	3	8	8 gpm	1	5
Leaks	5	12	0.008 gpm	continuous	1440
Total	100	230	0.16	120	100

Notes: (1) "gpm" is gallons per minute.
(2) Average demand at busy server, $\alpha = (218 \text{ gal/day})/(100 \text{ min/day}) = 2.18$ gpm (excludes leaks).
(3) Average duration at a busy server, $\tau = (100 \text{ min/day})/(120 \text{ uses/day}) = 0.83$ min/use.
(4) Average pulse volume at a busy server, $\phi = \alpha\tau = (2.18 \text{ gpm})(0.83 \text{ min/use}) = 1.82$ gal/use.
(5) Average daily customer arrival rate, $\lambda = (120 \text{ uses/day})/(24 \text{ hrs/day}) = 5.00$ uses/hr.
(6) Average daily utilization factor, $\rho = \lambda\tau = (100 \text{ min/day})/(1440 \text{ min/day}) = 0.069$.

4.3.2 Assigning Water Demands

The pattern of water demands in municipal distribution systems is quite unpredictable, changing with time and location. The inherent randomness in residential water use makes it difficult to assign accurate water demands to network nodes. The inaccuracies in nodal demands constitute one of the primary sources of uncertainty in the calibration and application of network models (Walski et al., 2003).

Nonetheless, for purposes of network modeling, representative estimates of residential water demand must be specified. As a starting point, it is important to select values appropriate for the circumstance under consideration. For instance, a hydraulic investigation into the adequacy of system pressure during a fire emergency would adopt high demands during peak hours. In contrast, a water quality study on the concentration of a disinfectant residual expected in a remote service area would likely be based on average demands during a routine day.

One approach often used to assign nodal demands adopts a "top-down convention." Under this strategy, the overall system-wide average demand is assumed known, usually from utility records of water production. Base demands (i.e., daily average use) are allocated among individual demand nodes in proportion to water usage registered at metered accounts or according to the number of households assigned to a node.

Most water distribution systems experience leaks (AWWA, 2003). In the absence of any information about the type or location of leakage, such losses can be distributed uniformly among the demand nodes. Table 4.2 illustrates a simple application of the top-down convention for a network of 1000 homes with a total demand of 120 gpm including a 15 percent water loss by leakage. The homes are assigned to six demand nodes, each assumed to represent a relatively homogeneous neighborhood.

The total demand listed in the right-most column of Table 4.2 is the sum of the assumed leakage and allocated use. It represents the average *base demand* for each node. Over the course of a typical day, the actual water demand will vary, in some cases dramatically (Anderson and Watson, 1967). In residential settings, the greatest water use usually occurs during the breakfast and dinner periods with lowest demands in the early morning hours (Bowen et al., 1993; Mayer et al., 1999).

To capture diurnal variations in water demand, network models employ a series of concatenated steady-state analyses. This approach, also known as extended-period simulation (EPS), is common in network water quality studies where unsteady flow patterns are important (Rossman et al., 1994). In a conventional EPS, the nodal demands change at a set time interval (usually every hour) according to a pattern of user-specified demand multipliers.

There can be considerable latitude in selecting the hourly multipliers, provided that they preserve the base demand over the course of the day. Hence, during the calibration of the network hydraulic model, it is not unusual to fine-tune the hourly multipliers and/or

TABLE 4.2 Allocating Network Water Demand with Assumed Uniform Leakage

Network node number	Number of homes	Percent of total	Uniform leakage(gpm)	Allocated demand (gpm)	Total demand (gpm)
1	80	8	3	8.2	11.2
2	200	20	3	20.4	23.4
3	150	15	3	15.3	18.3
4	120	12	3	12.2	15.2
5	300	30	3	30.6	33.6
6	150	15	3	15.3	18.3
Total	1000	100	18	102	120

base demands in order to better match model predictions against field observations (Walski et al., 2003).

4.3.3 Poisson Rectangular Pulse (PRP) Premise

Residential water use can be analyzed with basic principles from the queuing theory developed by Erlang (1917–18) in his seminal study of telephone calls to operator banks. In the context of municipal distribution systems, residential water demands at single family homes behave as a time dependent Poisson Rectangular Pulse (PRP) process. In the jargon of queueing theory, home occupants are customers while water fixtures and appliances are servers. When occupants draw water from the distribution network, the home is considered "busy"; otherwise the home is considered "idle."

Under the PRP hypothesis, the frequency of residential water use is assumed to follow a Poisson arrival process with a time dependent rate parameter (Buchberger and Wu, 1995). When a water use occurs, it is represented as a single rectangular pulse of random duration and random but steady intensity as shown in Fig. 4.1. Buchberger and Wells (1996) found that over 80 percent of indoor residential water demands occur as single pulses. They demonstrated that more complex demand patterns are easily converted to an equivalent single pulse.

By virtue of the Poisson assumption, it is unlikely that more than one pulse will start at the same instant. Owing to the finite duration of each water pulse, however, it is possible that two or more pulses with different starting times will overlap for a limited period. When this occurs, the total water use at the residence is the sum of the joint intensities from the coincident pulses. Buchberger and co-workers (2003) present an extensive body of compelling evidence based on detailed field studies to corroborate the validity of the PRP model for residential water demands.

The PRP model requires five parameters (α, β^2, τ, ω^2, λ) to characterize residential water use. Each has a clear physical interpretation:

1. α is the average rate of water use at a *busy* fixture in the home,
2. β^2 is the variance of the rate of water use at a *busy* fixture,
3. τ is the average duration of water use at a *busy* fixture,
4. ω^2 is the variance of the duration of water use at a *busy* fixture, and
5. λ is the average customer arrival rate in the home.

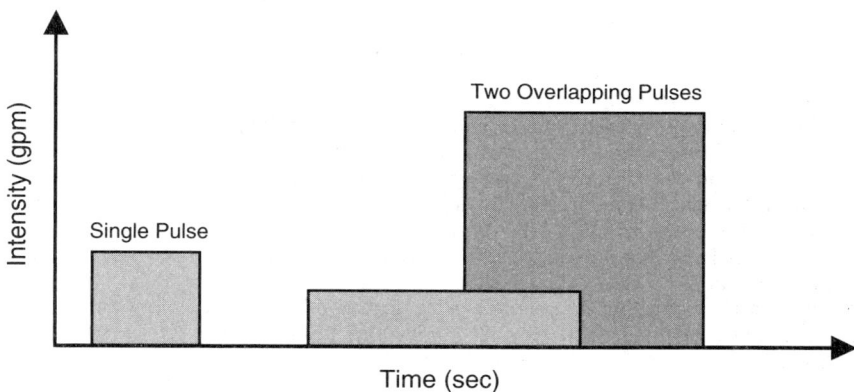

FIGURE 4.1 Definition sketch for Poisson Rectangular Pulse process.

TABLE 4.3 Typical Range of Parameter Values for PRP Model of Residential Water use

Pulse location and feature	Typical range for the mean value	Typical range for the standard deviation	Typical range for the coef variation	Representative probability distribution
Indoor intensity	α = 1.0 to 4.0 gpm	β = 0.5 to 2.0 gpm	Θ_q = 0.3 to 0.7	Log-normal
Indoor duration	τ = 0.5 to 2.0 min	ω = 0.5 to 3.0 min	Θ_d = 1.0 to 3.0	Log-normal
Outdoor intensity	α = 3.0 to 6.0 gpm	β = 1.0 to 2.0 gpm	Θ_q = 0.3 to 0.7	Log-normal
Outdoor duration	τ = 15 to 30 min	ω = 30 to 60 min	Θ_d = 2.0 to 4.0	Log-normal

Note: Indoor values compiled from Appendix 6.7 in Buchberger et al. (2003).

For example, Table 4.1 suggests that some representative PRP parameter values for indoor water use are: mean pulse intensity, $\alpha \approx 2.2$ gpm; mean pulse duration, $\tau \approx 0.8$ min per use, and daily mean arrival rate, $\lambda \approx 5.0$ uses per hour (in a single-family residence of four occupants). There is not enough information in Table 4.1 to estimate the pulse variability, β^2 or ω^2. However, recent field studies (Buchberger et al., 2003) suggest that β can range from 0.5 to 2.0 gpm while ω can range from 0.5 to 3.0 min for indoor water use in a single-family home.

It is important to stress that these PRP parameter estimates are nominal values averaged over the course of a typical day. Actual daily values at individual households may deviate significantly from these estimates (see Table 4.3). Furthermore, at any given home, values of the PRP parameters may change from hour to hour. In most cases, hourly variations in PRP parameter values will be relatively minor, except for the arrival rate where a strong diurnal pattern is common (see top portion of Fig. 4.2).

Central to the PRP concept for residential water use is a dimensionless parameter, known as the "utilization factor" ($\rho = \lambda\tau$), that incorporates the joint effects of customer arrival rates (λ) and busy fixture durations (τ). In a single family home, ρ approximates the percentage of time that the household uses water. The example under Table 4.1 gives $\rho \approx 0.069$ as an *average* daily value for a household of four people. As shown on the "ρ-profile" in the bottom portion of Fig. 4.2, the behavior of ρ during the course of a day closely mimics the unsteady pattern for customer arrivals. The hourly variations in the λ and ρ profiles are the primary source of time dependence in the PRP model of residential water use.

Both the arrival rate (λ) and the utilization factor (ρ) are additive across homes in the neighborhood. This is a very important property, because it implies that the PRP concept for residential water use is scalable. The PRP model can be readily applied to arbitrarily large networks (including the entire municipal water distribution system) simply by aggregating the utilization factors from all individual residences.

Besides painting a simple picture of residential water demands at network nodes, the PRP model provides several useful results for flows in branching pipelines. In contrast to a looping pipe where flow reversals are possible, flow in a branching pipe does not reverse direction under normal operating conditions. In branching pipes, water moves only toward the terminal end of the line in response to downstream demands. Branching pipes comprise a significant portion (up to 25 percent and more) of the water mains in most municipal distribution systems. A dead-end trunkline is the most common example of a branching pipe.

Suppose a branching pipeline supplies water to a neighborhood of N homes where there is a constant leakage rate, $L \geq 0$. From PRP principles it can be shown that the mean of the instantaneous flow rate $Q(t)$ at time t in the branching supply pipe is

$$E[Q(t)] = N\alpha\rho(t) + L \qquad (4.26)$$

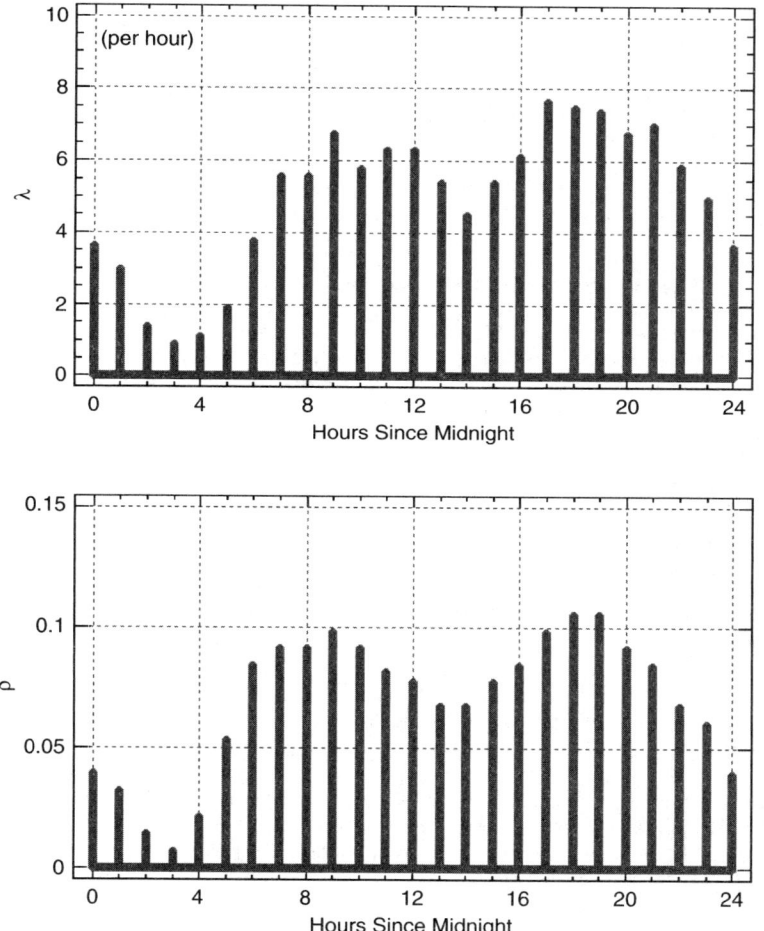

FIGURE 4.2 Typical hourly patterns for customer arrivals (top, $\bar{\lambda} = 5.0$ per hour) and the utilization factor (bottom, $\bar{\rho} = 0.069$) at a single-family residence of four (based on observations at 21 homes over a 30-day period as reported in Buchberger et al., 2003).

The variance of the instantaneous flow rate at time t is

$$\mathrm{Var}[Q(t)] = N[\alpha^2 + \beta^2]\rho(t) \tag{4.27}$$

and the probability of stagnation at time t in the pipeline is

$$\mathrm{Prob}[Q(t) = 0] = \frac{\exp[-N\rho(t)]}{\exp[1/\delta(L)]} \tag{4.28}$$

Here $\delta(L)$ is the Dirac delta function with property $\delta(L = 0) = \infty$; otherwise, $\delta(L \neq 0) = 0$. The result in Eq. (4.28) says there can be no stagnation in the supply line when leakage is present (i.e., Prob[$Q = 0$] = 0 if $L > 0$), but stagnation might occur when leakage is absent (i.e., Prob[$Q = 0$] = exp[$-N\rho$] if $L = 0$). Constant leakage in the neighborhood increases the mean flow through the supply line [see Eq. (4.26)], but has no effect on the variance of the pipe flow [see Eq. (4.27)].

The values for α and β used in Eqs. (4.26) and (4.27) are representative of water demand pulses for the entire neighborhood. $\rho(t)$ signifies that the utilization factor may change with time, as depicted in Fig. 4.2. The theoretical results for pipe flows given in Eqs. (4.26)–(4.28), in conjunction with the additive property of the utilization factor, make the PRP model a very convenient and tractable tool for simulating residential water demands.

4.3.4 PRP Model for Demand Simulation

An interactive C++ computer code called PRPsym has been developed at the University of Cincinnati to simulate residential water demands that follow a Poisson rectangular pulse process (Li and Buchberger, 2003). The PRPsym code generates *instantaneous* (i.e., second-by-second basis) water demands for each node in a pipe network. Resulting demand values are then integrated over time to match the averaging interval selected for the particular network application.

For example, suppose it is necessary to run a one-day extended period simulation using 15-min time steps. The PRPsym code generates 86,400 water demands (one value for each second of the one-day EPS) for each node in the network. The demands are grouped into nonoverlapping strings of 900 consecutive 1-s flows and then averaged to provide a random demand at each node for each quarter-hour of the EPS.

The general algorithm to simulate residential water demands with the PRP model is a three-step process, as outlined below:

Step 1. Generate random customer arrivals from an exponential distribution.

Step 2. Generate random pulse intensities from a lognormal distribution.

Step 3. Generate random pulse durations from a lognormal distribution.

In step 1, the arrival of a customer signals the occurrence of a water demand. The exponential distribution for waiting times between water demands follows directly from the premise of a Poisson arrival process. The arrival rate $\lambda(j, k)$ for the Poisson process at node j during time step k is given by

$$\lambda(j, k) = \frac{\pi(k) \times Q(j)}{\phi(j)} \quad (4.29)$$

In Eq. (4.29), $\pi(k)$ is a user-specified arrival rate multiplier for time step k, $Q(j)$ is the base demand for node j (e.g., see column 6 in Table 4.2), and $\phi(j)$ is average volume of a residential water demand pulse at node j.

The arrival rate multiplier is analogous to the demand multiplier now used in EPS studies. On average, the arrival rate multiplier $\pi(k)$ is unity. During high-use periods, $\pi(k) > 1.0$; conversely, during low-use periods $\pi(k) < 1.0$. The average pulse volume is the product of the mean pulse intensity and the mean pulse duration, $\phi = \alpha\tau$. The example with Table 4.1 gives $\phi = 1.82$ gal. Use of the lognormal distribution for pulse intensity and pulse duration in steps 2 and 3 is based on field observations of residential water use (Buchberger and Wells, 1996; Buchberger et al., 2003).

During the summer months, a significant fraction of residential water use may occur outdoors. Indoor water use is a high frequency, short duration process. By contrast, outdoor water use tends to occur as low frequency, long duration pulses. Due to their incommensurate time scales and incompatible PRP parameters, it is necessary to simulate indoor and outdoor water use separately (Lee and Buchberger, 1999). Then, the total instantaneous residential demand at a network node is found as the sum of the joint indoor and outdoor water demands.

4.3.5 Parameter Estimation

The PRPsym program needs five input parameters to generate residential water demands, namely, mean and variance for pulse intensity (α and β^2), mean and variance for pulse duration (τ and ω^2), and an arrival rate multiplier (π). Owing to strong similarities among water use fixtures and household appliances in homes across North America, the patterns and properties of indoor water use are fairly consistent from coast to coast (Mayer et al., 1999). Hence, the values summarized in Table 4.3 provide reasonable initial estimates of the key PRP parameters for indoor residential water use. On the other hand, outdoor water use depends strongly on the weather and the season. As a consequence, outdoor use varies with location and time of year. Discretion is essential when adopting values from Table 4.3 for simulation of outdoor water use.

In some cases it may be necessary to estimate site-specific PRP parameter values from field measurements. Few utilities have the resources to continuously record water consumption at individual households in order to customize PRP parameters. Many utilities, however, are able to monitor flow along a branching pipeline. Careful pipe flow measurements can yield explicit estimates of PRP parameters for indoor residential water use, especially if the target neighborhood is small and the flow readings are frequent.

To illustrate, suppose that flow measurements are obtained for an entire day on a branching pipe that feeds a small DMA ($N < 200$ homes) having negligible leakage and no outdoor water use. If leaks are a problem, then the rate of leakage should be estimated and deducted from the flow measurements using the leak detection algorithm proposed by Buchberger and Nadimpalli (in press). If necessary, the pipe flow measurements should be taken during the winter season when outdoor water use is absent.

The pipe flow readings are split into 24 one-hour subsets. For instance, if flows $Q(t)$ were recorded with a frequency of one reading per second, then each of the 24 hourly data sets will contain $M = 3600$ flow observations. For each data subset, the sample mean $\overline{Q}_k$, the sample variance $S_{Q_k}^2$, the lag-one autocorrelation coefficient $r_k(1)$, and the stagnation percentage $P_k[0]$ for the kth hour are computed using the following expressions:

$$\overline{Q}_k = \left(\frac{1}{M}\right)\sum_{t=1}^{M} Q_k(t) \tag{4.30}$$

$$S_{Q_k}^2 = \left(\frac{1}{M-1}\right)\sum_{t=1}^{M}\left[Q_k(t) - \overline{Q}_k\right]^2 \tag{4.31}$$

$$r_k(1) = \frac{\sum_{t=1}^{M}\left[Q_k(t) - \overline{Q}_k\right]\left[Q_k(t-1) - \overline{Q}_k\right]}{\sum_{t=1}^{M}\left[Q_k(t) - \overline{Q}_k\right]^2} \tag{4.32}$$

$$P_k[0] = 1 - \left(\frac{1}{M}\right)\sum_{t=1}^{M} w_k(t) \tag{4.33}$$

where $w_k(t)$ is an indicator variable: $w_k(t) = 1$, if $Q_k(t) > 0$ and $w_k(t) = 0$, if $Q_k(t) = 0$.

Provided that pipe flow readings are frequent (e.g., 1 per second) and that the mean customer arrival rate is nearly constant during the hourly period, Eqs. (4.26)–(4.28) give an estimate of the mean pulse intensity for the kth hour as

$$\hat{\alpha}_k = \frac{-\overline{Q}_k}{\ln(P_k[0])} \qquad \text{if } P_k[0] > 0 \tag{4.34a}$$

$$\hat{\alpha}_k = \frac{S_{Q_k}^2}{(1+\Theta_q^2)\overline{Q}_k} \qquad \text{if } P_k[0] = 0 \tag{4.34b}$$

In Eq. (4.34b), $\Theta_q = \beta/\alpha$ is the assumed coefficient of variation of pulse intensity at a busy server (see Table 4.3). As an expedient, if it is further assumed that pulse durations have an exponential distribution, then the mean pulse duration for hour k is given by

$$\hat{\tau}_k = \frac{\Delta T}{2\left[1-\sqrt{r_k(1)}\right]} \tag{4.35}$$

where ΔT is the time interval between flow readings (note that ΔT must be very small, say under 2 s). The hourly arrival rate multipliers are given by

$$\hat{\pi}_k = \frac{\overline{Q}_k}{\sum_{k=1}^{24} \overline{Q}_k} \tag{4.36}$$

Although possible, it is not practical to replicate hourly variations in parameter values with the PRPsym code. Hence, the recommended PRP parameter estimates for water use at the DMA are taken as the average of the 24 hourly values,

For pulse intensity: $\hat{\alpha} = \left(\frac{1}{24}\right)\sum_{k=1}^{24} \hat{\alpha}_k$ and $\hat{\beta} = \Theta_q \hat{\alpha}$ (4.37a,b)

For pulse duration: $\hat{\tau} = \left(\frac{1}{24}\right)\sum_{k=1}^{24} \hat{\tau}_k$ and $\hat{\omega} = \Theta_d \hat{\tau}$ (4.38a,b)

Here Θ represents the assumed coefficient of variation (see Table 4.3). In the PRPsym program, the mean and variance of pulse intensity and pulse duration are held constant at a demand node, but their values can vary among nodes across the network.

The simple parameter estimation method outlined above works well when the water utility can isolate and monitor a small-scale DMA that is representative of residential consumers in large-scale service districts. The main drawback is that the utility must furnish an educated guess about the degree of variability (i.e., Θ values) for pulse intensity and pulse duration. An example will illustrate the procedure in the following section.

Guercio et al., (2001) present an alternate approach for PRP parameter estimation that is suitable for large service areas and time-averaged flow measurements. However, their method is computationally intensive, involving implicit optimization of nonlinear equations. Finally, it should be noted that the PRPsym code is not a water demand forecaster. It will not predict real-time water demands for a utility. This niche is covered by IWR-MAIN and other empirical regression models (Dziegielewski and Boland, 1989).

4.3.6 PRP Example Application

A brief example based on detailed field measurements is presented here to demonstrate estimation of PRP model parameters and simulation of residential water demands using PRPsym.

TABLE 4.4 Residential Water Demands at the Dead-End Loop on Monday September 15, 1997

Home number	Measured water demand			Computed pulse statistics		
	Daily pulse count	Daily volume (gal)	Daily duration (min)	Intensity α (gpm)	Duration τ (min)	Volume ϕ (gal)
1	86	60.7	35.3	1.72	0.41	0.71
2	135	288.4	98.5	2.93	0.73	2.14
3	42	74.9	37.1	2.02	0.88	1.78
4	**326**	166.3	89.8	1.85	**0.28**	**0.51**
5	33	129.3	65.8	1.97	1.99	**3.92**
6	119	121.0	55.2	2.19	0.46	1.02
7	97	198.9	65.5	3.04	0.68	2.05
8	116	315.5	93.6	**3.37**	0.81	2.72
9	52	178.9	64.4	2.78	1.24	3.44
10	97	119.2	76.3	1.56	0.79	1.23
11	58	96.5	46.2	2.09	0.80	1.66
12	106	271.4	108.1	2.51	1.02	2.56
13	93	**421.7**	169.2	2.49	1.82	4.53
14	58	**35.3**	24.1	**1.47**	0.42	0.61
15	68	186.4	74.8	2.49	1.10	2.74
16	84	106.8	38.5	2.77	0.46	1.27
17	24	44.3	**19.9**	2.22	0.83	1.85
18	136	270.4	**174.0**	1.55	1.28	1.99
19	**20**	76.4	44.9	1.70	**2.24**	3.82
20	82	267.4	82.5	3.24	1.01	3.26
21	109	88.7	36.7	2.42	0.34	0.81
Grand total	1,941	3,518	1,500			
Average	92.4	167.5	71.4	2.35	0.77	1.81

Notes: (1) α = Total volume/total duration (col 3/col 4).
(2) τ = Total duration/total count (col 4/col 2).
(3) $\phi = \alpha\tau$ = Total volume/total count (col 3/col 2).
(4) Maximum and minimum values are underlined in bold.

The study site is a small neighborhood with about 55 people residing in 21 single-family homes on a dead-end loop supplied by a 6-in cast iron main. Water demands at each home were monitored around the clock for a period of 7 months in 1997 (for details, see Buchberger et al., 2003). The example considered here is a snapshot from a "typical" workday, Monday September 15, 1997.

Table 4.4 shows the daily pulse count, daily water volume and daily pulse duration measured at each of the 21 homes during the 24-h monitoring period. Among the 21 homes there is considerable variability in water use, with nearly an order-of-magnitude spread between minimum and maximum entries in each column. For instance, the daily pulse count ranged from a low of 20 to a high of 326 water uses with a household average of 92.4 pulses per day (3.85 per hour). The daily water volume ranged from a low of 35.3 gal to a high of 421.7 gal with a household average of 167.5 gal. The daily demand duration ranged from a low of 19.9 min to a high of 174.0 min with a household average of 71.4 min.

From these measured water demand data, the mean pulse intensity, the mean pulse duration, and mean pulse volume were found for each home as summarized on the right side of Table 4.4. Here too, there is considerable variability in the household water pulse statistics,

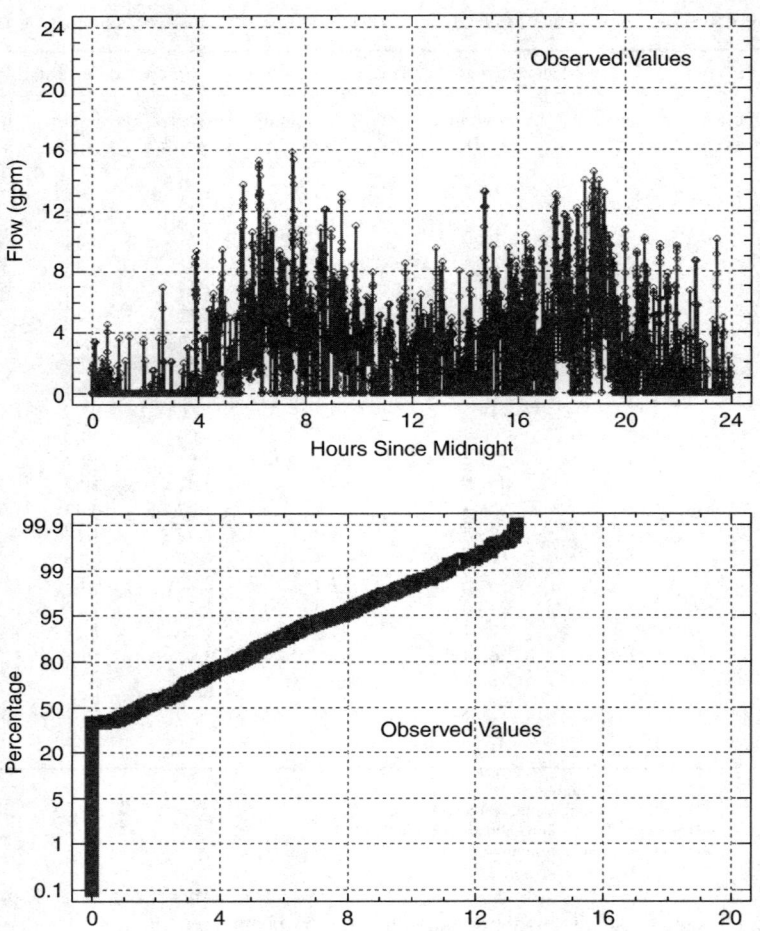

FIGURE 4.3 Time series plot (top) and normal probability plot (bottom) of reconstituted flows into a dead-end loop resulting from observed indoor water demands at 21 single family homes on Monday September 15, 1997.

though values are generally within the ranges listed in Table 4.3. The bottom row of Table 4.4 gives representative neighborhood values of the mean pulse intensity ($\alpha = 2.35$ gpm), the mean pulse duration ($\tau = 0.77$ min) and the mean pulse volume ($\phi = 1.81$ gal) computed from the grand totals. In most network modeling situations, these PRP parameter values would not be known a priori. Instead, they would need to be estimated from pipe flow measurements applied to Eqs. (4.34)–(4.38).

The record of instantaneous water demands at each home was used to reconstruct the flow in the pipeline feeding the dead-end loop. The top portion of Fig. 4.3 shows the time series of 86,400 reconstructed pipe flows, one for each second of the day, at the entrance to the study site on Monday, September 15, 1997. The time series of pipe flows exhibits the well-known diurnal pattern of residential water use with a slow period in the early morning hours and peak periods during breakfast and dinner times. The maximum instantaneous flow into the study site reached about 16 gpm just before 8 a.m. The minimum flow was zero, and this occurred numerous times. The bottom portion of Fig. 4.3 shows the normal probability plot of the reconstructed pipe flows. Here, the vertical spike at zero indicates that stagnation (i.e., no demand, $Q = 0$) occurred about 40 percent of the time in the dead-end loop on Monday, September 15, 1997.

The hourly statistics of the reconstructed pipe flows are summarized in Table 4.5. Stagnation in the mainline occurred every hour, ranging from a low of about 4 percent during breakfast to a high of about 92 percent in the predawn hours. Not surprisingly, stagnation and the pipe flow are inversely related; high stagnation implies low flows and vice versa. The mean hourly flow ranged from a low of 0.24 gpm (predawn) to a high of 5.27 gpm (breakfast). The variance of the hourly flows tends to follow a pattern similar to the mean flow with highest values during the busiest hourly periods. The lag-one autocorrelation of the hourly flows was consistently very high as shown in column 5.

Using the flow statistics in columns 2 to 5 of Table 4.5, hourly estimates of the mean pulse intensity, the mean pulse duration and the arrival multiplier were computed using Eqs. (4.34a), (4.35), and (4.36), respectively. Results, listed in Table 4.5 show that the hourly mean pulse intensity ranges from $\hat{\alpha}_k = 1.3$ to 2.8 gpm while the hourly mean pulse duration ranges from $\hat{\tau}_k = 0.4$ to 2.0 min. Taking column averages, the overall neighborhood values of these PRP parameters are estimated to be $\hat{\alpha} = 2.13$ gpm and $\hat{\tau} = 1.15$ min. These imply a mean pulse volume, $\hat{\phi} = 2.13 \times 1.15 = 2.45$ gal.

Comparing these estimated PRP parameters against the known neighborhood PRP parameters (values listed at the bottom of Table 4.4) reveals mixed results. The estimated mean intensity ($\hat{\alpha}$) is reasonably close to the actual value (2.13 gpm versus 2.35 gpm, a 9 percent difference). However, the estimated mean duration ($\hat{\tau}$) is 50 percent too high (1.15 min versus 0.77 min). This discrepancy likely indicates that predictions from Eq. (4.35) are poor when pulse durations are not exponentially distributed. This issue needs more investigation. The bias in the mean duration inflates the mean volume (2.45 gal versus 1.81 gal, a 35 percent difference). The net effect is that, in order to meet the neighborhood daily water demand, the number of pulses simulated by computer will be considerably less than the number of pulses observed in the field.

To proceed with the water demand simulation, the variability of the pulse intensity and pulse duration must be estimated. In this exercise, it is assumed that $\Theta_q = 0.5$ (so that $\hat{\beta} = 1.06$ gpm) and $\Theta_d = 1.5$ (so that $\hat{\omega} = 1.73$ min). Both Θ values are taken from the midrange of the coefficients of variation for indoor use listed in Table 4.3.

Using the arrival multipliers ($\hat{\pi}_k$, see column 8 of Table 4.5) and estimates of the four pulse parameters ($\hat{\alpha}, \hat{\beta}, \hat{\tau}, \hat{\omega}$, see column 4 of Table 4.6), the PRPsym program generated instantaneous (i.e., second by second) indoor water demands at the 21 homes along the dead-end study site for a 24-h period. Table 4.6 summarizes key statistics of the water demand pulses for three cases:

1. Statistics computed from 1941 observed residential water demands.
2. Statistics computed from reconstituted pipe flows [Eqs. (4.34)–(4.38)].
3. Statistics computed from 1465 simulated residential water demands.

The PRPsym program preserved the *estimated* values of the mean and standard deviation of the pulse intensity and duration (compare columns 4 and 5 of Table 4.6). As expected, owing to the mean volume discrepancy, the number of simulated pulses is much smaller than the number of observed pulses (1465 versus 1941, a 25 percent difference).

The simulated water demands were then used to reconstruct flows into the entrance of the dead-end loop. Table 4.7 provides a comparison of the statistics of the reconstructed flows

TABLE 4.5 Flows and PRP Parameters at the Dead-End Loop on Monday September 15, 1997

		Computed pipe flow quantities				Estimated PRP parameters		
Hour k	Stagnation probability $P_k[0]$	Mean $\overline{Q}_k$ (gpm)	Variance $S^2_{Q_k}$ (gpm)^2	Lag-1 ACF $r_k(1)$	Mean intensity $\hat{\alpha}_k$ (gpm)	Mean duration $\hat{\tau}_k$ (min)	Arrival multiplier $\hat{\pi}_k$
1	0.733	0.402	0.616	0.960	1.29	0.42	0.20
2	0.918	0.237	0.732	0.981	2.77	0.88	0.10
3	0.896	0.268	0.933	0.984	2.44	1.05	0.10
4	0.789	0.422	1.302	0.989	1.78	1.48	0.20
5	0.422	2.173	5.590	0.992	2.52	2.03	0.90
6	0.165	2.474	5.400	0.988	1.37	1.40	1.00
7	0.039	5.272	9.891	0.987	1.63	1.32	2.20
8	0.102	4.154	9.071	0.987	1.82	1.25	1.70
9	0.244	3.467	8.192	0.986	2.46	1.20	1.40
10	0.179	3.445	6.850	0.985	2.00	1.12	1.40
11	0.370	1.745	3.300	0.981	1.76	0.88	0.70
12	0.544	1.510	4.017	0.983	2.48	0.98	0.60
13	0.347	2.128	3.769	0.985	2.01	1.08	0.90
14	0.418	2.031	5.180	0.989	2.33	1.52	0.80
15	0.431	2.119	6.229	0.987	2.52	1.25	0.90
16	0.164	3.505	5.713	0.982	1.94	0.92	1.40
17	0.320	2.808	7.366	0.984	2.46	1.05	1.10
18	0.073	4.954	10.28	0.987	1.89	1.28	2.00
19	0.136	4.561	11.12	0.984	2.29	1.05	1.90
20	0.142	4.511	10.62	0.987	2.31	1.28	1.80
21	0.353	2.344	5.645	0.980	2.25	0.83	1.00
22	0.343	2.418	6.933	0.988	2.26	1.35	1.00
23	0.627	1.079	3.900	0.986	2.31	1.18	0.40
24	0.756	0.616	1.849	0.981	2.20	0.90	0.30
Full 24 h	0.396	2.443	7.860	0.990			
Average	0.396	2.443	5.604	0.984	2.13	1.15	1.00

TABLE 4.6 Statistics of Observed, Estimated and Simulated Water Demand Pulses at 21 Homes Along the Dead-End Loop

Parameter	Units	Observed value[*]	Estimated value[†]	Simulated value[‡]
No. of demand pulses	—	1941	—	1465
Water pulse intensity				
Mean, α	(gpm)	2.35	2.13	2.17
Standard deviation, β	(gpm)	1.30	1.06	1.03
Coef of var, β/α	—	0.55	0.50	0.47
Water pulse duration				
Mean, τ	(min)	0.77	1.15	1.13
Standard deviation, ω	(min)	1.55	1.73	1.50
Coef of var, ω/τ	—	2.01	1.50	1.33
Water pulse volume				
Mean, ϕ	(gal)	1.81	2.45	2.45
Standard deviation, ψ	(gal)	4.45	4.29	3.80
Coef of var, ψ/ϕ	—	2.45	1.75	1.55

[*]*Observed values* are computed from residential water demands recorded in the field.
[†]*Estimated values* are computed from reconstituted pipe flows using Eqs. (4.34)–(4.38).
[‡]*Simulated values* are computed from residential water demands generated with PRPsym.

into the dead-end loop based on observed and simulated water demands. Despite the difference in mean pulse volumes, there is a remarkably good agreement between the statistics of the flows based on observed water demands and those based on simulated water demands. In general, the differences between flow statistics are in the range of 2 to 20 percent. The agreement between observation and simulation is especially good for the mean flow, the coefficient of variation, autocorrelation, and the predicted frequencies of stagnant flow and laminar flow. The largest difference (17.7 percent) occurred between the maximum observed pipe flows, with simulation results indicating that higher flow values than were observed could be expected. It is very interesting to note that either a stagnant or laminar

TABLE 4.7 Statistics of Observed and Simulated Flows at Entrance to Dead-End Link on Monday September 15, 1997

Parameter/characteristic	Units	Observed value	Simulated value	Percent difference
Number of flow readings	—	86,400	86,400	—
Mean of pipe flow	(gpm)	2.44	2.50	2.4
Variance of pipe flow	(gpm)^2	7.86	8.76	11.4
Coefficient of variation	—	1.15	1.18	2.6
Minimum pipe flow	(gpm)	0	0	0
Maximum pipe flow	(gpm)	15.85	18.65	17.7
Lag-1 autocorrelation	—	0.990	0.989	0
Stagnation ($Q = 0$ gpm)	(%)	39.6	40.5	2.2
Laminar ($0 < Q < 4.6$ gpm)	(%)	38.6	39.0	1.0
Transition ($4.6 < Q < 8.1$ gpm)	(%)	16.6	14.5	12.6
Turbulent ($Q > 8.1$ gpm)	(%)	5.2	6.0	15.4

Note: Flow regime is based on Reynolds number (Re) for pipe with 6-in diameter and water temperature of 65°F.

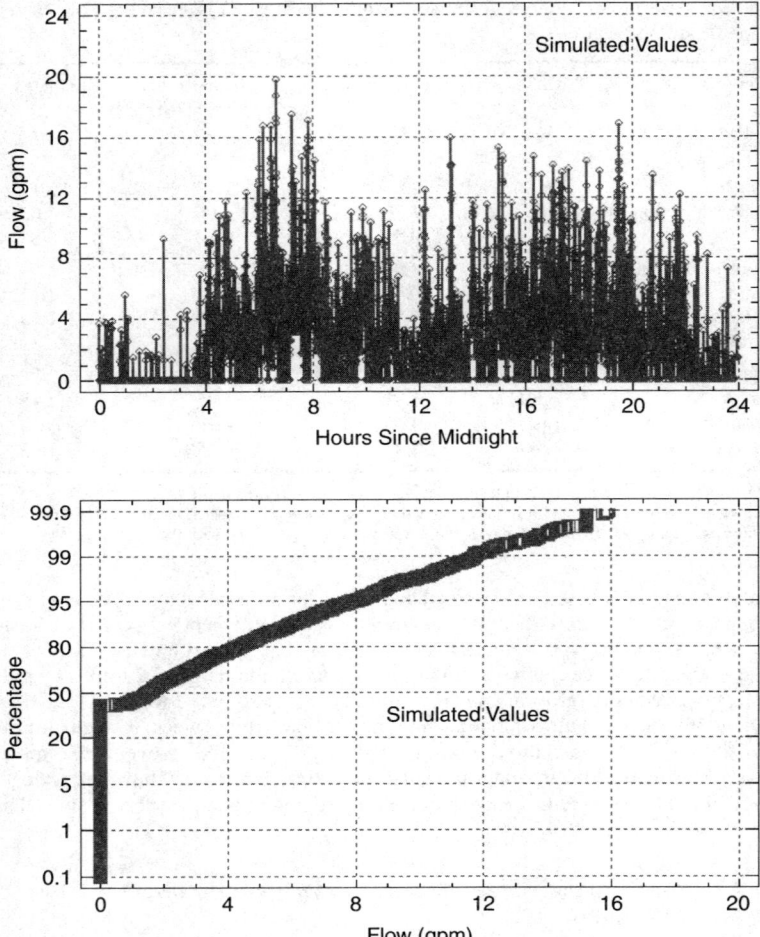

FIGURE 4.4 Time series plot (top) and normal probability plot (bottom) of reconstituted flows into a dead-end loop resulting from simulated indoor water demands at 21 single family homes on Monday September 15, 1997.

flow condition (Re < 2300) prevailed over 80 percent of the time in the dead-end mainline on Monday, September 15, 1997.

Figure 4.4 shows the corresponding times series (top) and normal probability plot (bottom) for the 86,400 pipe flows reconstructed from simulated water demands. The pronounced diurnal pattern present in the simulated time series matches well with the pipe flows from observed demands, shown in Fig. 4.3. The probability plot (bottom) based on simulated water demands is nearly indistinguishable from the probability plot based on observed water demands (Fig. 4.3). Both have prominent vertical spikes at $Q = 0$ (signifying stagnation) and a continuous tail of random flows ranging from 0 to 16 gpm. Considering the excellent agreement between the probability plots and the close comparison between flow statistics, the simulated demands from the PRP model appear to replicate quite well the behavior of the flows observed at the study site.

4.3.7 Summary

The Poisson Rectangular Pulse (PRP) model provides a rigorous framework for describing residential water demands in municipal pipe networks. A simple example based on field observations was presented to demonstrate how key parameters for the PRP model can be estimated from routine pipe flow measurements. The calibrated PRP model was then used to simulate realistic residential water demands.

What is the future of demand simulation in water supply? The introduction presented here just hints at the potential versatility of network simulation. Simulating water demands with the PRP process offers a direct way to incorporate the random nature of water use into problems that deal with design, operation and protection of water distribution systems. This allows water utilities to investigate a broad array of operational scenarios and thereby develop insight about the expected norms and the possible extremes of the network response. Such information complements traditional deterministic models, corroborates field experience and is essential for risk-based assessments. Considering the evolution of efficient network solvers, growth in desktop computational power, and advances in large-scale database management, it seems likely that demand simulation will emerge as an important element in water distribution system analyses.

4.4 CONDUCTING A TRACER STUDY

4.4.1 Introduction and Background

A tracer study is a mechanism for determining the movement of water within a water supply system. In such a study, a conservative substance is injected into the water and the resulting concentration of the substance is measured over time as it moves through the system. Historically, tracer studies have been used to study the movement of water through water treatment plants and distribution systems. Though conceptually quite straightforward, a successful field tracer study requires careful planning and implementation in order to achieve useful results.

Tracer studies have been most commonly used as a means of determining the travel time through various components in a water treatment plant (Teefy and Singer, 1990; Teefy, 1996). Its most frequent usage is to ascertain that there is an adequate chlorine contact time in clearwells as required by U.S. EPA regulations. Generally a conservative chemical such as fluoride, rhodamine WT, or calcium, sodium or lithium chloride is injected into the influent of the clearwell as either a pulse or a step function. Subsequently, the concentration is monitored in the effluent and statistics such as the T10 value (time until 10 percent of the injected chemical reaches the outflow) are calculated.

Tracer studies have also been conducted in distribution system storage facilities to understand the mixing processes within the facility. Fluoride was injected into the influent of a reservoir in California and sampled in both the effluent and at an interior point to assess mixing (Grayman et al., 1996; Boulos et al., 1996). In a research study sponsored by AwwaRF, Grayman et al. (2000) used fluoride and calcium chloride as tracers to study mixing in a variety of underground, ground level and elevated tanks.

Within distribution systems, tracer studies provide a means to understand the movement of water throughout the distribution system. This can serve multiple purposes including (1) calculating travel time or water age; (2) calibrating a water distribution system hydraulic model; (3) defining zones served by a particular source and blending with water from other sources; and (4) determining the impacts of accidental or purposeful contamination of a distribution system. A variety of tracers have been used in distribution system studies including chemicals that are injected into the system (e.g., fluoride, calcium chloride), chemical feeds that are turned off (e.g., fluoride), and natural occurring constituents that differ in different sources serving a system.

Grayman et al. (1988) studied blending in a multiple source system by measuring hardness, chloroform and total trihalomethanes that differed significantly between the surface water source and the groundwater sources. Clark et al. (1993) turned off a fluoride feed to a water system and traced the resulting front of low fluoride water as it moved through the system and subsequently restarted the fluoride feed and traced the front of fluoridated water. In an AWWARF sponsored research project, Vasconcelos et al. (1996, 1997) conducted a series of tracer studies around the United States for use in developing models of chlorine decay. DiGiano and Carter (2001) turned off the fluoride feed at one treatment plant and changed the coagulant feed at another plant to simultaneously trace water from both plants and used the resulting feed measurements to calculate mean constituent residence time in the system. Grayman (2001) discusses the use of tracers in calibrating hydraulic models. The Stage 2 DBPR Initial Distribution System Evaluation Guidance Manual (U.S. EPA, 2003) recognized the use of tracers as a means of calibrating models and predicting residence time as a partial substitute for required field monitoring.

4.4.2 Procedures for Conducting a Distribution System Tracer Study

A tracer study can be divided into three phases—planning, conducting the actual field work, and analysis of the results. In the planning phase, the logistics of the study are determined including tracer selection and injection methods; system operation; personnel requirements, training and deployment; monitoring strategy; and assembling and testing all equipment. An important part of the planning step is the use of a distribution system model to simulate the behavior of the tracer during the course of the actual tracer study so as to have a good understanding of what is expected to occur. The outcome of the planning phase should be a detailed work plan for conducting the tracer study. The actual field study should be preceded by a small-scale field test of the tracer injection and monitoring program. This step will identify potential problems prior to the actual full-scale field study. Following the field program, the data are analyzed, methods and results documented, and the tracer study assessed in order to improve future such studies. Specific procedures are described below.

Tracer Selection. A tracer should be selected to meet the specific needs of a project. Criteria that can influence the selection of a particular tracer include the following:

- Regulatory requirements
- Analytical methods for measuring tracer concentration
- Injection requirements
- Chemical composition of the finished water
- Cost
- Public perception

The most commonly used tracers in distribution systems are fluoride, calcium chloride, sodium chloride, and lithium chloride.

Fluoride is a popular tracer for those utilities that routinely add fluoride as part of the treatment process. In this case, the fluoride feed can be shut off, and a front of low-fluoride water is traced. However, several states that regulate and require the use of fluoride will not permit the feed to be shut off, even for short periods. Availability of continuous online monitoring equipment for fluoride is limited so that measurements are generally performed manually in the field or laboratory. Fluoride can interact with coagulants that have been added during treatment and in some circumstances can interact with pipe walls leading to nonconservative behavior.

Calcium chloride has been used in many tracer studies throughout the United States. Generally a food grade level of the substance is required. It can be monitored by measuring conductivity, or by measuring the calcium or chloride ion. The upper limit for concentration of calcium chloride is generally limited by the secondary MCL for chloride (250 mg/L).

Sodium chloride has many similar characteristics to calcium chloride in that it can be traced by monitoring for conductivity or the chloride or sodium ion. The allowable concentration for sodium chloride is limited by the secondary MCL for chloride and potential health impacts of elevated sodium levels.

Lithium chloride, a popular tracer in the U.K., is less frequently used in the United States, in part because of the public perception of lithium as a medical pharmaceutical. Additionally, concentrations of lithium must be determined as a laboratory test.

The general characteristics of the tracers are presented in Table 4.8 adapted from Teefy (1996). In addition to injection of these popular tracer chemicals, other methods have been used to induce a change in the water quality characteristics of the water that can then be traced through the distribution system. These include the following:

- Switching coagulants (i.e., switch from $FeCl_3$ to $(Al)_2(SO_4)_3$)
- Monitoring differences in source waters.

Injection Procedures. The goal of a distribution system tracer study is to create a change in water quality in the distribution system that can be traced as it moves through the system. This is accomplished by injecting a tracer over a period of time (i.e., a pulse) and observing the concentrations at selected locations within the distribution system at some interval of time. Best results are usually obtained by creating a near-instantaneous change in concentration in the receiving pipe at the point of injection and maintaining this concentration at a relatively constant value for a period of time (typically several hours). A variation on this approach has been to inject a series of shorter pulses and monitor the resulting concentration in the distribution system. However, if the pulses are too short in duration or the distribution system is too large or complex, then the pulses may interact resulting in confusion as to the movement of water in the system.

Injection may be accomplished by pumping the tracer into the pressurized system at a connection to a pipe or by feeding the tracer into a clearwell. For the pumping situation, the pump should be selected to overcome the system pressure and to inject at the required rate. It is desirable to have a flow meter both on the injection pump and on the receiving pipe so that the rate of injection and the resulting concentration in the receiving pipe is known. When feeding the tracer into a clearwell, the tracer should be well mixed in the clearwell, and the concentration of the tracer should be monitored in the clearwell effluent. In general, the volume of water in the clearwell should be minimized in order to generate an effluent tracer front as sharp as possible.

Monitoring Program. Tracer concentrations can be monitored in the distribution system by taking grab samples and analyzing them in the field or laboratory, or through the use of on-line monitors that analyze a sample at a designated frequency. When manual sampling is employed, a route for visiting monitoring locations is generally defined and the circuit is repeated at a predefined frequency. If online monitoring is employed, the instruments should be checked at frequent intervals during the tracer study in order to ensure that they are operating properly and to take a grab sample that can later be compared to the automated measurement.

TABLE 4.8 Tracer Characteristics (Adapted from Teefy, 1996)

	Tracers				
	Fluoride	Calcium	Sodium	Lithium	Chloride
Commonly available forms	H_2SiF_6, NaF, Na_2SiF_6	$CaCl_2$	NaCl	dry LiCl	$CaCl_2$, KCl, NaCl
Analytical methods	IC, ISE, SPADNS methods	AA, IC, ICP, EDTA titration	AA, IC, ICP, FEP	AA, IC, ICP, FEP	IC, ISE, $AgNO_3$ titration, $Hg(NO_3)_2$ titration
Typical chemical cost	$2 per gallon for H_2SiF_6	$20–30 per pound for lab grade $CaCl_2$	$2–5 per pound for lab grade NaCl	$50–150 per kilogram	$2–5 per pound for lab grade NaCl
Typical analytical cost per sample	$10–30	$15–30	$15–30	$20–45, AA more expensive than ICP	$15–30
Typical background levels	0–4 mg/L	Vary greatly (1–300 mg/L), use only when low	Vary greatly (1–500 mg/L)	Usually below 5 µg/L	Vary greatly (1–250 mg/L)
Regulatory limits	4 mg/L federal MCL, 2 mg/L secondary MCL	None known	20 mg/L for restricted diet (EPA recommendation)	None known	250 mg/L secondary standard

Note: Tracers
$CaCl_2$ calcium chloride
H_2SiF_6 hydrofluosilicic acid
KCl potassium chloride
LiCl lithium chloride
NaF sodium fluoride
NaCl sodium chloride
Na_2SiF_6 sodium silicofluoride

Analytical methods
AA atomic absorption spectrometry
$AgNO_3$ silver nitrate
EDTA ethylenediaminetetraacetic acid
FEP flame emission photometric method
$Hg(NO_3)_2$ mercuric nitrate
IC ion chromatography
ICP inductively coupled plasma
ISE ion selective electrode

Since sampling is generally performed around the clock during tracer studies, monitoring locations should be selected to provide for accessibility at all times. Dedicated sampling taps and hydrants are frequently used as sampling sites. Alternatively, buildings such as fire stations or 24 h-businesses can be used. When using hydrants, in order to facilitate grab samples, a faucet is frequently installed on one of the hydrant ports.

For both, grab samples and online monitoring, precautions should be taken to assure that the sample is representative of water in the main serving the sampling site. The travel time from the main can be calculated based on the flow rate, the distance from the main to the sample site, and the diameter of the connecting pipe. For grab samples, either water should be allowed to run continuously at a reasonable rate during the study or the sampling tap should be flushed prior to each sample for long enough to ensure water from the main. For online monitors, a constant flow rate should be maintained and the travel time from the main to the monitor calculated and considered when analyzing results.

System Operation. The operation of the water system will have a significant effect on movement of water in the distribution system during a tracer study. Therefore, information on the system operation should be collected both on a real-time basis and following the completion of the study. Real-time information can be used to ensure that the system is being operated in accord with the planning and to make modifications in monitoring in case of significant variations in system operation. Additional system operation data can usually be obtained from a SCADA system following the completion of the study and used as part of the modeling process.

One category of data that are especially useful for both real-time assessment and post study analysis are flow measurements in key pipes. For example, if flow rates in the pipe receiving the tracer are much higher or lower than planned, then travel times will likely be faster or slower than anticipated. This may necessitate changes in monitoring schedules. Flow rates in the receiving pipe, and other inflows and outflows to the study area are also very useful when modeling the distribution system following the study.

4.4.3 Example Tracer Studies

Tank Tracer Study. A tracer study was conducted as part of a research project at a 2-million-gallon Hydropillar tank to determine the mixing characteristics in the facility (Grayman et al., 2000; Hartman et al., 1998). The tank was constructed with sampling taps located at three elevations on the central access core that runs vertically through the center of the tank, and on the inflow and outflow lines. A diagram of the facility is presented in Fig. 4.5.

A liquid, 29 percent, food grade calcium chloride solution with a calcium concentration of 123,000 mg/L was used as the tracer chemical. The calcium chloride was tested by a certified laboratory to verify that it met all food grade specifications under Food and Drug Administration guidelines. The water utility obtained permission from the State regulatory agency to add calcium chloride to its water supply prior to the tracer study. Fluoride was considered as a tracer chemical but ruled out because of state regulations that did not allow utilities that currently fluoridate their water supply to shut off their fluoride treatment.

A total of 330 gal of calcium chloride was pumped into the influent pipe of the tank over a 9.5-h period during the fill period. A metering pump with a maximum capacity of 70 gal/h was used. The objective was to approximately double the background concentration (33 mg/L) of calcium in the tank at the end of its fill period. An ultrasonic flow meter was used to monitor inflow and the tracer feed rate was periodically adjusted in order to maintain a relatively constant tracer concentration in the influent.

Sampling began just prior to the start of the fill period and continued for approximately 40 h covering two fill periods and one-and-a-half draw periods. Samples were taken from the sampling taps and analyzed for both conductivity and calcium. Calcium concentrations

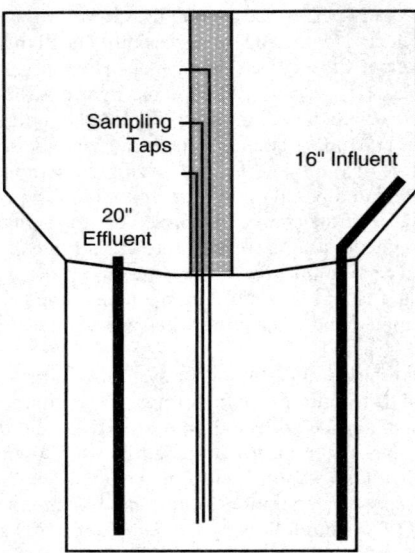

FIGURE 4.5 Schematic diagram of tank.

were measured with a pH/Ion/Conductivity meter, a calcium ion selective electrode (ISE), and a single junction reference electrode. Conductivity was also measured on the same pH/Ion/Conductivity meter using a conductivity probe. Temperature and chlorine residual were also measured as part of a water quality study. Samples were taken from the inlet line only during the fill period and from the outlet line during the draw period. Sampling from the tap located at the highest water level was curtailed when the water level dropped below that elevation. Generally samples were taken from each tap at approximately 15- to 30-min intervals.

The data that were collected during the tracer study were used to construct a time history of tracer concentrations at each of the sampling points. An example of a plot of conductivity data is shown in Fig. 4.6. Plots of calcium concentration displayed similar patterns.

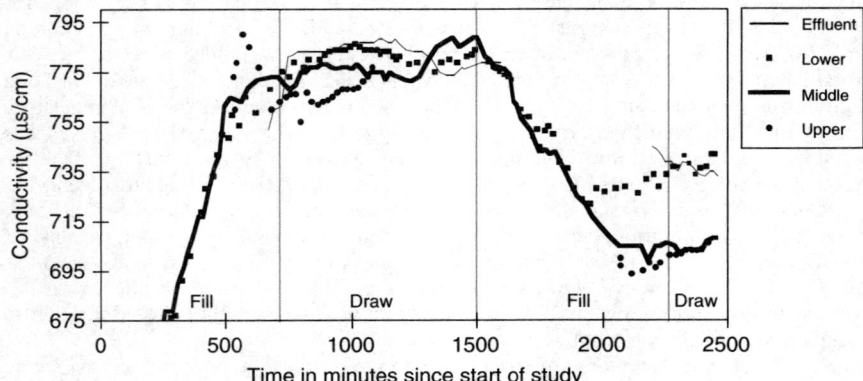

FIGURE 4.6 Plot of conductivity at different sampling locations in tank.

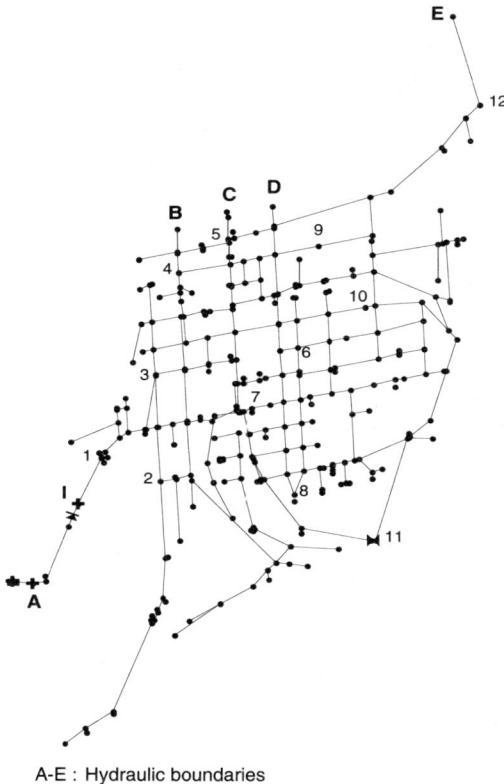

A-E : Hydraulic boundaries
I : Injection point
1-12 : Monitoring locations

FIGURE 4.7 Network map of distribution system in tracer study area.

Examination of the data indicates a relatively well mixed tank but also indicates some short term variations. For example, during the first draw period, higher conductivity concentrations are seen in the effluent and at the lower tap indicating that the newer water containing the tracer chemical is more prevalent in the lower portions of the tank. However, by the end of this draw period, this pattern has largely dissipated. At the start of the second draw period, the conductivity at the lower tap and in the effluent is again noticeably higher than at the other taps indicating continued local variations from a completely mixed tank.

Distribution System Tracer Study. A tracer study was performed to calibrate a hydraulic model of part of a distribution system (see Fig. 4.7). The area has a history of high customer complaints of red water which are due to adverse local system conditions. The primary cause of red water is the presence of unlined cast iron pipes which are prevalent in a major part of the area with installation occurring in the early 1900s. The secondary cause may be older water permitting the release of iron from the pipe wall deposits which have formed over time. A well-calibrated hydraulic model is needed to diagnose and address areas, such as this where adverse conditions exist.

Planning. In the planning stage, food grade calcium chloride solution (29 percent) was selected, and approved for a tracer study in the distribution system. The application used similar procedures as the tank tracer study. An existing hydraulic model was used to provide hydraulics of the study area during the planning stage. The preliminary simulation showed that there were five nodes which supplied water into and out of the study area. Identifying these nodes as boundaries A–E in Fig. 4.7, an all-pipe submodel was created for the tracer study. Boundary A was the primary source of water which supplies most of the demand for the study area. Boundary E was the primary point where water exited the study area. Boundaries B, C, and D showed only minor exchanges of water into and out of the study area. An imaginary reservoir was added to the upstream of boundary A since EPANET requires at least one fixed grade node (reservoir or tank) as a source. The submodel was run to ensure that the flow velocities and pressures matched the existing hydraulic model.

The model used three types of demand patterns to simulate water use in the study area—residential, industrial (including commercial), and unaccounted uses. To assist in the mass-balance process of demand in the study area, pipes connecting boundaries A and E were selected to measure velocity using ultrasonic flow meters. Injection of the tracer was simulated as a step input of 200 mg/L of chloride for 9 h using the hydraulic model. The point of injection was node I which was located downstream of boundary A. Based on the simulation run, monitoring locations were selected to measure temporal progress of the tracer at fire hydrants in the system. Conductivity was selected as the surrogate parameter to be recorded in the field since it had an approximately linear relationship to chloride concentration of the tracer. Using an inexpensive conductivity probe with a data logger, continuous readings were recorded.

Preliminary Run. A preliminary run was performed in the field before the actual tracer test. An access ditch was dug in advance to uncover the pipe, and to install a 2 in tap with a ball valve for injection of the tracer. An ultrasonic flow meter was also installed upstream of the injection point with proper separation from the injection point for accurate measurement. The calcium chloride tracer was injected into the main using a metering pump, and the injection rate was monitored using a rotameter. The injection assembly, consisting of HDPE pipe, fittings, metering pump, and rotameter, was disinfected, and checked for presence of any coliform to make sure that it was microbiologically safe before use in the field. To maintain chloride concentrations below the secondary MCL (250 mg/L), grab samples were taken immediately downstream of the injection point for chloride measurements every 30 min. The background chloride concentration of the system was around 35 mg/L. The preliminary run revealed two important things. First, the flow rate at boundary A was lower than the model prediction. Second, it took considerable time for the tracer to travel from the connection point of the main to the fire hydrant conductivity monitor. Based on these two observations, two modifications were implemented for the field tracer study. First, the injection rate of the tracer was decreased for the actual study. Second, it was suggested and implemented for the tracer study that the flow rate through the fire hydrants was increased to reduce the travel time. This modification of opening the hydrants for continuous conductivity reading during the study period would create additional demand in the system, and it was suggested to install a totalizer at the hydrant effluent to record the amount of flush water.

Field Tracer Study. Following the preliminary run, the field tracer study was performed. A pumping station was scheduled to supply water at a constant rate to the study area during the injection period. The injection of the calcium chloride tracer took place over a 9-h period. The injection flow rate was closely monitored, using a Rotameter, to maintain the chloride concentration at 190 mg/L based on chloride concentrations from grab samples measured on-site, and conductivity readings from the continuous monitor at the immediate downstream monitoring location. Conductivity was continuously recorded at the monitoring locations for 48 h after the injection started. Grab samples were also collected from the monitoring locations regularly to analyze the chloride concentration. The volume from the totalizer on the hydrants to sampling locations was recorded, and time-stamped when the grab samples were taken. This flow volume data was converted into the average rate over the measured duration.

Calibration: Preparation. The model was modified to better represent the study area. To accomplish this, first, the tracer injection was simulated by the option "mass boost" using the actual flow pattern during the injection period to allow the concentration to change according to flow rate in the pipe. Next, hydrants and their stub pipes were added to the existing model to simulate the additional demand and travel time required for tracer monitoring. Finally, calibration of the model was processed using the conductivity data collected at the monitored locations. Taking the linear regression between continuous conductivity reading and chloride concentration from grab samples, conductivity was converted into chloride concentration for the purpose of calibration. At each location, chloride values which trace temporal progress of the tracer were selected to build the calibration data file for the EPANET calibrated model.

Calibration: Global Adjustment with Demand Pattern. Initial comparison between field results and model predictions before the calibration showed that the model predicted well in all areas except those close to the boundary areas (Fig. 4.8). To improve the prediction in these areas, several adjustments were incorporated into the calibration either globally and/or locally. Actual measured flow rates were applied to the pipes at boundaries A and E, and water demand patterns were modified to balance the mass flow in the system. There has been no significant change in the study area since the model was originally developed 5 years ago. Upon review, it was felt that the industrial (and commercial), and unaccounted

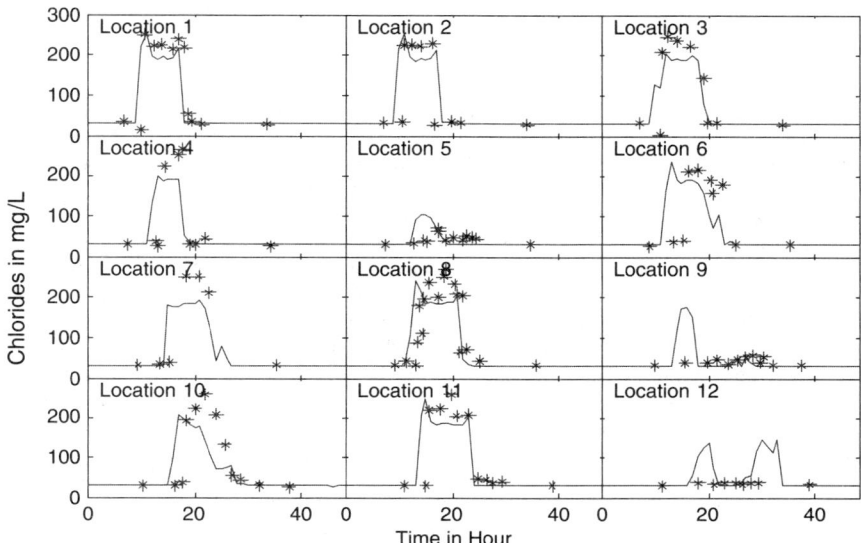

FIGURE 4.8 Comparison of model predictions and calibration points at monitoring locations before calibration.

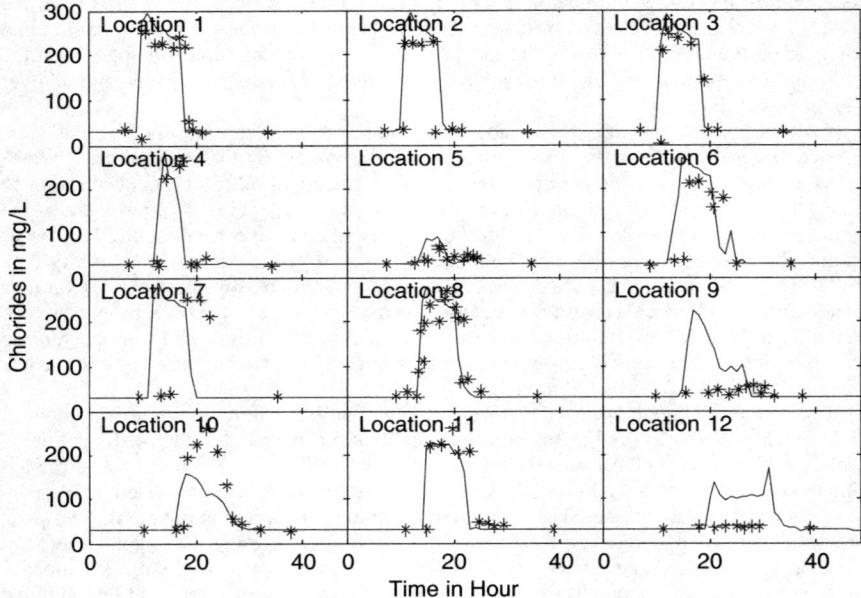

— Model prediction, ∗ Calibration points from chloride readings

FIGURE 4.9 Comparison of model predictions and calibration points at monitoring locations after global calibration of adjusting residential demand pattern.

demand patterns were still reliable. However, the residential demand pattern may have changed due to demographic changes in the area. Thus, the residential demand pattern was adjusted during the calibration process. Demand patterns were exported to a spreadsheet with flow data at boundaries A–E and demand data at nodes. Then, the residential demand was adjusted until mass-balance was achieved in the system for each hour. Unfortunately after this adjustment of mass-balance of the system demand, the predicted results still showed a mismatch at locations primarily downstream of boundaries B, C, and D (Fig. 4.9).

Calibration: Local Adjustment with Boundary Flow. It was suggested that the flow at boundaries B, C, and D was critical in the calibration process. Since no field flow data were available, a sensitivity analysis was carried out by changing the demand multiplier at each boundary node. This analysis showed these boundaries affected several locations. One boundary node showed a dilution effect at nearby locations, and the other two resulted in a change in travel time of the tracer to some locations. The demand multiplier was changed at all three nodes for various combinations. Then, the model was run to compare the calibration of these combinations after the mass-balance was achieved among the system demand. The predicted concentration after the final calibration is shown in Fig. 4.10. Locations downstream of boundaries B, C, and D have improved in calibration, but the adjustment resulted in reduced accuracy of predictions at the mid region (locations 6 and 7).

Calibration: Further Improvement? Calibration can be further improved by adjusting travel time and/or tracer concentrations through changing demands at a specific node, altering the C value of a pipe, and/or changing the status of a certain valve to modify direction

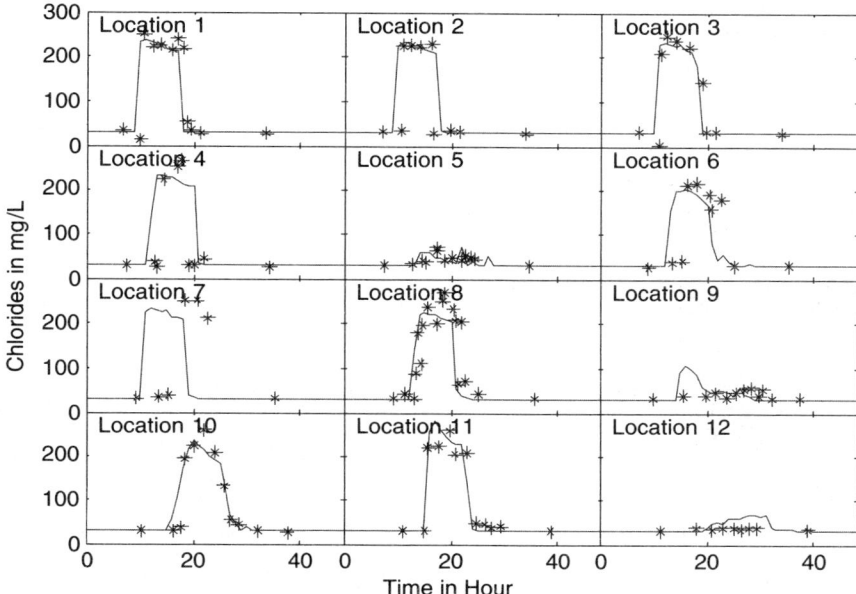

FIGURE 4.10 Comparison of model predictions and calibration points at monitoring locations after local calibration of adjusting demand multiplier at hydraulic boundaries B, C, and D.

of flow, travel time, and/or water source. However, these adjustments should be verified with field status before confirming any calibration results.

4.5 SUMMARY AND CONCLUSIONS

Water systems are complex and difficult to characterize. It has become conventional wisdom that water quality can change significantly as water moves through a water system. This awareness has led to the development of water quality/hydraulic models which can be used to understand the factors that affect these changes and to track and predict water quality changes in drinking water networks. Recent events have focused the water industries attention on the issue of water system vulnerability and it is apparent that the most vulnerable portion of a water utility is the network itself. Therefore interest has grown rapidly in the potential use of water quality/hydraulic models for assessing water system security and for their potential for assisting in the protection of water systems from deliberate contamination by biological and chemical threat agents.

In order to use water quality/hydraulic models correctly there are two features that must be understood. These are the characterization of system demands and the proper calibration of these models using tracer tests. Recent research in both of these areas indicates that a better understanding of both of these aspects could substantially improve the use of models for system protection.

REFERENCES

American Water Works Association, "Emergency Planning for Water Utilities" (M19), ISBN 1-58321-204-3, Catalog No. 20019. 169 pp., 2001.

American Water Works Association, AWWA Manual M1, Distribution System Requirements for Fire Protection. Denver, CO, American Water Works Association, 1989.

Anderson, J. S., and K. S. Watson, "Patterns of Household Usage," *J. Am. Water Works Assoc.* 59(10): 1228–1237, 1967.

AWWA Water Loss Control Committee, "Applying Worldwide BMPs in Water Loss Control," *J. Am. Water Works Assoc.* 95(8): 65–79, 2003.

Biswas, P., C. Lu, and R. M. Clark, "A Model for Chlorine Concentration Decay in Drinking Water Distribution Pipes," *Water Res.* 27(12): 1715–1724, 1993.

Boulos, P. F., T. Altman, P. A. Jarrige, and F. Collevati, "Discrete Simulation Approach for Network-Water-Quality Models," *J. Water Res. Planning Management* 121(1): 49–60, 1995.

Boulos, P. F., W. M. Grayman, R. W. Bowcock, J. W. Clapp, L. A. Rossman, R. M. Clark, R. A. Deininger, and A. K. Dhingra, "Hydraulic Mixing and Free Chlorine Residual in Reservoirs," *J. AWWA* 88(7): 48–59, 1996.

Boulos, P. F., W. M. Grayman, R. W. Bowcock, J. W. Clapp, L. A. Rossman, R. M. Clark, R. A. Deininger, and A. K. Dhingra, "Hydraulic Mixing and Free Chlorine Residuals in Reservoirs," *J. Am. Water Works Assoc.* 88(7): 48–59, 1996.

Bowen, P. T., J. F. Harp, J. W. Baxter, and R. D. Shull, *Residential Water Use Patterns*, AWWA Research Foundation, Denver, CO, 105 pp., 1993.

Buchberger, S. G., and G. J. Wells, "Intensity, Duration and Frequency of Residential Water Demands," *ASCE J. Water Resour. Planning Management* 122(1): 11–19, 1996.

Buchberger, S. G., and G. Nadimpalli, "Leak Estimation in Water Distribution Systems by Statistical Analysis of Flow Readings," *ASCE J. Water Resour. Planning Management*, in press.

Buchberger, S. G., and L. Wu, "A Model for Instantaneous Residential Water Demands," *ASCE J. Hydraul. Eng.* 121(3): 232–246, 1995.

Buchberger, S. G., and T. G. Schade, "Poisson Rectangular Pulse Model for Residential Water Use," *Proceedings XXVII IAHR Congress*, San Francisco, CA, August 10–15, 1997.

Buchberger, S. G., and Y. Lee, "Evidence Supporting the Poisson Pulse Hypothesis for Residential Water Demands," *Proceedings CCWI: International Conference on Computing and Control for the Water Industry*, Exeter, UK, September 13–15, 1999.

Buchberger, S. G., J. T. Carter, Y. H. Lee, and T. G. Schade, *Random Demands, Travel Times, and Water Quality in Deadends*, American Water Works Association, Denver, CO, 470 pp., 2003.

Burrows, W. D., and S. E. Renner, "Biological Warfare Agents as Potable Water Threats," U.S. Army Combined Arms Support Command, Fort Lee, VA, 1998.

Cesario, L., *Modeling Analysis and Design of Water Distribution Systems*, American Water Works Association, Denver, CO, 1995.

Characklis, W. G., *Bacterial Regrowth in Distribution Systems*, American Water Works Association Research Foundation and American Water Works Association, Denver, CO, 1988.

Chaudhry, M. H., *Applied Hydraulic Transients*, 2nd ed. Van Nostrand Reinhold, New York, 1987.

Clark, R. M., "Assessing the Etiology of a Waterborne Outbreak: Public Health Emergency or Covert Attack," in Jennifer Hatchett (ed.), *Proceedings of the First Water Security Summit*, Haested Press, Haested Methods, Inc., Waterbury, CT, pp. 170–179, 2002.

Clark, R. M., and D. L. Tippen, "Water Supply" in Robert A. Corbitt (ed.), *Standard Handbook of Environmental Engineering*, 1st edn., McGraw-Hill, New York, pp. 5.173–5.220, 1990.

Clark, R. M., and R. A. Deininger, "Minimizing the Vulnerability of Water Supplies to Natural and Terrorist Threats," in *Proceedings of the American Water Works Association's IMTech Conference*, Atlanta, GA, April 8–11, pp. 1–20, 2001.

Clark, R. M., and R. A. Deininger, "Protecting the Nation's Critical Infrastructure: The Vulnerability of U.S. Water Supply Systems," *J. Contingen. Crisis Management* 8(2): 76–80, 2000.

Clark, R. M., E. E. Geldreich, K. R. Fox, E. W. Rice, C. H. Johnson, J. A. Goodrich, J. A. Barnick, and F. Abdesaken, "A Tracking a *Salmonella* Serovar *Typhimurium* Outbreak in Gideon, Missouri: Role of Contaminant Propagation Modeling," *J. Water Supply Res. Technol. - Aqua*. 45(4): 171–183, 1996.

Clark, R. M., F. Abdesaken, P. F. Boulos, and R. E. Mau, "Mixing in Distribution System Storage Tanks: Its Effects on Water Quality," *ASCE J. Environ. Eng.* 122(9): 814–821, 1996.

Clark, R. M., L. Rossman, and L. Wymer "Modeling Distribution System Water Quality: Regulatory Implications," *J. Water Resour. Planning Management, ASCE*. 121(6): 423–428, 1995.

Clark, R. M., W. M. Grayman, J. A. Goodrich, R. A. Deininger, and A. F. Hess, "Field Testing Distribution Water Quality Models," *J. AWWA* 83(7): 67–75, 1991.

Clark, R. M., W. M. Grayman, R. M. Males, and J. Coyle, "Modeling Contaminant Propagation in Drinking Water Distribution Systems," *Aqua* 37(3): 137–151, 1988.

Clark, R. M., W. M. Grayman, R. Males, and A. Hess, "Modeling Contaminant Propagation in Drinking-Water Distribution Systems," *J. Environ. Eng. ASCE*, 119(2): 349–364, 1993.

Danckwerts, P. V., "Continuous Flow Systems. Chemical Engineering Science," 2(1): 1–18, 1958.

DiGiano, F. A., and G. Carter, "Tracer Studies to Measure Water Residence Time in a Distribution System Supplied by Two Water Treatment Plants," *Proceedings, AWWA Annual Conference*, 2001.

Dziegielewski, B., and J. J. Boland, "Forecasting Urban Water Use: The IWR-MAIN Model," *Water Resour. Bull.* 25(1): 101–109, 1989.

Erlang, A. K., "Solution of Some Problems in the Theory of Probabilities of Significance in Automatic Telephone Exchanges," *Post Office Electrical Engrs. J.* 10: 189–197, 1917–18.

Flack, J. E., *Urban Water Conservation: Increasing Efficiency-in-Use Residential Water Demand*, ASCE, New York, 99 pp., 1982.

Germeles, A. E., "Forced Plumes and Mixing of Liquids in Tanks," *J. Fluid Mech.* 71 (Part 3) 21–26, 1975.

Grayman, W. M., "Use of Tracer Studies and Water Quality Models to Calibrate a Network Hydraulic Model," *Curr Methods*, vol. 1, no. 1, Haestad Press, Waterbury, CT, pp. 38–42, 2001.

Grayman, W. M., and R. M. Clark, "Using Computers to Determine the Effect of Storage on Water Quality," *J. AWWA* 85(7): 67–77, 1993.

Grayman, W. M., and R. M. Clark, "Water Quality Modeling in a Distribution System," in *Proceedings of the AWWA 1991 Annual Conference*, American Water Works Association, Denver, CO, 1991.

Grayman, W. M., L. A. Rossman, and E. E. Geldreich, "Water Quality," *Water Distribution Systems Handbook*, Larry W. Mays (ed.), McGraw-Hill, New York, NY, pp. 9.1–9.22, 2000.

Grayman, W. M., L. A. Rossman, C. Arnold, R. A. Deininger, C. Smith, J. F. Smith, and R. Schnipke, *Water Quality Modeling of Distribution System Storage Facilities*, AWWA Research Foundation and American Water Works Association, Denver, CO, pp. 230, 2000.

Grayman, W. M., R. A. Deininger, A. Green, P. F. Boulos, R. W. Bowcock, and C. C. Godwin, "Water Quality and Mixing Models for Tanks and Reservoirs," *J. Am. Water Works Assoc.* 88(7): 60–73, 1996.

Grayman, W. M., R. A. Deininger, and R. M. Clark, "Vulnerability of Water Supply to Terrorist Activities," *CE News* 14(2): 34–38, 2002.

Grayman, W. M., R. M. Clark, and J. A. Goodrich, "The Effects of Operation, Design and Location of Storage Tanks on the Water Quality in a Distribution System," in *Proceedings of the Water Quality Modeling in Distribution Systems Conference*, American Water Works Association Research Foundation and American Water Works Association, Denver, CO, 1991.

Grayman, W. M., R. M. Clark, and R. M. Males, "Modeling Distribution-System Water Quality: Dynamic Approach," *J. Water Resour. Planning Management, ASCE*, 114(3): 295–312, 1988.

Guercia, R., R. Magini, and I. Pallavicini, "Instantaneous Residential Water Demand as Stochastic Point Process," *Proceedings of First International Conference on Water Resources Management*, Thessaloniki, Greece, pp. 129–138, 2001.

Hartman, D. J., H. H. Jiang, J. DeMarco, and F. Cossins, "Impact of Storage Tank Design and Operation on Maintaining Water Quality," in: AWWA Annual 1998 Conference Proceedings, Dallas, Texas, 1998.

Holloway, M. B., "Dynamic Pipe Network Computer Model," Ph.D. thesis, Washington State University, Pullman, WA, 1985.

Islam, M. R. "Modeling of Chlorine Concentration in Unsteady Flows in Pipe Networks," Ph.D. thesis, Washington State University, Pullman, WA, 1995.

Jeppson, R. W., *Analysis of Flow in Pipe Networks*, Ann Arbor Science, Ann Arbor, MI, 1976.

Karney, B. W., "Hydraulics of Pressurized Flow," in Larry W. Mays (ed.), *Water Distribution Systems Handbook*, McGraw-Hill, New York, pp. 2.1–2.43, 2000.

Kennedy, M. S., S. S. Moegling, and K. Suravallop, "Assessing the Effects of Storage Tank Design," *J. AWWA* 85(7): 78–88, 1993.

Kennedy, M. S., S. Sarikelle, S. Moegling, and K. Suravallop, "Mixing Characteristics in Distribution System Storage Reservoirs," in *Proceedings of the AWWA 1991 Annual Conference*, American Water Works Association, Denver, CO, 1991.

Larock, B. E., R. W. Jeppson, and G. Z. Watters, *Hydraulics of Pipeline Systems*, CRC Press LLC, Boca Raton, FL, pp. 12–26, 2000.

Lee, Y. H., and S. G. Buchberger, "Modeling Indoor and Outdoor Residential Water Use as the Superposition of Two Poisson Rectangular Pulse Processes," *Proceedings of ASCE 26th National Conference on Water Resources Planning and Management*, Tempe, Arizona, June 6–9, 1999.

Li, Z., and S. G. Buchberger, *PRP Simulator Users Manual*, University of Cincinnati, Cincinnati, OH, 2003.

Linaweaver, Jr., F. P., J. C. Geyer, and J. B. Wolff. *Residential Water Use, Final and Summary Report*, Johns Hopkins University, 79 pp., with appendices, 1966.

Litke, D. W., and L. F. Kauffman, *Analysis of Residential Use of Water in the Denver Metropolitan Area, Colorado 1980-87*, USGS-WRI Report 92–4030, 1993.

Lu, C., "Theoretical Study of Particle, Chemical, and Microbial Transport in Drinking Water Distribution Systems," Ph.D. thesis, University of Cincinnati, Cincinnati, OH, 1991.

Maddaus, W. O., *Water Conservation*, American Water Works Association, Denver, Colorado, 93 pp., 1987.

Mau, R., P. Boulos, R. Clark, W. Grayman, R. Tekippe, and R. Trussell, "Explicit Mathematical Models of Distribution System Storage Water Quality," *J. Hyd. Eng.* 121(10): 699–709, 1995.

Mayer, P. W., W. B. DeOreo, E. M. Opitz, J. C. Kiefer, W. Y. Davis, B. Dziegielewski, and J. O. Nelson, *Residential End Uses of Water*, American Water Works Association, Denver, CO, 310 pp., 1999.

Opitz, Eva, "Demand and Management Models," in Larry W. Mays (ed.), *Urban Water Supply Handbook*, McGraw-Hill, New York, pp. 5.3–5.55, 2002.

PL107-188. Public Health Security and Bioterrorism Preparedness and Response Act of 2002.

President, "White Paper," The Clinton Administration Policy on Critical Infrastructure Protection: Presidential Decision Directive 63, 1998, available at: http//www.ciao.gov/CIAO_Document_Library/paper598.htm

Rossman, L. A., *EPANET User's Manual*, U.S. EPA, Cincinnati, OH, 1994.

Rossman, L. A., R. M. Clark, and W. M. Grayman, "Modeling Chlorine Residuals in Drinking Water Distribution Systems," *J. Environ. Eng.* 120(4): 803–820, 1994.

Rossman, L. A., R. M., Clark, and W. M. Grayman, "Modeling Chlorine Residuals in Drinking Water Distribution Systems," *ASCE J. Environ. Eng.* 120(4): 803–820, 1994.

Skov, K. R., A. F. Hess, and D. B. Smith, "Field Sampling Procedures for Calibration of a Water Distribution System Hydraulic Model," in *Water Quality Modeling in Distribution Systems*, American Water Works Association Research Foundation/U.S. EPA and American Water Works Association, Denver, CO, 1991.

Solley, W. B., R. R. Pierce, and H. A. Perlman, *Estimated Use of Water in the United States in 1995*, U.S. Geological Survey Circular 1200, Denver, CO, 71 pp., 1998.

Teefy, S. M., and P. C. Singer, "Performance Testing and Analysis of Tracer Tests to Determine Compliance of a Disinfection Scheme with the SWTR." *J. Am. Water Works Assoc.* 82(12): 88–98, 1990.

Teefy, S. M., *Tracer Studies in Water Treatment Facilities: A Protocol and Case Studies*, American Water Works Association Research Foundation and American Water Works Association, Denver, CO, 1996.

U.S. EPA, "25 Years of the Safe Drinking Water Act: History and Trends," EPA 816-R-99-007, Office of Water, December 3, 1999.

U.S. EPA, "Baseline Threat Information for Vulnerability Assessments of Community Water Systems," 2002.

U.S. EPA, "National Primary Drinking Water Regulations: Total Coliforms (Including Fecal Coliforms and *E. coli*); Final Rule," *Federal Register* 54: 27544, 1989a.

U.S. EPA, "National Primary Drinking Water Regulations; Filtration, Disinfection; Turbidity, *Giardia lamblia*, viruses, *Legionella*, and Heterotrophic Bacteria; Final Rule," *Federal Register* 54(124): 27486, 1989b.

U.S. EPA, *The Stage 2 DBPR Initial Distribution System Evaluation Guidance Manual*, U.S. EPA, 2003.

Vasconcelos, J. J., L. A. Rossman, W. M. Grayman, P. F. Boulos, and R. M. Clark, "Kinetics of Chlorine Decay," *J. Am. Water Works Assoc.* 89(7): 54–65, 1997

Vasconcelos, J., P. Boulos, W. Grayman, L. Kiene, O. Wable, P. Biswas, A. Bhari, L. Rossman, R. Clark, and J. Goodrich, *Characterization and Modeling of Chlorine Decay in Distribution Systems*, American Water Works Association Research Foundation and American Water Works Association, Denver, CO, 1996.

Walski, T. M., D. V. Chase, D. A. Savic, W. M. Grayman, S. Beckwith, and E. Koelle, *Advanced Water Distribution Modeling and Management*, Haestad Press, Waterbury, CT, 751 pp., 2003.

CHAPTER 5
CYBER THREATS AND IT/SCADA SYSTEM VULNERABILITY

Srinivas Panguluri
Shaw Environmental, Inc., Cincinnati, Ohio.

William R. Phillips, Jr.
CH2M-Hill, Gainesville, Florida

Robert M. Clark
Environmental Engineering and Public Health Consultant, Cincinnati, Ohio.

5.1 INTRODUCTION

Advances in computer technologies, especially during the 1990s, have helped water utilities automate several aspects of their water supply, distribution, and information management systems. Post 9/11, security vulnerabilities and threat assessments have led water utilities to invest in improving the physical security of their infrastructure. Although the improvements in physical security are essential, they can lead to a false sense of safety. In the age of digital interconnections, operations at even the most physically secure site can be disrupted with relative ease through digital backdoors. A cyber attack has been defined as a computer-to-computer attack that undermines the confidentiality, integrity, or availability of a computer or information resident in it (O'Shea, 2003). This strict definition excludes direct system *attacks* by an unauthorized individual who has *obtained* physical access to that machine. Protection is just as important against the most obvious means such as an individual using the keyboard of an unattended, but logged-on main computer, or by inserting a CD with an embedded virus launches an attack directly on the main system.

The computer system infrastructure of a medium to large water utility typically includes its financial system, Human Resource (HR) system, Laboratory Information Management System (LIMS), Supervisory Control and Data Acquisition (SCADA) system, Computerized Maintenance Management System (CMMS), etc. The financial, HR, LIMS, and CMMS systems are considered to be a part of the utility's Information Technology (IT) infrastructure run by a utility or local government IT group on a daily 8- to10-h schedule.

The SCADA systems are typically run by a separate core utility operations and maintenance group on a 24 × 7 basis.

During the late 1990s, medium to large utilities integrated their IT and SCADA systems for economic reasons. For example, one of the largest utilities in the world, the Water Corporation of Western Australia (WCWA), integrated its IT and SCADA infrastructure (Wiese, 1999). With utility or local government IT systems typically connected to the Internet, integration of the IT SCADA infrastructures could increase the potential damage of cyber attack. The Internet Storm Center run by the SANS Institute[*] currently gathers more than 3 million cyber-attack reported entries every day and displays the latest trends and origins of cyber attacks throughout the world. It is estimated that there are many more unreported attacks occurring on a daily basis. A cyber-attack could disrupt the general IT and/or the SCADA infrastructure and cause significant damage. However, careful planning of IT/SCADA integration can minimize any damage from such an attack.

There is a popular misconception that if the SCADA system is not connected to the Internet it is safe. The Maroochy Sewage Spill in Australia (Tagg, 2001) shatters this myth. 49-year old Vitek Boden of Brisbane was sentenced to two years in prison for hacking into a local council's computerized waste management system. He was able to cause an environmental disaster, where millions of liters of raw sewage spilled into rivers, parks and the grounds of a Hyatt Regency hotel (during March and April 2000). Boden used a two-way radio to alter the pump station operations. The police also found a program on his laptop that allowed him to *hack* into the council computers and control the sewage management system. There was evidence that his laptop had been used at the same times that the pumps were illegally discharging raw sewage.

In the United States the infamous *computer outlaw*, Kevin Mitnick, eluded the police, U.S. Marshalls, and FBI for over 2 years. While he was on the run, he broke into countless computers, intercepted private electronic communications, and copied personal and confidential materials. His activities included altering information, corrupting system software, and eavesdropping on users. Mitnick used a number of tools to commit his crimes, including *social engineering*,[†] cloned cellular telephones, *sniffer* programs placed on victims' computer systems, and hacker software programs (U.S. DOJ, 1999). While these may appear to be isolated incidents, they shed light on the potential vulnerabilities of SCADA systems.

Although there is no "panacea" for eliminating cyber and other attacks to a utility's computer system infrastructure, the information presented in this chapter can be used to minimize and mitigate the impacts of such attacks. Section 5.2 of this chapter provides an overview of a utility's computer system infrastructure. Section 5.3 presents an overview of the ongoing initiatives and standards being developed to protect the SCADA infrastructure. Section 5.4 exposes the most common vulnerabilities observed for computer system infrastructure. Section 5.5 contains information on the various vulnerability assessment and planning tools available to protect against cyber attacks. Section 5.6 describes typical cyber-attack scenarios and Sec. 5.7 showcases the tools and methods employed by cyber attackers. Section 5.8 identifies methods for mitigating such attacks and Sec. 5.9 presents a brief summary of procedures for incident response and business continuity. Section 5.10 presents summary and conclusions and also provides a compilation of useful web links that can be used to extend the reader's

---

[*]The System Administration, Networking and Security Institute.

[†]Social Engineering is the practice of using people such as an employee or someone familiar with a company or firm to provide access to the company's computer systems or files. This practice might include the relatively unsophisticated act of looking over someone's shoulder while they are keying in their access codes, or searching through their garbage to obtain passwords and/or access codes. A more sophisticated approach might be to call a company and pretend to be a vendor, consultant, or contractor in order to gain access to the computer systems. An especially insidious approach is to send an e-mail that contains a seemingly harmless attachment but in fact contains a "Trojan horse" that opens a connection to the outside through the firewall.

knowledge on some of the topics presented in this chapter. The chapter concludes with a listing of the various references used throughout this chapter.

The information we present in this chapter is derived from generally available sources. No data or information utilized was obtained from restricted or classified sources. In many cases the suggestions we have made are based on common sense and good practice. Nevertheless, we believe that it is important for the water utility industry to pay serious consideration to the importance of cyber safety.

5.2 OVERVIEW OF UTILITY'S COMPUTER SYSTEM INFRASTRUCTURE

The applications in use at a typical utility and the information flow between them are depicted in Fig. 5.1. This diagram presents a robust argument for the IT/SCADA integration work done in the 1990s to improve utility operations and management. A utility's network architecture can be simplified into logical blocks as represented in Fig. 5.2. Rather than depicting a typical network, this figure represents a composite of the network vulnerabilities found in completing vulnerability assessments of more than 20 utility SCADA systems. Figure 5.2 illustrates that the integration, which eliminated so-called "islands of automation" often provided paths for attacking the entire computer system from any point on the system.

Cyber-attacks to the IT infrastructure may cause significant financial damage and disruptions of the utility's internal operations but they are not expected to cause any immediate water supply disruptions. However, cyber attacks on the SCADA system could have an immediate detrimental impact on the water supply. Also, a prolonged disruption of the IT infrastructure could also lead to water supply disruptions. The vulnerabilities, threats and mitigation methodologies for both IT and SCADA systems are similar but this chapter focuses on the potential cyber-attacks on the SCADA systems that pose a greater threat to the user community. In order to understand the potential impact of a cyber-attack it is important to understand the various elements of a typical SCADA system.

In the water industry today, the term SCADA is often used to include both in-plant computer-based process control systems and computer-based systems providing monitoring and control of geographically distributed (remote) raw water production and treated water distribution facilities. In fact, the distinction between the more traditional SCADA systems used for central monitoring and control of remote facilities, such as wells and pump stations, versus in-plant computer-based process control systems is blurred. Monitoring and control of in-plant and remote equipment is often provided by a single system with a common computer-based operator interface generically referred to as a Human Machine Interface (HMI).

The sole remaining distinction, response time (the time from when a command is issued or a measurement made and it is executed or displayed), is also fading as more and more broadband options are becoming affordably available to utilities for communications with remote facilities. Local area networks (LAN) are predominantly used to provide reliable and high-speed in-plant communications. Communications with remote facilities were, in the past, provided by narrow-band radio systems or leased analog phone lines. Narrow-band radios are still in use today, though broadband alternatives are emerging. However, where fiber optic or leased digital phone lines are available, broadband alternatives allow remote facilities to communicate with in-plant LANs over wide area networks (WAN) at data rates high enough to support the use of Internet protocols and provide remote programming of controllers. These WANs create, a double-edged sword. A distinction between in-plant and remote control strategies and procedures is fading. However, the use of WANs subjects remote facilities to the same kinds of cyber attacks faced by in-plant systems.

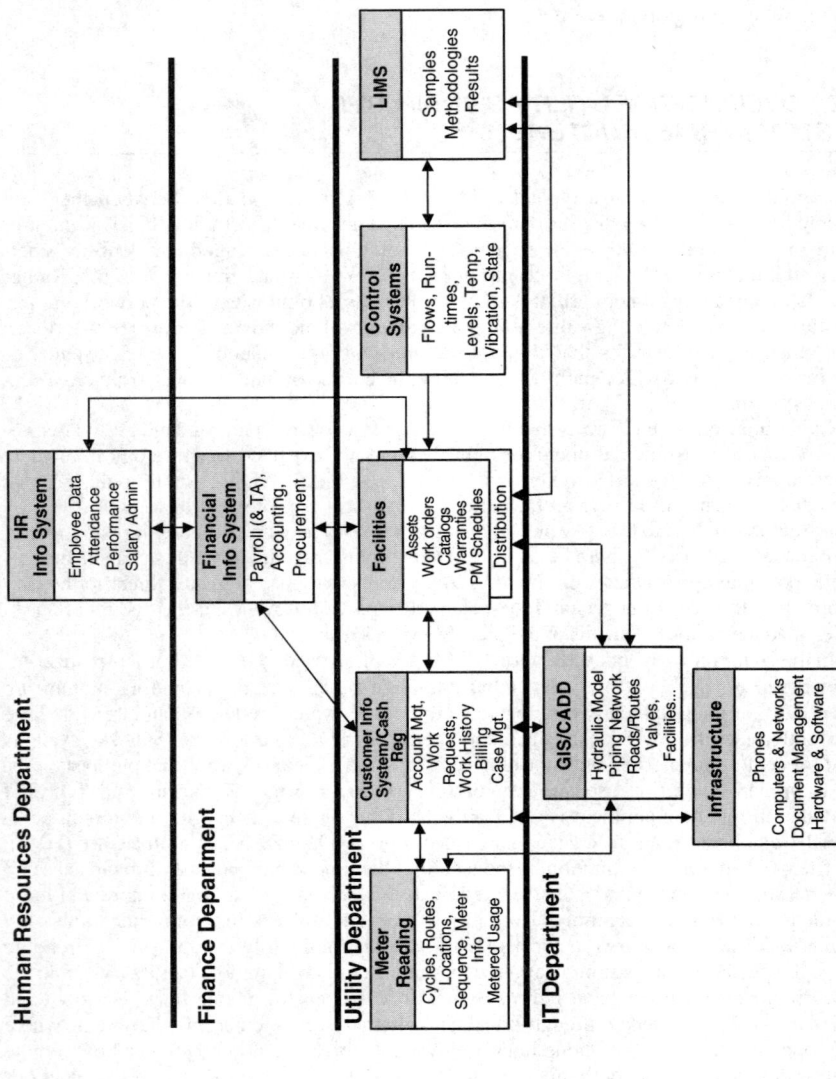

FIGURE 5.1 Typical mid-to-large water utility information flow requirements (*Phillips Jr., 2003*).

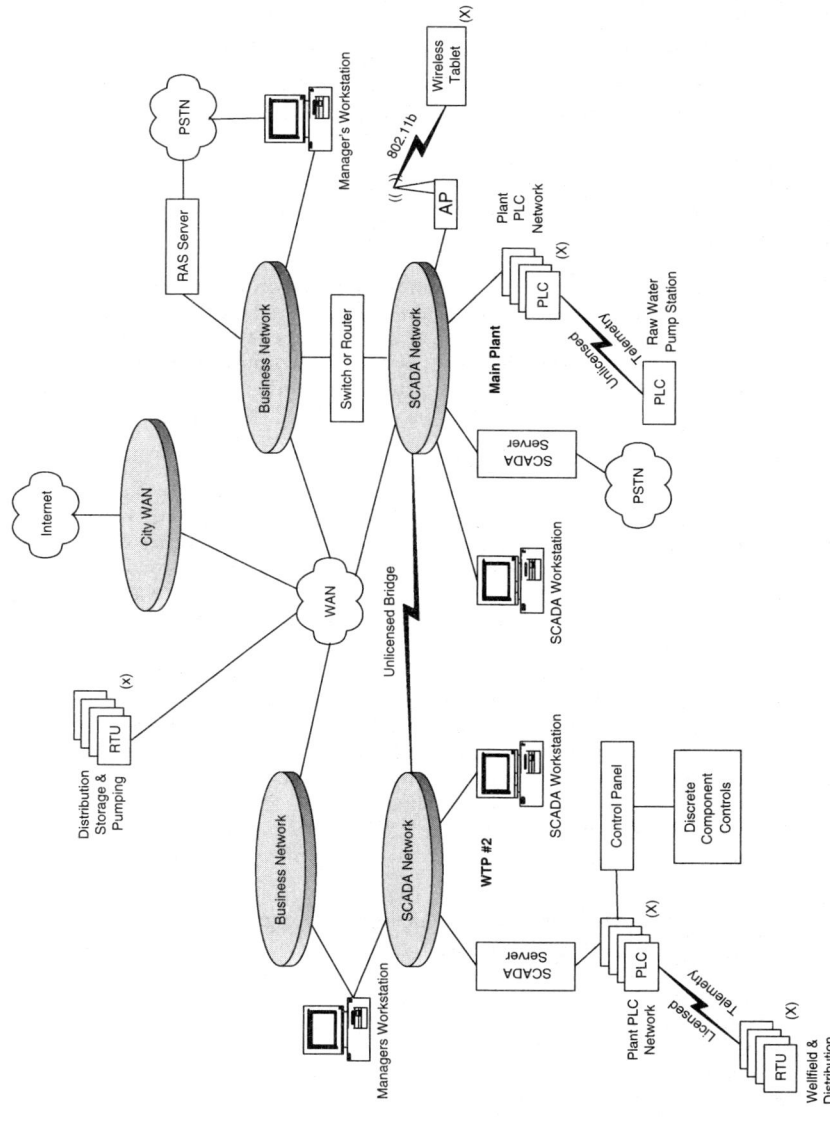

FIGURE 5.2 Typical mid-to-large water utility's composite network infrastructure (*Phillips Jr., 2003*).

Unlike a handful of IT communications protocols, there are 100s of different SCADA communication protocols and application program interfaces (APIs) that are used by the SCADA systems. The common industrial communication protocols (based on their native application domain) can be categorized as follows (Pratt, 2003):

- Sensor Networks-protocols initially designed to support discrete I/O
- Device Networks-protocols originally focused on process instrumentation
- Control Networks-protocols typically used to connect controllers and I/O systems
- Enterprise Networks-protocols that focus on IT applications

The sensor network protocols handle discrete inputs and outputs that relay simple information such as an "on/off" position. An example is an indicator light or a motor starter which may be reported as, or positioned to, an on or off condition. These are simple and fast protocols, the popular sensor level networks include: AS-i, CAN, DeviceNet, Interbus, and LON.

Device networks are a product of process automation, and grew from the need to accurately measure and control continuous physical processes. Unlike the sensor network, the device networks handle predominantly analog data and tend to be more complex than those in sensor networks. The devices are more complex, the data and the status information of the variables are richer and more varied. The device network protocols prevalent in the process automation industry include Foundation Fieldbus H1, HART, and PROFIBUS-PA (Pratt, 2003).

Control networks provide a communication backbone that allows integration of controllers, I/O, and subnetworks. Control networks stand at the crossroads between the growing capabilities of industrial networks and the penetration of IT enterprise networks into the control system. At this level, technology and networks are in a state of flux and, in many cases, the technology is relatively new. The popular type of control network protocols include BACnet, ControlNet, Industrial Ethernet, Ethernet/IP, FF-HSE, MODBUS, and PROFIBUS-DP (Pratt, 2003). It is expected that some type of Ethernet and IP addressing will become the standard in coming years.

A vast majority of the SCADA communication protocols are simple by design and hence easy to break. However, the impact of such intrusions are expected to be localized and limited due to the nature of the master/slave communication design of SCADA networks. Some of the existing SCADA hardware use electronic chips that do not even have the computing horsepower to encrypt the transmission for security. Also, the chips that have the computing power to encrypt, when retrofitted with encryption software can have performance related issues. The following subsection presents an overview of the ongoing collaborative initiatives to develop standards to protect the SCADA systems from cyber attacks.

5.3 ONGOING SCADA ENCRYPTION INITIATIVES AND STANDARDS DEVELOPMENT

Presidential Decision Directive 61 raised the issue of Critical Infrastructure Protection (CIP). In response, the Gas Technology Institute (GTI) was awarded a contract by the Technical Support Working Group (TSWG) to identify encryption algorithms that would protect gas SCADA systems from cyber-attack. GTI found that the least-cost approach to protecting SCADA systems was to incorporate the Digital Encryption Standard (DES), the Rivest, Shamir, Adelman (RSA) public key algorithm, and the Diffie-Hillman number generating algorithm as a suite of algorithms into SCADA units by adding a special purpose $5 encryption chip at the time the SCADA unit was zured. It was found that these same algorithms could be installed on existing SCADA units, but at a higher cost and with reduced communication speeds. GTI presented this work to several manufacturers, gas companies, etc., upon its completion in 1999. However, there was little interest in using this code, given that few gas companies asked for security to be included in their SCADA units. The events of 9/11 have changed the perspective of all gas utilities and SCADA vendors. In October of 2001, the American Gas Association (AGA), GTI, the Institute of Electrical and Electronics Engineers (IEEE), National Institute of Standards & Technology (NIST), gas and electric utilities operators, SCADA and cryptographic vendors, and security industry experts have taken an accelerated approach to developing and implementing these encryption algorithms on existing and new SCADA units.

The AGA now plans to issue a series of AGA 12 documents to incorporate lessons learned and to expand the scope to address new SCADA/DCS designs (AGA 12-1 Working Group, 2003). AGA 12-1's goal is to "get the plain text messages off the wire as quickly as possible." To achieve this goal AGA 12-1 focuses attention on the retrofit market, which encompasses 80 percent of the existing SCADA systems. New SCADA systems that are a part of a truly distributed processing architecture of networked Intelligent Electronic Devices (IEDs[*]) will be the focus of AGA 12-2. In addition to other challenges AGA 12-2 will provide additional cryptographic protocol specifications to enhance interoperability of cryptographic components produced by different vendors. Retrofit solutions based on AGA 12-1 will be field tested, and lessons learned will require changes to current specifications. These changes will be subject to version control, and published as updates to AGA 12-1.

The NIST has a separate, ongoing initiative on CIP, to support the development and dissemination of standards for process control security. NIST has established the Process Control Security Requirements Forum (PCSRF), a working group comprised of vendors and users of process control automation. The PCSRF is applying the ISO 15408 Common Criteria methodology to develop Protection Profiles for process control. As a part of this effort NIST is conducting meetings to collect industry-specific requirements. At the time of this writing, several of the industry-specific meetings have been completed; scheduling is underway for a water industry meeting. At the time of publication the NIST PCSRF has released a first draft of the "System Protection Profile for Industrial Control Systems." NIST is also releasing a series of special publications addressing information system security certification and accreditation for federal government agencies.

Another major initiative is by the Instrumentation, Systems, and Automation Society (ISA), which is developing the standard ISA-SP99 for Manufacturing and Control Systems Security. The SP99 Committee will establish standards, recommended practices, technical reports, and related information that will define procedures for implementing electronically secure manufacturing and control systems and security practices, and assessing electronic security performance. Guidance is directed toward those responsible for designing, implementing, or managing manufacturing and control systems and shall also apply to users, system integrators, security practitioners, and control systems manufacturers and vendors. The committee's focus is to improve the confidentiality, integrity, and availability of components or systems used for manufacturing or control and provide criteria for procuring and implementing secure control systems. Compliance with the committee's guidance will not only improve manufacturing and control system electronic security, but will also help identify vulnerabilities and address them, thereby reducing the risk of compromising confidential information or causing Manufacturing Control Systems degradation or failure. The committee is currently generating technical content for two reports planned to be published soon.

The major vendors of the SCADA system have their own initiatives to collaborate with the aforementioned groups to improve the security of the SCADA infrastructure.

---

[*]IED is a general term that refers to any device that incorporates one or more processors with the capability to receive or send data/control from or to an external source.

Furthermore, the industry leader in IT networking equipment, Cisco, has formed a Critical Infrastructure Assurance Group (CIAG) to work with the various aforementioned forums to perform vulnerability research and provide future enhancements (Franz, 2003).

5.4 TOP 10 VULNERABILITIES OBSERVED IN THE COMPUTER SYSTEM INFRASTRUCTURE

The network block diagram (Fig. 5.2) shows a composite of many of the network vulnerabilities found in conducting EPA-mandated vulnerability assessments. The recent assessments conducted on the SCADA systems of a number of large water utilities in the United States have indicated the following list of vulnerabilities:

1. Operator station logged on all the time even when the operator is not present at the workstation, thereby rendering the authentication process useless
2. Physical access to the SCADA equipment relatively easy
3. Unprotected SCADA network access from remote locations via Digital Subscriber Lines (DSL) and/or dial-up modem lines
4. Insecure wireless access points on the network
5. Most of the SCADA networks directly or indirectly connected to the Internet
6. No firewall installed or the firewall configuration is weak or unverified
7. System event logs not monitored
8. Intrusion detection systems not used
9. Operating and SCADA system software patches not routinely applied
10. Network and/or router configuration insecure; passwords not changed from manufacturers default

At minimum, individual water utilities should periodically review and address these vulnerabilities. A detailed vulnerability assessment approach and various planning tools available to the utility are listed in the following section. Outsourcing some of these audit functions to specialists is recommended.

5.5 VULNERABILITY ASSESSMENT APPROACH AND PLANNING TOOLS

Several vulnerability assessment methodologies and tools are available for a utility to evaluate potential weaknesses in its computer system infrastructure. These methodologies help a utility evaluate the susceptibility to potential cyber-attacks and identify corrective actions that can mitigate the risk and seriousness of the consequences of such attacks. An independent assessment can play a vital role is developing an unbiased plan.

The assessment should cover both the IT and the SCADA infrastructure. A Risk Assessment Methodology for Water (RAM-W) was developed specifically for the water industry by the American Waterworks Association Research Foundation (AWWARF) and Sandia National Laboratories. One of the components of the RAM-W process is, "SCADA fault trees." The SCADA fault trees are graphical representations that show how each point of vulnerability can be used by a cyber-attacker to destroy and/or disable critical SCADA system components, or interfere with the normal utility operation. The "SCADA fault trees" are then used in conjunction with risk calculations to rank and select the security improvements. Fault trees are discussed in chapter 7.

The IT system vulnerability assessment is not addressed at the same level as SCADA assessments by RAM-W. However, it is essential to secure IT systems along with the SCADA systems especially if they are connected. Weaknesses in IT systems can be exploited to disable or disrupt the SCADA system operations.

The U.S. Department of Energy (2002) has developed a 21-step guide to improve cyber-security of SCADA networks. The specific 21 steps are listed below:

1. Identify all connections to SCADA networks.
2. Remove unnecessary connections to the SCADA network.
3. Evaluate and strengthen the security of any remaining connections to the SCADA network.
4. Harden SCADA networks by removing or disabling unnecessary services.
5. Do not rely on proprietary protocols to protect your system.
6. Implement security features provided by device and system vendors.
7. Establish strong controls over any medium that is used as a backdoor into the SCADA network.
8. Implement internal and external intrusion detection systems and establish 24-h-a-day incident monitoring.
9. Perform technical audits of SCADA devices and networks, and any other connected networks, to identify security concerns.
10. Conduct physical security surveys and assess all remote sites connected to the SCADA network to evaluate their security.
11. Establish SCADA "Red Teams" to identify and evaluate possible attack scenarios.
12. Clearly define cyber security roles, responsibilities, and authorities for managers, system administrators, and users.
13. Document network architecture and identify systems that serve critical functions or contain sensitive information that require additional levels of protection.
14. Establish a rigorous, ongoing risk management process.
15. Establish a network protection strategy based on the principle of defense in depth.
16. Clearly identify cyber security requirements.
17. Establish effective configuration management processes.
18. Conduct routine self-assessments.
19. Establish system backups and disaster recovery plans.
20. Senior organizational leadership establish expectations for cyber security performance and hold individuals accountable for their performance.
21. Establish policies and conduct training to minimize the likelihood that organizational personnel will inadvertently disclose sensitive information regarding the SCADA system design, operations, or security controls.

DYONYX (Blume, 2002) and PlantData Technologies (Pollet, 2002) have a methodology called "Rings of Defense." In this methodology the IT network and SCADA security are broken down into several layers. The appropriate configuration of the "rings" is considered to be flexible and the employment of an integrated and coordinated set of layers is key in the design of a security approach.

Another method of vulnerability assessment (Munshi, 2003) recommends the evaluation of the four levels of the SCADA system. Level 1 evaluation checks for vulnerabilities of the field equipment such as the PLCs, RTUs, and flowmeters. Level 2 evaluation assesses the vulnerabilities of the communication media such as LAN media, dial-up, and DSL. Level 3 evaluation scrutinizes the SCADA host subsystem. Finally, Level 4 evaluation looks at the whole enterprise for weaknesses.

All of the methodologies presented in this sub-section should lead to the development, documentation, and enforcement of an effective security policy that is unique to each system. A one-size-fits-all approach will not be effective.

Even if a great deal of time and money is spent on assessing the vulnerability of a system to cyber-attacks it is often almost impossible to secure all the vulnerabilities in a system. Therefore, it is prudent to assess the consequences of specific types of cyber threats and secure a system for those specific threats. Schneier (1999) has developed a formal "attack tree" methodology which is very similar to the fault tree analysis used by RAM-W for analyzing the security of systems based on varying attacks. Basically, attacks against a system are represented in a tree structure, with the goal as the root node and different ways of achieving that goal as leaf nodes. This methodology is purported to provide a different way of thinking about security, to capture and reuse expertise about security, and to respond to changes in security. Security is not a product—it is a process. Attack trees form the basis of understanding that process.

5.6 TYPICAL ATTACK SCENARIOS

Cyber-attackers focus on known IT/SCADA system vulnerabilities. The tools and methods used depend upon the characteristics of the vulnerability being exploited. The modes of attacks can be broadly categorized as follows—unauthorized access and/or authentication, data interception, data modification/destruction, and system disruption. Furthermore, cyber-attacks can originate from inside or outside the utilities network. Whatever the mode and origin of an attack, once a system is compromised, data can then be altered or destroyed, and communications can be blocked or rerouted. Additionally, settings can be changed deliberately or randomly such that equipment either fails to operate when needed, or operates when it should not. These types of intrusions can both damage the equipment and/or disrupt the service. The following five example attack scenarios illustrate how cyber-attackers can exploit system vulnerabilities.

5.6.1 Scenario 1

Statistically most cyber-attacks are "inside jobs." A disgruntled current or former staff member can sabotage systems and settings. Similarly, an insider with privileged information can be approached by a cyber-attacker and be bribed or duped into sabotaging systems and settings or creating access mechanisms for the attacker to gain future access.

5.6.2 Scenario 2

For systems equipped with modems or Internet based remote access points a "war dialer" or "port scanning tool" can be used to automate an attack and gain access to the system. A "war-dialer" is a program that can scan hundreds of phone numbers typically in the vicinity of the utility's publicly available phone numbers, looking for answering modems. Similarly, a port scan or ping-sweep program can be used to identify active system ports and/or network IP addresses belonging to a public utility. When a connection is found, multiple returns, question marks, "HELP," and "HELLO" are entered to probe the connection and look for clues as to the kind of connection. Once a login dialog is acquired the intruder can use "social engineering" to determine login information, or launch a dictionary-based or brute-force password attack (Oman et al., 2001). When the connection is made the intruder has access to the SCADA system.

5.6.3 Scenario 3

This scenario applies to systems equipped with wireless access points implementing 802.11b technology. The 802.11b devices have a serious security flaw that compromises the wireless encryption key. Widely available free software, such as AirSnort and NetStumbler, give hackers free tools to crack wireless codes within minutes. Anyone in the vicinity of the access point with a laptop PC and the aforementioned software equipped with a $60 wireless network card, and a directional antenna (which can be made from a Common Cardboard can) can launch an effective attack. A simple Internet search would provide a cyber-attacker step-by-step instructions, pictures, and even videos on how to hack into such a system. Once they steal the wireless encryption key, they can use a freely available protocol analyzer such as. Ethereal or Sniffit to spy on the network. The next step for the attacker is to wait until a maintenance engineer signs onto a PLC and then capture the password. Furthermore, it is common for individuals to use the same passwords for multiple systems, in which case the attacker has just obtained passwords for other secure devices and networks. (Brown, 2002).

5.6.4 Scenario 4

In this scenario a staff member with access to the utility's IT/SCADA systems is duped into installing or running a computer game or otherwise seemingly innocuous application by a current or former associate, or virtually anyone on the Internet. The installed application contains a "Trojan horse program," that opens a backdoor into the computer network (Oman et al., 2001). Once the application is installed, the cyber-attacker is automatically notified that the backdoor is open and gains access to the system.

5.6.5 Scenario 5

A cyber-attacker can access the network from inside or outside and run a program that starts flooding the network with useless traffic, thus jamming the links and denying service to users and SCADA devices causing random disruption of service.

5.7 POPULAR TOOLS USED BY CYBER ATTACKERS

The tools used for cyber-attacks depend upon many things such as the operating system platform, SCADA software, network type, etc. Water utilities typically use a recent version of Windows[*] or a flavor of the UNIX operating system. By far, Windows is the most popular

---

[*]Windows is a registered trademark of Microsoft Corporation.

type of operating system used by water utilities. However, there are a number of market leading SCADA HMI applications in use by utilities. The popular applications include Wonderware InTouch, Intellution FIX, Citect, and RSView.[†] A cyber-attacker can potentially exploit the weakness of the operating system, the network, and/or the SCADA software to gain unauthorized access.

There is a popular myth that somehow non-Windows operating systems such as UNIX and VMS are more secure. The SANS website publishes a top 20 list that includes the top 10 Windows and the top 10 UNIX vulnerabilities. VMS, though less common than UNIX, is commonly used in integrated SCADA systems and the vulnerabilities associated with it are as well known as those for the more common Windows and UNIX operating systems. There are hundreds of cyber-attack tools that are freely available on the Internet for download. A full listing of such tools is impossible to compile as the list grows almost on a daily basis. Chick (2003) maintains a website that provides a listing of popular cyber attack tools. A partial listing, most of which is extracted from his website, is presented below.

5.7.1 Trojan horse

A destructive program that masquerades as a benign application. Unlike viruses, Trojan horses do not replicate themselves but they can be just as destructive. One of the most insidious types of Trojan horse is a program that claims to rid your computer of viruses but instead introduces viruses onto your computer. The popular Trojans include Netbus 2.0 pro, Bo2k, Back Orifice, and Admin. Troj. Kikzyurarse.

5.7.2 Worms and Viruses

The terms virus and worm are often used synonymously to describe malicious, autonomous computer programs. Most contemporary computer viruses are, in fact, worms. The worm epidemic of recent months, enabled by a common "buffer overflow[‡]" exploit, allows cyber-attackers to hijack legitimate computer programs for illicit purposes. These were once the dominion of only the most elite programmers. However, in recent years, buffer overflow attacks have become more and more popular, and they are now the favorite among cyber-attackers of all skill levels. Popular recent examples include Code Red II worm, Nimda worm, etc. (Vatis, 2001).

5.7.3 Scanners

The scanners scan for open and vulnerable IP ports that can be attacked. The popular scanners include ChaoScanner, Port Scanner, NetGhost Domain Scanner, FTP Scan Anonymous port scanner, Mirror Universe 2.1, NetCop 1.6, Site Scan, etc.

5.7.4 Windows NT Hacking Tools

A variety of Windows NT system tools are available on the Internet. Popular hacking tools include Unsecure NT password cracker, Red Button NT exploit, Get Admin, NT Crack, CPU Hog, Crash, NTFS Dos Access, etc.

[†]Wonderware is a trademark of the Production Management Division of Invensys plc, Intellution FIX is a trademark of GE Fanuc, Citect is a trademark of Citect Pty Ltd, and RSView is a trademark of Rockwell Software.

[‡]Buffer overflow: an event in which more data is put into a buffer than the buffer has been allocated. This results in the excess data replacing other data or program code. Left uncorrected, buffer overflows can be exploited to crash the system, gain operating system access or execute inserted code all leading to unauthorized system access.

5.7.5 ICQ Hacking tools

ICQ or "I Seek You" programs were designed to make it easy to get in touch with people. The program could communicate by email, chat, SMS, phone or pager, and makes the process as straightforward as calling across a room to start a friendly conversation. As with any other program hacking tools, various ICQ attacking tools are now available on the Internet. The popular ICQ attack tools include ICQ Port Scanner, ICQ IP Sniffer, ICQ Password Stealer, ICQ 99 UIN Cracker Beta, ICQ Pager, ICQ Flood 95, ISoaQ, Dark ICQ, etc.

5.7.6 Mail Bombs

Although they are much easier to detect, trace and terminate, they can cause damage to the IT infrastructure that may lead to SCADA disruptions. The popular mail bombs include UpYours3, MailBomber, KaBoom v3.0, AnonyMail MailBomber, etc.

5.7.7 Nukers

These are used to "nuke" or crash computers and networks. The popular nukers include Win Nuke, Blood Lust, Muerte, etc.

5.7.8 Key Loggers

Key loggers log keystrokes and can be used to access passwords. The popular keyloggers include Invisible Keylogger, Key Copy 1.01, KeyLog Windows, KeyLog 2.0, KeyLog 2.5, Dos Keylogger, etc.

5.7.9 Hackers' Swiss Knife

These programs contain a wide collection of hacking tools. The popular tools include Computer Warfare, Agressor, Genius 2.6, Hackers Utility, etc.

5.7.10 Password Crackers

Most password crackers are brute-force crackers. They try passwords until the target file or program is cracked. Brute-force is not an efficient way of cracking especially if the password is well-chosen. The popular password crackers include Killer Cracker, PGP Cracker, XIT Quick, Cracker Jack, etc.

5.7.11 BIOS Crackers

These tools are designed to crack a computer's BIOS password. The popular tools include AMI BIOS Cracker, Kill CMOS, Award BIOS Cracker, etc.

For some of these tools to be used, the attacker would need to have physical access to the host computer and/or requires the operator to be duped to install the programs. However, an attacker with social engineering skills can expose the system without having physical access to the various systems. The examples presented demonstrate the range and availability of tools available to hackers. These tools can also be used to conduct system penetration testing to see how easy or difficult it is to hack your systems. Penetration testing allows specific vulnerabilities to be identified and fixed, thereby minimizing exposure to potential cyber attacks.

5.8 IDENTIFYING METHODS FOR MITIGATING SUCH ATTACKS

Once an evaluation of the IT/SCADA system is completed the first step is to disable or remove programs that are not necessary for the SCADA systems to operate. Then the utility should address the "Top 10 Vulnerabilities" (listed in Sec. 5.4). Furthermore, the utility at a minimum should look at improving the following infrastructure components:

5.8.1 Physical Security

Disable devices not required for operation, locate equipment in confined areas, limit and monitor physical access to the servers and SCADA workstations via doors and access restriction.

5.8.2 Access and Authentication Methods

Change username and passwords from manufactures provided defaults. Improve password lengths, encourage the use of non-dictionary words with numbers and non-alpha-numeric characters, and require routine changes. If necessary consider introducing strong authentication and access methods such as the RSA two-factor authentication, biometric authentication, virtual private network (VPN) and public key infrastructure (PKI) for remote access.

5.8.3 Software Improvements

Apply operating system and SCADA software patches routinely. Upgrade the software in a cyclic manner, typically 3 to 5 years. Install software patches on routers, firewalls and switches, if available. Install virus protection and intrusion detection software. Use various software encryption schemes where possible. Install ICSA certified firewalls with Stateful Packet Inspection on connections between IT and SCADA systems, and on wireless and Internet connections.

5.8.4 Privacy Improvements

Add IPSec or other security layer to any wireless access point running 802.11b. Also, for all wireless access points enable the highest level of wireless encryption protocol (WEP), change the default service set ID (SSID) that ships with the router, set up specific network card (MAC) address authentication, disable *ad-hoc* (that allows wireless peer-to-peer connections), and "broadcast" (that periodically transmits the SSID) mode of operations (Netgear, 2003). Either disconnect analog modems, temporarily connect them when they are needed, or install secure analog modems for remote access. Secure analog modems typically employ an authentication mechanism (username/password) in the dial string and are configured to automatically dial back to programmed locations. Adding VPN or PKI will improve security for any remote access option used. Consider reducing communications media vulnerabilities by introducing fiber optic or other types of more secure media that make eavesdropping more difficult. Radio and leased line communications should use encryption. For low data-rate telemetry applications, some encrypted radios are already becoming available and the GTI initiative discussed earlier in this chapter is seeking to define requirements for and promote development of broadly applicable telemetry encryption devices.

5.8.5 Network Topology Improvements

Remove any network hub (use switches). Consider RAS and VPN options for remote access. Redesign SCADA subnet configurations and remove any workstation with dual network interfaces or other insecure connections to other "networks." Provide appropriately sized uninterruptible power supply (UPS) to all critical components of the SCADA network.

5.9 INCIDENT RESPONSE AND BUSINESS CONTINUITY

Cyber and/or SCADA attacks are inevitable. Therefore it is critical that utilities prepare, keep current and practice incident response and business continuity plans so that they can respond quickly when attacks do occur. Although there is no one-size-fits-all solution, there are some common guidelines that can be applied in developing utility-specific plans. The general steps in any plan are preparation, identification, containment, eradication, recovery, and follow-up.[*] Likewise, defining goals is a necessary first step in preparing the plan. Though each utility's goals will be different, common themes include protection of human life and safety; protection of mission-critical systems and data; minimizing and mitigating damage and service disruptions; maintaining the public's confidence; restoring normal operation as quickly as possible; and providing credible, accurate evidence of the attack. This last goal is critical for preventing similar attacks and identifying and prosecuting criminal activity.

5.9.1 Preparation

Preparation includes preparing the plan, training, and routine practice. The plan should be comprehensive and should be based on guidelines published by government agencies. One such guideline is the SANS Institute's "Computer Security Incident Handling Guide" published under the SANS step-by-step series. This comprehensive guide presents a step-by-step approach to incident response that can be used as a starting point in developing a plan tailored to each utility's needs. Legal review of the plan is also critical to improve the chances of being able to prosecute criminal activity without infringing on anyone's rights. Everyone responsible for implementing the plan must be trained. Role-playing and simulation exercises can be used to test, refine, and practice the plan.

5.9.2 Response Process Phases

The response process can be divided into three phases. Phase 1 includes detection, identification, assessment, and triage. Phase 2 includes containment, evidence collection and analysis, and mitigation. Phase 3 includes eradication, recovery, and follow-up.

---

[*]FCC Computer Security Incident Response Guide, Integrated Management Services, Inc. December 2001.

Phase 1. The detect-delay-respond physical security model also applies to cyber/SCADA security. The better a utility's cyber security, the sooner an attack will be detected and the longer it will take the attacker to inflict damage. Preparation should include implementing and maintaining cyber/SCADA security measures to delay attackers and promote early detection while minimizing false alarms.

Because false alarms do occur, first responders need to be able to quickly, quietly, and accurately determine if the anomaly detected is in fact an incident. They need to do this while protecting the evidence, and preserving the evidence chain of custody.

Phase 2. Once an anomaly is identified as an incident, containment without contaminating the environment is the next priority. Evidence is also collected and initial analysis performed to determine if affected system(s) can continue operation or must be shut down.

Phase 3. Once the incident is contained, eradication can begin. Eradication includes removing the cause of the incident to prevent recurrence. The attack evidence can also be used to improve defenses. Also, a vulnerability analysis should be performed to determine if related vulnerabilities have appeared. Finally, affected systems can be restored using the latest "clean" backup. Once the normal operation is restored, affected systems should be more closely monitored than normal and a follow-up report should be prepared while incident memories and evidence are fresh.

5.9.3 SCADA Specific Incident Response

The incident response steps outlined above are general and can be applied to any computer system. Because SCADA is usually critical to utility operation, SCADA specific incident response procedures are usually recommended. The SCADA specific procedures are generally in addition to the general computer incident handling procedures.

Unlike most computer systems, which are centrally focused, SCADA systems are focused on the network edge. Critical IT resources and information are generally at the core of the network. However, SCADA systems are usually designed to distribute intelligence to the network edge device, the PLC or RTU. Therefore, the first priority for SCADA incident response is to determine which process systems are affected and to place those systems in LOCAL/MANUAL control mode. Also, SCADA systems are usually isolated from business networks and even telemetry systems during the containment step. Finally, SCADA incident response must also include evaluation, containment and restoration of PLC and RTU programs in addition to those of the SCADA computers.

5.9.4 Business Continuity

In preparing the business continuity portion of the plan, it is important to include a range of options that can be combined to tailor the response to the specific incident. Here again, there is no simple formula that works for all utilities. Every utility needs to analyze its specific situation and vulnerabilities. However, in general, incidents can be characterized into Levels 1, 2, and 3, with Level 3 being the most severe.

Level 1 incidents could be characterized as those affecting day-to-day operations but easily mitigated through the use of readily available alternatives such as redundant systems or alternative methods. Level 2 incidents could be characterized as involving more complicated solutions requiring staff time to resolve. For example, there may be a need to reconfigure a server or to identify or trace a virus in the IT system. Level 2 incidents do not have ready available mitigation alternatives and require active intervention. Failure to resolve Level 2 incidents, when they arise, might escalate the problem to Level 3. Level 3 incidents could be characterized as directly affecting a utility's operations such as the ability to deliver water, or posing a threat to the utility's revenue stream or having staff and/or public safety implications.

The response time for taking corrective action to maintain business continuity would vary with incident level. Taking up to a week to finish restoring systems affected by a Level 1 incident(s) might be acceptable. However, the impact of a Level 3 incident must be mitigated as quickly as possible to maintain the water supply and public confidence. Likewise the level of effort and interagency coordination required for mitigation would be exponentially higher for a Level 3 event than for a Level 1 event. The practice of including simulated incidents is crucial in refining business continuity planning and in preparing those who must execute the plan.

5.10 SUMMARY AND COMPILATION OF USEFUL INFORMATION LINKS

Information presented in this chapter illustrates the potential vulnerabilities in a water utility's electronic infrastructure that can be exploited by a cyber-attacker. The information presented is intended to guide the utilities to assess vulnerabilities and minimize the risk of exposure to a cyber attack. As technology to defend against a cyber-attack evolves, so do the attack tools. The best defense against cyber-attacks is to be prepared and to stay current. The following is a listing of important websites, where the most current information can be obtained and used to minimize and mitigate cyber-attacks on critical infrastructure.

- http://www.cert.org/—The Carnegie Mellon Computer Emergency Response Team
- http://www.fedcirc.gov/—The Federal Computer Incident Response Center
- http://ists.dartmouth.edu/—The Institute for Security Technology
- http://www.nipc.gov/—The National Infrastructure Protection Center
- http://www.sans.org/—The System Administration, Networking, and Security Institute
- http://isc.incidents.org/—Internet Storm Center
- http://www.cve.mitre.org/—Common Vulnerabilities and Exposures
- http://www.epa.gov/ogwdw/security/index.html—EPA's Water Infrastructure Security
- http://www.waterISAC.org/—Water Information Sharing and Analysis Center
- http://www.isd.mel.nist.gov/projects/processcontrol/—NIST Process Control Security Requirements Forum (PCSRF)
- http://www.isa.org/MSTemplate.cfm?MicrositeID=988&CommitteeID=6821—ISA-SP99, Manufacturing and Control Systems Security
- http://www.sandia.gov/CIS/—Critical Infrastructure Surety at Sandia National Laboratories
- http://www.cisco.com/security_services/ciag/—Cisco's Critical Infrastructure Assurance Group (CIAG)

ACKNOWLEDGMENTS

The authors would like to acknowledge Mr. Roy C. Haught and the U.S Environmental Protection Agency for their leadership, advice, and guidance on the application of information systems for controlling and managing water treatment and distribution systems.

REFERENCES

AGA 12-1 Working Group, Cryptographic Protection of SCADA Communications—DRAFT 1—, AGA Report No. 12-1, March 24, 2003, available at: http://www.gtiservices.org/security/

Brown, A., "SCADA vs. the hackers," *Mechanical Engineering Online*, 2002, available at: http://www.memagazine.org/backissues/dec02/features/scadavs/scadavs.html

Blume, R., Mitigating Security Risks in SCADA/DCS System Environments, 2002.

Chick, D., "The Bitter Network Administrator—Hacking, Cracking and Attacking Tools," 2003, available at: http://www.thenetworkadministrator.com/hackertools.html

DYONYX, Mitigating Security Risks in SCADA/DCS System Environments available at: http://www.dyonyx.com/website_pdfs/SCADA_Security.pdf

FCC Computer Security Incident Response Guide, Integrated Management Services, 2001.

Franz, M., "A future of SCADA and Control System Security—API Industry Security Forum," 2003, available at: http://www.io.com/~mdfranz/papers/franz_API_future_of_scada_security.apn

Munshi, D., "Pipelines Have Help Available to Safeguard Their SCADA Systems," *Pipeline Gas J.* 2003, available at: http://www.pipelineandgasjournal.com

Netgear, "10 Simple Steps to Wireless Security—Simple, Low-Cost Ways to Optimize the Security of Your Wireless LANs," 2003, available at: http://www.netgear.com/pdf_docs/10StepsWirelessSecurity.pdf

Oman, P., E. O. Schweitzer, and J. Roberts, *Safeguarding IEDS, Substations, and SCADA Systems Against Electronic Intrusions*, Schweitzer Engineering Laboratories, 2001, available at: http://www.selinc.com/techppros/6118.pdf

O'Shea, K., *Cyber Attack Investigative Tools and Technologies*, Institute for Security Technology Studies at Dartmouth College, Hanover, NH, 2003, available at: http://htcia_siliconvly.org/contacts.htm

Phillips Jr., W. R., "Solving the Puzzle of Providing Appropriate Cyber Security While Maintaining Operations Effectiveness and Efficiency," *presented at the Florida Water Resources Conference by Bill Phillips*, PE, 2003.

Pollet, J., *SCADA Security Strategy*, PlantData Technologies, 2002, available at: http://www.plantdata.com/scada_security.htm

Pratt, W., "Evaluating Fieldbus Networks—Choose the Right Tool for the Job," 2003, available at: http://www.hartcomm.org/develop/network/compnet.html

Schneier, B., "Attack Trees: Modeling security threats," *Dr. Dobb's J.* 1999, available at: http://www.counterpane.com/attacktrees-ddj-ft.html

Tagg, L., "Aussie Hacker Jailed for Sewage Attacks," 2001, available at: http://cooltech.iafrica.com/technews/837110.htm

U.S. Department of Energy, President's CIP Board, "21 Steps to Improve Cyber Security of SCADA Networks," 2002, available at: http://www.counterterrorismtraining.gov/updates_102002.html

U.S. Department of Justice, United States Attorney's Office Central District of California, 1999, available at: http://www.usdoj.gov/criminal/cybercrime/mitnick.htm

Vatis, M., *Cyber Attacks During the War on Terrorism: A Predictive Analysis*, Institute for Security Technology Studies at Dartmouth College, Hanover, NH, 2001, available at: http://www.ists.dartmouth.edu/ISTS/counterterrorism/cyber_a1.pdf

Wiese, I., SCADA Talks—The Flow of Information Aids the Flow of Water Down Under, Industrial Computing, December ed., 1999, available at: http://www.isa.org/isaolop/journals/pdf/ic/991234.pdf

CHAPTER 6
ASSESSING THE RISKS TO DRINKING-WATER SUPPLIES FROM TERRORISTS ATTACKS

Malcolm S. Field
National Center for Environmental Assessment
Office of Research and Development
U.S. Environmental Protection Agency
Washington, D.C.

6.1 INTRODUCTION

Terrorist threats to the nation's potable water supplies have recently become a major concern for the country. The events of September 11, 2001 (9/11) and subsequent anthrax attacks have proven the vulnerability of basic civilian infrastructures to terrorists. While the past attacks included the physical destruction of large structures housing significant populations by detonation and aerosol attacks on a smaller scale, the potential for a biological or chemical attack on important potable water supplies cannot be discounted (Burrows and Renner, 1999). Developing a preparedness for and response to a terrorist attack is an essential aspect to countering terrorist attacks (Lane et al., 2001).

Current efforts, intended to protect potable water supplies, tend to focus on early warning systems (EWS) (Foran and Brosnan, 2000) to detect initial arrival of hazardous biological and chemical agents. While an improvement over conventional methods of tracking contaminated-water outbreaks (MacKenzie et al., 1994), EWSs may be regarded as inadequate in and by themselves.

Predicting when, and at what concentration a toxic substance released in the respective source area will reach a water-production facility is essential for water managers. The concern of managers for water-supply systems was recently aggravated by the realization that terrorists could deliberately release a toxic poison into their system with the potential to cause widespread illnesses, deaths, and panic before adequate protection measures could be activated. Whereas EWSs will certainly be beneficial when available, knowledge of the source area(s) as well as knowledge of the transport and survivability of the suspect agent are critical.

Disclaimer: The views expressed in this paper are solely those of the author and do not necessarily reflect the views or policies of the U.S. Environmental Protection Agency.

Knowledge of source area(s) and identification of their particular vulnerabilities can lead to the installation of the most appropriate security apparatus. Source area(s) security apparatuses commonly considered include typical security measures (e.g., fences) and/or atypical security measures (e.g., armed security guards) which are most likely to be applied at large water-supply systems (Reed, 2001). Unfortunately, it appears that much of the country is of the misconception that only very large water-supply systems are threatened (U.S. EPA, 2001) and then really only at the downstream end, beyond the treatment system. While this situation is certainly worthy of concern, it is not a seriously realistic attack scenario.

Smaller municipal water systems where source areas are known to occur at some significant distance from the actual supply are more likely to be severely threatened. For example, a karst spring that is used as a municipal water supply and is known to be directly connected to a sinking stream, karst window, or sinkhole several kilometers away should be of significant concern to the local water managers. While these smaller supplies may only serve a few thousand people at most, the potential for a terrorist attack causing illnesses, deaths, and widespread terror are very realistic.

The threats can be better assessed, however, if water managers have a general sense of potential solute-transport rates and likely receptor concentrations for a given release in a given source area. Reasonable predictions of solute-transport rates and concentrations will allow water managers to (1) provide for a higher level of physical security, (2) enhance detection systems, (3) develop alternative strategies to deal with microbial or toxic substance attack(s), and (4) have time to implement the appropriate strategy when an attack does occur.

The purpose of this paper is to outline a method for a proactive approach to protecting water supplies. A hypothetical scenario is used to illustrate the value of the approach presented.

6.2   CHEMICAL AND BIOLOGICAL AGENTS

Numerous toxic agents are known to be available to terrorists. Most of the basic literature regarding toxic agents available to terrorists fall in the category of chemical and biological weapons which must be distinguished from chemical and biological materials. CB weapons refers to the use of warfare agents specifically designed, developed, and secured by the military to support specific military doctrines of a state while CB materials refers to the use of any toxic substance or pathogen (Zanders, 1999). Most of the literature focusing on terrorist attacks on basic infrastructures focus on CB weapons. Information on CB materials must necessarily be obtained from the basic human health and environmental literature.

6.2.1   Biological Agents

Biological agents as water pathogens can be separated out as biological warfare agents and nonwarfare-specific agents. While a significant concern regarding biological warfare agents is warranted, nonwarfare agents warrant equal concern.

Biological Warfare Agents.   A basic listing of potential replicating biological warfare agents appears in Table 6.1 (Burrows and Renner, 1999, 1998, p. 2) and potential warfare biotoxins appears in Table 6.2 (Burrows and Renner, 1999).

TABLE 6.1 Summary of Threat Potential of Replicating Biological Warfare Agents

Agent/ disease	Weaponized	Water threat	Infective-* dose	Stable in water	Chlorine[†] tolerance
Bacteria					
Anthrax	Yes	Yes	6×10^3 spores (inh)	2 years (spores)	Spores resistant
Brucellocis	Yes	Probable	10^4 organisms (uns)	20–72 days	Unknown
Cholera	Unknown	Yes	10^3 organisms (ing)	Survives well	Easily killed
Clostridium perfringens	Probable	Probable	10^8 organisms (ing)	Common in sewage	Resistant
Glanders	Probable	Unlikely	3.2×10^6 spores (uns)	Up to 30 days	Unknown
Melioidosis	Possible	Unlikely	Unknown	Unknown	Unknown
Plague	Probable	Yes	500 organisms (inh)	16 days	Unknown
Salmonella	Unknown	Yes	10^4 organisms (ing)	8 days, fresh water	Inactivated
Shingellosis	Unknown	Yes	10^4 organisms (ing)	2–3 days	Inactivated[‡]
Tularemia	Yes	Yes	10^8 organisms (ing)	Up to 90 days	Inactivated[¶]
Rickettsia					
Q fever	Yes	Possible	25 organisms (uns)	Unknown	Unknown
Typhus	Probable	Unlikely	10 organisms (uns)	Unknown	Unknown
Bacteria-like					
Psittacosis	Possible	Possible	Unknown	18–24 h, seawater	Unknown
Virus					
Encephalomyelitis	Probable	Unlikely	25 particles (aer)	Unknown	Unknown
Enteric viruses	Unknown	Yes	6 particles (ing)	8–32 days	Inactivated[§]
Hemorrhagic fever	Probable	Unlikely	10^5 organisms (ing)	Unknown	Unknown
Smallpox	Possible	Possible	10 particles (uns)	Unknown	Unknown
Protozoa					
Cryptosporidiosis	Unknown	Yes	132 oocysts (ing)	Stable days or more	Resistant

Abbreviations: aer = aerosol; ing = ingestion; inh = inhalation; uns = unspecified.
*Total infective dose
[†]Ambient temperature, ≤1 ppm free available chlorine, 30 min or as indicated.
[‡]0.05 ppm, 10 min
[¶]1.00 ppm, 5 min
[§]Rotavirus
Source: After (Burrows and Renner, 1999, 1998, p. 2).

TABLE 6.2 Summary of Threat Potential of Warfare Biotoxins

Biotoxin	Weaponized	Water threat	NOAEL* (2 L · d⁻¹)	Stable in water	Chlorine[†] tolerance
Aflatoxin	Yes	Yes	75 µg · L⁻¹	Probably stable	Probably tolerant
Anatoxin A	Unknown	Probable	Unknown	Inactivated in days	Probably tolerant
Botulinum toxins	Yes	Yes	0.0004 µg · L⁻¹	Stable	Inactivated, 6 ppm, 20 min
Microcystins	Possible	Yes	1.0[‡] µg · L⁻¹	Probably stable	Resistant at 100 ppm
Ricin	Yes	Yes	15 µg · L⁻¹	Stable	Resistant at 10 ppm
Saxitoxin	Possible	Yes	0.4 µg · L⁻¹	Stable	Resistant at 10 ppm
Staphylococcal enterotoxins	Probable	Yes	0.1 µg · L⁻¹	Probably stable	Unknown
T-2 mycotoxin	Probable	Yes	65[¶] µg · L⁻¹	Stable	Resistant
Tetrodotoxin	Possible	Yes	1.0 µg · L⁻¹	Probably stable	Inactivated, 50 ppm

*NOAEL = no-observed-adverse-effect level. Estimated as 7.5 times the NOAEL calculated for consumption of 15 L · d⁻¹.
[†]Ambient temperature, ≤1 ppm free available chlorine, 30 min or as indicated.
[‡]World Health Organization drinking water standard for NOAEL.
[¶]NOAEL derived from short-term U.S. Department of Defense Tri-Service standard.
Source: After (Burrows and Renner, 1999).

The biological agents and biotoxins listed in Tables 6.1 and 6.2 are deadly in their own right but may also become weaponized at a biological weapons research facility. For example, the anthrax (*Bacillus anthracis*) attacks that occurred in the United States right after 9/11 consisted of weaponized anthrax. Anthrax spores are typically weaponized by reducing the size of the spores and preventing them from flocculating to keep them airborne for long periods of time where they are more likely to be inhaled. Flocculation of anthrax spores is generally prevented by simply drying the spores and combining them with an appropriate agent such as bentonite or silica (Weiss and Warrick, 2001) to cause repulsive charges. Additionally, the anthrax spores virulence may be increased (Nass, 2002).

Nonwarfare Agents. Tables 6.1 and 6.2 refer specifically to modification of pathogens for use by states in warfare, but some that are listed and others that are not listed could readily be used by terrorists to contaminate drinking-water supplies. Although these agents may not necessarily be weaponized, they are still deadly in their own right. Some biological agents that terrorists might more likely consider using are listed in Table 6.3 (Ford, 1999, p. 2). Some obvious overlaps, such as *Cryptosporidium* spp. occur in the Tables 6.1 and 6.3, but others appear to be unique to military activities and may not be so readily available to terrorists. The fact that some organisms were not identified as biological warfare agents should not be construed to imply that the nonwarfare agent is not of significant concern. For example, though the biological agents *Eschericia coli* and *Campylobacter jejuni* (Table 6.3) are not identified as biological warfare agents (Table 6.1) or biological warfare biotoxins (Table 6.2), these organisms were responsible for the deaths of seven individuals

TABLE 6.3 Summary of Nonwarfare-Specific Pathogens in Drinking Water[a]

Agent	Infectious[b] dose	Estimated[c] incidents	Survival in drinking water, days	Survival[d] strategies
Bacteria				
Vibrio cholerae	10^3	(very few)[d]	30	vnc, ic
Salmonella spp.	10^6–10^7	5.9×10^4	60–90	vnc, ic
Shingella spp.	10^2	3.5×10^4	30	vnc, ic
Toxigenic *Eschericia coli*	10^2–10^9	1.5×10^5	90	vnc, ic
Campylobacter spp.	10^6	3.2×10^5	7	vnc, ic
Leptospira spp.	3	?[f]	?	?
Francisella tularensis	10	?	?	?
Yersinia enterocolitica	10^9	?	90	?
Aeromonas spp.	10^8	?	90	?
Heliobacter pylori	?	High	?	?
Legionella pneumophila	>10	1.3×10^{4e}	Long	vnc, ic
Microbacterium avium	?	?	Long	ic
Protozoa				
Giardia lamblia	1–10	2.6×10^5	25	cyst
Cryptosporidium parvum	1–30	4.2×10^5	?	oocyst
Naegleria fowleri	?	?	>365[h]	cyst
Acanthamoeba spp.	?	?	?	cyst
Entamoeba histolica	10–100	?	25	cyst
Cyclospora caytanensis	?	?	>365[h]	oocyst
Isospora belli	?	?	>365[h]	oocyst
The microsporidia	?	?	?	spore, ic[f]
Ballantidium coli	25–100	?	20	cyst
Toxoplasma gondii	10–100[h,i]	>4000[h,j]	>365[h]	oocyst
Viruses[g]				
Total estimates	1–10	6.5×10^6	5–27[k]	adsorption/ absorption

Abbreviation: ? = unknown; ic = intracellular survival and/or growth; vnc = viable but not culturable.

[a]Except where noted, data are compiled from Morris and Levin (1995); WHO (1999); Hazen and Toranzos (1990); Geldreich (1996).
[b]Infectious dose is the number of infectious agents that produce symptoms in 50% of tested volunteers. Volunteers are not usually susceptible individuals, and therefore these numbers are not useful for risk estimates.
[c]U.S. point estimates.
[d]Very few outbreaks of cholera occur in the U.S. and these are usually attributable to imported foods (Breiman and Butler, 1998).
[e]Data from Breiman and Butler (1998).
[f]Possible ic with microsporidia-like organisms (Hoffman et al., 1998).
[g]Includes Norwalk virus, poliovirus, coxsackievirus, echovirus, reovirus, adenovirus, HAV, HEV, rotavirus, SRSV, astrovirus, coronavirus, calicivirus, and unknown viruses.
[h]Data from Marquea King (*pers. comm.*).
[i]Based on pigs.
[j]Congenital cases.
[k]Estimated for HAV, Norwalkvirus, and rotavirus (Weber et al., 1994).
Source: After (Ford, 1999).

TABLE 6.4 U.S. Environmental Protection Agency Contaminant Candidate List (CCL)

Protozoa	Bacteria	Viruses
Ancanthamoeba	Aeromonas	Adenoviruses
Microsporidia	Cyanobacteria	Caliciviruses
(Enterocytozoon & Septata)	Heliobacter pylori	Coxsackie Viruses
	Mycobacterium Avium intracellulare	Echoviruses

and the sickening of more than 2000 people in Walkerton, Ontario (Canada) when released into the community drinking water (O'Connor, 2000, p. 42). A much more comprehensive list and review of potential drinking-water pathogens may be found in Rose and Grimes (2001).

As an example of the amount of research needed on microbial pathogens in drinking water, the U.S. Environmental Protection Agency's microbiological contaminant candidate list (CCL) currently considers only a small number of pathogens (Table 6.4). The small number of microbial pathogens listed in Table 6.4 is a direct consequence of the need to keep the initial research efforts to a manageable size.

6.2.2 Threat Posed by Genetic Engineering

Genetically engineered organisms are a relatively recent advance in the development of biological warfare agents. Knowledge of existing biological agents may be very transient with the recent advances in genetic engineering. By modifying the genetic structure of toxic and nontoxic organisms, infectivity may be enhanced, time from infection to disease may be shortened, and severity of disease outcome may be increased. Production of large quantities of replicating microorganisms for weaponization, through recombinant methodologies, is now relatively inexpensive with the possibility of creating new agents having desirable properties for biological warfare such as increased virulence, hardiness, resistance to antibiotics, disinfectants, and the natural immune system (Birks, 1990) being expected (Takafuji et al., 1997, pp. 679–682). Some animal and human pathogens may need no more enhancement than to be provided with an effective delivery system to overcome traditional water treatment methodology and expose immunologically naïve populations. Infectious agents delivered by a primary water route may continue to spread by secondary human-to-human contact. An infectious epidemic will continue to spread as long as a nonimmune population is exposed and the disease does not outpace its host organisms (e.g., hemorrhagic fever viruses such as Ebola). Although genetically engineered oganisms for warfare purposes were previously considered to be too difficult to be taken very seriously, recent advances in biotechnology have demonstrated the feasibility of bioengineering (Couzin, 2002).

As an example of the possibilities associated with genetic engineering, consider cholera (*Vibrio cholerae*). A very toxic bacterial pathogen in drinking water, it is easily killed by simple chlorination (Table 6.1). Modern genetic engineering techniques could be utilized to reduce or even eliminate cholera's susceptibility to chlorine disinfection which could result in serious outbreaks of cholera wherever this bioengineered form of the pathogen was released.

6.2.3 Chemical Agents

Unlike biological agents, the number of potential chemical agents that might be used by terrorists for attacks on drinking-water supplies appears to be much more limited. While this might be true for chemical warfare agents, it is not necessarily a valid perception for non-warfare agents.

Chemical Warfare Agents. Several chemical agents originally developed for aerosol dispersal during warfare may also be effective when dispersed in water (Table 6.5). The effectiveness can be through direct ingestion, dermal absorption, and inhalation during showering.

Chemical Nonwarfare Agents. While chemical warfare agents warrant the greatest concern in terms of potential chemical agents that could be used by terrorists, several chemical nonwarfare agents pose a significant threat to human health if released in drinking-water supplies. Table 6.6 lists some chemical nonwarfare agents that may be regarded as potential threats to drinking-water supplies if released by terrorists.

Some of the chemicals listed in Table 6.6 are highly toxic (e.g., Compound 1080, sodium cyanide) and pose a major concern for water managers, but others (e.g., sodium azide) are somewhat less toxic. In the case of sodium azide, ingestion of sufficient quantities can result in individuals developing decompression sickness (commonly known as the bends) similar to scuba divers who ascend too rapidly.

6.3 SOURCE-WATER PROTECTION

Perhaps the most critical element necessary for protecting the nation's drinking-water supplies is source-water protection. Source-water protection consists of delineating the sources of water, inventorying potential sources of contamination in those areas, and making susceptibility determinations (U.S. EPA, 1997, p. 1-11).

The most basic aspect of source-water protection is source-water delineation which is nothing more than mapping out the drainage basin from which the water is derived. This can take the form of mapping surface-water divides on a topographic map to comprehensive quantitative-tracing studies. Detailed guidances on source-water delineation may be found in U.S. EPA (1987), Bradbury et al. (1991), and Schindel et al. (1997), but only Schindel et al. (1997) provides a detailed guidance on the use of tracer test methods for source-water delineation. Field (2002d) and Mull et al. (1988) provide detailed discussions on quantitative tracer testing for more comprehensive evaluations of hydrologic systems. Understanding the conditions of the source and identifying environmental stressors such as pH, temperature, chemical constituents, turbidity, and various nutrients (e.g., from sewage source), potential intermediate hosts, and seasonality of pathogens provides a more comprehensive perspective of the source water.

Quantitative Tracer Testing. Quantitative tracer testing is the most reliable method for source-water delineation. The basic methodology consists of releasing a known quantity of tracer material into a source-water location (e.g., karst sinkhole) and recovering the tracer at a downstream location (e.g., karst spring). Careful mass-balance analyses provide insights into the nature of the transport mechanisms operative in the system. By repeating this procedure at several tracer-injection sites and recovering the tracer at all possible recovery locations, a clear delineation of source water is established.

To establish the ability of the flow system of interest to transport different types of materials at different times and different flow conditions, tracer tests covering a range of hydrologic conditions need to be conducted. For this reason, Mull et al. (1988) conducted numerous tracer tests within the same flow system over one year which may or may not be adequate depending upon the overall range of conditions that may develop.

TABLE 6.5 Summary of Threat Potential of Chemical Warfare Agents

Common name	Chemical name	CAS no.	LD_{50} (mg · kg^{-1})	NOAEL (µg · kg^{-1} · d^{-1})
		Incapacitants		
Agent BZ	3-quinuclidinyl benzilate	6581-06-2	18–25	0.5
Lysergide (LSD)	9,10-didehyro-N,N-diethyl-6-methylergoline-8β-carboxamide	50-37-3	46	0.5–2.0
LSD Based BZ	—	—	—	—
Mescaline	3,4,5-trimethoxy-β-phenethylamine	—	—	—
Benzilates	—	—	—	—
		Nerve agents		
Agent GA (tabun)	ethyl N,N-dimethylphosphoramidocyanidate	77-81-6	9.3*	—
Agent GB†(sarin)	isopropyl methylphosphonofluoridate	107-44-8 50642-23-4	43–158	—
Agent GD‡ (soman)	pinacolyl methyl phosphonofluoridate	96-64-0 50642-24-5	20–165	—
Agent GF	—	—	—	—
Agent VE	o-ethyl-s-[2-(diethylamino)ethyl-]ethylphosphonothiolate	—	—	—
Agent VG	o,o-ethyl-s-[2-(diethylamino)ethyl-]phosphorothiolate	—	—	—
Agent VK	—	—	—	—
Agent VM	o-ethyl-s-[2-(diethylamino)ethyl-]methylphosphonothiolate	—	—	—
Agent VR-55	o-isobutyl-s-[2-(diethylamino)ethyl-]methylphosphonothiolate	—	—	—
Agent VX‖,§	o-ethyl-s-[2-(diisopropylamino)ethyl-]methylphosphonothiolate	50782-69-9 51848-47-6 53800-40-1 70938-84-0	—	—

TABLE 6.5 Summary of Threat Potential of Chemical Warfare Agents (*Continued*)

Common name	Chemical name	CAS no.	LD_{50} (mg · kg^{-1})	NOAEL (µg · kg^{-1} · d^{-1})
Vessicants				
Sulfur mustard (H or HD)	bis(2-chloroethyl)sulfide	505-60-2	—	—
Distilled mustard (DM)	—	—	—	—
Nitrogen mustard (HN)	—	—	—	—
Lewisite (L)	2-chlorovinyldichloroarsine	541-25-3	—	—
Phosgene oxime (CX)	dichloroformoxime	—	25	—
Mustard lewisite (HL)	—	—	—	—
Lung irritants				
Phosgene (CG)	—	—	—	—
Diphosgene (DP)	trichloromethylchloroformate	—	—	—
PS Chloropicrin	trichloronitromethane nitrochloroform	76-06-2	—	—
Chlorine gas	—	—	—	—
Perfluoroisobutene	1,1,1,3,3-pentafluoro-2-(trifluoromethyl)propene	382-21-8	—	—
Blood gases				
Hydrogen cyanide (AC)	hydrocyanic acid (HCN)	74-90-8	—	—
Cyanogen chloride (CK)	CNCl	506-77-4	—	—

Note: The principal chemical warfare agents are mostly limited to cyanide.
*Monkey, percutaneous.
†GB2 represents binary chemical Agent GB.
‡GD2 represents binary chemical Agent GD.
¶Russian equivalent, V-Gas.
§VX2 represents binary chemical Agent VX.
Source: After (NAP, 1995; Cordesman, 2001; WHO, 2001).

TABLE 6.6 Summary Threat of Some Potential Chemical Nonwarfare Agents

Common name	Chemical name	CAS no.	LD_{50} (mg · kg^{-1})	NOAEL (mg · kg^{-1} · d^{-1})
Compound 1080	sodium fluoroacetate	62-74-8	2–5	0.05
Sodium cyanide	NACN	143-33-9	2.2	20.4
Potassium cyanide	KCN	151-50-8	—	27
Cyanogen bromide	cyanogen bromide (CNBr)	506-68-3	20	44
Aldicarb	2-methyl-2-(methylthio) propionaldehyde *o*-(methylcarbamoyl)oxime	116-06-3	—	—
Strychnine		57-24-9	2.35	—
Sodium azide	sodium azide (NaN$_3$)	26628-22-8	—	3.57
Potassium silver cyanide	potassium silver cyanide (KAg(CN)$_2$)	506-61-6	21	82.7
Paris green	copper acetoarsenite	12002-03-8	22	—

Additional testing using differing materials, such as sorbing tracers, light and dense-phase tracers, and particulate matter, is essential.

Particulate matter as tracer material is essential for understanding the ability of pathogens to migrate in the flow system. Fluorescent microspheres have been shown to be very valuable in demonstrating the ability of microscopic particles to move through supposedly very tight porous-media systems (Harvey et al., 1989, 1993). Conventional perspectives about smaller-sized particles (e.g., virus) being the most readily transportable particles may not always be true. Fluorescent microspheres are limited in the sense that they do not die off as do living pathogens. Microspheres are further limited in that they are incapable of replicating exponential growth in the flow systems that may occur with bacteria under the right conditions and possibly with protozoa with intermediate hosts. This latter problem is not expected when approximating the behavior of viruses in the environment.

6.4 EFFICIENT HYDROLOGIC TRACER-TEST DESIGN

To better facilitate tracer testing in hydrologic systems, a new Efficient Hydrological Tracer-test Design (EHTD) methodology has been developed (Field, 2002a). Application of EHTD to a study site resulted in successful tracer tests and showed that a good tracer-test design can be developed prior to initiating a tracer test (Field, 2000, p. 26). Subsequent comparison analyses documented the ability of EHTD to predict tracer-test results (Field, 2002b).

6.4.1 Basic Design of EHTD

EHTD is based on the theory that field-measured parameters (e.g., discharge, distance, cross-sectional area) can be combined in functional relationships that describe solute-transport processes related to flow velocity and times of travel. EHTD applies these initial estimates for times of travel and velocity to a hypothetical continuous stirred tank reactor (CSTR) as

ASSESSING THE RISKS TO DRINKING-WATER SUPPLIES

an analog for the hydrological-flow system to develop initial estimates for tracer concentration and axial dispersion D based on a preset average tracer concentration $\overline{C}$. The one-dimensional advection-dispersion equation (ADE)

$$R_d \frac{\partial C}{\partial t} = D \frac{\partial^2 C}{\partial z^2} - v \frac{\partial C}{\partial z} - \mu C + \gamma(z) \tag{6.1}$$

is solved for it's root for preset $\overline{C}$ where $\gamma(z)$ was originally taken as zero and was not a part of the original form of EHTD (Field, 2002a). Using the preset $\overline{C}$ then provides a theoretical basis for an estimate of necessary tracer mass. Application of the predicted tracer mass with the hydraulic and geometric parameters in the ADE allows for an approximation of initial sample-collection time and subsequent sample-collection frequency where 65 samples have been empirically determined to best describe the predicted tracer-breakthrough curve (BTC).

6.4.2 Range of Capabilites of EHTD

Recognizing that solute-transport processes operative in hydrological systems all follow the same basic theoretical principles suggests that an appropriate model for estimating tracer mass would function effectively for all hydrological systems. However, such a model would need to be able to account for slight differences in the nature of the flow systems (e.g., effective porosity) and the manner in which the tracer test is conducted (e.g., tracer-release mode).

Breakthrough curves, predicted using the tracer-test design program, EHTD, for various hydrological conditions have been shown to be very reliable (Field, 2002b). The hydrological conditions used to evaluate EHTD ranged from flowing streams to porous-media systems so that the range of capabilities of EHTD could be assessed. The flowing streams used to evaluate EHTD included tracer tests conducted in small and large surface-water streams, a solution conduit, and a glacial-meltwater stream. The porous-media systems used to evaluate EHTD included natural-gradient, forced-gradient, injection-withdrawal, and recirculation tracer tests. Comparisons between the actual tracer tests and the results predicted by EHTD showed that EHTD adequately predicted tracer breakthrough, hydraulic characteristics, and sample-collection frequency in most instances.

6.5 PREDICTING THE OUTCOMES OF A TOXIC RELEASE

The effect of accidental and deliberate releases of toxic substances to drinking-water supplies needs to be predicted if water managers are to initiate appropriate actions should a release occur. EHTD can be used to predict the effects of a toxic-substance release once source-water areas have been established. By using the same measured or estimated parameters intended for tracer-mass estimation and entering a solute mass for EHTD to use, solute-transport parameters and downstream arrival concentrations are predicted.

A similar prediction methodology had been previously developed (Kilpatrick and Taylor, 1986; Taylor et al., 1986; Mull et al., 1988), but these previous methods required considerably more time and effort. Worse, these previous methods failed to reproduce known results; downstream concentrations were greatly overestimated when these methods were applied. EHTD, however, reliably reproduces known results.

6.5.1 Methodology

EHTD predicts the effects of a toxic-substance release by initially predicting solute-transport parameters and estimated solute mass as described in Field (2002a). EHTD was modified to solve the ADE as a boundary value problem (BVP) for a third-type inlet condition which conserves mass (Toride et al., 1995, p. 5). Additional modifications allow for consideration of an initial value problem (IVP) for uniform background concentration and production value problem (PVP) for exponential production (Toride et al., 1995, pp. 9–12). In dimensionless form the ADE now appears as (Toride et al., 1995, p. 4)

$$R_d \frac{\partial C_r}{\partial T} = \frac{1}{P_e} \frac{\partial^2 C_r}{\partial Z^2} - \frac{\partial C_r}{\partial Z} - \mu^E C_r + \gamma^E(Z) \tag{6.2}$$

where C_r represents the reduced volume-averaged solute concentration.

The solution to the ADE for resident concentration and third-type inlet condition is given as (Toride et al., 1995, p. 6)

$$C^R(Z,T) = C^B(Z,T) + C^I(Z,T) + C^P(Z,T) \tag{6.3}$$

where the R superscript denotes resident concentration, and the B, I, P superscripts denote boundary, initial, and production value problems, respectively (Toride et al., 1995, p. 6). All other parameters are described in the Notations section.

Boundary Value Problem. The BVP may be solved for an impulse release as a Dirac (δ) function by (Toride et al., 1995, p. 8)

$$C^B(Z,T) = M_B \, \Gamma_1^E(Z,T) \tag{6.4}$$

and for a pulse release for the case where $\mu^E = 0$ by (Toride et al., 1995, p. 8)

$$C^B(Z,T) = \sum_{i=1}^{2} (g_i - g_{i-1}) \, \Gamma_2^E(Z, T - \hat{T}_i) \tag{6.5}$$

and for the case where $\mu^E \neq 0$ by (Toride et al., 1995, p. 8)

$$C^B(Z,T) = \sum_{i=1}^{2} (g_i - g_{i-1}) \, \Gamma_3^E(Z, T - \hat{T}_i) \tag{6.6}$$

where the E superscript denotes dimensionless equilibrium concentration. The auxiliary functions Γ_1^E, Γ_2^E, and Γ_3^E are defined in the appendix.

Initial Value Problem. The IVP may be solved for uniform initial concentration by (Toride et al., 1995, p. 10)

$$C^I(Z,T) = C_i \, \Gamma_4^E(Z,T) \tag{6.7}$$

The auxiliary function Γ_4^E is defined in the appendix.

Production Value Problem. The PVP may be solved for solute production that changes exponentially with distance by (Toride et al., 1995, p. 12)

$$\gamma^E(Z) = \gamma_1 + \gamma_2 e^{-\lambda^P Z} \tag{6.8}$$

which gives the solution as (Toride et al., 1995, p. 12)

$$C^P(Z, T) = \frac{1}{R_d} \int_0^T \gamma_1 \Gamma_4^E(Z, T; 0) + \gamma_2 \Gamma_5^E(Z, T; \lambda^P) dT \qquad (6.9)$$

$$= \begin{cases} \frac{\gamma_1}{\mu^E}[1 - \Gamma_4^E(Z, T; 0) - \Gamma_3^E(Z, T; \mu^E)] + \frac{\gamma_2}{R_d}\int_0^T \Gamma_5^E(Z, T; \lambda^P) dt & (\mu^E > 0) \quad (6.10) \\ \frac{\gamma_1}{R_d} \Gamma_6^E(Z, T) + \frac{\gamma_2}{R_d}\int_0^T \Gamma_4^E(Z, T; \lambda^P) dt & (\mu^E = 0) \quad (6.11) \end{cases}$$

where the auxiliary functions Γ_5^E and Γ_6^E are defined in the appendix. Exponential production and decay will be highly dependent on effects imposed by environmental stressors. For example, high levels of turbidity and nutrient loading are necessary to ensure adequate growth of the pathogens.

Upon entering a solute mass, EHTD proceeds using the measured parameters and calculated functional relationships (Field, 2002a). Entering a solute mass directly, causes the preset average concentration $\overline{C}$ to be overridden and a new $\overline{C}$ to be predicted. A typical breakthrough curve representing the downstream effects of the release is then produced. For pathogen releases, simple conversions for mass and concentration need to be undertaken, however.

6.6 EXPERIMENTAL EXAMPLE

To evaluate the ability of EHTD to predict the effects of a deliberate release of a chemical or biological agent, a karstic aquifer in which a relatively small spring is used for drinking water is investigated. Tracer tests have established the connection between a distant karst window [depression revealing a part of a subterranean river flowing across its floor, or an unroofed part of a cave (Field, 2002c, p. 110)] and the spring which serves to illustrate the vulnerability of such a water supply to a terrorist attack.

6.6.1 Hydrologic System

The example system considered here consists of a karst window in which the stream flowing at the base of the window has been connected through tracing tests to a spring used by a small city for drinking-water supplies. The karst window is not far from a major thoroughfare and is easily accessible. Basic measured field parameters necessary for EHTD prediction are shown in Table 6.7 with associated functional relationships and related transport parameters. All measured parameters listed in Table 6.7 were calculated directly from one tracer test in the flow system which may or may not be representative of the system at different times and hydrologic conditions. Also, it is very unlikely that the solution conduit maintains a straight line so that a sinuosity factor $S_f \approx 1.3$ is usually multiplied to the straight-line distance, but was not done so here. Tracer retardation R_d and tracer decay μ were subsequently developed to account for delayed arrival times and significant tracer loss (≈ 35 percent).

Cursory examination of the measured parameters and functional relationships shows that transport rates are quite high, dispersion is significant, and transport is strongly dominated by advective forces rather than diffusive forces. The impact of these parameters is that a release of a toxic substance in this area will arrive quickly at the water-supply spring with relatively little dispersion. Discrepancies between the measured and predicted transport parameters relate mostly to the effect of $R_d = 1.05$ which was applied to EHTD, but not considered as part of the original data analysis.

TABLE 6.7 Tracer-Test Design Parameters

Parameter	Measured	Predicted
Field parameters		
Release Mode	Impulse	Impulse
Q, m$^3 \cdot$ H^{-1}	1.16×10^2	1.16×10^2
L^a, m	9.14×10^2	9.14×10^2
A, m^2	1.84×10^0	1.84×10^0
$\overline{C}$, µg $\cdot$ L^{-1}	4.10×10^0	4.10×10^0
C_p, µg $\cdot$ L^{-1}	4.20×10^0	4.20×10^0
Functional relationships		
t_p, h	1.45×10^1	1.39×10^1
$\bar{t}$, h	1.72×10^1	1.45×10^1
v_p, m $\cdot$ h^{-1}	6.30×10^1	6.57×10^1
v, m $\cdot$ h^{-1}	5.31×10^1	6.30×10^1
V, m^3	1.99×10^3	1.77×10^3
Axial dispersion		
D, m$^2 \cdot$ h^{-1}	3.28×10^2	6.60×10^2
P_e	1.50×10^2	8.73×10^2
Tracer reaction		
R_d	1.00×10^0	1.05×10^0
μ, h^{-1}	0.00×10^0	1.80×10^{-2}

^{a} Transport distance = straight-line distance.

Applying the measured parameters and tracer reaction values to EHTD resulted in a visually acceptable fit between the predicted BTC and the actual measured BTC (Fig. 6.1) when the actual mass of tracer released ($M = 3.57$ g) is matched.

6.6.2 Chemical/Biological Release Examples

Consider two possible toxic releases into a water-supply system. A potentially deadly pathogen might include *Vibrio cholerae* (Table 6.3) while a potentially deadly chemical substance that could be released might include Compound 1080 (fluoroacetic acid [CAS NUMBER: 62-74-8]). *Vibrio cholerae* is well documented in history for its known toxicity although modern society has managed to control this pathogen (Table 6.1). Referring to Table 6.6 for a toxic chemical, Compound 1080 stands out. Compound 1080 is a highly toxic pesticide (NOAEL = 0.05 mg $\cdot$ kg$^{-1} \cdot$ d^{-1}; LOAEL = 0.2 mg $\cdot$ kg$^{-1} \cdot$ d^{-1}; and human LD$_{50}$ = 2 to 5 mg $\cdot$ kg^{-1}) used to control rodents and coyotes.

Acquisition of a toxic chemical is not difficult. On May 10, 2002 7.6 tons (6895 kg) of sodium cyanide were hijacked in Hidalgo State, Mexico (Jordan, 2002). Although a majority of the NaCN was recovered and was probably stolen by mistake, this instance serves to illustrate the likelihood that highly toxic compounds may easily fall into the hands of would-be terrorists. Other potentially deadly chemicals, such as Compound 1080 may just as easily fall into the wrong hands.

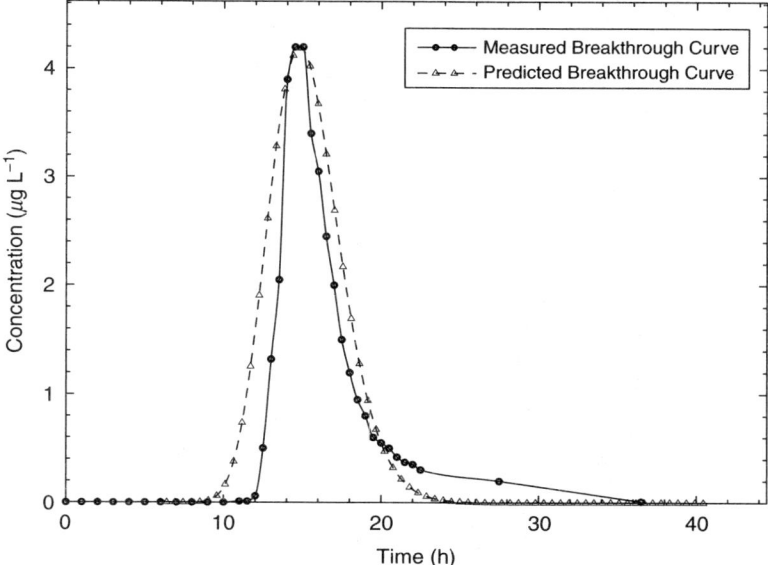

FIGURE 6.1 Comparison of measured data for the site tracer test with EHTD predicted results. Circles represent actual sample-collection times and triangles represent EHTD-recommended sample-collection times.

6.6.3 Release of Compound 1080

Suppose just 1 kg of Compound 1080 was to be deliberately released into the flow system. Such a release would result in a significant downstream peak concentration (Fig. 6.2). A peak concentration of $1.18 \text{ mg} \cdot \text{L}^{-1}$ is sufficiently large enough to warrant an acute risk assessment be conducted.

Compound 1080 Risk Assessment. A release of 1 kg of Compound 1080 resulting in a peak concentration downstream of the release site equal to $1.18 \text{ mg} \cdot \text{L}^{-1}$ can be assessed for its impact on human health by conducting a standard risk assessment. An acute exposure assessment for ingestion E_1 is estimated from (Field, 1997; U.S. EPA, 1989b, p. 6-35)

$$E_1 = \frac{\int_{t_0}^{t_1} C(t)dt}{(t_1 - t_0)} \frac{I_g}{B_w} \qquad (6.12)$$

where $C(t)$, the temporally averaged concentration = $1.48 \times 10^{-1} \text{ mg} \cdot \text{L}^{-1}$, $I_g = 1.4 \text{ L} \cdot \text{d}^{-1}$ (Field, 1997; U.S. EPA, 1989a, pp. 2-1–2-10), $t_1 - t_0 = 45.0$ h is the exposure time, and $B_w = 70$ kg (Field, 1997; U.S. EPA 1989a, pp. 5-1–5-7). Its associated hazard quotient H_{Q_1} is estimated by (Field, 1997; U.S. EPA, 1989b, p. 8-11)

$$H_{Q_1} = \frac{E_1}{RfD} \qquad (6.13)$$

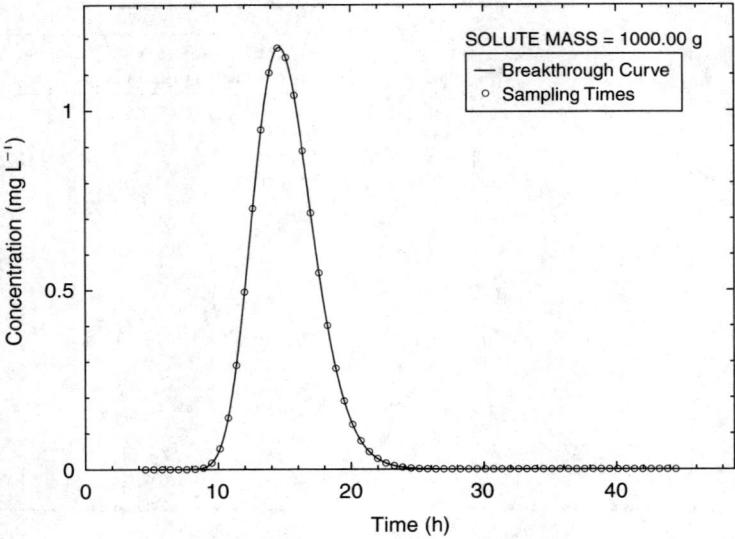

FIGURE 6.2 Breakthrough curve results from release of 1 kg of Compound 1080. Circles represent EHTD-recommended sample-collection times.

An acute exposure assessment for inhalation E_2 is estimated by (Field, 1997; U.S. EPA, 1989b, p. 6-44)

$$E_2 = \frac{\int_{t_0}^{t_1} C(t)dt}{(t_1 - t_0)} \frac{I_h W_u S_d}{B_w V_a F_r} \quad (6.14)$$

where $I_h = 0.6$ m$^3 \cdot$h^{-1} (Field, 1997; U.S. EPA, 1989a, pp. 3-1–3-8), $W_u = 719$ L (Field, 1997; U.S. EPA, 1989a, pp. 5-34–5-36), $S_d = 0.17$ h, $V_a = 2$ m^3 (Field, 1997; U.S. EPA, 1989a, pp. 5-34–5-36), and $F_r = 1$ each day (Field, 1997; U.S. EPA, 1989a, pp. 5-34–5-36). Its associated hazard quotient H_{Q_2} is estimated from (Field, 1997; U.S. EPA, 1989b, pp. 8-5 and 8-11)

$$H_{Q_2} = \frac{E_2 B_w}{RfC\, I_h} \quad (6.15)$$

Lastly, an acute exposure assessment for dermal contact E_3 is estimated from (Field, 1997; U.S. EPA, 1989b, p. 6-37)

$$E_2 = \frac{\int_{t_0}^{t_1} C(t)dt}{(t_1 - t_0)} \frac{S_a P_c S_d K_f}{B_w F_r} \quad (6.16)$$

where $S_a = 1.82 \times 10^4$ cm^2 (Field, 1997; U.S. EPA, 1989a, pp. 4-1–4-16), $P_c = 0.074$ cm$\cdot$h^{-1} (Field, 1997; U.S. EPA, 1992a, p. 5-43), and $K_f = 10^{-3}$ L$\cdot$cm^{-3} (Field, 1997; U.S. EPA, 1989a, p. 6-37). Its associated hazard quotient H_{Q_3} is estimated from (Field, 1997; U.S. EPA, 1989b, pp. 8-11 and A-2)

$$H_{Q_3} = \frac{E_3}{RfD\, A_e} \quad (6.17)$$

TABLE 6.8 Acute Risk Assessment for Compound 1080

Pathway	Exposure assessment (mg·kg^{-1}·d^{-1})	Hazard quotient (dimen.)
Ingestion*	2.96×10^{-3}	1.48×10^{2}
Inhalation†	7.75×10^{-5}	1.88×10^{2}
Dermal	1.16×10^{-5}	2.91×10^{0}
	Hazard Index, $H_I = 3.39 \times 10^2$	

*$RfD = 2.0 \times 10^{-5}$ mg·kg^{-1}·d^{-1} (U.S. EPA, 1992b).
†$RfC = 2.0 \times 10^{-6}$ mg·m^{-3} (assumed).

where $A_e = 20\%$. The hazard index H_i is then obtained by summing all the previously estimated hazard quotients (Field, 1997; U.S. EPA, 1989b, p. 8-13)

$$H_I = \sum_{i=1}^{n} H_{Q_i} \tag{6.18}$$

Table 6.8 shows the basic exposure and risk numbers associated with the Compound 1080 release. The resulting hazard index H_I of 3.39×10^2 is high enough to warrant significant concern by a water manager. An $H_I > 1$ is reason for concern so an $H_I > 10^2$ should probably prompt the water manager to issue a no-use warning.

6.6.4 Release of *Vibrio cholerae*

It has been suggested that it would be very difficult for terrorists to release a deadly pathogen into drinking-water supplies in sufficient quantities to cause serious illness because of the large volume necessary. Consider a one quart thermos [suggested by Hickman (1999) as a logical container] with a 2.5-percent concentration of *V. cholerae* (enteric gram-negative rod bacteria ≈ 2 to 4 μm). The actual concentration N_p of cholera in the thermos may be calculated using Euclidean geometry by

$$N_p = \frac{2\omega \, 10^{12}}{\rho_p \pi a b^2} \tag{6.19}$$

where $a = 3.0$ μm, $b = 0.5$ μm, and $\rho_p = 1.05$ g·cm^{-3} for *V. cholerae*. For a $\omega = 2.5\%$ concentration, $N_p = 2.02 \times 10^{10}$ mL^{-1}. The mass M_p for an individual cholera particle may be calculated by

$$M_p = \frac{\rho_p \pi 10^{-12} a b^2}{2} \tag{6.20}$$

Using Eq. (6.20) $M_p = 1.24 \times 10^{-12}$ g. The total mass of all the cholera particles M_p^T is then calculated by

$$M_p^T = V N_p M_p \tag{6.21}$$

where $V \approx 946$ mL (one quart thermos) resulting in $M_p^T = 23.70$ g, which appears to be a relatively small amount. Applying this value for M_p^T to EHTD with an initial *V. cholerae* concentration of 5.08×10^{-1} mL^{-1} and allowing for exponential growth results in a $C_p = 27.94$ μg·L^{-1} (Fig. 6.3),

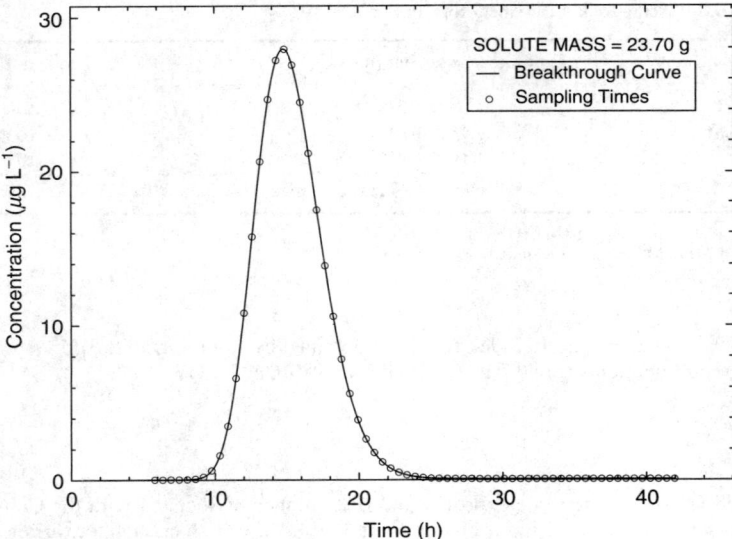

FIGURE 6.3 Breakthrough curve results from release of 23.70 g of *V. cholerae*. Circles represent EHTD-recommended sample-collection times.

which translates into a downstream particle concentration $N_p = 2.26 \times 10^4$ mL^{-1}. Temporally averaging results in $C(t) = 3.75$, which translates into an average downstream particle concentration $N_p = 3.03 \times 10^3$ mL^{-1}.

***Vibrio cholerae* Risk Assessment.** A release of 23.70 g of *V. cholerae* resulting in a peak concentration downstream of the release site equal to 27.94 µg · L^{-1} can be assessed for its impact on human health by determining the probability of infection P_I by use of the beta-Poisson model (Haas, 1983)

$$P_I(d) = 1 - \left[1 + \frac{d}{N_{50}}(2^{1/\alpha} - 1)\right]^{-\alpha} \quad (6.22)$$

where $\alpha = 0.25$ and $N_{50} = 243$ for *V. cholerae* (Haas et al., 1999, p. 430). Equation (6.22) may be related to the probability of morbidity $P_{D:I}$ (clinical illness) by (Haas et al., 1999, p. 306)

$$P_{D:I}(d) = 1 - \left[1 + \frac{d}{N_{50}^*}(2^{1/\alpha^*} - 1)\right]^{-\alpha^*} \quad (6.23)$$

where $\alpha^* = 0.495$ and $N_{50}^* = 3364$ for *V. cholerae* (Hass et al., 1999, p. 308). The probability of mortality $P_{M:D}$ may then be estimated by (Gerba et al., 1996)

$$P_{M:D}(d) = P_{D:I} F_a \quad (6.24)$$

although evidence in support of Eq. (6.24) is conspicuously lacking. Table 6.9 shows the risks associated with the *V. cholerae* release. While the values for P_I, $P_{D:I}$, and $P_{M:D}$ for a

TABLE 6.9 Acute Risk Assessment for *V. cholerae*

Population	Risk of infection, P_I	Risk of illness, $P_{D:I}$	Risk of death,[*] $P_{M:D}$	Risk of death,[†] $P_{M:D}$
Individual	5.65×10^{-2}	1.90×10^{-3}	1.90×10^{-7}	1.90×10^{-5}
10,000	5.65×10^{2}	1.90×10^{1}	1.90×10^{-3}	1.90×10^{-1}

[*]Healthy population ($F_a = 0.01\%$).
[†]Immunocompromised and elderly population ($F_a = 1.0\%$).

single individual may be taken as fairly low, they are significant given the very small volume (946 mL) released. For an exposed population of 10,000 people, the severity of the risks will become apparent where nearly 6 percent of the population will become infected, 3 percent of which will exhibit morbidity (Table 6.9). Accordingly, only 0.01 percent of the healthy population contracting the illness would exhibit mortality while 1 percent of the immunocompromised population contracting the illness would exhibit mortality. The probabilities listed in Table 6.9 are misleading in that healthy individuals and immunocompromised individuals are separated into two distinct groups for the modeling process. In addition, many individuals will receive a much higher dose while others will receive a much lower dose depending on when water ingestion actually occurs relative to the arrival time of the coliform bacteria.

Equations (6.22) to (6.24) are acceptable for pathogens such a *V. cholerae* because cholera, the prototype toxigenic diarrhea, is secretory in nature. It is not easily spread from human-to-human and is therefore a disease factor in drinking-water supplies primarily. However, a large portion of the population may be infected from primary water-borne exposures; immunocompromised and the elderly may face a greater risk for mortality (1 in 500) than would be a healthy population (1 in 5 million) (Table 6.9). Supportive therapies may be intensive and require prolonged hospitalization.

The reported rate of symptomatic-to-asymptomatic cases during exposures is 1:400 (Dire and McGovern, 2002). Diseases that may also spread by person-to-person contact after infection occurs may be more effectively modeled using an infection-transmission model similar to those developed by Eisenberg et al. (1996) and Chick et al. (2001). For example, typhoid fever, a type of enteric fever caused by the pathogen *Salmonella typhosa* (see Tables 6.1 and 6.3), is readily communicable and requires more sophisticated modeling if the risks are to be quantified.

Typical water chlorination readily kills *V. cholerae*, but weaponization of *V. cholerae* could make it resistant to chlorination and other disinfectants. *Vibrio cholerae* is well-documented to form biofilms with individual organisms taking a rugose form which increases its resistance to typical chlorination and possibly other disinfectants (Sánchez and Taylor, 1997; Wai et al., 1999; Reidl and Klose, 2002). Cholera survival in drinking water may be further exacerbated by inadequate water-treatment plant chlorination (O'Connor, 2000, pp. 106–107) or water-treatment plant sabotage (Hickman, 1999).

6.7 CONCLUSIONS

Terrorist attacks on drinking-water supplies must be regarded as inevitable. While basic security efforts are useful, it is beneficial to conduct simulation studies of possible releases of toxic substances so as to gain insights into the nature of potential threats.

The tracer-test design program, EHTD was modified for use in conducting model simulations of potential attacks. Modifications consisted of conversion to a third-type inlet condition for resident concentrations, inclusion of routines for uniform initial (background) concentration and exponential production (growth) parameters, and the ability to bypass preset $\overline{C}$. When run with user-selected solute-mass as input, EHTD bypasses the preset $\overline{C}$ to allow for prediction of downstream concentrations. Initial solute concentrations and/or exponential production may significantly affect the final concentration estimates if these entered values are substantial. Additionally, secondary transmission following initial infection should be considered for the final population disease outcome estimate.

By conducting basic model simulation studies, water managers can also develop standard risk assessments for chemical and biological attacks on their drinking-water supplies. By developing basic risk assessments, water managers can gain a general sense as to how vulnerable their respective water supplies are to various types of toxic contaminants and release amounts. Assessment of the vulnerabilities can then be used to develop human health protection-strategies (e.g., boil water or don't drink health advisories) for use in the event of a terrorist attack.

While not a preventative counterterrorist tool similar to the posting of armed guards, the methodology described is useful for predicting events and for developing protection plans. It is expected, however, that this methodology will be just one small piece in the arsenal of tools available to water managers as they continue to develop protection programs for the nation's drinking-water supplies.

APPENDIX: DEFINITIONS FOR THE AUXILIARY FUNCTIONS Γ_i^E

A1. Γ_1^E

$$\Gamma_1^E(Z,T) = \exp\left(\frac{-\mu^E T}{R_d}\right)\sqrt{\frac{P_e}{\pi R_d T}} \exp\left[\frac{-P_e(R_d Z - T)^2}{4 R_d T}\right] - \frac{P_e}{2R_d} e^{PZ} \text{erfc}\left(\frac{R_d Z + T}{\sqrt{4 R_d T/P_e}}\right) \quad (A1)$$

A2. Γ_2^E

For $\mu^E = 0$, Γ_2^E is given as

$$\Gamma_2^E(Z,T) = \frac{1}{2}\text{erfc}\left(\frac{R_d Z - T}{\sqrt{4 R_d T/P_e}}\right) + \sqrt{\frac{P_e}{\pi R_d}} \exp\left[\frac{-P_e(R_d Z - T)^2}{4 R_d T}\right]$$
$$-\frac{1}{2}\left(1 + P_e Z + \frac{P_e T}{R_d}\right)\exp(PeZ)\,\text{erfc}\left(\frac{R_d Z + T}{\sqrt{4 R_d T/P_e}}\right) \quad (A2)$$

A3. Γ_3^E

For $\mu^E \neq 0\ (\mu^E > -P_e/4)$, Γ_3^E is given as

$$\Gamma_3^E(Z,T) = \frac{1}{1+u}\exp\left[\frac{P_e(1-u)Z}{2}\right]\mathrm{erfc}\left(\frac{R_d Z - uT}{\sqrt{4R_d T/P_e}}\right)$$

$$+\frac{1}{1-u}\exp\left[\frac{P_e(1+u)Z}{2}\right]\mathrm{erfc}\left(\frac{R_d Z + uT}{\sqrt{4R_d T/P_e}}\right)$$

$$-\frac{2}{1-u^2}\exp\left[P_e Z + \frac{P_e(1-u^2)T}{4R_d}\right]\mathrm{erfc}\left(\frac{R_d Z + uT}{\sqrt{4R_d T/P_e}}\right) \quad (A3)$$

with

$$u = \sqrt{1 + \frac{4\mu^E}{P_e}}$$

A4. Γ_4^E

$$\Gamma_4^E(Z,T) = \exp\left(\frac{-\mu^E T}{R_d}\right)\left\{1 - \frac{1}{2}\mathrm{erfc}\left[\frac{R_d(Z-Z_i)-T}{\sqrt{4R_d T/P_e}}\right]\right.$$

$$-\sqrt{\frac{P_e T}{\pi R_d}}\exp\left\{P_e Z - \frac{P_e[R_d(Z+Z_i)+T]^2}{4R_d T}\right\}$$

$$\left.+\frac{1}{2}\left[1 + P_e(Z+Z_i) + \frac{P_e T}{R_d}\right]\exp(P_e Z)\mathrm{erfc}\left[\frac{R_d(Z+Z_i)+T}{\sqrt{4R_d T/P_e}}\right]\right\} \quad (A4)$$

A5. Γ_5^E

$$\Gamma_5^E(Z,T) = \exp\left(\frac{-\mu^E T}{R_d} + \frac{\lambda^2 T}{R_d P_e} + \frac{\lambda T}{R_d} - \lambda Z\right)\left\{1 - \frac{1}{2}\mathrm{erfc}\left[\frac{R_d Z - (1+2\lambda/P_e)T}{\sqrt{4R_d T/P_e}}\right]\right.$$

$$+\frac{1}{2}\left(1 + \frac{P_e}{\lambda}\right)\exp(P_e Z + 2\lambda Z)\mathrm{erfc}\left[\frac{R_d Z + (1+2\lambda/P_e)T}{\sqrt{4R_d T/P_e}}\right]$$

$$\left.-\frac{P_e}{2\lambda}\exp\left(-\frac{\mu^E T}{R_d} + P_e Z\right)\mathrm{erfc}\left[\frac{R_d Z + T}{\sqrt{4R_d T/P_e}}\right]\right\} \quad (A5)$$

A6. Γ_6^E

$$\Gamma_6^E(Z,T) = T + \frac{1}{2}\left(R_d Z - T + \frac{R_d}{P_e}\right) \text{erfc}\left(\frac{R_d Z - T}{\sqrt{4R_d T/P_e}}\right)$$

$$- \sqrt{\frac{P_e T}{4\pi R_d}}\left(R_d Z + T + \frac{2R_d}{P_e}\right)\exp\left[-\frac{P_e(R_d Z - T)^2}{4R_d T}\right]$$

$$+ \left[\frac{T}{2} - \frac{R_d}{2P_e} + \frac{P_e(R_d Z - T)^2}{4R_d}\right]\exp(PeZ)\text{erfc}\left(\frac{R_d Z + T}{\sqrt{4R_d T/P_e}}\right) \quad \text{(A6)}$$

NOTATION

- a = long dimension of rod-shaped particle (L)
- α = slope parameter for median infection estimate
- α^* = slope parameter for median morbidity estimate
- A = cross-sectional area of flow system (L^2)
- A_e = adsorption efficiency (dimen.)
- b = short dimension of rod-shaped particle (L)
- b_f = one half fracture width (L)
- $\overline{C}$ = averaged tracer concentration (M·L^{-3})
- $\overline{C}^E$ = mean volume-averaged tracer concentration (M·L^{-3})
- C_p = peak tracer concentration (M·L^{-3})
- d = dose = $N_p I_g$ (# T^{-1})
- D = axial dispersion (L^2·h^{-1})
- E_1 = acute exposure for ingestion of a chemical (M·M^{-1}·T^{-1})
- E_2 = acute exposure for inhalation of a chemical (M·M^{-1}·T^{-1})
- E_3 = acute exposure for dermal contact with a chemical (M·M^{-1}·T^{-1})
- F_a = fatality rate (dimen.)
- F_r = frequency of showers (T)
- g_i = input concentrations for pulse injection; ($i = 1, 2; g_0 = g_2 = 0$)
- $g(t)$ = function of values such that $g(t_i) = C_i$
- γ_i = dimensionless exponential production (growth) constants for the PVP [$i = 1, 2$]
- γ^E = dimensionless production = $\frac{L(n_e\gamma_i + \rho_b K_d \gamma_s)}{n_e v c_0}$; $\frac{L(b_f \gamma_i + 2 K_a \gamma_s)}{b_f v c_0}$; $\frac{L(r\gamma_i + 2 K_a \gamma_s)}{r v c_0}$
- Γ_i^E = auxiliary functions for equilibrium transport [see Appendix]
- H_I = hazard index for all pathways (dimen.)
- H_{Q_1} = hazard quotient for ingestion (dimen.)
- H_{Q_2} = hazard quotient for inhalation (dimen.)
- H_{Q_3} = hazard quotient for dermal contact (dimen.)

I_g = amount of water ingested per day ($L^3 \cdot T^{-1}$)
I_h = inhalation rate ($L^3 \cdot T^{-1}$)
K_f = volumetric conversion for water ($L^3 \cdot L^{-3}$)
L = characteristic distance from point of injection to point of recovery (L)
M = solute mass (M)
M_p = particle mass (M)
M_p^T = mass of total number of particles (M)
N_{50} = median infectious dose (# T^{-1})
N_{50}^* = median morbidity dose (# T^{-1})
N_p = concentration of particles (# L^{-3})
P_c = skin permeability constant ($L \cdot T^{-1}$)
P_e = Péclet number (dimen.)
P_I = probability of infection (dimen.)
$P_{D:I}$ = probability of morbidity (dimen.)
$P_{M:D}$ = probability of mortality (dimen.)
ρ_p = particle density ($M \cdot L^{-3}$)
Q = flow system discharge ($L^3 \cdot T^{-1}$)
r = solution conduit radius (L)
R_d = solute retardation (dimen.)
RfC = reference concentration ($M \cdot M^{-1}$)
RfD = reference dose ($M \cdot M^{-1} \cdot T^{-1}$)
S_a = skin surface area (L^2)
S_d = shower duration (T)
S_f = sinuosity factor (dimen.)
t = time (T)
$\bar{t}$ = mean residence time (T)
t_p = peak time of arrival (T)
T = dimensionless time = $\frac{vt}{L}$
$\hat{T}_i$ = dimensionless pulse time = $\frac{vt_0}{L}$; $(i = 1, 2; \hat{T}_1 = 0)$
v = average time of travel ($L \cdot h^{-1}$)
μ = solute decay (T^{-1})
μ^E = dimensionless equilibrium decay = $\frac{L(n_e\mu_l + \rho_b K_d \mu_s)}{n_e v}$; $\frac{L(b\mu_l + 2K_a\mu_s)}{bv}$; $\frac{L(r\mu_l + 2K_a\mu_s)}{rv}$
μ_l = liquid phase solute decay (T^{-1})
μ_s = sorbed phase solute decay (T^{-1})
V = volume (L^3)
V_a = shower stall volume (L^3)
ω = concentration of particulate matter for a concentrated volume (%)
W_u = water usage (L^3)

ACKNOWLEDGMENTS

The author would like to thank Marquea King and Brenda Boutin of the U.S. Environmental Protection Agency, National Center for Environmental Assessment for contributing critical information and data and for providing useful comments used in Table 6.3 and other parts of this manuscript. Their careful review of the manuscript, assistance and contributions greatly improved the manuscript and are gratefully acknowledged.

REFERENCES

Birks, J. W., "Weapons Forsworn: Chemical and Biological Weapons," in A. H. Ehrlich, and J. W. Birks (eds.), *Hidden Dangers: Environmental Consequences of Preparing for War*. Sierra Club, San Francisco, pp. 161–189, 1990.

Bradbury, K. R., M. A. Muldoon, A. Zaporozec, and J. Levy, "Delineation of Wellhead Protection Areas in Fractured Rocks," Tech. Rep. EPA 570/9-87-009, U.S. Envir. Prot. Agency, Washington, DC, 144, pp., 1991.

Breiman, R. F., and J. C. Butler, "Legionnaires' Disease: Clinical, Epidemiological, and Public Health Perspectives," *Semin. Respir. Infect.* 13: 84–89, 1998.

Burrows, W. D., and S. E. Renner, "Medical Issues Information Paper: Biological Warfare Agents as Threats to Potable Water," Tech. Rep. IP-31-017, U.S. Army, Aberdeen, MD, 1998.

Burrows, W. D., and S. E. Renner, "Biological Warfare Agents as Threats to Potable Water, *Environ. Health Pers.* 107(12): 975–984, 1999.

Chick, S. E., J. S. Koopman, S. Soorapanth, and M. E. Brown, "Infection Transmission System Models for Microbial Risk Assessment," *The Science of the Total Environment*. 274: 197–207, 2001.

Cordesman, A. H., "Defending America: Asymmetric and Terrorist Attacks with Chemical Weapons," Tech. Rep. Center for Strategic and International Studies (CSIS), Washington, DC, 52 pp., 2001, available at: http://www.csis.org/burke/hd/reports/chemTerr010923.pdf [accessed 3 October 2002]

Couzin, J., "Active Poliovirus Baked from Scratch," *Science* 297: 174–175, 2002.

Dire, D. J., and T. McGovern, "CBRNE—Biological Warfare Agents," Tech. Rep. eMedicine: Instant Access to the Mind of Medicine, 34 pp., 2002, http://www.emedicine.com/emerg/topic853.htm [accessed 3 October 2002].

Eisenberg, J. N., E. Y. Seto, A. W. Olivieri, and R. C. Spear, "Quantifying Water Pathogen Risk in an Epidemiological Framework," *Risk Analysis* 16(4): 549–563, 1996.

Field, M. S., "Risk Assessment Methodology for Karst Aquifers: (2) Solute-Transport Modeling," *Environ. Monit. Assess.* 47, 23–37, 1997.

Field, M. S., "Ground-Water Tracing and Drainage Basin Delineation for Risk Assessment Mapping for Spring Protection in Clarke County, Virginia," Tech. Rep. NCEA-W-0936, U.S. Environmental Protection Agency, Washington, DC, 36 pp., 2000.

Field, M. S., "Efficient Hydrologic Tracer-Test Design for Tracer-Mass Estimation and Sample-Collection Frequency. 1. Method Development," *Environ. Geol.* 42(7): 827–838, 2002a.

Field, M. S., "Efficient Hydrologic Tracer-Test Design for Tracer-Mass Estimation and Sample-Collection Frequency. 2. Experimental Results," *Environ. Geol.* 42(7): 839–850, 2002b.

Field, M. S., "A Lexicon of Cave and Karst Terminology with Special Reference to Environmental Karst Hydrology," Tech. Rep. EPA/600/R-02/003, U.S. Envir. Prot. Agency, Washington, DC, 214 pp., 2002c.

Field, M. S., "The QTRACER2 Program for Tracer-Breakthrough Curve Analysis for Hydrological Tracer Tests," Tech. Rep. EPA/600/R-02/001, U.S. Envir. Prot. Agency, Washington, DC, 179 pp., 2002d.

Foran, J. A., and T. M. Brosnan, "Early Warning Systems for Hazardous Biological Agents in Potable Water," *Environ. Health Pers.* 108(10): 979–982, 2000.

Ford, T. E., "Biological Warfare Agents as Threats to Potable Water," *Environ. Health Pers. Supp.* 107(1): 191–206, 1999.

Geldreich, E., "The Worldwide Threat of Waterborne Pathogens," in G. F. Craun (ed.), *Water Quality in Latin America: Balancing the Microbial and Chemical Risks from Drinking Water Disinfection*, ILSI, Washington, pp. 19–43, 1996.

Gerba, C. P., J. B. Rose, C. N. Haas, and K. D. Crabtree, "Waterborne Rotavirus: A Risk Assessment," *Water Res.* 30(12): 2929–2940, 1996.

Haas, C. N., "Estimation of Risk Due to Low Doses of Microorganisms: A Comparison of Alternative Methodologies," *Am. J. Epidemiol.* 118: 573–582, 1983.

Haas, C. N., J. B. Rose, and C. P. Gerba, *Quantitative Microbial Risk Assessment*. Wiley, New York, 1999.

Harvey, R. W., L. H. George, R. L. Smith, and D. R. LeBlanc, "Transport of Microspheres and Indigenous Bacteria Through a Sandy Aquifer: Results of Natural- and Forced-Gradient Tracer Experiments," *Environ. Sci. Technol.* 25: 51–56, 1989.

Harvey, R. W., N. E. Kinner, D. MacDonald, D. W. Metget, and A. Bunn, "Role of Physical Heterogeneity in the Interpretation of Small-Scale Laboratory and Field Observations of Bacteria, Microbial-Sized Microspheres, and Bromide Transport Through Aquifer Sediments," *Water Resour. Res.* 29(8): 2713–2721, 1993.

Hazen, T. C., and G. A. Toranzos, "Tropical Source Water," in G. A. McFeters (ed.), *Assessing and Managing Health Risks from Drinking Water Contamination: Approaches and Applications*, Drinking Water Microbiology, New York, pp. 32–53, 1990.

Hickman, D. C., "A Chemical and Biological Warfare Threat: USAF Water Systems at Risk," Tech. Rep. U.S. Air Force Counterproliferation Center, Air War College, Air University, Maxwell Air Force Base, Ala., 28 pp., 1999, available at: http://www.au.af.mil/au/awc/awcgate/cpc-pubs/hickman.htm [accessed 7 June 2002].

Hoffman, R., R. Michel, E. N. Schmid, and K.-D. Muller, "Natural Infection with Microsporidium Organisms (KW19) in *vanella* spp. (Gymnamoebia) Isolated from a Domestic Tap-Water Supply," *Parasitol. Res.* 84: 164–166, 1998.

Jordan, M., "Mexicans Search for Lost Cyanide," *Washington Post* (May 28, 2002), A13.

Kilpatrick, F. A., and K. R. Taylor, "Generalization and Applications of Tracer Dispersion Data," *Water Resour. Bull.* 22(4): 537–548, 1986.

Lane, H. C., J. La Montagne, and A. S. Fauci, "Bioterrorism: A Clear and Present Danger," *Nature Med.* 7(12): 1271–1273, 2001.

MacKenzie, W. R., N. J. Hoxie, M. E. Proctor, M. S. Gradus, K. A. Blair, D. E. Peterson, J. J. Kazmierczak, D. G. Addiss, K. R. Fox, J. B. Rose, and J. P. Davis, A Massive Outbreak in Milwaukee of *Cryptosporidium* Infection Transmitted Through the Public Water Supply," *N. Engl. J. Med.* 331(3): 161–167, 1994.

Morris, R. D., and R. Levin, "Estimating the Incidence of Waterborne Infectious Disease Related to Drinking Water in the United States," in E. G. Reichard, and G. A. Zapponi (eds.), *Assessing and Managing Health Risks from Drinking Water Contamination: Approaches and Applications*, IAHS Publ. No. 233, International Association of Hydrological Sciences, Wallingford, U.K., pp. 75–88, 1995.

Mull, D. S., T. D. Liebermann, J. L. Smoot, and L. H. Woosley, Jr., "Application of Dye-Tracing Techniques for Determining Solute-Transport Characteristics of Ground Water in Karst Terranes," Tech. Rep. EPA/904/9-88-001, U.S. Envir. Prot. Agency, Region IV, Atlanta, GA, 103 pp., 1988.

NAP, "Guidelines for Chemical Warfare Agents in Military Field Drinking Water," National Academy Press, Washington, DC, 1995.

Nass, M., "In Search of the Anthrax Attacker: Following Valuable Clues," Tech. Rep., 8 pp., 2002, available at: http://www.redflagsweekly.com/nassanthrax3.html, redflagsweekly.com, [accessed 3 January 2003].

O'Connor, D. R., "Report of the Walkerton Inquiry: The Events of May 2000 and Related Issues," Tech. Rep. Part One, Ministry of the Attorney General, Toronto, ON, Canada, 504 pp., 2000.

Reed, C., "Security Measures on Tap," *Geotimes* 46(12): 26–28, 2001.

Reidl, J., and K. E. Klose, "*Vibrio cholerae* and Cholera: Out of the Water and into the Host," *FEMS Microbiol. Lett.* 741: 1–15, 2002.

Rose, J. A., and D. J. Grimes, "Reevaluation of Microbial Water Quality: Powerful New Tools for Detection and Risk Assessment," Tech. Rep. Amer. Acad. of Micro., Washington, DC, 19 pp., 2001.

Sánchez, J. L., and D. N. Taylor, "Cholera," *The Lancet* 349: 1825–1830, 1997.

Schindel, G. M., J. F. Quinlan, G. Davies, and J. A. Ray, "Guidelines for Wellhead and Springhead Protection Area Delineation in Carbonate Rocks," Tech. Rep. EPA/904/B-97/003, U.S. Envir. Prot. Agency, Washington, DC, 1997.

Takafuji, E. T., A. Johnson-Winegar, and R. Zajtchuk, "Medical Challenges in Chemical and Biological Defense for the 21st Century," in F. R. Sidell, E. T. Takafuji, and D. R. Franz (eds.), *Textbook of Military Medicine: Medical Aspects of Chemical and Biological Warfare*. Office of the Surgeon General, Department of the Army, United States of America, Washington, DC, pp. 677–685, 1997.

Taylor, K. R., R. W. J. James, and B. M. Helinsky, "Traveltime and Dispersion in the Shenandoah River and its Tributaries, Waynesboro, Virginia to Harpers Ferry, West Virginia," Tech. Rep. WRI 86-4065, U.S. Geological Survey, Towson, MD, 60 pp., 1986.

Toride, N., F. J. Leij, and M. T. van Genuchten, "The CXTFIT Code for Estimating Transport Parameters from the Laboratory or Field Tracer Experiments," Version 2.0. Tech. Rep. 137, U.S. Salinity Lab., Riverside, California, 121 pp., 1995.

U.S. EPA, "Guidelines for Delineation of Wellhead Protection Areas," Tech. Rep. EPA 4401/6-87-010, U.S. Envir. Prot. Agency, Washington, DC, 209 pp., 1987.

U.S. EPA, "Exposure Factors Handbook," Tech. Rep. EPA/600/8-89/043, U.S. Environ. Protect. Agency, Washington, DC, 1989a.

U.S. EPA, "Risk Assessment Guidance for Superfund: Human Health Evaluation Manual, Part A (Interim Final)," Tech. Rep. EPA 9285.701A, U.S. Environ. Protect. Agency, Washington, DC, 1989b.

U.S. EPA, "Dermal Exposure Assessment: Principles and Applications (Interim Report)," Tech. Rep. EPA/600/8-91/011B, U.S. Environ. Protect. Agency, Washington, DC, 1992a.

U.S. EPA, "Integrated Risk Information (IRIS)," Version 1.0. Tech. Rep., U.S. Environ. Protect. Agency, Washington, DC, 1992b.

U.S. EPA, "State Source Water Assessment and Protection Programs," Tech. Rep. EPA 816-R-97-009, U.S. Environ. Protect. Agency, Washington, DC, 1997.

U.S. EPA, "Protecting the Nation's Water Supplies from Terrorist Attack: Frequently Asked Questions," Tech. Rep., U.S. Environ. Protect. Agency, Washington, DC, 2001, available at: http://www.epa.gov/safewater/security/ secqanda.html [accessed 4 March 2002].

Wai, S. N., Y. Mizunoe, and S. Yoshida, "How *Vibrio cholerae* Survive During Starvation," *FEMS Microbiol. Lett.* 180: 123–131, 1999.

Weber, J. T., W. C. Levine, D. P. Hopkins, and R. V. Tauxe, "Risks at Home and Abroad," *Arch. Intern. Med.* 154: 551–556, 1994.

Weiss, R., and J. Warrick, "Army Working Weapons-Grade Anthrax," Washington Post (December 13, 2001), A16, 2001.

WHO, "Guidelines for Drinking-Water Quality: Recommendations," World Health Organization, Geneva, 1999.

WHO, "Public Health Response to Biological and Chemical Weapons," World Health Organization, Geneva, 2001.

Zanders, J. P., "Assessing the Risk of Chemical and Biological Weapons Proliferation to Terrorists," *Nonprolif. Rev.* Fall, 17–34, 1999.

CHAPTER 7
METHODOLOGIES FOR RELIABILITY ANALYSIS

Yeou-Koung Tung
Hong Kong University of Science and Technology
Clear Water Bay, Kowloon, Hong Kong

Larry W. Mays
Arizona State University, Tempe, Arizona

7.1 INTRODUCTION

There is a need for the reliability analysis of water supply/water distribution systems. The ideas of using event/fault tree analysis for this purpose were discussed in Mays (1989). Unfortunately, these methods have not been implemented by water utilities for reliability assessment. The purpose of this section is to reintroduce some very valuable methodologies that can be applied in the reliability computations for vulnerability assessment.

A formal quantitative reliability analysis for an engineering system involves a number of procedures as illustrated in Fig. 7.1. First, the system domain is defined, the type of system is identified, and the conditions involved in the problem are defined. Second, the kind of failure is identified and defined. Third, factors that contribute to the working and failure of the system are identified. Fourth, uncertainty analysis is performed for each of the contributing component factors or subsystems. Fifth, based on the characteristics of the system and the nature of the failure, a logic tree is selected to relate the failure modes and paths involving different components or subsystems. Fault tree, event tree, and decision tree are the logic trees often used. Sixth, appropriate method or methods that can combine the components or subsystems, following the logic of the tree to facilitate computation of system reliability, are identified and selected. Seventh, computation following the methods selected in the sixth step is performed to determine the system failure probability and reliability. Eighth, if the risk-cost associated with the system failure is desired and the failure damage cost function is known or can be determined, it can be combined with the system failure probability function determined in step 7 to yield the expected risk cost.

Water supply systems are often so large and complex that teams of experts of different disciplines are required for performing the reliability analysis and computation. The logic trees are tools that permit division of team works and subsequent integration for the system results.

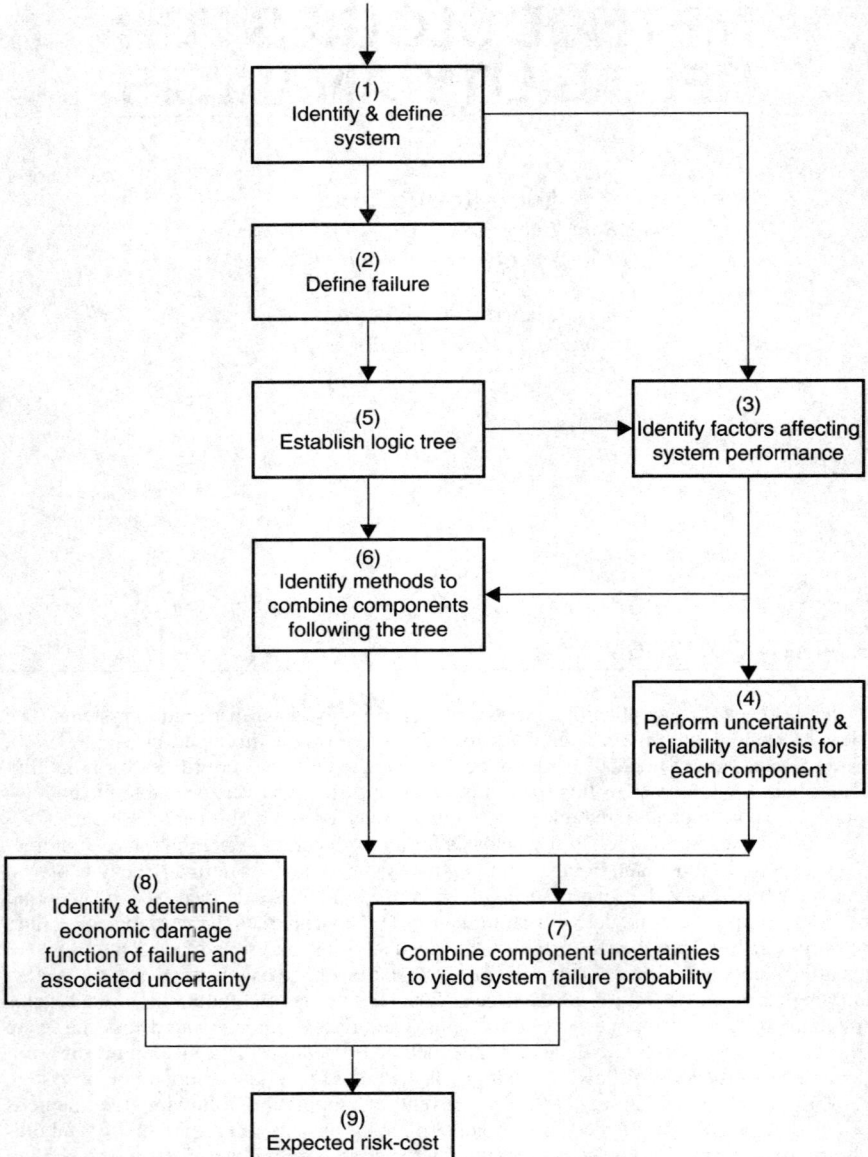

FIGURE 7.1 Procedure for infrastructural engineering system reliability assessment.

Event/fault tree calculations have been used in the chemical/petroleum and nuclear industries for some time (Fullwood and Hall, 1988; Peplow et al., 2003; Sulfredge et al., 2003). Software tools used for event/fault tree analysis typically rely on cut set approaches. These packages, designed for highly reliable systems, base on the low failure probabilities nature of the problem to use several approximations that greatly speed up the calculations. One example is the Visual Interactive Site Analysis Code (VISAC), developed at Oak Ridge National Laboratory (ORNL), which uses a geometric model of a facility, coupled with an event/fault tree model of plant systems, to evaluate the vulnerability of a nuclear power plant as well as other infrastructure objects. The event/fault tree models, associated with facility vulnerability calculations, typically involve systems with high-component failure probabilities resulting from an attack scenario. Such methodologies could be employed for water distributions systems. VISAC is a Java-based graphical user interface (GUI) that can analyze a variety of accidents/incidents at nuclear or industrial facilities ranging from simple component sabotage to an attack with military or terrorist weapons. A list of damaged components from a scenario is then propagated through a set of event/fault trees to determine the overall facility failure probability, the probability of an accompanying radiological release, and the expected facility downtime.

7.2 GENERAL VIEW OF SYSTEM RELIABILITY COMPUTATION

As mentioned previously, the reliability of a system depends on the component reliabilities, as well as interaction and configuration of components. Consequently, computation of system reliability requires knowing what constitutes the system being in a failed or satisfactory state. Such knowledge is essential for system classification and dictates the methodology to be used for system reliability determination.

7.2.1 Classification of Systems

From the reliability computation viewpoint, the classification of the system primarily depends on how the system performance is affected by its components or modes of operation. A multiple-component system called a *series system* requires that all of its components must perform satisfactorily to allow satisfactory performance of the entire system. Similarly, for a single-component system involving several modes of operation, it is also viewed as a series system if satisfactory performance of the system requires satisfactory performance of all its different modes of operation.

A second basic type of system is called the *parallel system*. A parallel system is characterized by the property that the system would serve its intended purpose satisfactorily as long as at least one of its components or modes of operation perform satisfactorily.

For most real-life water supply systems, system configurations are complex in which the components are arranged as a mixture in series–parallel or in the form of a loop. In dealing with the reliability analysis of a complex system, the general approach is to reduce the system configuration, based on the arrangement of its components or modes of operation, to a simpler situation for which the reliability analysis can be performed easily. However, this goal may not always be achievable for which a special procedure would have to be devised.

7.2.2 Basic Probability Rules for System Reliability

The solution approaches to system reliability problems can broadly be classified into the *failure-mode approach* and the *survival-mode approach* (Bennett and Ang, 1983).

The failure-mode approach is based on the identification of all possible failure modes for the system, whereas the survival-mode approach is based on all possible modes of operation under which the system will be operational. The two approaches are complementary. Depending on the operational characteristics and configuration of the system, a proper choice of one of the two approaches can often lead to significant simplification in reliability computation.

Consider that a system has M components or modes of operation. Let event F_i represent that the ith component or mode of operation is in the failure state. If the system is a series system, the failure probability of the system is the probability that at least one of the M components or modes of operation fails, namely,

$$p_{f,\text{sys}} = P(F_1 \cup F_2 \cup \cdots \cup F_M) = P\left(\bigcup_{i=1}^{M} F_i\right) \tag{7.1}$$

in which $p_{f,\text{sys}}$ is the failure probability of the system. On the other hand, the system reliability, $p_{s,\text{sys}}$, is the probability that all its components or modes of operation perform satisfactorily,

$$p_{s,\text{sys}} = P(F_1' \cap F_2' \cap \cdots \cap F_M') = P\left(\bigcap_{i=1}^{M} F_i'\right) \tag{7.2}$$

in which F_i' is the complementary event of F_i representing that the ith component or mode of operation does not fail.

In general, failure events associated with system components or modes of operation are not mutually exclusive. Therefore, the failure probability for a series system can be computed as

$$p_{f,\text{sys}} = P\left(\bigcup_{i=1}^{M} F_i\right) = \sum_{i=1}^{M} P(F_i) - \sum\sum_{i \neq j} P(F_i, F_j) + \sum\sum\sum_{i \neq j \neq k} P(F_i, F_j, F_k) - \cdots$$
$$+ (-1)^M P(F_1, F_2, \ldots, F_M) \tag{7.3}$$

Hence, the reliability for a series system is

$$p_{s,\text{sys}} = P(F_1') \times P(F_2' \mid F_1') \times P(F_3' \mid F_1', F_2') \times \cdots \times P(F_M' \mid F_1', F_2', \ldots, F_{M-1}') \tag{7.4}$$

In the case that failure events are mutually exclusive, or the probability of joint occurrence of multiple failures is negligible, the failure probability of a series system can be easily obtained as

$$p_{f,\text{sys}} = \sum_{i=1}^{M} P(F_i) \tag{7.5a}$$

with the corresponding system reliability

$$p_{s,\text{sys}} = 1 - p_{f,\text{sys}} = 1 - \sum_{i=1}^{M} P(F_i) \tag{7.5b}$$

Under the condition that all failure events are statistically independent, the reliability of a series system can be computed as

$$p_{s,\text{sys}} = \prod_{i=1}^{M} P(F_i') = \prod_{i=1}^{M} [1 - P(F_i)] \tag{7.6a}$$

with the corresponding system failure probability

$$p_{f,\text{sys}} = 1 - \prod_{i=1}^{M} [1 - P(F_i)] \tag{7.6b}$$

The component failure probability $P(F_i)$ can be determined by an appropriate method that accounts for the involved factors affecting the performance of the component.

For a parallel system, the system would fail if all its components or modes of operation failed. Hence, the failure probability of a parallel system is

$$p_{f,\text{sys}} = P(F_1 \cap F_2 \cap \cdots \cap F_M) = P\left(\bigcap_{i=1}^{M} F_i\right) \tag{7.7}$$

The reliability of a parallel system, on the other hand, is the probability that at least one of its components or modes of operation is functioning, that is,

$$p_{s,\text{sys}} = P(F_1' \cup F_2' \cup \cdots \cup F_M') = P\left(\bigcup_{i=1}^{M} F_i'\right) \tag{7.8}$$

Hence, under the condition of independence for all failure events, the failure probability of a parallel system simply is

$$p_{f,\text{sys}} = \prod_{i=1}^{M} P(F_i) \tag{7.9a}$$

with the corresponding system reliability being

$$p_{s,\text{sys}} = 1 - \prod_{i=1}^{M} P(F_i) = 1 - \prod_{i=1}^{M} [1 - P(F_i')] \tag{7.9b}$$

For mutually exclusive failure events, reliability of a parallel system can be computed as

$$p_{s,\text{sys}} = \sum_{i=1}^{M} P(F_i') = M - \sum_{i=1}^{M} P(F_i) \tag{7.10a}$$

with the corresponding system failure probability

$$p_{f,\text{sys}} = 1 + \sum_{i=1}^{M} P(F_i) - M \tag{7.10b}$$

Unfortunately, for a real-life system involving multiple components or modes of operation, the corresponding failure events are neither independent nor mutually exclusive.

Consequently, the computations of exact values of system reliability and failure probability would not be a straightforward task as the number of system components or the system complexity increases. In practical engineering applications, bounds on system reliability are computed based on simpler expressions with less computational effort. As can be seen in the next subsection, to achieve tighter bounds on system reliability or failure probability, a more elaborate computation would be required. Of course, the required precision on the computed system reliability is largely dependent on the importance of the satisfactory performance of the system under consideration.

7.2.3 Bounds for System Reliability

Despite the system under consideration being a series or parallel system, the evaluation of system reliability or failure probability involves probabilities of union or intersection of multiple events. Knowing the bounds of system failure probability, that is,

$$\underline{p}_{f,\text{sys}} \leq p_{f,\text{sys}} \leq \overline{p}_{f,\text{sys}} \tag{7.11a}$$

with $\underline{p}_{f,\text{sys}}$ and $\overline{p}_{f,\text{sys}}$ being the lower and upper bounds of system failure probability, respectively, the corresponding bounds for system reliability can be obtained as

$$1 - \overline{p}_{f,\text{sys}} = \underline{p}_{s,\text{sys}} \leq p_{s,\text{sys}} \leq \overline{p}_{s,\text{sys}} = 1 - \underline{p}_{f,\text{sys}} \tag{7.11b}$$

Similarly, after the bounds on system reliability are obtained, the bounds on system failure reliability can be easily computed.

First-Order (or Unimodal) Bounds. It has been shown by Ang and Tang (1984) that the first-order bounds on the probability of joint occurrence of several positively correlated events are

$$\prod_{i=1}^{M} P(A_i) \leq P\left(\bigcap_{i=1}^{M} A_i\right) \leq \min_{j}\{P(A_j)\} \tag{7.12}$$

Hence, the first-order bounds for reliability of a series system with positively correlated nonfailure events can be computed as

$$\prod_{i=1}^{M} P(F_i') \leq p_{s,\text{sys}} \leq \min_{j}\{P(F_i')\} \tag{7.13a}$$

or in terms of failure probability as

$$\prod_{i=1}^{M} [1 - P(F_i)] \leq p_{s,\text{sys}} \leq \min_{j}\{1 - P(F_j)\} \tag{7.13b}$$

Similarly, by letting $A_i = F_i$, the first-order bounds on the failure probability of a parallel system with positively correlated failure events can be immediately obtained as

$$\prod_{i=1}^{M} P(F_j) \leq p_{f,\text{sys}} \leq \min_{j}\{P(F_j)\} \tag{7.14a}$$

and the corresponding bounds for the system reliability are

$$1 - \min_{j}[P(F_j)] \leq p_{s,\text{sys}} \leq 1 - \left[\prod_{i=1}^{M} P(F_j)\right] \tag{7.14b}$$

In the case that all events A_i's are negatively correlated, the first-order bounds for the probability of joint occurrence of several negatively correlated events are

$$0 \leq P\left(\bigcap_{i=1}^{M} A_i\right) \leq \prod_{i=1}^{M} P(A_i)$$

The bounds for reliability of a series system, with $A_i = F_i'$, containing negatively correlated events are

$$0 \leq p_{s,\text{sys}} \leq \prod_{i=1}^{M} [1 - P(F_i)] \tag{7.15}$$

whereas for a parallel system, with $A_i = F_i$, are

$$1 - \left[\prod_{i=1}^{M} P(F_i)\right] \leq p_{s,\text{sys}} \leq 1 \tag{7.16}$$

It should be pointed out that the first-order bounds for system reliability may be too wide to be meaningful. Under such circumstances, tighter bounds might be required which can be obtained at the expense of more computations.

Second-Order (or Bi-Modal) Bounds. The second-order bounds are obtained by retaining the terms involving joint probability of two events. As the probability of union of several events is

$$P\left(\bigcup_{i=1}^{M} A_i\right) = \sum_{i=1}^{M} P(A_i) - \sum\sum_{i \neq j} P(A_i, A_j) + \sum\sum\sum_{i \neq j \neq k} P(A_i, A_j, A_k) - \cdots$$
$$+ (-1)^M P(A_1, A_2, \ldots, A_M) \tag{7.17}$$

Probability computation using partial terms of the right-hand-side provides only an approximation to the exact probability of union. Notice the alternating signs in Eq. (7.17) as the order of the terms increases. It is evident that the inclusion of only the first-order terms, that is, $P(A_i)$, produces an upper bound for $P(A_1 \cup A_2 \cup \cdots \cup A_M)$. The consideration of only the first two-order terms yields a lower bound, the first three-order terms again an upper bound, and so on (Melcher, 1987).

A simple bound for the probability of a union is

$$\sum_{i=1}^{M} P(A_i) - \sum_{i}\sum_{j} P(A_i, A_j) \leq P\left(\bigcup_{i=1}^{M} A_i\right) \leq \min\left[1, \sum_{i=1}^{M} P(A_i)\right] \tag{7.18}$$

It should be pointed out that the above bounds produce adequate results only when the values of $P(A_i)$ and $P(A_i, A_j)$ are small. Ditlevsen (1979) developed a better lower bound as

$$P(A_1) + \sum_{i=2}^{M} \max\left\{ \left[P(A_i) - \sum_{j=1}^{i-1} P(A_j, A_i) \right], 0 \right\} \leq P\left(\bigcup_{i=1}^{M} A_i \right) \qquad (7.19)$$

Notice that the above lower bound for the probability of a union depends on the order in which the events are labeled. A useful rule of thumb is to order the event in the order of decreasing importance (Melchers, 1987). In other words, events are ordered such that $P(A_{[1]}) > P(A_{[2]}) > \cdots > P(A_{[M]})$ with $[i]$ representing the rank of event according to its probability of occurrence.

As for the upper bound, it can be derived as

$$P\left(\bigcup_{i=1}^{M} A_i \right) \leq \min\left\{ 1, \sum_{i=1}^{M} P(A_i) - \sum_{i=2}^{M} \max_{j<i}[P(A_j, A_i)] \right\} \qquad (7.20)$$

Again, this upper bound value is also dependent on the order of events.

As can be seen, the computation of second-order bound for the probability of a union requires determination of the joint probability for combinations of all possible pairs of events involved. The second-order bounds for the failure probability of a series system can be obtained as

$$p_{f,\text{sys}} \leq \sum_{i=1}^{M} P(F_i) - \sum_{i=2}^{M} \max_{j<i}[P(F_j, F_i)]$$

$$\geq P(F_1) + \sum_{i=2}^{M} \max\left\{ \left[P(F_i) - \sum_{j=1}^{i-1} P(F_j, F_i) \right], 0 \right\} \qquad (7.21)$$

Similarly, the second-order bounds for the reliability of a parallel system are

$$p_{f,\text{sys}} \leq \sum_{i=1}^{M} P(F_i') - \sum_{i=2}^{M} \max_{j<i}[P(F_j', F_i')]$$

$$\geq P(F_1') + \sum_{i=2}^{M} \max\left\{ \left[P(F_i') - \sum_{j=1}^{i-1} P(F_j', F_i') \right], 0 \right\} \qquad (7.22)$$

As will be seen in the next section, the evaluation of system reliability, by any method employed, involves the determination of probability of union or intersection. For a real-life system, in general, the task of determining the exact probability of union or intersection involving large number of events is not trivial. The use of an approximation, such as that described in this section, may be needed. However, one should be cautioned that the above approximations, especially the first-order approximations, are suitable for problems in which the probability of individual event is small and that of joint occurrence of events is negligible. This condition may not be satisfied for an

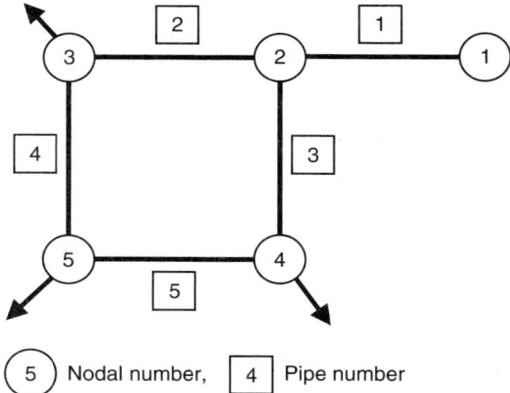

FIGURE 7.2 Example water distribution network.

unreliable system whose component failure probabilities are high and this would be the case when considering the water supply system is under terrorist attacks or deliberate sabotage.

7.3 RELIABILITY ASSESSMENT METHODS

7.3.1 State Enumeration Method

The *state enumeration method* lists all possible mutually exclusive states of the system components that define the state of the entire system. In general, for a system containing M components, each of which can be classified into N operating states, there will be N^M possible states for the entire system. For example, if the state of each of the M components is classified into failed and operating states, the system has 2^M possible states.

Once all the possible system states are enumerated, the states that result in successful system operation are identified and the probability of the occurrence of each successful state is computed. The last step is to sum all of the successful state probabilities, which yield the system reliability.

Consider the simple water distribution network in Fig. 7.2. The tree diagram such as Fig. 7.3 is called an *event tree* and the analysis involving the construction of an event tree is referred to as *event-tree analysis*. As can be seen, an event tree simulates not only the topology of a system, but also, more importantly, the sequential or chronological operation of the system.

Example 7.1 Consider a simple water distribution network consisting of five pipes and one loop, as shown in Fig. 7.2. Node 1 is the source node and nodes 3, 4, and 5 are demand nodes. The components of this network subject to possible failure are the five pipe sections. Within a given time period, each pipe section has an identical failure probability of 5 percent due to breakage or other causes that require it to be removed from service. System reliability is defined as the probability that water can reach all three demand nodes from the source. Furthermore, it is assumed that the states of serviceability of each pipe are independent.

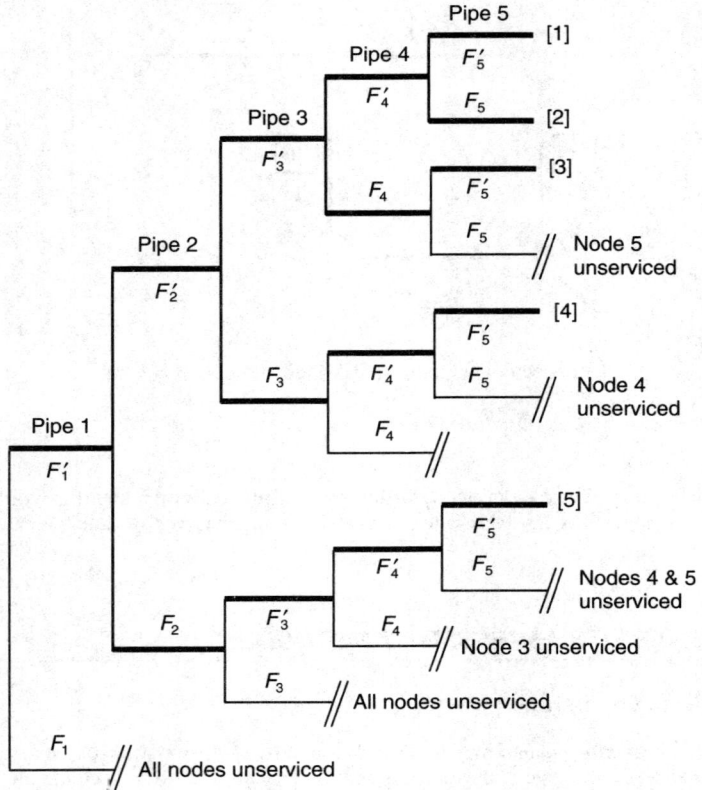

F'_i = Nonfailure of pipe i
F_i = Failure of pipe i
// = Branch associated with unserviceability of one or more node

FIGURE 7.3 Event tree for the reliability of example water distribution network.

Solution Using the state enumeration method for system reliability evaluation, the event tree, as shown in Fig. 7.3, can be constructed to depict all possible combinations of component states in the system. Since each pipe has two possible states, that is, failure (F) or nonfailure (F'), the tree, if fully expanded, would have $2^5 = 32$ branches. However, knowing the role that each pipe component plays in the network connectivity, exhaustive enumeration of all possible states is not necessary.

For example, referring to Fig. 7.3, one realizes that when pipe 1 fails, all demand nodes cannot receive water, indicating a system failure, regardless of the state of the remaining pipe sections. Therefore, branches in the event tree beyond this point do not have to be constructed. Applying some judgment in event tree construction in this fashion can generally lead to a smaller tree. However, for a complex system, this may not be a trivial task.

The system reliability can be obtained by summing up the probabilities associated with all of the nonfailure branches. In this example, there are five branches, as indicated by the heavy lines in the tree, for which all users can have the water delivered by the system. Therefore, the system reliability is

$$p_{s,\text{sys}} = P\left[\bigcup_{i=1}^{5} B_{[i]}\right] = \sum_{i=1}^{5} P(B_{[i]})$$

where $P(B_{[i]})$ = the probability that the branch $B_{[i]}$ of the event tree provides full service to all users. The probability associated with each branch resulting in satisfactory delivery of water to all users can be calculated, due to independence of serviceability of individual pipe, as the following:

$$P(B_{[1]}) = P(F_1') P(F_2') P(F_3') P(F_4') P(F_5')$$
$$= (0.95)(0.95)(0.95)(0.95)(0.95) = 0.77378$$
$$P(B_{[2]}) = P(F_1') P(F_2') P(F_3') P(F_4') P(F_5)$$
$$= (0.95)(0.95)(0.95)(0.95)(0.05) = 0.04073$$
$$P(B_{[3]}) = P(F_1') P(F_2') P(F_3') P(F_4) P(F_5')$$
$$= (0.95)(0.95)(0.95)(0.05)(0.95) = 0.04073$$
$$P(B_{[4]}) = P(F_1') P(F_2') P(F_3) P(F_4') P(F_5')$$
$$= (0.95)(0.95)(0.05)(0.95)(0.95) = 0.04073$$
$$P(B_{[5]}) = P(F_1') P(F_2) P(F_3') P(F_4') P(F_5')$$
$$= (0.95)(0.05)(0.95)(0.95)(0.95) = 0.04073$$

Therefore, the system reliability is the sum of above five probabilities associated with the operating state of the system, which is

$$p_{s,\text{sys}} = 0.77378 + 4(0.04073) = 0.93668.$$

7.3.2 Path Enumeration Method

This is a very powerful method for system reliability evaluation. A *path* is defined as a set of components or modes of operation, which lead to a certain outcome of the system. In system reliability analysis, the system outcomes of interest are those of failed state or operational state. A *minimum path* is one in which no component is traversed more than once in going along the path. Under this methodological category, *tie-set analysis* and *cut-set analysis* are the two well-known techniques.

Cut-Set Analysis. A *cut set* is defined as a set of system components or modes of operation which, when failed, cause the failure of the system. It is powerful for evaluating system reliability for two reasons: (1) it can be easily programmed on digital computers for fast and efficient solutions of any general system configuration, especially in the form of a network, and (2) cut sets are directly related to the modes of system failure and therefore, identify the distinct and discrete ways in which a system may fail. For example, in a water distribution system, a cut set will be the set of system components including pipe sections, pumps, storage facilities, etc., which when failed jointly, would disrupt the service to certain users.

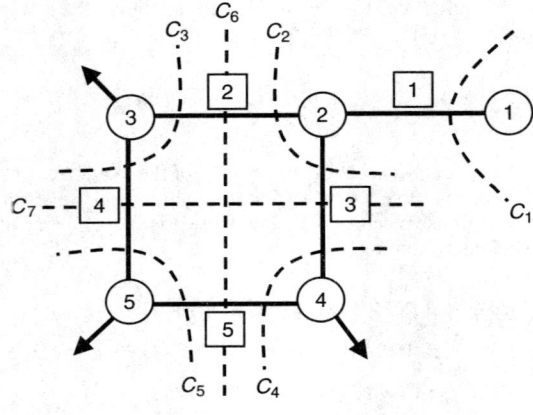

5 Nodal number, 4 Pipe number, C_i = ith cut set

FIGURE 7.4 Minimum cut sets for the example water distribution network.

The cut set method utilizes *minimum cut sets* for calculating the system failure probability. A minimum cut set is a set of system components which, when all failed, causes failure of the system, but when any one component of the set does not fail, does not cause system failure. A minimum cut set implies that all components of the cut set must be in the failure state to cause system failure. Therefore, the components or modes of operation involved in a minimum cut set are effectively connected in parallel and each minimum cut set is connected in series. Consequently, the failure probability of a system can be expressed as

$$P_{f,\text{sys}} = P\left[\bigcup_{i=1}^{I} C_i\right] = P\left[\bigcup_{i=1}^{I}\left(\bigcap_{j=1}^{J_i} F_{ij}\right)\right] \quad (7.23)$$

in which C_i is the ith of the total I minimum cut sets; J_i is the total number of components or modes of operation in the ith minimum cut set; and F_{ij} represents the failure event associated with the jth components or mode of operation in the ith minimum cut set. In the case that the number of minimum cut sets I is large, computing the bounds for probability of a union can be applied. The bounds on failure probability of the system should be examined for their closeness to ensure adequate precision is obtained. Figure 7.4 shows the cut sets for the simple water distribution system.

EXAMPLE 7.2 Refer to the simple water distribution network (shown in Fig. 7.2) in Example 7.1. Evaluate the system reliability using the minimum cut set method.

Solution Based on the system reliability, as defined, seven minimum cut-sets for the example pipe network are identified:

$C_1: F_1$ $\qquad$ $C_2: F_2 \cap F_3$ $\qquad$ $C_3: F_2 \cap F_4$ $\qquad$ $C_4: F_3 \cap F_5$

$C_5: F_4 \cap F_5$ $\qquad$ $C_6: F_2 \cap F_5$ $\qquad$ $C_7: F_3 \cap F_4$

where C_i = the ith cut set and F_k = the failure state of pipe link k. The above seven cut-sets for the example network are shown in Fig. 7.4. The *system unreliability* $p_{f,\text{sys}}$ is the probability of occurrence of the union of the cut-set, that is,

$$p_{f,\text{sys}} = P\left[\bigcup_{i=1}^{7} C_i\right]$$

The system reliability can be obtained by subtracting $p_{f,\text{sys}}$ from 1. However, the computation, in general, will be very cumbersome for finding the probability of the union of large numbers of events, even if they are independent. In this circumstance, it is computationally easier to compute the system reliability as

$$p_{s,\text{sys}} = 1 - P\left[\bigcup_{i=1}^{7} C_i\right] = P\left[\bigcap_{i=1}^{7} C_i'\right]$$

Since all cut sets behave independently, all their complements also behave independently. The probability of the intersection of a number of independent events is

$$p_{s,\text{sys}} = P\left[\bigcap_{i=1}^{7} C_i'\right] = \prod_{i=1}^{7} P(C_i')$$

where

$$P(C_1') = 0.95, \quad P(C_2') = P(C_3') = \cdots = P(C_7') = 0.9975$$

Hence, the system reliability of the example water distribution network is

$$p_{s,\text{sys}} = (0.95)(0.9975)^6 = 0.9360$$

Tie-Set Analysis. As the complement of cut set, a *tie set* is a minimal path of the system in which system components or modes of operation are arranged in series. Consequently, a tie set fails if any of its components or modes of operation fail. All tie sets are effectively connected in parallel, that is, the system will be in the operating state if any of its tie sets are functioning. Therefore, the system reliability can be expressed as

$$p_{s,\text{sys}} = P\left[\bigcup_{i=1}^{I} T_i\right] = P\left[\bigcup_{i=1}^{I}\left(\bigcap_{j=1}^{J_i} F_{ij}'\right)\right] \tag{7.24}$$

in which T_i is the ith tie set of all I tie sets; J_i is the total number of components or modes of operation in the ith tie set; and F_{ij}' represents the nonfailure state of the jth component in the ith tie set. Again, when the number of tie sets is large, computation of exact system reliability by Eq. (7.23) could be computationally cumbersome. In such a condition, bounds for system reliability could be computed.

The main disadvantage of the tie-set method is that failure modes are not directly identified. Direct identification of failure modes sometimes is essential if a limited amount of a resource is available to focus on a few dominant failure modes. Figure 7.5 shows the tie sets for the simple water distribution system.

1. Find all minimum paths. In general, this has to be done with the aid of computer when the number of components is large and the system configuration is complex.
2. Find all required unions of the paths.

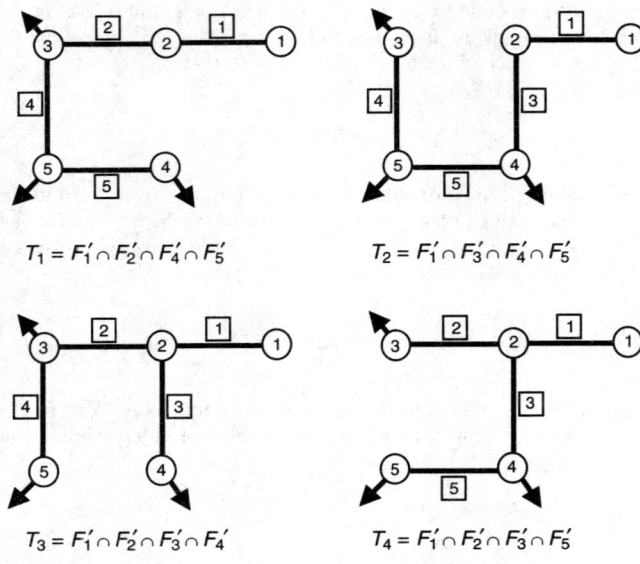

T_i = the ith tie set
F'_k = Nonfailure of pipe section k

FIGURE 7.5 Minimum tie sets for the example water distribution network.

3. Give each path union a reliability expression in terms of module reliability.
4. Compute the system reliability in terms of module reliabilities.

EXAMPLE 7.3 Refer to the simple water distribution network as shown in Fig. 7.2. Use tie-set analysis to evaluate the system reliability.

Solution The minimum tie sets (or path), based on the definition of system reliability given previously, for the example network are

$$T_1: F'_1 \cap F'_2 \cap F'_4 \cap F'_5;$$
$$T_2: F'_1 \cap F'_3 \cap F'_4 \cap F'_5;$$
$$T_3: F'_1 \cap F'_2 \cap F'_3 \cap F'_4;$$
$$T_4: F'_1 \cap F'_2 \cap F'_3 \cap F'_5.$$

where T_i = the ith minimum tie set and F'_k = the nonfailure of the kth pipe link in the network. The four minimum tie sets are shown in Fig. 7.5. The system reliability is

$$\begin{aligned}p_{s,\text{sys}} &= P(T_1 \cup T_2 \cup T_3 \cup T_4) \\ &= [P(T_1) + P(T_2) + P(T_3) + P(T_4)] \\ &\quad - [P(T_1,T_2) + P(T_1,T_3) + P(T_1,T_4) + P(T_2,T_3) + P(T_2,T_4) + P(T_3,T_4)] \\ &\quad + [P(T_1,T_2,T_3) + P(T_1,T_2,T_4) + P(T_1,T_3,T_4) + P(T_2,T_3,T_4)] \\ &\quad - P(T_1,T_2,T_3,T_4)\end{aligned}$$

METHODOLOGIES FOR RELIABILITY ANALYSIS

Since all pipes in the network behave independently, all minimum tie sets behave independently. In such circumstances, the probability of the joint occurrence of multiple independent events is simply equal to the multiplication of the probability of the individual event. That is,

$$P(T_1) = P(F_1')\,P(F_2')\,P(F_4')\,P(F_5') = (0.95)^4 = 0.81451$$

Similarly,

$$P(T_2) = P(T_3) = P(T_4) = 0.81451$$

Note that, in this example, the intersections of more than two minimum tie sets are the intersections of the nonfailure state of all five pipe sections. For example,

$$T_1 \cup T_2 = (F_1' \cap F_2' \cap F_4' \cap F_5') \cup (F_1' \cap F_3' \cap F_4' \cap F_5') = (F_1' \cap F_2' \cap F_3' \cap F_4' \cap F_5').$$

The system reliability can be reduced to

$$\begin{aligned} p_{s,\text{sys}} &= [P(T_1) + P(T_2) + P(T_3) + P(T_4)] - 3\,P(F_1' \cap F_2' \cap F_3' \cap F_4') \\ &= 4(0.81451) - 3\,(0.95)^5 \\ &= 0.9367 \end{aligned}$$

In summary, the path enumeration method involves the following steps (Henley and Gandhi, 1975):

7.3.3 Conditional Probability Approach

The approach starts with a selection of key components and modes of operation whose states (operational or failure) would decompose the entire system into simple series and/or parallel subsystems for which the reliability or failure probability of subsystems can be easily evaluated. Then, the reliability of the entire system is obtained by combining those of the subsystems using the conditional probability rule as:

$$p_{s,\text{sys}} = p_{s|F_i'} \times p_{s,i} + p_{s|F_i} \times p_{f,i} \tag{7.25}$$

in which $p_{s|F_i'}$ and $p_{s|F_i}$ are the conditional system reliability given that the ith component is operational, F_i', and failed, F_i, respectively; and $p_{s,i}$ and $p_{f,i}$ are the reliability and failure probabilities of the ith component, respectively.

Except for very simple and small systems, a nested conditional probability operation is inevitable. Efficient evaluation of system reliability of a complex system hinges largely on a proper selection of key components, which generally is a difficult task when the scale of the system is large. Furthermore, the method cannot be easily adapted to computerization for problem solving. Figure 7.6 illustrates the conditional probability method for the simple water distribution system.

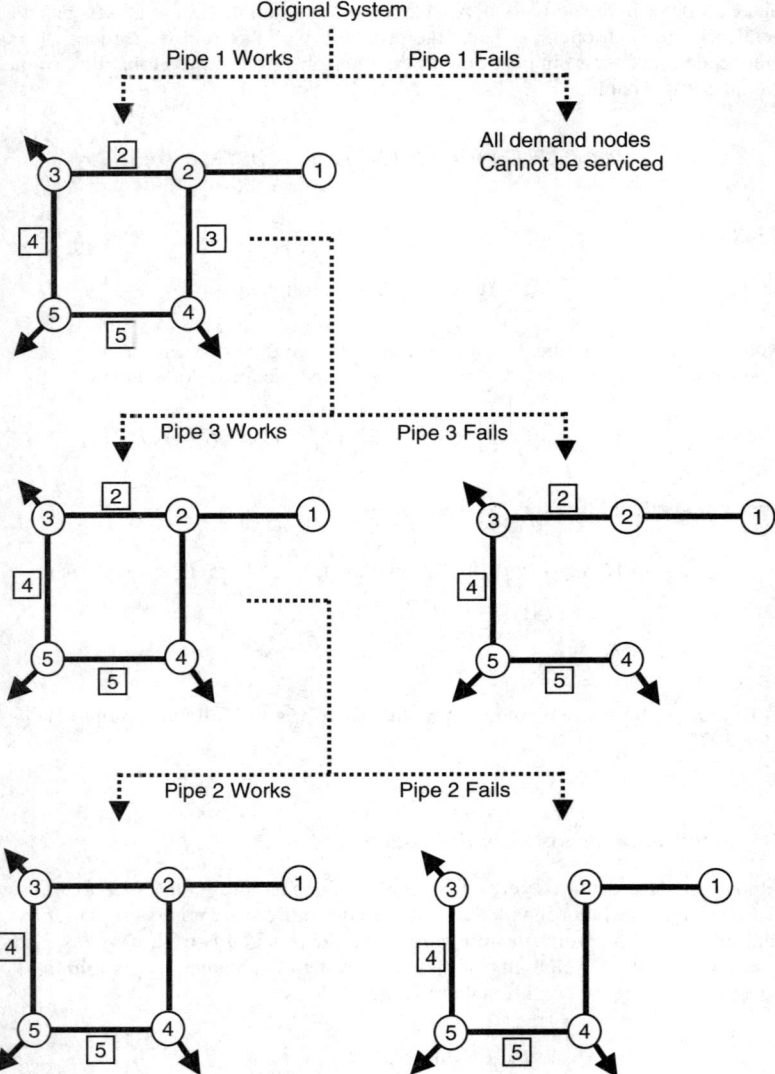

FIGURE 7.6 Decomposition of the example water distribution network.

Example 7.4 Find the system reliability of the water distribution network in Fig. 7.2 using conditional probability approach.

Solution Using the conditional probability approach for system reliability evaluation, first select pipe section 1 as the key element, which decomposes the system into a simpler configuration, as shown in Fig. 7.6. After the entire system is decomposed into a simple system configuration, the conditional probability of the decomposed systems can be easily evaluated. For example, the conditional system reliability, after imposing F'_1 and F_3 for pipes 1 and 3, respectively, can be expressed as

$$p_{s,\text{sys} \mid F'_1, F_3} = P(F'_2 \cap F'_4 \cap F'_5) = (0.95)^3 = 0.8574$$

where $p_{\text{sys} \mid F_2, F_3}$ = conditional system reliability. Conditional system reliabilities for other imposed conditions are shown in Fig. 7.6. After the conditional system reliabilities for the decomposed systems are calculated, the system reliability is computed as

$$p_{s,\text{sys}} = p_{s,\text{sys} \mid F'_1, F_3} \times P(F'_1, F_3) + p_{s,\text{sys} \mid F'_1, F'_3, F'_2} \times P(F'_1, F'_3, F'_2)$$
$$+ p_{s,\text{sys} \mid F'_1, F'_3, F_2} \times P(F'_1, F'_3, F_2)$$
$$= (0.8574)(0.95)(0.05) + (0.9975)(0.95)^3 + (0.9025)(0.95)^2(0.05)$$
$$= 0.9367$$

7.3.4 Fault-Tree Analysis

The major objective of fault tree construction is to represent the system condition, which may cause system failure, in a symbolic manner. In other words, the fault tree consists of sequences of events that lead to system failure. Fault-tree analysis, unlike event-tree analysis, is a backward analysis, which begins with a system failure and traces backward, searching for possible causes of the failure. Fault tree analysis was initiated at Bell Telephone Laboratories and Boeing Aircraft Company (Barlow et al., 1975). Since then it has been used for evaluating the reliability of many different engineering systems. In hydrosystem engineering design, fault-tree analysis has been applied to evaluate the risk and reliability of earth dams (Cheng et al., 1993), underground water control systems (Bogardi et al., 1987), and water-retaining structures including dikes and sluice gates (Vrijling, 1993).

Dhillon and Singh (1981) pointed out the advantages and disadvantages of the fault-tree analysis technique. Advantages include the following:

1. Provides insight into the system behavior.
2. Requires engineers to understand the system thoroughly and deal specifically with one particular failure at a time.
3. Helps to ferret out failures deductively.
4. Provides a visible and instructive tool to designers, users and management to justify design changes and trade-off studies.
5. Provides options to perform quantitative or qualitative reliability analysis.
6. Technique can handle complex systems.
8. Commercial codes are available to perform the analysis.

Disadvantages include the following:

1. Can be costly and time-consuming.
2. Results can be difficult to check.
3. Technique normally considers that the system components are in either working or failed state; therefore, the partial failure states of components are difficult to handle.
4. Analytical solutions for fault trees containing standbys and repairable components are difficult to obtain for the general case.
5. To include all types of common-cause failure requires considerable effort.

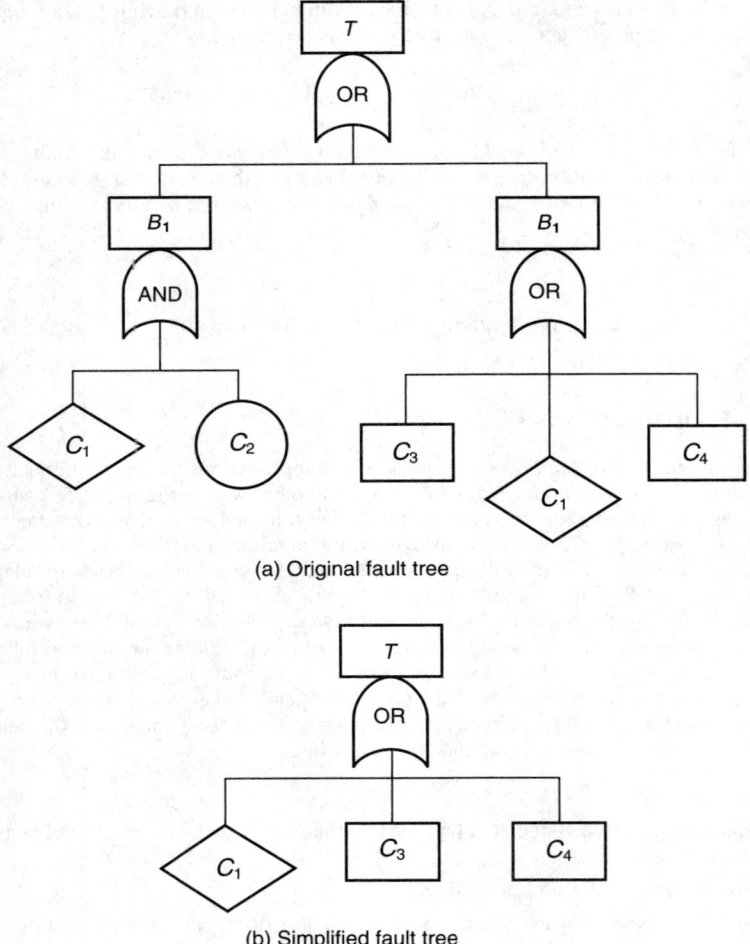

FIGURE 7.7 Example fault tree (a) and (b).

A *fault tree* is a logical diagram representing the consequence of the component failures (basic or primary failures) on the system failure (top failure or top event). A simple fault tree is given in Fig. 7.7 as an example. There are two major types of combination nodes (or gates) used in a fault tree. The *AND* node implies that the output event occurs only if all the input events occur jointly, corresponding to the intersection operation in probability theory. The *OR* node indicates that the output event occurs if any one or more of the input events occur, that is, a union. System reliability $p_{s,\text{sys}}(t)$ is the probability that the top event does not occur over the time interval $(0, t)$.

Before constructing a fault tree, engineers must thoroughly understand the system and its intended use. One must determine the higher order functional events and continue the

METHODOLOGIES FOR RELIABILITY ANALYSIS 7.19

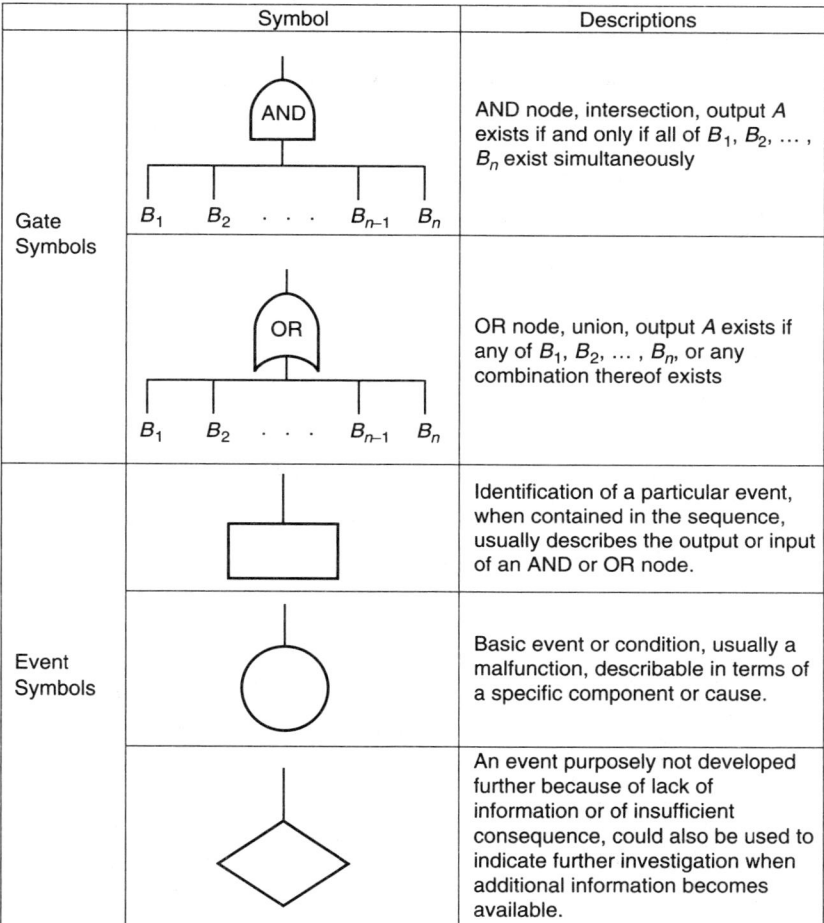

FIGURE 7.8 Basic fault tree symbols.

fault event analysis to determine their logical relationships with lower level events. Once this is accomplished, the fault tree can be constructed. A brief description of fault tree construction is presented below. The basic concepts of fault-tree analysis are presented in Henley and Kumamoto (1981) and Dhillon and Singh (1981).

There are actually two types of building blocks—*gate symbols* and *event symbols* as illustrated in Fig. 7.8. Gate symbols connect events according to their casual relation such that they may have one or more input events but only one output event. A *fault event*, denoted by a rectangular box, results from a combination of more basic faults acting through logic gates. A circle denotes a basic component failure that represents the limit of resolution of a fault tree. A diamond represents a fault event whose causes have not been fully developed. For more complete descriptions on other types of gate and event symbols, readers are referred to Henley and Kumamoto (1981).

7.20 WATER SUPPLY SYSTEMS SECURITY

Henley and Kumamoto (1981) present heuristic guidelines for constructing fault trees, which are summarized below:

1. Replace abstract events by less abstract events.
2. Classify an event into more elementary events.
3. Identify distinct causes for an event.
4. Couple trigger event with "no protection actions."
5. Find cooperative causes for an event.
6. Pinpoint component failure events.
7. Develop component failure.

Boolean algebra operations are used in fault tree analysis. Thus, for the fault tree shown in Fig. 7.7(a),

$$B_1 = C_1 \cap C_2; B_2 = C_3 \cup C_4 \cup C_1$$

Hence, the top event is related to the component events as

$$T = B_1 \cup B_2 = (C_1 \cap C_2) \cup (C_3 \cup C_4 \cup C_1) = C_1 \cup C_3 \cup C_4$$

Thus, the probability of the top event occurring can be expressed as

$$P(T) = P(C_1 \cup C_3 \cup C_4)$$

If C_1, C_3, and C_4 are mutually exclusive, then

$$P(T) = P(C_1) + P(C_3) + P(C_4)$$

The original fault tree can be reduced to an equivalent, but simpler, tree as shown in Fig. 7.7(b). In the case that the component's reliability or failure probability is a function of time, system reliability $p_{z,sys}(t)$ is the probability that the top event does not occur over the time interval $(0, t)$.

Another example of a fault tree construction is given for the pumping system in Fig. 7.9. In this pumping system, the tank is filled in 10 min and empties in 50 min, having a cycle time of 60 min. After the switch is closed, the time is set to open the contacts in 10 min. If the mechanism fails, then the horn sounds and the operator opens the switch to prevent pressure tank rupture. The fault tree for the pumping system is shown in Fig. 7.10.

Evaluation of Fault Trees. The basic steps used to evaluate fault trees include the following:

1. Construct the fault tree.
2. Determine the minimal cut sets.
3. Develop primary event information.
4. Develop cut set information.
5. Develop top event information.

To evaluate the fault tree, one should always start from the minimal cut sets, which in essence, are *critical paths*. Basically, the fault tree evaluation comprises two distinct processes: (a) the determination of the logical combination of events that cause top event failure expressed in the minimal cut sets and (b) the numerical evaluation of the expression.

Cut sets, as described previously, are collections of basic events such that if all these basic events occur, then the top event is guaranteed to occur. The tie set is a similar concept

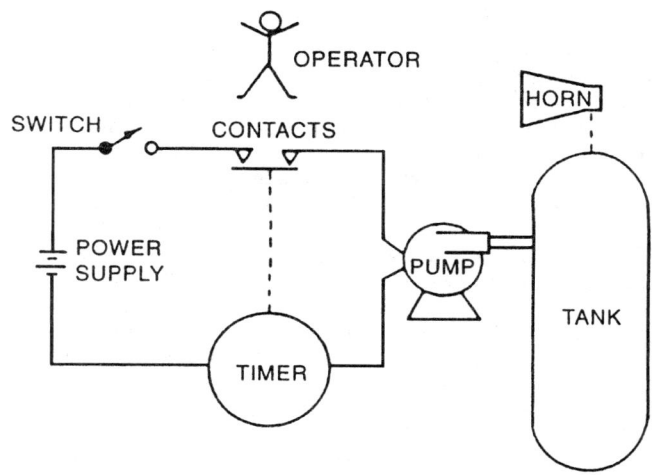

FIGURE 7.9 Schematic diagram for an example pumping system. (*Source: Henley and Kumamoto, 1981.*)

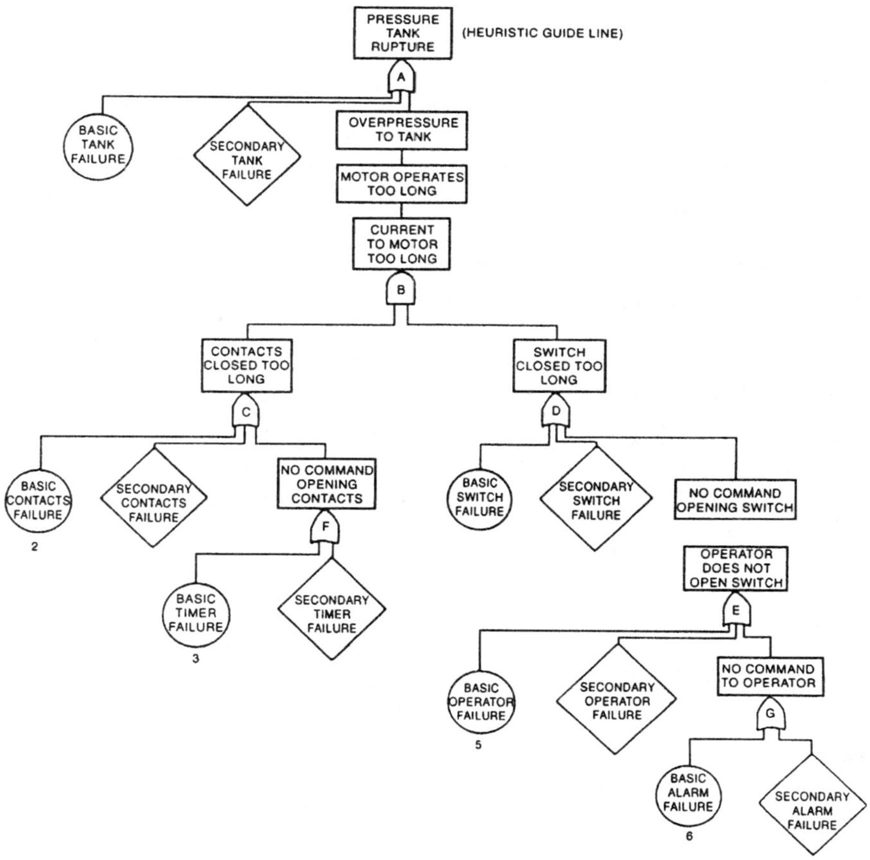

FIGURE 7.10 Fault tree for the example pumping system. (*Source: Henley and Kumamoto, 1981.*)

to the cut set in that it is a collection of basic events such that if the top event does not occur, then none of the events occur. As one could imagine, a large system has an enormous number of failure modes. A minimal cut set is one that if any basic event is removed from the set, the remaining events collectively are no longer a cut set. By the use of minimum cut sets, the number of cut sets and basic events are reduced in order to simplify the analysis.

System availability, $A_{sys}(t)$, is the probability that the top event does not occur at time t, which is the probability of the systems operating successfully when the top event is an OR combination of all system hazards. System unavailability, $U_{sys}(t)$, on the other hand, is the probability that the top event occurs at time t, which is either the probability of system failure or the probability of a particular system hazard at time t.

System reliability $p_{s,sys}(t)$ is the probability that the top event does not occur over time interval $(0, t)$. System reliability requires continuation of the nonexistence of the top event and its value is less than or equal to the availability. On other hand, system unreliability $p_{f,sys}(t)$ is the probability that the top event occurs before time t and is complementary to system reliability. Also, system unreliability, in general, is greater than or equal to system unavailability. From system unreliability the system failure density $f_{sys}(t)$ can be obtained.

REFERENCES

American Water Works Association, *Emergency Planning for Water Utilities (M19)*, Denver, CO, 2001.

American Water Works Association, *Water System Security: A Field Guide*, Denver, CO, 2002.

Association of State Drinking Water Administrators and the National Rural Water Association, *Security Vulnerability Self-Assessment Guide for Small Drinking Water Systems Serving Between 3,300 and 10,000*, 2002.

Barlow, R. E., J. B. Fussell, and N. D. Singpurwalla, *Reliability and Fault Tree Analysis: Theoretical and Applied Aspects of System Reliability and Safety Assessment*, SIAM, Philadelphia, 1975.

Bogardi, I., L. Duckstein, and F. Szidaroviszky, "Reliability Estimation of Underground Water Control Systems Under Natural and Sampling Uncertainty," in L. Duckstein and E. J. Plate (eds.), *Engineering Reliability and Risk in Water Resources*, Martinus Nijhoff Publishers, Drodrecht, The Netherlands, pp. 423–441, 1987.

Cheng, S.-T., B. C. Yen, and W. H. Tang, "Stochastic Risk Modeling of Dam Overtopping," in Ben C. Yen and Y.-K. Tung (eds.), *Reliability and Uncertainty Analyses in Hydraulic Design*, American Society of Civil Engineers, New York, pp. 123–132, 1993.

Dhillon, B. S., *Engineering Reliability: New Techniques and Applications*, Wiley, New York, 1981.

Fullwood, R. R., and R. E. Hall, *Probabilistic Risk Assessment in Nuclear Power Industry*, Pergamon Press, Oxford, 1988.

Henley, E. J. and H. Kumamoto, *Reliability Engineering and Risk Assessment*, Prentice-Hall, Englewood Cliffs, NJ. 1981.

Henley, E. J., and S. L. Gandhi, "Process Reliability Analysis," *Am. Inst. Chem. Eng. J.* 21(4): 677–686, July 1975.

Mays, L. W. (ed.), *Reliability Analysis of Water Distribution Systems*, American Society of Civil Engineers, New York. 1989.

Mays, L. W. (ed.), *Urban Water Supply Handbook*, McGraw-Hill, New York, 2002.

Mays, L. W. (ed.), *Water Distribution Systems Handbook*, McGraw-Hill, New York, 2000.

National Rural Water Association, "Rural and Small Water and Wastewater System Emergency Response Plan Template," available at: www.nrwa.org, Duncan, OK.

Peplow, D. E., C. D. Sulfredge, R. L. Saunders, R. H. Morris, and T. A. Hann, *Calculating Nuclear Power Plant Vulnerability Using Integrated Geometry and Event/Fault Tree Models*, Oak Ridge National Laboratory, Oak Ridge, TN, 2003.

METHODOLOGIES FOR RELIABILITY ANALYSIS

President's Commission on Critical Infrastructure Protection, Appendix A, Sector Summary Reports, *Critical Foundations: Protecting America's Infrastructure*: A-45, available at: http://www.ciao.gov/PCCIP/PCCIP_Report.pdf.

Sulfredge, C. D., R. L. Saunders, D. E. Peplow, and R. H. Morris, *Graphical Expert System for Analyzing Nuclear Facility Vulnerability*, Oak Ridge National Laboratory, Oak Ridge, TN, 2003.

U.S. EPA, "Guidance for Water Utility Response, Recovery, and Remediation Actions for Man-Made and/or Technological Emergencies," available at: http://www.epa.gov/safewater/security/erguidance.pdf.

U.S. EPA, "Guidance for Water Utility Response, Recovery, and Remediation Actions for Man-Made and/or Technological Emergencies," EPA 810-R-02-001, Office of Water (4601), available at: www.epa.gov/safewater, April 2002.

U.S. EPA, "Instructions to Assist Community Water Systems in Complying with the Public Health Security and Bioterrorism Preparedness and Response Act of 2002," EPA 810-R-02-001, Office of Water, available at: www.epa.gov/safewater/security, January 2003.

U.S. EPA, "Large Water System Emergency Response Plan Outline: Guidance to Assist Community Water Systems in Complying with the Public Health Security and Bioterrorism Preparedness and Response Act of 2002," available at: www.epa.gov/safewater/, June 2003.

U.S. EPA, "Vulnerability Assessment Fact Sheet 12-19," EPA 816-F-02-025, Office of Water, available at: www.epa.gov/safewater/security/va fact sheet 12-19.pdf, also at www.epa.gov/ogwdw/index.html, November 2002.

U.S. EPA, information available at: http://www.epa.gov/swercepp/cntr-ter.html

U.S. EPA, *Water Security Strategy for Systems Serving Populations Less than 100,000/15 MGD or Less*, July 9, 2002.

Vrijling, J. K., "Development in Probabilistic Design of Flood Defenses in the Netherlands," in Ben C. Yen and Y.-K. Tung (eds.), *Reliability and Uncertainty Analyses in Hydraulic Design*, American Society of Civil Engineers, New York, pp. 133–178, 1993.

Washington State Department of Health, *Emergency Response Planning Guide for Public Drinking Water Systems*, DOH PUB. #331-211, Olympia, Washington, May 2003.

CHAPTER 8
DEVELOPING AN EARLY WARNING SYSTEM FOR DRINKING WATER SECURITY AND SAFETY

Robert M. Clark,[*] Nabil Adam,[†] Vijayalakshmi Atluri,[†]
Milton Halem,[†] Eric Vowinkel,[‡] Pen C. Tao,[§]
Laura Cummings,[¶] and Eva A. Ibraham[**]

8.1 INTRODUCTION

The events of September 11, 2001 have raised concerns over the safety and security of the Nation's critical infrastructure including water and waste water systems. The U.S. Environmental Protection Agency (U.S. EPA) responded to these events by establishing a Water Protection Task Force (WPTF) composed of members of the U.S. EPA's Office of Water (OW), Regional Office staff and liaisons from other U.S. EPA programs. In addition, EPA's Office of Research and Development (ORD) established a National Center for Homeland Security Research (NCHSR) to develop a program devoted to water security research. Both OW and ORD have been given the responsibility for developing plans and conducting research to improve the security of the nation's drinking water and wastewater infrastructure.

Security of water systems is not a new issue and the potential for natural, accidental, and purposeful contamination of water supplies has been the subject of many studies. For example, in May 1998, President Clinton issued Presidential Decision Directive (PDD) 63 that outlined a policy on critical infrastructure protection, including our nation's water

---

[*]Public Health and Environmental Engineering Consultant, Cincinnati, Ohio
[†]The Center for Information Management, Integration and Connectivity (CIMIC), Rutgers University, Newark, New Jersey
[‡]U.S. Geological Survey, Edison, New Jersey
[§]North Jersey District Water Supply Commission, Wanaque, New Jersey
[¶]Passaic Valley Water Commission, Totowa, New Jersey
[**]American Water, Voorhees, New Jersey

supplies (President, 1998). However, it wasn't until after September 11, 2001 that the water industry truly focused on the vulnerability of the nation's water supplies to security threats. In recognition of current water security concerns President George Bush signed The Bioterrorism Act into law in June 2002 (HR 3448). Under the requirements of The Bioterrorism Act, drinking water utilities have been required to prepare Vulnerability Assessments and Emergency Response Plans. Vulnerability Assessments and Emergency Response Plans will be unique.

In addition to the development of new technology, responses to perceived and real water security threats will require the establishment of new types of institutions. For example, the Urban Security Group at the University of Michigan has developed recommendations for the establishment of institutions to implement Emergency Response Plans for public water systems. Both the Safe Drinking Water Act and Emergency Planning Community Right to Know Act of 1986 require the establishment of State Emergency Response Commissions (Burns et al., 2002).

As a part of the process of developing mechanisms for responding to water security threats, Rutgers University and EPA's Region II Office convened, in June 2002, a workshop entitled "Monitoring and Modeling Drinking Water Systems for Security and Safety." Attendees from Industry, Local, State, Federal Agencies, and members from academia discussed the state-of-the-art in the area of water system security. As a consequence of these discussions, a consortium was formed and a memorandum of understanding (MOU) was drafted, proposing a framework for establishing a Regional Drinking Water Security and Safety Consortium (RDWSSC). The goal of the MOU is to implement the drinking water security recommendations from the workshop. Members of the consortium include Rutgers University's Center for Information Management, Integration and Connectivity (CIMIC), the U.S. Geological Survey (USGS), American Water (AW), the Passaic Valley Water Commission (PVWC), the North Jersey District Water Supply Commission (NJDWSC), the New Jersey Department of Environmental Protection (NJDEP), and the U.S. Environmental Protection Agencies (USEPAs), Region II Office. Development of an early warning system for source and distributed water was identified as being of critical importance by the Consortium. In this context, an early warning system (EWS) is an integrated system of monitoring stations located at strategic points in a water utility's source waters or in its distribution system, designed to warn against contaminants that might threaten the health and welfare of drinking water consumers. As will be discussed later, an EWS should be integrated or packaged with appropriate sensors and predictive modeling capability.

In December 2002 a follow-up to the previous workshop was held to refine the needs for research as related to early warning systems for security in drinking water. After this meeting the consortium representatives signed the MOU. As the organizational research arm for implementation of the recommendations, Rutgers University has established a laboratory for water security (LWS), under CIMIC.

This chapter discusses the nature of the security threat to water supplies, the recommended elements of an EWS and relevant plans for developing and implementing an EWS.

8.2 NATURE OF THE THREAT

There are nearly 60,000 community water supplies in the United States serving over 226 million people. Over 53 percent of these systems supply water to less then 2.4 percent of the population and 5.4 percent supply water to 78.5 percent of the population. Most of these systems provide water to less than 500 people. In addition, there are 140,000 noncommunity water systems that serve schools, recreational areas, trailer parks, etc. (U.S. EPA, 1999). Some of the common elements associated with water supply systems in the United States are as follows:

- *Water source*, which may be a surface impoundment such as a lake, reservoir, river, or ground water from an aquifer
- *Conventional treatment facilities* including filtration, which removes particulates and potentially pathogenic microorganisms, followed by disinfection, primarily for surface supplies.
- *Transmission systems*, which include tunnels, reservoirs, and/or pumping facilities, and storage facilities
- *Distribution system* carrying finished water through a system of water mains and subsidiary pipes to consumers

8.2.1 Vulnerability of Water Systems

Water systems are spatially diverse and therefore, have an inherent potential to be vulnerable to a variety of threats—physical, chemical, and biological—that may compromise the system's ability to reliably deliver safe water. Community water supplies are designed to deliver water under pressure and generally supply most of the water for fire fighting purposes. Loss of water or a substantial loss of pressure could disable fire-fighting capability, interrupt service, and disrupt public confidence. This loss might result from sabotaging pumps that maintain flow and pressure, or disabling electric power sources that could cause long-term disruption. Many of the major pumps and power sources in water systems have custom designed equipment and could take months or longer to replace (Clark and Deininger, 2001).

These areas of vulnerability include:

1. raw water source (surface or groundwater)
2. raw water channels and pipelines
3. raw water reservoirs
4. treatment facilities
5. connections to the distribution system
6. pump stations and valves, and
7. finished water tanks and reservoirs.

Each of these system elements present unique challenges to the water utility in safeguarding the water supply (Clark and Deininger, 2000).

8.2.2 Physical Disruption

The ability of a water supply system to provide water to its customers can be compromised by destroying or disrupting key physical elements of the water system. These elements include raw water facilities (dams, reservoirs, pipes, and channels), treatment facilities, and distribution system elements (transmission lines and pump stations).

Physical disruption may result in significant economic cost, inconvenience, and loss of confidence by customers, but has a limited direct threat to human health. Exceptions to this generalization include (1) destruction of a dam that causes loss of life and property in the accompanying flood wave and (2) an explosive release of chlorine gas at a treatment plant.

Water utilities should examine their physical assets, determine areas of vulnerability, and increase security accordingly. An example of such an action might be to switch from chlorine gas to liquid hypochlorite, especially in less secure locations which decrease the risk of exposure to poisonous chlorine gas. Redundant system components would provide backup capability in case of accidental or purposeful damage to facilities.

8.2.3 Contamination

Contamination is generally viewed as the most serious potential terrorist threat to water systems. Chemical or biological agents could spread throughout a distribution system and result in sickness or death among the consumers and for some agents, the presence of the contaminant might not be known until emergency rooms report an increase in patients with a particular set of symptoms (Clark, 2002). Even without serious health impacts, just the knowledge that a group had breached a water system could seriously undermine customers' confidence in the water supply (Grayman et al., 2002).

Accidental contamination of water systems has resulted in many fatalities as well. Examples of such outbreaks include cholera contamination in Peru (Clark et al., 1995), *Cryptosporidium* contamination in Milwaukee, Wisconsin (U.S.) (Fox and Lytle, 1996). and *Salmonella* contamination in Gideon, Missouri (U.S.). In Gideon, the likely culprit was identified as pigeons infected with *Salmonella* that had entered a tank's corroded vents and hatches (Clark et al., 1996).

The U.S. Army has conducted extensive testing and research on potential biological agents (Burrows and Renner, 1998 and 1999). Table 8.1 summarizes information on the agents most likely to have an impact on water systems. Though much is known about these agents, as is evident in the table, there is still research needed to fully characterize the impacts, stability, and tolerance to chlorine of many of these agents.

TABLE 8.1 Potential Threat of Selected Biological Agents to Water Systems

Agent	Type	Stable in water	Chlorine* tolerance
Anthrax	Bacteria	2 years (spores)	Spores resistant
Cholera	Bacteria	"Survives well"	"Easily killed"
Plague	Bacteria	16 days	Unknown
Salmonella	Bacteria	8 days, fresh water	Inactivated
Shigellosis	Bacteria	2–3 days	Inactivated 0.05 ppm, 10 min
Tularemia	Bacteria	Up to 90 days	Inactivated 1 ppm, 5 min
Aflatoxin	Biotoxin	Probably stable	Probably tolerant
Botulinum toxins	Biotoxin†	Stable	Inactivated 6 ppm, 20 min
Cryptosporidiosis	Protozoan‡	Stable days or more	Oocysts resistant
Microcystins	Biotoxin	Probably stable	Resistant at 100 ppm
Ricin	Biotoxin	Unknown	Resistant at 10 ppm
Staph enterotoxins	Biotoxin	Probably stable	Unknown
Tetrodotoxin	Biotoxin	Unknown	Inactivated 50 ppm
T-2 mycotoxin	Biotoxin	Stable	Resistant
Hepatitis A	Virus	Unknown	Inactivated 0.4 ppm, 30 min
Saxitoxin	Biotoxin	Stable	Resistant at 10 ppm

*Ambient temperature; < 1 ppm free available chlorine for 30 min or as indicated.
‡Consisting of one cell or of a colony of like or similar cells.
†Toxic to humans.
Source: Based on Burrows and Renner (1998, 1999).

Characteristics that would enhance the potential for an agent to contaminate drinking or recreational water include the following:

- Resistance to disinfectants at normal concentrations
- Resistance to boiling for 1 to 3 minutes
- A low oral infectious dose
- Easy availability
- Easy to culture without sophisticated equipment
- Survival in water for long periods of time
- Difficult to remove by common water treatment practices

Although all of the above agents (and many others) could result in very significant health impacts, the risks vary considerably. For example, Botulinum toxin, because of its lethality in very small doses, is considered to be among the most serious threats. There are many factors that contribute to the relative risk of the various agents including availability, lethality, stability, and tolerance to chlorine or other disinfectants.

Deininger and Meier (2000) ranked some agents and compounds in terms of their relative factor of effectiveness, R, based on lethality and solubility, using the following equation:

$$R = solubility\ in\ water\ (\text{in mg/L}) / [1000 \times lethal\ dose\ (\text{in mg/human})]$$

Table 8.2 lists values of R for various biological agents and chemicals by decreasing level of effectiveness, that is, decreasing degree of lethality in water (Deininger and Meier, 2000).

Many locations within the overall water supply system are vulnerable to the introduction of chemical or biological agents. In many cases, the most accessible location is in the raw surface water source. An agent introduced in a surface water source is subject to dilution, exposure to sunlight, and treatment therefore it follows that the most serious threats are posed by an agent introduced into the finished water at a treatment facility or within the distribution system. Possible points of entry include the treatment plant clear well, distribution system storage tanks, and reservoirs, pump stations, and direct connections to distribution system mains.

TABLE 8.2 Relative Toxicity of Some Poisons in Water

Compound	R
Botulinum toxin A	10000
VX	300
Sarin	100
Nicotine	20
Colchicine	12
Cyanide	9
Amiton	5
Fluoroethanol, sodium, fluoroacetate	1
Selenite	1
Arsenite, Arsenate	1

Source: Based on Burrows and Renner (1998, 1999).

8.3 THE EARLY WARNING SYSTEM CONCEPT

Early warning systems (EWSs) are intended to reliably identify low probability/high-impact contamination events in source or distributed water. The following requirements were identified by an International Life Sciences Institute (ILSI) working group report on EWSs for hazardous events in water (ILSI, 1999):

- Provide warning in sufficient time to respond to the contamination event and prevent exposure of the public to the contaminant.
- Is capable of detecting all potential contamination threats.
- Can be operated remotely.
- Can identify the point at which the contaminant was introduced.
- Has a low rate of false positive and false negative results.
- Provides continuous, year-round surveillance.
- Produces results with acceptable accuracy and precision.
- Requires low skill and training.
- Is affordable to the majority of public water systems.

Although there isn't an EWS in existence that meets all of the above criteria, there are some conventional monitoring systems that might satisfy some of these elements. However, these types of systems will provide the only opportunity to prevent the public from being exposed to a covert attack on the water supply using biological, chemical, or radiological agents. Since it is unlikely that an EWS would be capable of detecting all potential contamination threats, it should have a strong tie to public health surveillance monitoring. An EWS is only one of many tools including physical security, infrastructure hardening, and optimized treatment that can be implemented to protect public water supplies.

8.3.1 Characteristics of Early Warning Systems

A contamination event in source or distributed water must be identified in time to allow for an appropriate response that mitigates or eliminates its adverse impact (ILSI, 1999). The features of an ideal EWS include the following:

Rapid Response Time. The response time for an EWS, is the time period from the point at which the contaminant contacts the sensor to the point when a result is reported and a response is initiated. Depending on the nature of the threat, in order to prevent damage, the response time must be such that it allows measures to be taken to prevent or minimize exposure of the public to the contaminant. The required response time will depend on a number of factors, such as the point at which the contaminant is introduced into the system; treatment plant characteristics and retention times; and the nature of the contaminant itself. In some cases, a response time of several hours is sufficient, while in other cases, the response time will have to be on the order of minutes. In most cases, an EWS will likely utilize field deployable units or biomonitoring systems.

Fully Automated. Ideally, an EWS would require little or no operator intervention and would allow for 24-h operation. The system should also allow for remote operation such that a response at a remote sensor would be immediately relayed to a central data management system. In the event that a contaminant is detected, the EWS should immediately trigger an alarm and contact a plant operator via page, phone, and/or fax.

Screens for a Range of Contaminants. As discussed above there are a number of agents that could pose a potential threat if introduced into a water supply. It is, however, impossible to know in advance which agent would be used, therefore EWSs should have the capability to screen for a range of potential agents. The problem that is often encountered is that methods that are effective for screening for a large number of agents aren't specific enough to distinguish between harmful and benign substances. Thus, the ability to screen for a range of contaminants must be balanced against the need for specificity.

Specific for the Contaminants of Concern. Specificity refers to the ability of an EWS to identify specific biological, chemical or radiological agents. Ideally, an EWS would positively identify specific agents that pose a threat to public health, and would be capable of differentiating between these substances and closely related, yet benign substances.

Sufficient Sensitivity. Sensitivity refers to the lowest level of detection and quantification that an EWS can achieve. An EWS should be sufficiently sensitive to provide quantification of a specific agent at the lowest level that poses a threat to public health.

Low Occurrence of False Positives and False Negatives. False positives occur when an EWS indicates the presence of a contaminant that is actually absent from the sample. False negatives occur when a contaminant is present at levels of concern, but is not detected by the EWS. Both types of errors are problematic and, ideally, the rates of false positives and false negatives would be zero. However, given the fact that there is always a probability of either type of error occurring, it is necessary to characterize the rate at which they occur and to minimize their impact on decision making.

High Rate of Sampling. Assuming that a monitoring device is online, it is desirable to have the device monitor for contaminants as frequently as is possible. A high rate of sampling would be crucial if the agent is introduced in one defined event and is entrained in the water system in plug flow. Detection of a "contaminant plug" would offer an excellent way to distinguish sabotage from natural changes in water quality.

Reliable and Rugged. The systems must be able to withstand field conditions and still perform reliably. Challenges that a remote monitoring system may face include power outages, extreme environmental conditions, potential vandalism or theft, and fouling of sensor components (biological, chemical, or particulate fouling).

Requires Minimal Skill and Training. The equipment used in the EWS should not require excessive skill or training to operate and maintain the equipment or to interpret the results.

Affordable Cost. The more affordable the monitoring system the more available it will be to plants with limited resources, and the more widely it can be deployed within a system.

8.4 THE CONSORTIUM WATER SYSTEMS

The goal of the consortium research is to achieve the afore-mentioned goals in developing an EWS. In addition, the consortium will provide a framework where Federal, state, and local government agencies' representatives, highly talented scientists, water utility professionals, and leaders in the area of water security can share their expertise and resources. It is intended to provide a test bed for the rapid prototyping of advanced and still evolving technologies to monitor drinking water resources and distribution networks in order to better protect the public.

An immediate step toward achieving this goal is the development and implementation of an end-to-end real-time monitoring and modeling early warning pilot system that will:

- Consists of currently available state-of-the-art biochemical sensors, predictive modeling tools, and information management infrastructure
- Provides decision makers and the public with reliable and timely assessments
- Satisfies the consortium members' requirements for reliability, scalability, and accuracy under operational field conditions
- Ensure the continued safety and security of drinking water at the source and distribution network within our region and within our nation for future generations

It is envisioned that this consortium may grow in the future to incorporate other universities, utilities and possibly other agencies as appropriate and mutually acceptable. The work of the consortium will include, but not be limited to organizing and conducting workshops; convening seminars on relevant technology applications; developing training opportunities for consortium-affiliated personnel; and conducting real time pilot studies. Several utilities have inquired about the possibility of joining the Consortium at some future date.

The Consortium water systems are located in one of the most complex water systems in the United States (Fig. 8.1), consisting of three watershed management areas. Water is withdrawn from the various surface sources in the watersheds and then either used directly by a utility or distributed through a number of interties to other utilities. As can be seen in Fig. 8.1, there are numerous surface water supplies in the three watersheds. This interconnectivity makes these particular water systems excellent candidates for inclusion in a research program to develop an EWS.

The North Jersey District Water Supply Commission (NJDWSC) of the State of New Jersey (NJDWSC) is one of the largest drinking water purveyors in the State of New Jersey. NJDWSC provides treated water on a wholesale basis to the Passaic Valley Water Commission, and also supplies wholesale treated water to its many contracting municipalities and raw water to United Water New Jersey. NJDWSC's Wanaque/Monksville Reservoirs System is located in northern Passaic County, New Jersey. The Reservoir System beginning with Upper and Lower Greenwood lakes drains a 95-mi^2 watershed covering parts of Orange and Rockland counties in New York, and Passaic County in New Jersey. The Wanaque/Monksville Reservoirs System is New Jersey's largest (in terms of population served) water supply system. The system provides a safe yield of 173 million gallons per day to approximately 2 million people in 90 towns and cities in the northern half of the state. Additionally, there are two pump stations designed to pump up to 250 million gallons per day of water from the Pompton/Passaic Rivers and 150 million gallons per day from the Ramapo River into the Wanaque Reservoir. These rivers are the supplementary sources of water supply to the Wanaque Reservoir. NJDWSC's Water Treatment Plant/Operations Center is located in Wanaque, NJ, where the raw water is pumped from the Wanaque Reservoir then purified and filtered to ensure its safety and potability.

The Passaic Valley Water Commission (PVWC) owns and operates the Little Falls Water Treatment Plant (LFWTP) located in Totowa, New Jersey. Three raw water sources are available to the LFWTP, including Point View Reservoir water, Pompton River water and Passaic River water. In addition, finished water supplied to PVWC by NJDWSC is also distributed to the Totowa facility. The two finished water supplies, either directly or as a blend, are distributed to PVWC's retail and wholesale customers, which represent a population of over 750,000. American Water is one of PVWC's major customers.

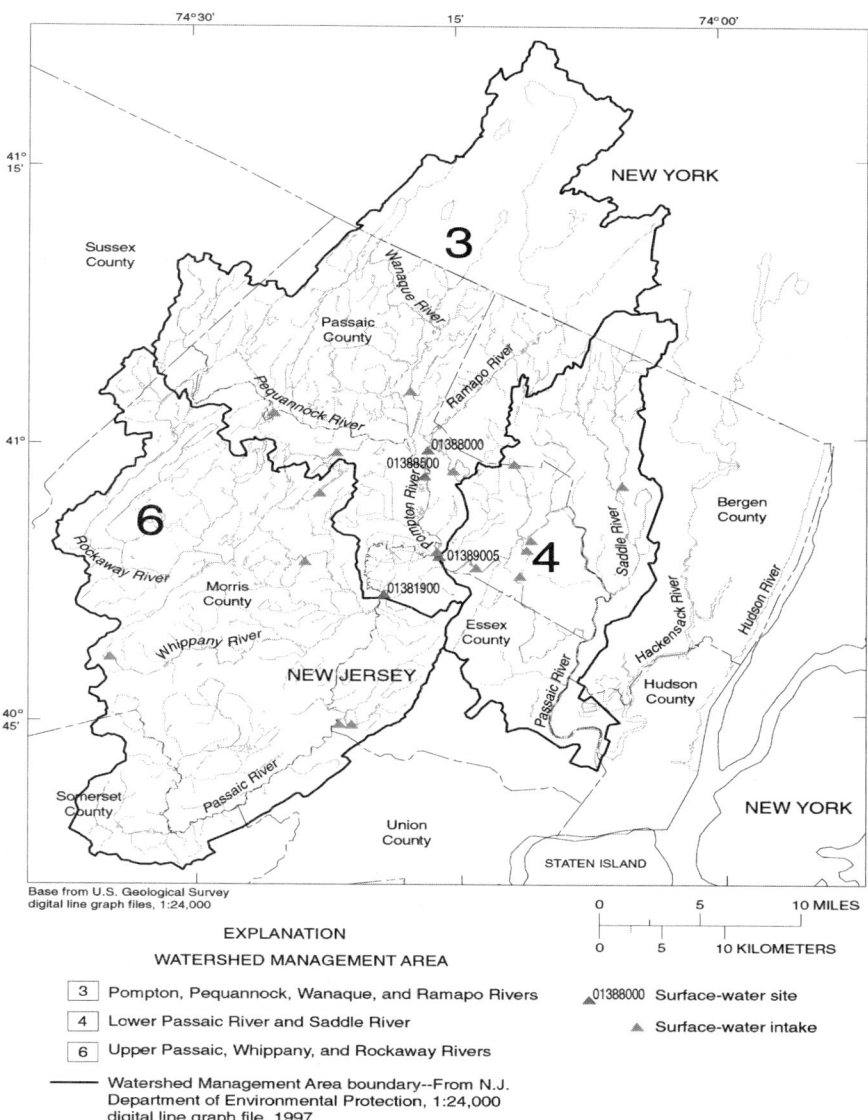

FIGURE 8.1 Proposed and current continuous Water-Quality Monitoring Sites and Surface Water Intakes in the Passaic River basin, New Jersey.

The Haddon Heights water system is part of the New Jersey-American Water Company, which is a utility subsidiary of the American Water System that owns and operates more than 85 surface water plants in 27 states and three Canadian provinces. American Water is the largest private water utility in the United States devoted exclusively to the business of water supply. The New Jersey-American Water Company serves 600,000 customers

(2 million people) and provides an average of 43.55 mgd to approximately 91,200 customers in the combined Burlington, Camden and Haddon service areas in 2001. Approximately 90 percent of the customer base is residential, 8.4 percent is commercial, 0.5 percent is comprised of the other public authority category, and nearly 0.1 percent is industrial. The fire service classification amounts to approximately 1 percent of the customer base. At the end of 2001, there were 13 resale customers in the Burlington, Camden and Gloucester Counties area that purchased water from NJAWC. The Burlington and Haddon systems are interconnected via a regional pipeline from the Delaware Valley Regional Water Treatment Plant. The Camden system is also connected to the regional pipeline via 24-in and 30-in transmission main that is routed through a part of the Haddon system. Therefore, the three systems can be considered as one combined system with a service area of approximately 100 mi^2 (16,100 m^2). Most of the projected growth in the system is expected to occur in a few of the municipalities in Burlington County, and in Voorhees, Gloucester, and Cherry Hill townships in Camden County. A significant portion of the projected growth in usage is likely to occur as additional resale usage to the growing municipalities outside of the NJAWC service area.

All three water utilities have monitoring systems and procedures currently in place. For example, NJDWSC has a comprehensive ongoing monitoring and modeling reservoir water quality management plan. The PVWC and the Haddon Heights systems have ongoing compliance monitoring programs.

8.5 DEVELOPING AN EARLY WARNING DECISION SUPPORT SYSTEM

As mentioned, the consortium's objective is to encourage collaboration and cooperation among the member water utilities, and state, local, and Federal government to enhance water security. One of the first efforts the consortium plans to undertake is the design, development, and implementation of an end-to-end early warning decision support system (EWDSS) that can evolve over time incorporating advanced sensor technologies as they become available. This effort will consist of the integration of three separate but closely related research components, namely, monitoring, modeling, and information management systems. The issues that must be addressed in the design of an EWS include system characterization, vulnerability/threat characterization, establishment of appropriate alarm levels, availability of appropriate monitoring technologies, location of sensors, and data management and analysis. As a first step in the design of an early warning system, it is necessary to characterize the study area including source waters, intake structures, treatment facilities, chemical storage facilities, and distribution system infrastructure. In addition, access points, flow and demand patterns, and pressure zones within each distribution system must be determined. For example, high usage rates that are largely driven by industry have significantly different public health implications as compared to high usage rates in an urban center with a high population density. These types of usage patterns must be considered in the design of an EWS intended to protect public health. Hydraulic models will be or have been developed for each distribution system.

Some of the general issues to be considered in setting up an EWS are given below.

Alarm Levels. It is necessary to identify the concentrations at which the agents pose a threat to human health so that alarms can be triggered at appropriate levels. The response that is initiated by an alarm must be established before the EWS is deployed. A tiered approach to an EWS may allow a broad screening approach at the first tier to detect a number of potential contaminants, followed by specific, confirmatory analyses. If a false alarm leads to a decision to issue a notice to the public to stop using the water, public health as well as confidence could be impacted.

Method Performance and Data Quality. Once target agents for the EWS have been identified, it is necessary to select a monitoring technology for the particular agent or class of agents.

Fate and Transport of Pathogens and Chemicals. Chemical and microbial agents can behave in a variety of ways as they migrate through a water system. Environmental conditions, the presence of oxidants or other treatment chemicals, and the hydraulic characteristics of the system will affect the concentration and characteristics of these agents. If information is available on agent characteristics that affect their fate and transport, it should be factored into the design of an EWS. For example, if a target agent is known to chemically degrade at a certain rate in the presence of free chlorine, it may be possible to use a hydraulic/chemical model of the distribution system to predict the concentration profile through the system. This in turn can be used to select optimal locations for sensors as discussed further below.

Sensor Location. The location and density of sensors in an early warning system is significantly impacted by the result of the system characterization, vulnerability assessment, threat analysis, and usage considerations. Placement of sensors in a distribution system is appropriate because it is one of the most vulnerable points in a water system. Contaminants introduced directly into the distribution system would bypass all treatment barriers except the disinfectant residual. The size, complexity, and dynamic nature of distribution systems also complicate the selection of sensor locations. Proper characterization of the distribution system, usage patterns, and identification of critical system nodes (e.g., hospitals, law enforcement and emergency response agencies, etc.) is necessary to design an effective monitoring network with a distribution system. Due to their complexity and dynamic nature, it may be beneficial to develop a hydraulic/water quality model of the system to assist in the placement of sensors. Some of these issues have been explored in the literature (Lee and Deininger, 1992). However, even if sensors can be optimally located within a distribution system, there will probably be insufficient time to prevent exposure of a portion of the public to the contaminated water. Monitoring conducted within the distribution system will provide time to limit exposure, isolate the contaminated water, and initiate mitigation/remediation actions.

The initial research phase will focus on the rapid deployment of a prototype EWDSS. This prototype will be based on integrating currently available water sensor monitoring systems, existing models, and the current information infrastructure for the PVWC and the AW drinking water distribution networks and the source water for all three utilities. In succeeding years, it is planned that, state-of-the-art monitoring sensors will be acquired and advanced models coupled to enhanced information system architecture with interactive management decision support capabilities. In conjunction with CIMIC the members of the consortium will develop advanced information and decision support systems which will integrate the outputs of as yet undeveloped sensors and "cutting edge" modeling systems using the prototype EWS data configuration. The following three subsections discuss the three specific tasks to be addressed.

8.5.1 Sensor and Monitoring

It is anticipated that the sensor and monitoring research component of this effort will be maintained and managed by the USGS in collaboration with the three water utilities. It will provide the basis for a prototype EWS and will augment the existing monitoring program for the three utilities. The USGS will work with each of the utilities to install the platforms and maintain the sensors and employing the NJDWSC, the PVWC, and the AWS laboratories for analysis of the grab sample data. Each of the utilities involved in the consortium has slightly different objectives for participating in the cooperative agreement.

NJDWSC's goals for the proposed research are to install and operate state-of-the-art continuous monitoring systems at field locations in their source water reservoirs. PVWC's and AWS's primary goals for this project are to participate in field tests to determine viable options for online sensor technology that can be utilized to predict chemical and/or biological anomalies either in the source or finished water supplies.

Specific tasks to be undertaken are as follows:

- Determination of the optimal location of EWS stations in conjunction with the consortium.
- Select an initial station in order to test the proof of concept for collection of data.
- Design and construct or upgrade gauging stations to house EWS equipment.
- Install, operate, and maintain real-time monitoring equipment using available and state-of-the-art sensors to determine background water-quality characteristics.
- Field-scale testing of new sensor probes developed by Federal, state, and private laboratories.
- Generate and manage a stream of real-time water-quality data that would benefit the water companies in making critical management decisions regarding treatment plant and distribution system operations.
- Test and evaluate different sampling and installation techniques for reliability and maintenance including (a) in situ and (b) flow through methods.
- Evaluate the optimal locations for the placement of the probes in streams and distribution systems to ensure that the water supply is being protected from accidental and intentional contamination.
- Assist in the evaluation of previously conducted time of travel studies to determine optimal locations of upstream EWS stations in source waters and in distribution systems.
- In conjunction with consortium members, assist in the statistical interpretation of real-time data and other sensor data to predict contamination in source water and in finished water.
- Provide technical guidance on the installation, operation, and maintenance of the real time monitoring equipment for deployment in other areas of the United States.

8.5.2 Modeling

It is planned that CIMIC's LWS, working collaboratively with the three water utilities and with the USGS, will conduct modeling and simulation studies to optimize the design of an early warning decision support system (EWDSS). The intent is to better protect source water and distribution systems from bio-chemical agents. The modeling component will focus both on the simulation of system design and predicting the movement of contaminants in source water and distribution systems (Wagner et al., 2000). Water in streams and reservoirs and in the distribution system moves in complex temporal and spatial patterns under varying demands. Modeling provides a mechanism for both understanding these complexities and a quantitative means for determining the optimal placement of sensors. In addition, the model outputs will provide data input to the information management system in a timely manner for decision support management displays. These modeling studies will initially use the current operational source water and distribution system models of the respective utilities as the core predictive model. Owing to the proprietary nature of these models, and the sensitivity of the data configuration expressed within the models, the experiments will be designed by the LWS but conducted by the utilities on their computational facilities in collaboration with LWS.

The USGS will also assist the LWS in determining the optimum location and design of the placement of the sensors from considerations of costs of implementation and maintenance of a multipurpose EWDSS for drinking water safety and security for the three utilities.

Advanced modeling will utilize some of the system simulation techniques developed by NASA, NOAA, and the DOD (Jastrow and Halem, 1970; Rohaly and Krishnamurti, 1993) for optimizing the design of global space observing systems. Such studies have played an important role in planning for future-integrated multiagency satellite observing systems of the earth's atmosphere, oceans, and land surfaces (Arnold and Dey, 1986). The studies proposed will constitute the first application and test of this important simulation methodology to optimize the design of an EWS for use by the consortium water utilities.

The following two tasks will be critical for the design and deployment of an optimal prototype sensor monitoring system and subsequent real data predictive studies.

Source Water Modeling. Initially, NJDWSC will collect daily stream flow records (from the USGS), water quality records (from USGS and NJDWSC), and daily climatic records such as rainfall, air temperature, solar radiation, cloud cover, wind speed, dew point, etc. (from U.S. Weather Service). Then, NJDWSC will use an existing two-dimensional water quality simulation model, called LAWATERS (Lateral Averaged Wind and Temperature Enhanced Reservoir Simulation, a propriety model owned jointly by Wanaque South Project co-owners [NJDWSC and United Water New Jersey] and Najarian Associates), to generate spatial (from surface to bottom for entire length of the Wanaque Reservoir) and temporal (60-min interval continuously from January 1, 1993 to December 31, 2002) water quality profiles.

Watershed climatic conditions and reservoir model predicted natural watershed runoff and off-stream pumped storage will serve to produce input conditions of water that feed the reservoir. A reservoir model called Wanaque South Management Program (WMSP) will be used to generate a "nature" record that will form the basis of the complete water state of the reservoir from which synthetic data will be generated to simulate the sensor data with appropriate error and noise limits added. The reservoir models will then be started from random initial states taken from the nature run with appropriate errors introduced into the fields and used to test the impact of configuring various numbers and types of sensors at different locations. The simulated instrument data at the different locations will be taken from the nature run interpolated to the position of the sensors with systematic instrument errors added. The nature run will also be used to verify the influence of the instruments to indirectly infer the reservoir water parameters. In this manner, quantitative estimates will be derived to determine the optimum locations for the sensors under different climate conditions.

Distribution System Modeling. One of the major goals of the study is to integrate distribution system hydraulic and water quality modeling with data captured from sensors and monitoring systems. This task will be undertaken within the PVWC and Haddon Heights water distribution systems. Both systems currently use proprietary water quality/hydraulic models. These models will be used to perform initial simulation studies to assist in locating three monitoring stations in each distribution system. An EPANET file will be developed from these data. EPANET (Rossman et al., 1994) is a public sector water quality/hydraulic water quality model and will be used to provide a common platform and to facilitate date transfer to other interested water utilities. EPANET is a full-featured water quality/hydraulic model that can be used to simulate, pressures, and contaminate transport in drinking water distribution systems. Note that it is anticipated that an EPANET model will not be *built* for the entire distribution system but for the study area only. An initial tracer calibration study will be conducted during a period when the operating conditions for the utility can be held relatively constant, most likely early fall when system operations are relatively stable.

After the utility model has been calibrated and an equivalent EPANET file has been created a second calibration run will be conducted under more dynamic operating conditions. This second run will be used to establish the feed back loops between the modeling predications and the outputs from the sensor/monitoring systems and will provide an independent check of the ability of the monitoring/sensing systems to self calibrate the quality/hydraulic models on a continuous basis.

Two research areas that will be investigated as spin-offs from this research are as follows:

- Development of techniques to differentiate a *real* signal above background when dealing with noisy data
- Development of feedback procedures for integrating sensor data with network modeling systems in drinking water distribution systems

8.5.3 Information Management

One of the challenges of a continuous, real-time monitoring system is management of the large amounts of data that are generated. Use of data acquisition software and a central data management center is critical. This will require that the individual sensor deployed in the system be equipped with transmitters, modems, direct wire, or some other means to communicate with the data acquisition and management system. Furthermore, the data management system must be capable of performing some level of data analysis and trending in order to assess whether or not an alarm level has been exceeded. The use of "smart" systems that evaluate trends and can distinguish between genuine excursions and noise could minimize the rate of false alarms.

A decision will have to be made regarding the action that is taken when the data management system detects an excursion above the alarm level. At a minimum, the system should notify system operators and/or emergency response officials. If possible, redundant communication should be used (e.g., notifying multiple individuals through multiple routes such as page and fax). In some cases, it may be appropriate to program the data management system to take initial response actions, such as closing valves or initiating additional sampling. However, these initial responses should be considered simple precautionary measures, and human decision officials should make the ultimate decisions regarding response actions.

A major difficulty in designing an EWS is balancing the ability to screen for a wide range of agents with the ability to positively identify a specific contaminant of concern. One way to address these conflicting objectives is through the use of tiered monitoring. In a two-staged monitoring approach, the first stage would utilize screening methods to detect a range of contaminants that could pose a threat to public health. A positive result from the screening analysis would trigger confirmatory analysis using more specific and sensitive techniques, and a positive result from the confirmatory analysis would trigger a response action. Such a staged approach could use less expensive sensors in the first stage to allow for greater coverage of the system. More involved and expensive techniques would be used for confirmatory analysis.

CIMIC will conduct requirement studies utilizing the above-mentioned consortium field testing sites to compare the specifications of the monitors as provided by the vendors, against actual field scale performance. Validation will be based on grab sample and other comparative techniques. A major task in this research effort will be the integration of distribution system models, with monitoring and sensing systems. An information system will be developed that collects the data from various sensors, validates, and stores it in a secure manner, as well as an "alert" management system that provides the utilities with decision support capabilities.

Sensor Data Management System. CIMIC will design and develop a system to manage the streaming data being collected in near real-time. Incoming data streams from a network of distributed sensor sources will supply data comprising of both spatial and temporal attributes. The sensor readings will form continuous data streams that are either to be archived or to be processed. The challenge here is to organize the multidimensional data in a single dimension storage device (such as a hard disk) in such a way that the temporal and spatial relationships are preserved. In other words, the indexes associated with the data should consider spatial and temporal indexes (Samet, 1990). Such an organization would result in efficient processing of queries. Techniques will be developed to couple the sensor data management system with each utility's SCADA system.

Security Enforcement. Security enforcement is a major issue in all EWSs and will be addressed by the consortium in the following manner:

Building a Security Policy Base. Security policies will be established in sharing the data, as well as the documents among the consortium members and public. For example, while a specific sensor location and the attribute values collected from it are sensitive, the aggregate information of the data from all the sensors may be less sensitive. In the same vein, a document reporting the results of the research can be shared with the public, but a map depicting the sensor locations is highly sensitive. In other words, parts of the document may need to be protected. This necessitates the need for fine-grained access control. These security policies will be established based on the need-to-know requirements and data quality and integrity objectives.

Development of a Security Module. The security module will have the following functionalities:

- *Authentication.* A password based authentication mechanism will be employed for users to access the sensor data and document databases.
- *Access control.* A fine-grained access control module, with perhaps role-based access control, if appropriate, will be implemented to enforce the security policies. The security model and mechanism will be built on prior work on content-based access control (Adam et al., 2002). This system will be capable of specifying access control permissions based on the users' credentials that can be verified through digital certificates, and based on the document or part of it.
- *Secure data transfer.* Confidentiality and integrity of the sensor data as well as the documents during their transit from the sensors to the system as well as from the system to the users will be ensured by employing encrypted communication using protocols such as SSL (secure socket layer).

Data Validation and Alert Management System. The fact that the sensors are placed in a rugged environment, may result in incorrect values. It is important to distinguish whether signals are due to a benign failure or due to a terrorist activity. Manual verification may not always be feasible due to the large amount of data being collected. Therefore, mechanisms must be in place to automate this process. To accomplish this, it is planned to develop a sensor data validation module, which relies on two underlying technologies.

Data Mining for Alert Generation. Essentially an outlier detection approach will be used to identify abnormal behavior of the sensors. In order to accomplish this, commercial off-the-shelf (COTS) data mining software, will be employed.

Exploiting Correlations Among the Sensor Data. Formal correlations among the data attributes (such as pH, temperature, dissolved oxygen, specific conductivity, turbidity, and chlorophyll), and their spatial and temporal attributes, will be exploited to determine the validity of the data. These formal relationships among data attributes will be extracted with the help of the models discussed previously.

A rule-based diagnostic reasoning approach will be developed to help in generating the alarms (attached to a sensor) and alerts (require human attention). Several alarms can be used to generate a higher level of alert.

Real-Time Data Acquisition Information Network Systems. Studies related to the automated remote data handling, sensor calibration, and validation including data quality and reliability under harsh environmental conditions for real-time data acquisition will be conducted. In addition, user data requirements will be collected for assessing real-time report generation, graphical and visual representation of information, and compliance reports for the EWS prototype system will be developed.

Research on the performance of predictive models, expert systems and knowledge management for coupling these systems with SCADA systems will also be studied. In particular, research will be conducted into the application of mobile field communication systems for acquisition of sensor data. The use of grid computing resources needed to support the assimilation of real-time monitoring information from a vast array of advanced sensors coupled in real time, will be explored.

8.6 SENSOR RESEARCH AND DEVELOPMENT

There is significant research effort being invested by private, nonprofit, and government laboratories in developing more robust and reliable sensors for use in water supply. The consortium in collaboration with the U.S. EPA/NCHSR and the LWS, will conduct workshops and symposia in an attempt to identify where this research is being conducted. The consortium will establish a series of study groups to evaluate the status of sensors that are close to field application. Promising technologies will be placed on a fast track to assess their potential application. Part of this effort will be working with vendors to develop prototype systems for testing in the selected field locations.

LWS will also focus on conducing research that is "on the cutting edge." For example, satellite technology may prove useful for identifying chemical and biological warfare agents in source water or for rapid transmission of data from monitoring stations. Similarly, bioanalysis on chip sensors is being developed at various labs both government and industrial that offer increased real-time automated data access for assimilation to an EWS.

A GIS interface will be developed to visualize the location and severity of an alert that may result from "triggering" the EWS (Atluri and Mazzoleni, 2002).

8.7 EWS AND PUBLIC HEALTH SURVEILLANCE

Although it will not be addressed initially, eventually the consortium intends to link its EWS to a Public Health Surveillance System.

Most terrorist attacks in the United States have been overt attacks and much has been discussed about the importance of first responders, including local public health agencies and their roles in dealing with these attacks. In the case of a covert attack, however, disease and illness would most likely propagate throughout the community before the attack is recognized. The first indication that an attack has, in fact, occurred will most likely be identified, first, by various elements of the public health system (Clark, 2001). The identification of an etiologic agent associated with a covert attack will depend on the timely recognition that a public health emergency exists, so that appropriate clinical and environmental samples can be obtained. An awareness of such an attack may also depend on the practices of individual laboratories which can influence whether the etiologic agent is identified. Clearly detecting a water security threat will require the development of sensitive public health surveillance and monitoring systems. Close communication among state and local health departments and water utilities will be essential. If evidence from various sources in the community suggests the possibility of a waterborne attack, affected water utilities should be contacted immediately. Utilities should be questioned about recent changes in source water quality and changes in treatment practices, and/or if there have been disturbances in the distribution system such as a main break, pump failures, or a reported intrusion into a tank.

Local and state environmental health specialists and water utility operators may be the first to observe and recognize an environmental problem potentially caused by an attack. Development of rapid communication methods to assist in the interaction of public health and environmental personnel at the local, state, and Federal levels is critical. The interaction between the output from an EWS and the public health surveillance system will be critical.

Detection of an attack will depend heavily on the level of interest and awareness of the local and state public health departments. It will no doubt vary greatly. Many states have established surveillance systems that will detect illnesses, based on syndrome, that might be related to an attack, including illnesses such as viral hemorrhagic fevers. Poison control centers, might be the first organizations to detect illnesses related to unusual biologic and chemical toxins. Unusual patterns of school absences or an increased level of morbidity or mortality in nursing homes might be indicators of an outbreak.

8.8 SUMMARY AND CONCLUSIONS

Emergency planning by water utilities is not new. The American Water Works Association publishes a manual on emergency planning and the State of California has developed extensive emergency guidance for water utilities because of the potential devastation associated with earthquakes. Presidential Decision Directive 63 requires that federal agencies implement plans to protect the nation's infrastructure and the Bioterrorism Act of 2002 requires that all drinking water utilities serving 100,000 people or more, must conduct vulnerability assessments. However, in times of extreme crisis, it is one thing to have plans in place to deal with a threat, and it is something else to have functioning systems that provide for communication with all levels of government and the resources to deal with it. The goal of the consortium is to develop a prototype institutional structure and robust hardware and software systems to assist in identifying a water security threat and then to respond and mitigate it. Beyond security issues, the Federal government is also placing increased emphasis on the need to manage water quality in distribution networks. The proposed monitoring and modeling systems will provide the potential for utilities to exercise real-time control on water quality from the source water to the delivery point in the network.

The consortium intends to pursue the following activities:

- Development of a guidance manual on the procedures, protocols, costs, and the operational procedures for implementing an early warning system both for source waters and distributed drinking water
- Development of information transfer and security protocols for the acquisition, storage, and transfer of real-time monitoring and operational data in drinking water utilities
- Testing and application of water quality/hydraulic modeling to drinking water distribution systems
- Testing and application of models for simulating water quality and quantity in source waters

This effort will assist in satisfying or providing data to support a number of other objectives identified in EPA's Homeland Research Strategy. These are as follows:

- Improvement of models for more accurately predicating the spread of contamination
- Comparison of the results of dynamic modeling to steady-state modeling, which is commonly used by most utilities
- Evaluation of the intrusion point in a network to cause maximum spread of a contaminant
- The application of advanced inversion techniques for identifying the point of contamination given "down stream" monitoring results
- Assist in valve location and in identifying the location of automated valves to minimize the threat of contamination during a water security incident
- Development of risk maps using water quality simulation models in conjunction with GIS

ACKNOWLEDGMENTS

The authors gratefully acknowledge the assistance of the following individuals for their assistance in completing the research reported in this paper: Rafael Jusino Atresino, Luke Ceberio, Yue He, and Dr. Kirk Barrett and Dr. Fransico Artigas of CIMIC, and Dr. Yelena Yesha of the University of Maryland, Baltimore County, Messers Bruce Kiselica, Water Supply Branch Chief, U.S. EPA, Region II, Steven Schmidt, Vice President, American Water, and Barker Hamill, Water Supply Branch Chief, New Jersey Department of Environmental Protection

REFERENCES

Adam, N. R., V. Atluri, E. Bertino, and E. Ferrari, "A Content-Based Authorization Model for Digital Libraries," *IEEE Transactions Knowledge and Data Engineering*, 13(4): 705–716, 2002.

Arnold, C. P., and C. H. Dey, "Observing System Simulation Experiments: Past, Present and Future," *Bull. Amer. Meteor. Soc.* 67: 687–695, 1986.

Burns, N. L., C. A. Cooper, D. A. Dobbins, J. C. Edwards, and L. K. Lampe, "Security Analysis and Response for Water Utilities," in L. W. Mays (ed.), *Urban Water Supply Handbook*, McGraw-Hill, New York, pp. 20.1–20-24, 2002.

Burrows, W. D., and S. E. Renner, "Biological Warfare Agents as Threats to Potable Water," *Environ. Health Perspect.* 107(12): 975–984, 1999.

Burrows, W. D., and S. E. Renner, *Biological Warfare Agents as Potable Water Threats*. U.S. Army Combined Arms Support Command, Fort Lee, VA, page 10, 1998.

Clark, R. M., "Assessing the Etiology of a Waterborne Outbreak: Public Health Emergency or Covert Attack," in Jennifer Hatchett (ed.), *Proceedings of the First Water Security Summit*, Haested Press, Haested Methods, Waterbury, CT, pp. 170–179, 2002.

Clark, R. M., and R. A. Deininger, "Minimizing the Vulnerability of Water Supplies to Natural and Terrorist Threats," in *Proceedings of the American Water Works Association's IMTech Conference held in Atlanta*, GA, pp. 1–20, April 8–11, 2001.

Clark, R. M., and R. A. Deininger, "Protecting the Nation's Critical Infrastructure: The Vulnerability of U.S. Water Supply Systems," *J. Contin. Crisis Management* 8(2): 73–80, 2000.

Clark, R. M., E. E. Geldreich, K. R. Fox, E. W. Rice, C. H. Johnson, J. A. Goodrich, J. A. Barnick, and F Abdesaken, "Tracking a Salmonella Serovar Typhimurium Outbreak in Gideon, Missouri: Role of Contaminant Propagation Modeling," *J. Water Supply Res. Technol. - Aqua*. 45(4): 171–183, 1996.

Clark, R. M., L. Rossman, and L. Wymer "Modeling Distribution System Water Quality: Regulatory Implications," *J. Water Resour. Planning Management ASCE*. 121(6): 423–428, Nov/Dec 1995.

Deininger, R. A., and P. G. Meier, "Sabotage of Public Water Supply Systems," in R. A. Deininger, P. Literathy, and J. Bartram (eds.), *Security of Public Water Supplies*, NATO Science Series, 2. Environment, Vol. 66, Kluwer Academic Publishers, Dordrecht; page 76–80, 2000.

Fox, K. R., and D. A. Lytle, "Milwaukee's Crypto Outbreak Investigation and Reccomendations," *J. Am. Water Works Assoc.* 88(9): 87–94, 1996.

Grayman, W. M., R. A. Deininger, and R. M. Clark, "Vulnerability of Water Supply to Terrorist Activities," *CE News*, 14: 34–38, March 2002.

HR 3448 Public Health Security and Bioterrorism Preparedness and Response Act

International Life Sciences Institute, Risk Science Institute, *Early Warning Monitoring to Detect Hazardous Events in Water Supplies*. ILSI PRESS, Washington DC, 1999.

Jastrow, R., and M. Halem, "Simulation Studies Related to the Global Atmospheric Research Program," *J.Atmos. Sci.* 26: 1705–1718, 1970.

Lee, J. Y., and R. A. Deininger, "Detecting Accidental Contamination in Municipal Water Networks," *J. Am. Water Works Assoc.* 88(9): 87–94, 1992.

PL107-188. Public Health Security and Bioterrorism Preparedness and Response Act of 2002.

President. White Paper. The Clinton Administration Policy on Critical Infrastructure Protection: Presidential Decision Directive 63, 1998.

Rohaly G. D., and T. N. Krishnamurti, "An Observing System Simulation Experiment for Laser Atmospheric Wind Sounder," *J. Appl. Meteor*. 32: 1453–1471, 1993.

Rossman, L. A., R. M. Clark, and W. M. Grayman, "Modeling Chlorine Residuals in Drinking Water Distribution Systems," *J. Environ. Eng*. 120(4): 803–820, 1994.

Samet, H., *Application of Spatial Data Structures*, Addison-Wesley, Reading, MA, 1990.

U.S. EPA, "25 years of the Safe Drinking Water Act: History and Trends," EPA 816-R-99-007, Office of Water, December 3, 1999.

Wagner, R. J., H. C. Mattraw, G. F. F. Ritz, and B. A. Smith, Guidelines and Standard Procedures for Continuous Water-Quality Monitors: Site Selection, Field Operation, Calibration, Record Computation, and Reporting: Water-Resources Investigations Report 00-4252, 2000, available at: http//www.ciao.gov/CIAO_Document_Library/paper598.htm

CHAPTER 9
RESPONDING TO A CONTAMINATION THREAT IN A DRINKING WATER NETWORK: THE POTENTIAL FOR MODELING AND MONITORING

Robert M. Clark
Environmental Engineering and Public Health Consultant, Cincinnati, Ohio

Steven G. Buchberger
University of Cincinnati, Cincinnati, Ohio

9.1 INTRODUCTION

In response to the events of September 11, 2001 the U.S. Environmental Protection Agency (U.S. EPA) established a Water Protection Task Force (WPTF) with the responsibility for developing plans for the safety and security of the Nation's water and waste water systems. Security of water systems is not a new issue and the potential for natural, accidental, and purposeful contamination of water systems has been the subject of many studies. For example, in May 1998, President Clinton issued Presidential Decision Directive (PDD) 63 that outlined a policy on critical infrastructure protection including our nation's water supplies (President, 1998). However, it was only after September 11, 2001 that the water industry truly focused on the vulnerability of the nation's water supplies to security threats. In recognition of current water security concerns, President George Bush signed The Bioterrorism Act into law in June 2002 (PL 107-188). The Bioterrorism Act requires drinking water utilities to prepare Vulnerability Assessments and Emergency Response Plans. Each Vulnerability Assessment and the Emergency Response Plan will be unique.

There are nearly 60,000 community water supplies in the United States serving over 226 million people. Over 63 percent of these systems supply water to less than 2.4 percent of the population and 5.4 percent supply water to 78.5 percent of the population. Most of these systems provide water to less then 500 people. In addition, there are 140,000 noncommunity water systems that serve schools, recreational areas, trailer parks, etc. (U.S. EPA, 1999)

Some of the common elements associated with water supply systems in the United States are as follows:

- A water source which may be a surface impoundment such as a lake, reservoir, river, or groundwater from an aquifer.
- Conventional treatment facilities (surface supplies) including filtration, which removes particulates and potentially pathogenic microorganisms, followed by disinfection.
- Transmission systems which include tunnels, reservoirs and/or pumping facilities, and storage facilities.
- A distribution system carrying finished water through a system of water mains and subsidiary pipes to consumers.

Community water supplies are designed to deliver water under pressure and generally supply most of the water for fire fighting purposes. Loss of water or a substantial loss of pressure could disable fire-fighting capability, interrupt service, and undermine public confidence. This loss might result from sabotaging pumps that maintain flow and pressure or from disabling electric power sources. Many of the major pumps and power sources in water systems have custom designed equipment and could take months or longer to replace (Clark and Deininger, 2000).

Water systems are spatially diverse and many of the system components such as tanks and pumps are in isolated locations. Water distribution networks, therefore, have an inherent vulnerability to a variety of threats—physical, chemical, and biological—that may compromise the system's ability to reliably deliver safe water. Areas of vulnerability include (1) the raw water source (surface or groundwater), (2) raw water channels and pipelines, (3) raw water reservoirs, (4) treatment facilities, (5) connections to the distribution system, (6) pump stations and valves, and (7) finished water tanks and reservoirs. Each of these system elements presents unique challenges to the water utility in safeguarding the water supply (Clark and Deininger, 2001).

The ability of a water supply system to provide water to its customers can be compromised by destroying or disrupting key physical elements of the water system. Key elements include raw water facilities (dams, reservoirs, pipes, and channels), treatment facilities, and distribution system elements (transmission lines and pump stations).

Physical disruption may result in significant economic cost, inconvenience, and loss of confidence by customers but has a limited direct threat to human health. Exceptions to this generalization include (1) destruction of a dam that causes loss of life and property in the accompanying flood wave and (2) an explosive release of chlorine gas at a treatment plant.

Contamination is generally viewed as the most serious terrorist threat to water systems and is the threat addressed in this chapter. Chemical or biological agents could spread throughout a distribution system and result in sickness or death among the consumers. For some agents, the presence of the contaminant might not be known until emergency rooms reported an increase in patients with a particular set of symptoms. Even without serious health impacts, just the knowledge that a group had breached a water system could seriously undermine customers' confidence in the water supply.

In this chapter, distribution system contamination threats will be characterized and the potential for monitoring in combination with water quality and hydraulic modeling, for responding to these threats, will be discussed. The fundamentals of monitoring and modeling are explained and two case studies are presented. One of the case studies illustrates the important interactions between monitoring, modeling, and data interpretation. The second case study discusses the use of public health and environmental models in combination with hydraulic modeling using a waterborne outbreak as an example. The use of water quality modeling to identify the cause of the outbreak will be discussed.

9.2 NETWORK CONTAMINATION THREATS

The distribution network is the most vulnerable part of the water system, and there are many opportunities to breach the network (Clark and Deininger, 2001). A terrorist with a small pump could target a section of the system and deliberately inject a contaminant into the network from a home, an office building, warehouse, fire hydrant, or pumping station. Backflow prevention devices which can provide protection against this type of attack are typically installed only in industrial and commercial facilities or can be easily disengaged. They are usually not installed in residences or at fire hydrants.

The advantage of using residential, commercial, or warehouse service connections as an entry point to the pipe network is the privacy which provides an opportunity for contaminant injection without interference. Using easily available equipment, it would be possible to connect a pump to force a contaminant into a distribution system. Injecting a contaminant directly into a larger distribution main would require a relatively small pump. Depending on the hydraulic design and demand conditions a relatively large portion of the distribution system could be impacted, potentially affecting a large number of individuals and possibly resulting in many casualties. Several studies have demonstrated this effect and it would be very difficult to track the spread of the contaminant and to find the source of the injection (Vasconcelos et al., 1996 and 1997). A contaminant also could be injected at multiple points simultaneously causing a great deal of confusion and giving the impression of massive contamination. If the agent were biological in nature, the injection event would be over before anyone was aware that an attack had occurred.

Although fire hydrants provide another access point and would provide the opportunity to tap into a distribution line such that a larger area could be contaminated than might occur with a residential service connection, there are limitations on using this approach. For example, fire hydrants are visible and injecting a contaminant into a water distribution system via a fire hydrant would require a pumper truck (fire engine) or some apparatus to reach the pressures and flow rates necessary to cause a significant impact.

One of the most important protections against an attack on a distribution system is the disinfectant residual maintained in the system. If chlorine is the disinfectant used, the chlorine residual of the water entering the distribution system is usually at least 1 mg/L. However, the residual will decay as water moves through the system. In many systems there may be only trace levels of chlorine at locations that are at the greatest distance from the treatment plant. Several studies have confirmed this fact Many systems exhibit very long residence times ranging from one to 10 days which compounds this loss of residual (Clark et al., 1991 and 1993; Buchberger et al., 2003).

The type of disinfectant being used in the network is important with regard to protection against security threats. Chlorine is a much more powerful disinfectant than monochloramine but many utilities are considering the substitution of chloramines for chlorine to reduce the formation of chlorinated DBPs in distributed water. Chloramines at the concentrations used in drinking water treatment are relatively ineffective against most viruses, parasites, and spores (Rice, 2001). Obviously this shift has increased the vulnerability of many systems to contamination with microbial agents, as well as some chemical agents, that would be effectively neutralized by free chlorine. In addition, there are many drinking water systems, mostly of smaller size, that do not practice chlorination at all. This includes unchlorinated groundwater distributed in national parks. Many military bases in the United States use local municipal water whenever feasible and both elevated and ground level storage tanks are especially vulnerable. Many storage tanks are vented to the atmosphere and contaminants could be poured into the storage tank through access hatches. All distribution systems that are not disinfected have to be considered as vulnerable to infectious agents that are stable in water. An example of this type of problem is discussed in this paper.

Another distribution system vulnerability is the possibility of creating an intentional cross connection between the drinking water and wastewater systems. This would provide a continuous supply of fecal contamination until the cross connection is discovered. Perhaps the greatest risk from a cross connection would occur at a remote location in the distribution system where disinfectant residuals are lowest.

If a contamination event occurs, one of the following scenarios may apply:

- The identity of the contaminant or the class of contaminant is known at the time of the event.
- The type of contaminant is suspected.
- Nothing is known about the contaminant.

The first objective would be to confirm that a contamination event has occurred, then identify the agent, and determine its concentration profile within the system. If the identity of the contaminant is known, as in the case of a major chemical spill in a source water, monitoring would be initiated immediately (Clark et al., 1990). Steps could be taken, such as closing the water intake when the slug goes by in order to protect the consumers and sampling would be initiated to characterize the contaminant profile.

Under the second scenario, initial monitoring may focus on the suspected agent. If the agent is not identified, then a screening analysis might be used. Under the third scenario, the only recourse is to apply screening methods.

Once a contamination event has been verified and characterized, it will be necessary to determine the fate of the contaminant in the system. Not only will this require sampling locations to be established in the system but will also require an understanding of the hydraulic behavior of the water in the distribution system. The final step will be to purge the contaminant from the system and confirm that the contaminant is no longer present.

In responding to a contamination threat it is important to effectively use monitoring to identify that a threat exists and then to understand movement of water in the system. Modeling techniques, in conjunction with monitoring data, could be used to identify the source of the threat, where contaminants may move in the system and to suggest remediation approaches. In the following sections the fundamental principles of effective monitoring and the application of hydraulic/water quality models are discussed.

9.3 DEVELOPMENT OF MONITORING SYSTEMS

Monitoring systems designed to warn of security threats, frequently called early warning systems (EWS), are intended to reliably identify low probability, high impact contamination events in source or distributed water. The following requirements were identified by an International Life Sciences Institute (ILSI) working group report on EWSs for hazardous events in water (ILSI, 1999):

- Provide warning in sufficient time to respond to the contamination event and prevent exposure of the public to the contaminant.
- Is capable of detecting all potential contamination threats.
- Can be operated remotely.
- Can identify the point at which the contaminant was introduced.
- Has a low rate of false positive and false negative results.
- Provides continuous, year-round surveillance.

- Produces results with acceptable accuracy and precision.
- Requires low skill and training.
- Is affordable to the majority of public water systems.

Although there is no EWS in existence that meets all of the above criteria, there are some conventional monitoring systems that might satisfy some of these elements. However, these types of systems will provide the only opportunity to prevent the public from being exposed to a covert attack on the water supply from biological, chemical, or radiological agents. It is unlikely that an EWS would be capable of detecting all potential contamination threats. However, any EWS in existence should have a strong relationship to public health surveillance monitoring. An EWS is only one of many tools including physical security, infrastructure hardening, and optimized treatment that can be implemented to protect public water supplies.

9.3.1 Water Quality Models

Modeling the movement of a contaminant within the distribution systems from various points of entry (e.g., treatment plants) to water users is based on the following three principles (Clark and Grayman, 1998):

- Conservation of mass within differential lengths of pipe
- Complete and instantaneous mixing of the water entering pipe junctions
- Appropriate kinetic expressions for the growth or decay of the substance as it flows through pipes and storage facilities

This change in concentration can be expressed by the following differential equation:

$$\frac{\partial C_{ij}}{\partial t} = -v_{ij}\frac{\partial C_{ij}}{\partial x} + k_{ij}C_{ij} \qquad (9.1)$$

where C_{ij} = substance concentration (g/m^3) at position x and time t in the link between nodes i and j
v_{ij} = flow velocity in the link (equal to the link's flow divided by its cross-sectional area) (m/s)
k_{ij} = rate at which the substance reacts within the link (s^{-1})

According to Eq. (9.1) the rate at which the mass of material changes within a small section of pipe equals the difference in mass flow into and out of the section plus the rate of reaction within the section. It is assumed that the velocities in the links are known beforehand from the solution to a hydraulic model of the network. Dispersive transport is assumed to be negligible, though this premise may not be a accurate in peripheral zones of the network where laminar and stagnant flow may occur (Buchberger et al., 2003). In order to solve Eq. (9.1) we need to know C_{ij} at $x = 0$ for all times (a boundary condition) and a value for k_{ij}.

Equation (9.2) represents the concentration of material leaving the junction and entering a pipe

$$C_{ij} = \frac{\Sigma_k Q_{ij}C_{kj}}{\Sigma Q_{kj}} \qquad (9.2)$$

where C_{ij} = the concentration at the start of the link connecting node i to node j (i.e., where $x = 0$)
C_{kj} = the concentration at the end of a link, in mg/L
Q_{kj} = the flow from node k to node j

Equation (9.2) implies that the concentration leaving a junction equals the total mass of a substance flowing into the junction divided by the total flow into the junction.

Storage tanks can be modeled as completely mixed, variable volume reactors in which the change in volume and concentration over time is as follows:

$$\frac{dV_s}{dt} = \sum_k Q_{ks} - \sum_i Q_{sj} \qquad (9.3)$$

$$\frac{dV_s C_s}{dt} = \sum_k Q_{ks} C_{ks} - \sum_i Q_{sj} C_{ij} + k_{ij}(C_s) \qquad (9.4)$$

where C_s = the concentration for tank s, in mg/L
dt = change in time, in seconds
Q_{ks} = flow from node k to s, in ft^3/s (m^3/s)
Q_{sj} = flow from node s to j, in ft^3/s (m^3/s)
dV_s = change in volume of tank at nodes, in ft^3 (m^3)
V = volume of tank at nodes, in ft^3 (m^3)
C_{ks} = concentration of contaminant at end of link, in mg/ft^3 (mg/L)
k_{il} = decay coefficient between node i and j, in s^{-1}

There are currently several models available for modeling both the hydraulics and water quality in a drinking water distribution system. Most of this discussion will focus on a U.S. EPA developed hydraulic/contaminant propagation model called EPANET (Rossman et al., 1994), which is based on mass transfer concepts (transfer of a substance through another on a molecular scale). Another approach to water quality contaminant propagation to be discussed is the approach developed by Biswas et al. (1993). This model uses a steady-state transport equation that takes into account the simultaneous advective transport of chlorine in the axial direction, diffusion in the radial direction, and consumption by first-order reaction in the bulk liquid phase. Islam (1995) developed a model called QUALNET, which predicts the temporal and spatial distribution of chlorine in a pipe network under slowly varying unsteady flow conditions. Boulos et al. (1995) proposed a technique called the event driven method (EDM), which is based on a "next-event" scheduling approach, which can significantly reduce computing times.

Solution Methods. There are several different numerical methods that can be used to solve contaminant propagation equations. Four commonly used techniques are Eulerian finite-difference method (FDM), Eulerian discrete volume method (DVM), Lagrangian time-driven method (TDM), and Lagrangian event-driven method (EDM).

The FDM approximates derivatives with finite-difference equivalents along a fixed grid of points in time and space. Islam (1995) used this technique to model chlorine decay in distribution systems.

The DVM divides each pipe into a series of equally sized, completely mixed volume segments. At the end of each successive water quality time step, the concentration within each volume segment is first reacted and then transferred to the adjacent downstream segment. This approach was used in the models that were the basis for early U.S. EPA studies.

The TDM tracks the concentration and size of a nonoverlapping segment of water that fills each link of a network. As time progresses, the size of the most upstream segment in a link increases as water enters the link. An equal loss in the size of the most downstream segment occurs as water leaves the link. The size of these segments remains unchanged.

The EDM is similar to TDM, except that rather than update an entire network at fixed time steps, individual link-node conditions are updated only at times when the leading segment in a link completely disappears through this downstream node.

EPANET. EPANET has been a key component in providing the basis for water quality modeling in the United States (Rossman, 1994). There are many commercially available hydraulic models that incorporate water quality models as well. EPANET is a computer program based on the extended period simulation (EPS) approach to solving hydraulic behavior of a network. In addition, it is designed to be a research tool for modeling the movement and fate of constituents within drinking water distribution systems. The model is available to be downloaded from the U.S. EPA website.

EPANET uses the Hazen-Williams formula, the Darcy-Weisbach formula, or the Chezy-Manning formula for calculating the head loss in pipes. All nodes have their elevations above sea level specified and tanks and reservoirs are assumed to have a free water surface. The hydraulic head is simply the elevation of the surface above sea level. The tank depth is assumed to change in accordance with the following equation:

$$dy = (q/A)\,dt \tag{9.5}$$

where dy = change in water level, in ft (m)
q = flow rate into (+) or out of (−) tank in ft³/s (m³/s)
A = crossectional area of the tank, in ft (m)
dt = time interval, in seconds

It is assumed that water usage rates, external water supply rates, and source concentrations at nodes remain constant over a fixed period of time, although these quantities can change from one period to another. Nodes are junctions where network components connect to one another (links), as well as tanks and reservoirs. The default period interval is 1 h but can be set to any desired value. Various consumption or water usage patterns can be assigned to individual nodes or groups of nodes.

EPANET solves a series of equations for each link using the gradient algorithm. Gradient algorithms provide an interactive mechanism for approaching an "optimal solution" by calculating a series of slopes that lead to better and better solutions. Flow continuity is maintained at all nodes after the first iteration. The method easily handles pumps and valves.

The set of equations solved for each link (between nodes i and j) and each node k is as follows:

$$h_i - h_j = f(q_{ij}) \tag{9.6}$$

$$\sum_i q_{ik} - \sum_j q_{kj} - Q_k = 0 \tag{9.7}$$

where q_{ij} = flow in link connecting nodes i and j, in ft³/s (m³/s)
h_j = hydraulic grade line elevation at node i (equal to elevation head plus pressure head), in ft (m))
Q_k = flow consumed (+) or supplied (−) at node k, in ft³/s (m³/s)
$f(\cdot)$ = functional relation between head loss and flow in a link

The set of equations for each storage node (tank or reservoir) in the system is as follows:

$$\frac{\partial y_s}{\partial t} = \frac{q_s}{A_s} \tag{9.8}$$

$$q_s = \sum_i q_{is} - \sum_j q_{sj} \tag{9.9}$$

$$h_s = E_s + y_s \tag{9.10}$$

where y_s = height of water stored at nodes, in ft (m)
A_s = cross-sectional area of storage nodes (infinite for reservoirs), in ft² (m²)
E_s = elevation of nodes, in ft (m)
q_s = flow into storage nodes, in ft³/s (m³/s)
t = time, in seconds
h_s = height of water in storage tanks, in ft (m)

Equation (9.7) expresses conservation of water volume at a storage node. Equations (9.8) and (9.9) express the same relationship for pipe junctions and Eq. (9.10) represents energy loss or gain due to flow within a link. For known initial storage node levels y_s, at time zero, Eqs. (9.10) and (9.7) are solved for all flows q_{ij} and heads h_i using Eq. (9.9) as a boundary condition. This is called hydraulically balancing the network and is accomplished by using an iterative technique to solve the resulting nonlinear equations.

After a network hydraulic solution is obtained, flow into (or out of) each storage node q_s is found using Eq. (9.8) and used in Eq. (9.7) to find new storage elevations after a time step dt. This process is then repeated for all the following time steps for the remainder of the simulation period.

9.3.2 Water Quality Simulation

EPANET uses the flows from the hydraulic simulation to track the propagation of contaminants through a distribution system. A conservation of mass equation is solved for the substance within each link between nodes i and j as follows:

$$\frac{\partial C_{ij}}{\partial t} = \frac{q_{ij}}{A_{ij}}\left[\frac{\partial C_{ij}}{\partial x_{ij}}\right] + \theta(C_{ij}) \tag{9.11}$$

where C_{ij} = concentration of substance in link i, j as a function of distance and time
x_{ij} = distance along link i, j, in ft (m)
q_{ij} = flow rate in link i, j at time t, in ft³/s (m³/s)
A_{ij} = cross-sectional area of link i, j, in ft² (m²)
$\theta(C_{ij})$ = rate of reaction of constituent within link i, j, in mass/ft³/s (mass/m³/s)

Equation (9.11) must be solved with known initial conditions at time zero. The following boundary condition at the beginning of the link, i.e., at node i where $x_{ij} = 0$ must hold

$$C_{ij}(0,t) = \frac{\sum_k q_{ki} C_{ki}(L_{ki}, t) + M_i}{\sum_k q_{ki} + Q_{si}} \tag{9.12}$$

where L_{ki} = length of link k, i
M_i = substance mass introduced by any external source at node i
Q_{si} = source's flow rate

and the summations are made over all links k, i that flow into the head node i of the link i, j.

The boundary condition for link k, i depends on the concentrations at the head of the nodes of all links k, i that flow into link i, j. Equations (9.11) and (9.12) form a coupled set of differential/algebraic equations over all links in the network. These equations are solved within EPANET using the DVM, which was described earlier.

Water quality time steps are chosen to be as large as possible without causing the flow volume of any pipe to exceed its physical volume. Therefore the water quality time step dt_{wq} source cannot be larger than the shortest time of travel through any pipe in the network, i.e.,

$$dt_{wq} = \text{Min}\left\{\frac{V_{ij}}{q_{ij}}\right\} \text{ for all pipes } i, j \qquad (9.13)$$

where V_{ij} is the volume of pipe i, j and q_{ij} is the flow rate of pipe i, j.

Pumps and valves are not a part of this determination because transport through them is assumed to occur instantaneously. Based on this water quality time step, the number of volume segments in each pipe (n_{ij}) is

$$n_{ij} = \text{INT}\left\{\frac{V_{ij}}{q_{ij} dt_{wq}}\right\} \qquad (9.14)$$

where INT (x) is the largest integer less than or equal to x. There is both a default limit of 100 for pipe segments or the user can set dt_{wq} to be no smaller than a user-adjustable time tolerance

Reaction Rate Model. Equation (9.1) provides a mechanism for evaluating the reaction of a substance as it travels through a distribution system (Rossman et al., 1994). Reactions can occur in the bulk phase or at the pipe wall. EPANET models both types of reactions using first-order kinetics. In general, within any given pipe, material in the bulk phase and at the pipe wall will decrease according to the following equation:

$$q(C) = -k_b C - \left[\frac{k_f}{R_h}\right](C - C_w) \qquad (9.15)$$

where $q(C)$ = total reaction rate
k_b = first-order bulk reaction rate, in s^{-1}
C = substance concentration in the bulk flow, in mass/ft^3 (mass/m^3)
k_f = mass transfer coefficient between the bulk flow and the pipe wall, in ft/s (m/s)
R_h = hydraulic radius of pipe (pipe radius/2), in ft (m)
C_w = substance concentration at the wall, in mass/ft^3 (mass/m^3)

Assuming a mass balance for the substance at the pipe wall yields:

$$k_f(C - C_w) = k_w^c C_w \qquad (9.16)$$

where k_w is the rate of chlorine demand at wall (wall demand), in ft/s (m/s).

Equation (9.16) pertains to the growth or decay of a substance, with mass transfer to or from the pipe wall depending on whether the sign of the equation is positive or negative. A negative sign means decay and a positive sign means growth.

There are three coefficients used by EPANET to describe reactions within a pipe. These are the bulk rate constant k_b and the wall rate constant k_w, which must be determined empirically and supplied as input to the model. The mass transfer coefficient is calculated internally by EPANET.

9.3.3 Other Features

EPANET can also model the changes in age of water and travel time to a node. The percentage of water reaching any node from any other node can also be calculated. Source tracing is a useful tool for showing the degree that water from a given source blends with that from other sources, and how this pattern changes over time.

The following input steps should be taken before using EPANET:

- Identify all network components and their connections. These components include pipes, pumps, valves, storage tanks, and reservoirs.
- Assign unique ID numbers to all nodes. These numbers must be between 1 and 2,147,483,647, but do not have to be in any specific order.
- Assign ID numbers to each link (pipe, pump, or valve). Both a link and a node can have the same ID number.
- Collect information on the following system parameters: diameter, length, roughness, and minor loss coefficient for each pipe.
- Characteristic operating curve for each pump.
- Diameter, minor loss coefficient, and pressure or flow setting for each control valve.
- Diameter and lower and upper water levels for each tank.
- Control rules that determine how pump, valve, and pipe settings change with time, tank water levels, or nodal pressures.
- Changes in water demands for each node over the time period being simulated.
- Initial water quality at all nodes and changes in water quality over time at source nodes.

Output from EPANET includes the following:

- Color-coded network maps
- Time series plots
- Tabular reports
- Duration and rate of flow
- Pressure at the node
- Concentration at the node

9.4 CASE STUDIES

In this section, two case studies are presented to illustrate how monitoring/modeling might be used to analyze data to determine whether or not a contamination event had occurred. The first case study was conducted at the North Marin Water Authority in California in order to track the impact of two water sources, one of very high quality and one of poor quality. Data were collected in the network and modeling applied to understand the mixing properties within the network itself (Clark, 2000). The second case study was a *Salmonella* outbreak in Gideon MO which resulted in a large percentage of the residents in a small community becoming very ill and in the death of seven nursing home residents. The community used ground water which was not disinfected (Clark, 2000).

9.4.1 North Marin Water Authority

At the time of the study, the North Marin Water District (NMWD) served a suburban population of 53,000 people who live in or near Novato, California. NMWD used two sources of water; Stafford Lake and the North Marin Aqueduct. The North Marin Aqueduct was a year-round source, but Stafford Lake was in use only during the warm summer months when precipitation was low and demand was high. Novato, the largest population center in the NMWD service area, was located in a warm inland coastal valley with a mean annual rainfall of 68.58 cm (27 in). There was virtually no precipitation during the growing season from May through September. Eighty-five percent of total water use was residential and the service area contained 13,200 single-family detached homes, which accounted for 65 percent of all water use (Clark, et al., 1994).

The water quality of the two sources differed greatly. The Stafford Lake water had a high humic content and was treated with conventional treatment and pre-chlorination doses of between 5.5 and 6.0 mg/L. The treated water had a residual of 0.5 mg/L when it left the treatment plant clearwell. Total trihalomethane formation potential (TTHMFP) levels in the Stafford Lake water were very high. The North Marin Aqueduct water was derived from a Raney Well Field along the Russian River. Although technically considered, groundwater, the source water, contained a high proportion of naturally surface filtered water. Disinfection was the only treatment received by the Aqueduct water and it was very low in precursor material with a correspondingly low TTHMFP. Both sources carried a residual chlorine level of approximately 0.5 mg/L when the water entered the system.

Figure 9.1 shows, a schematic of the major pipes in the service area distribution system, the major tanks and pumps in the North Marin network. Depending on the time of year and time of day, water entered the system from either one or both of the sources. The North Marin Aqueduct source operated year-round, 24 hours per day. The Stafford Lake source operated only during the peak demand period from 6:00 a.m. in the morning to 10:00 p.m. at night. EPANET was used to model the system hydraulics including the relative flow from each source, TTHMs and chlorine residual propagation (Rossman et al., 1994). The model was calibrated based on a comparison of simulated versus actual tank levels for the May 27 to 29, 1992 period of operation. The dynamic nature of the system led to both variable flow and quality conditions in the network. Flow directions frequently reversed within a given portion of the network during a typical operating day. The U.S. EPA designed a sampling protocol and sent a team of investigators to work with the NMWD staff for the period May 27 to 29, 1992. Figure 9.2 shows the time formation curves for TTHMs for the two source waters. These figures illustrate the significant difference in water quality from the two sources.

Figure 9.3 shows the TTHM sampling results at node 120. As can be seen, the sampling results show wide variation over a 48-h period. If one were attempting to identify a contamination event it would be very difficult to interpret these results. However, Fig. 9.4 shows the percentage of flow from the Stafford Lake source, based on model predictions for the same time period. When the predicted Stafford Lake source is superimposed on the sampling data, it is easy to see that the wide variations in sampling measurements are the result of the blending of the two sources in the system (Fig. 9.5). The same pattern is seen in Figs. 9.6 to 9.8 for node 105.

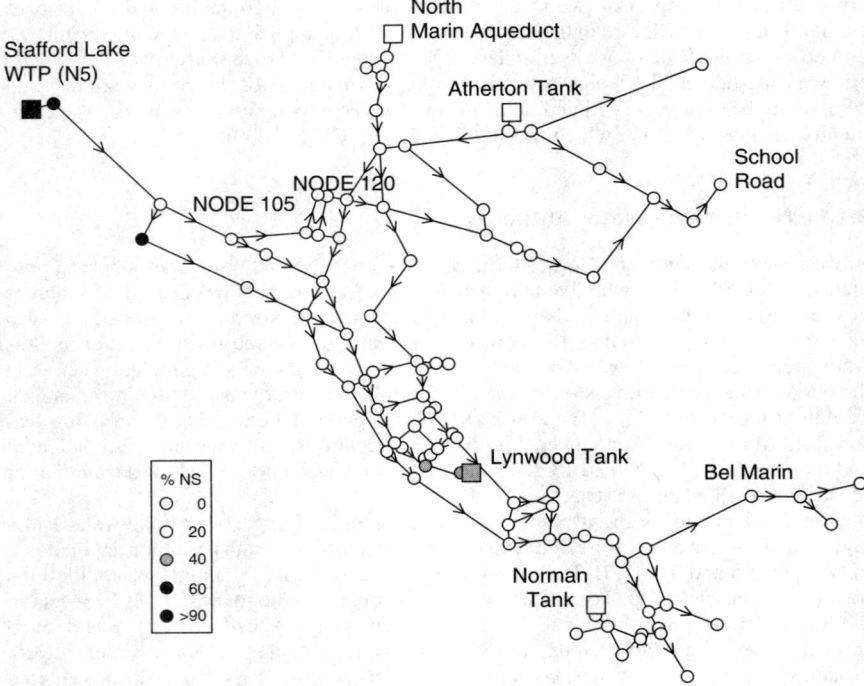

FIGURE 9.1 Schematic of North Marin system.

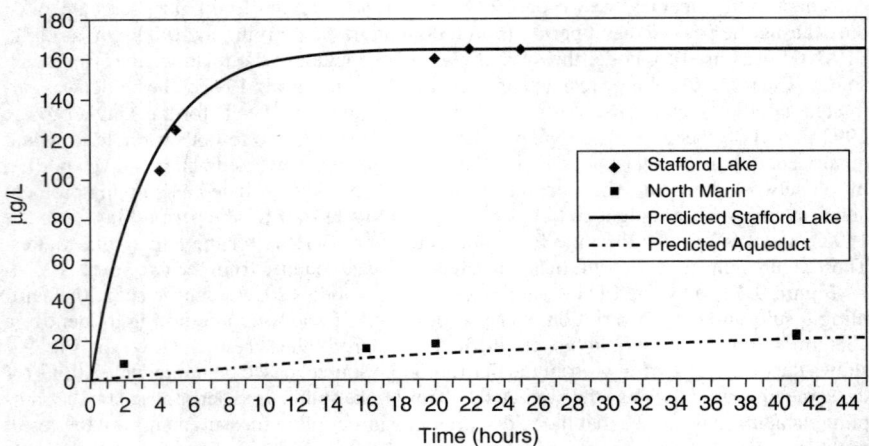

FIGURE 9.2 TTHM formation for Stafford Lake and North Marin aqueduct.

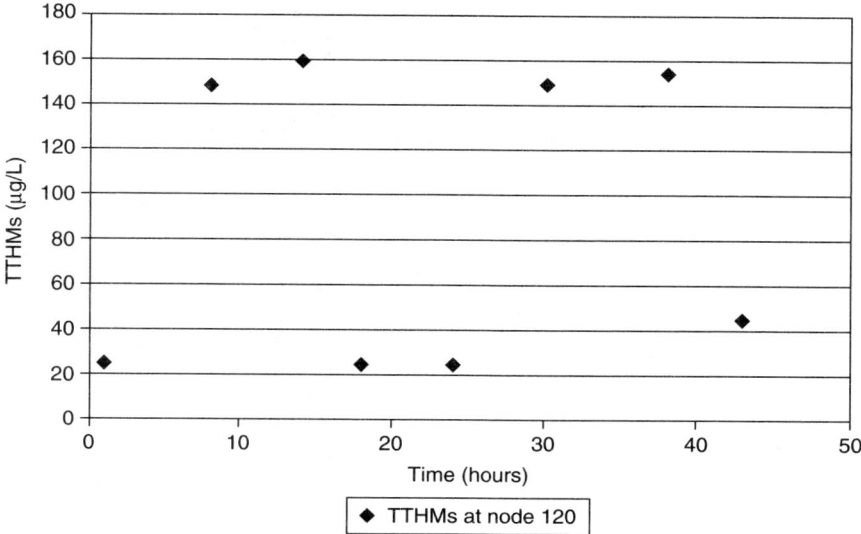

FIGURE 9.3 Actual TTHMs at node 120 versus time in hours.

If the data reviewed above were from a contamination threat, it is vital to link both the monitoring data and the network modeling results. This analysis would provide very useful information on the source of the contamination event and provide the ability to predict the profile of the contaminant throughout the network. In addition, the model might provide insight into potential remediation strategies such as shutting down or opening specific valves to flush the contaminant out of the system.

9.4.2 Gideon Waterborne Outbreak

Most terrorist attacks to-date in the United States have been overt attacks. These include the attack on the World Trade Center in 1993, the destruction of the Alfred P. Murrah office building in Oklahoma City in 1995, and of course the attack against the World Trade Center on September 11, 2001. However, in the case of a covert attack, disease and illness could propagate throughout the community before the attack was recognized. In fact, in the case of a covert attack, the first indication that an attack had occurred would be an increased demand on the public health system. The ability to identify an etiologic agent associated with a covert attack will depend on the timely recognition that a public health emergency exists, so that appropriate clinical and environmental samples can be obtained (Clark, 2001). Detecting such an attack will require the development of sensitive public health surveillance and monitoring systems.

The ability to differentiate between public health emergencies and routine disease in the general population and a covert attack will be critical for the identification and control of potential disease outbreaks. Water- and foodborne outbreak disease reporting provides an excellent example as to how a covert attack might be recognized. They also illustrate how effective, or ineffective, surveillance monitoring might be in recognizing the fact that there is an unexpected incidence of disease in the population. Close communication among state and local health departments and water utilities will be essential. If evidence from various

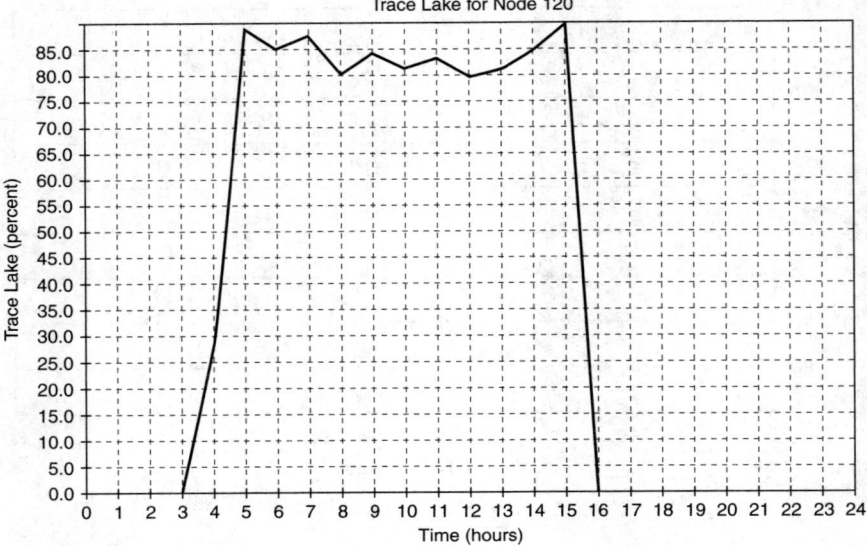

FIGURE 9.4 Percent of water from Stafford Lake at node 120.

sources in the community suggests the possibility of a waterborne attack, affected water utilities should be contacted immediately. Utilities should be questioned about recent changes in source water quality and changes in treatment practices, and/or if there have been disturbances in the distribution system such as a main break, pump failures or a reported intrusion into a storage tank.

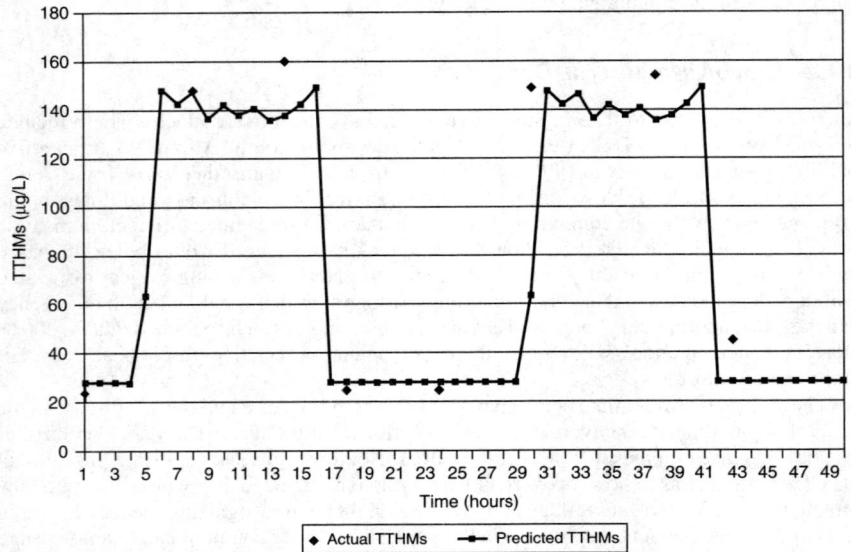

FIGURE 9.5 Predicted versus actual TTHMs at node 120.

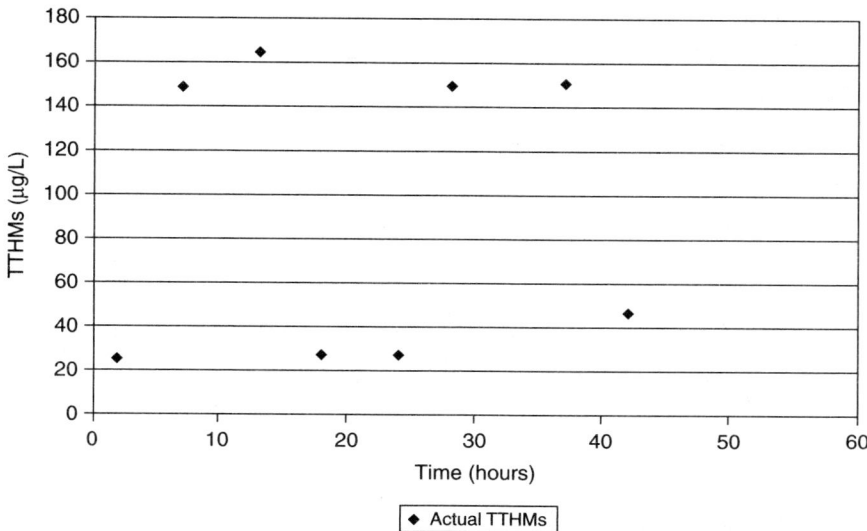

FIGURE 9.6 Actual TTHMs at node 105 versus time.

Detection of an attack will depend heavily on the level of interest and awareness of the local and state public health departments. Many states have established surveillance systems that will detect illnesses, based on syndrome, that might be related to an attack including illnesses such as viral hemorrhagic fevers. Poison control centers, might be the first organizations to detect illnesses related to unusual biologic and chemical toxins.

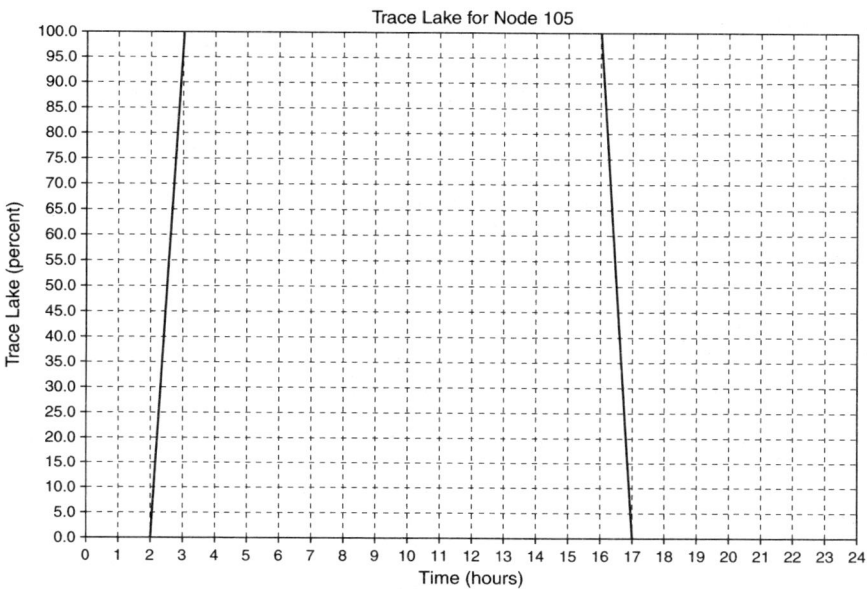

FIGURE 9.7 Percent of water from Stafford Lake at node 105.

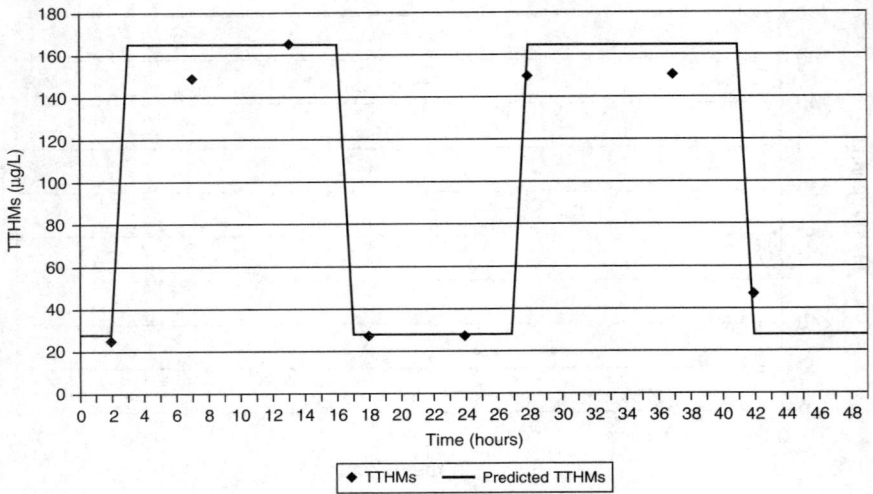

FIGURE 9.8 Predicted versus actual TTHMs at node 105.

Unusual patterns of school absences or an increased level of morbidity or mortality in nursing homes might be indicators of an outbreak.

Figure 9.9 shows the increased number of absentees in the Gideon Missouri schools from the period November 1 to December 18. As can be seen there is a spike of absences on November 12. If a covert attack were to be launched, these might be the only data available to warn that an attack had occurred.

The Gideon example also illustrates how water quality modeling might be used to help identify the source of an attack. Some of the same elements that existed in the Gideon outbreak may be present in a covert biological attack. With the increasing sophistication of water quality propagation modeling, it has now become possible to apply these types of models to waterborne disease outbreaks

Gideon, is located in Southeastern Missouri. In 1990, the population of Gideon was 1104 with a median income of $14,654 (25 percent of the population was below the poverty level) and with an unemployment rate of 11.3 percent. Major employers in Gideon were a nursing home with 68 residents and a staff of 62 and the Gideon schools with 444 students (kindergarten through 12th grade). The topography is flat and the predominant crop is cotton. The municipal water system had two elevated tanks. One tank was a 189 m^3 (50,000 gal) elevated tank and the other was a 378 m^3 (100,000 gal) elevated tank. Another tank in the area was privately owned by the Cotton Compress Corporation and had a volume of 378 m^3 (100,000 gal). The private tank was constructed in 1930 and was heavily rusted and in an obvious state of disrepair. It was connected via a backflow prevention valve to the city water system and used primarily for fire protection at the Cotton Compress, a local cotton baling industry. Both 378 m^3 (100,000 gal) tanks had broad flat roofs while the smaller municipal tank had a much steeper pitch.

The Gideon municipal water system was originally constructed in the mid-1930s and obtained water from two adjacent 396 m (1300 ft) deep wells. The well waters were not disinfected at the time of the outbreak. The distribution system consisted primarily of small diameter (5, 10, 15 cm (2, 4, and 6 in)) unlined steel and cast iron pipes. Tuberculation and corrosion were a major problem in the distribution pipes. Raw water temperatures were unusually high

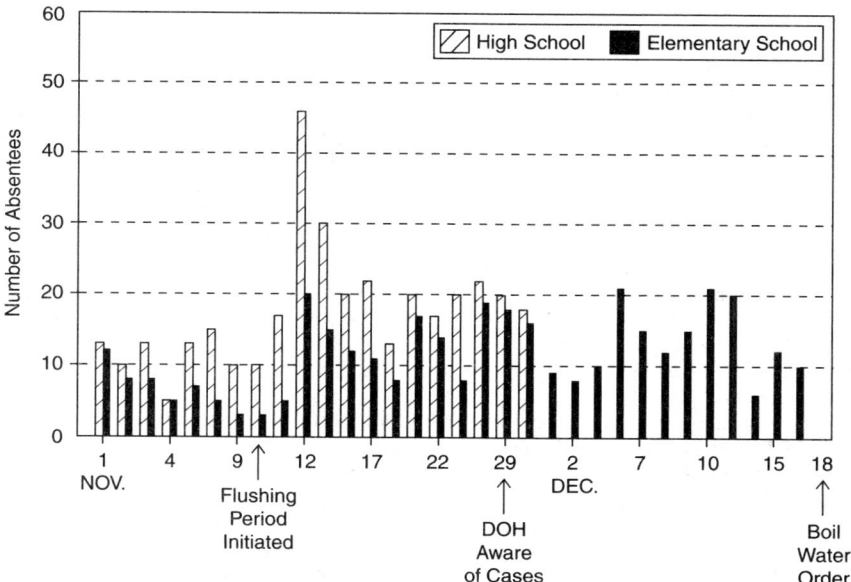

FIGURE 9.9 Number of absentees in Gideon Schools during outbreak.

for a ground water supply (14°C (58°F)) because the system overlies a geologically active fault. Under low flow or static conditions the water pressure was close to 3.5 kg/cm² (50 psi). However, under high flow or flushing conditions the pressure dropped dramatically. These sharp pressure drops were evidence of major problems, such as heavily tubercled pipe, in the Gideon distribution system. In the Cotton Compress yards, water was used for equipment washing, in rest rooms, and for consumption. The pressure gradient between the Gideon system and the Cotton Compress system was such that the private storage tank would overflow when the municipal tanks were filling. To prevent this from occurring a valve was installed in the influent line to the Cotton Compress tank. This same pressure differential kept water in the Cotton Compress tank unless there was a sudden demand in the warehouse area. The entire Cotton Compress water system was isolated from the Gideon system by a back flow prevention valve. There were no residential water meters in the Gideon system and residents paid a flat service rate ($11.50 per month) for both water and sewage service. The municipal sewage system operated by a gravity flow with two lift stations and served 429 households.

On November 10, in response to numerous taste and odor complaints, a sequential flushing program involving all 50 hydrants in the system was initiated. The flushing program was started in the morning and continued through the entire day. Each hydrant was flushed for 15 min at an approximate rate of 2.8 m³/min (750 gal/min).

On November 29, the Missouri Department of Health (DOH), became aware of two high school students from Gideon who were hospitalized with culture confirmed salmonellosis (Clark et al., 1996). Within 2 days five additional patients living in Gideon were hospitalized with salmonellosis (one student, one child from a day care, two nursing home residents, and one visitor to the nursing home). The State Public Health Laboratories identified the isolates as dulcitol negative *Salmonella* and the CDC laboratories identified the organism as serovar *typhimurium*. Interviews conducted by the DOH suggested that there were no food exposures common to a majority of the patients. All of the ill persons had consumed municipal water.

In December of 1993, 6 to 9 cases of diarrhea were reported at a local nursing home in Gideon, Missouri, raising the possibility of a waterborne outbreak (Clark et al., 1996). After an initial investigation by the Missouri DOH, the Missouri Department of National Resources (DNR) was contacted and water samples were taken at various points in the system (between December 17 and 21, 1993). Several samples were positive and yielded 1 to 6 total coliforms (TC) per 100 mL and a few samples were fecal coliform (FC) positive. Several other samples yielded results that were too numerous to count (TNTC) for coliforms and were also FC positive. Original speculation regarding the cause of the outbreak focused on a water tank located on private property.

The Missouri DNR was informed that the DOH suspected a water supply link to the outbreak. Water samples collected by the DNR on December 16 were positive for FC. On December 18, the city of Gideon, as required by the DNR, issued a boil water order. Signs were posted at the city hall, in the grocery store, and two area radio stations announced the boil water order.

Several water samples collected by the DNR on December 20 were also found to be FC positive. On December 23, a chlorinator was placed on line at the city well by the DNR, and nine samples were collected by the DOH and the DNR from various sites in the distribution system. None of the samples contained chlorine but one sample collected from a fire hydrant was positive for dulcitol-negative *Salmonella* serovar *typhimurium*. The Missouri DOH had informed the CDC about the outbreak in Gideon in early December and requested information about dulcitol negative *Salmonella* serovar *typhimurium*. On December 17, the DOH informed the CDC that contaminated municipal water was the suspected cause of the outbreak and (on December 22) invited the CDC to participate in the investigation. A flyer explaining the boil water order jointly produced by the DOH and the DNR, was placed in the mailbox of all of the homes in Gideon on December 29 and the privately owned water tower was physically disconnected from the municipal system on December 30. The DNR mandated that Gideon permanently chlorinate their water system. At the end of the study the EPA provided input to the DNR on the criteria necessary to lift the boil water order (Angulo et al., 1997).

Through January 8, 1994 the DOH had identified 31 cases with laboratory confirmed salmonellosis associated with the Gideon outbreak. The State Public Health Laboratories identified 21 of these isolates as dulcitol negative *Salmonella* serovar *typhimurium*. Fifteen of the 31 culture confirmed patients were hospitalized (including two patients hospitalized for other causes and who developed diarrhea while in the hospital). The patients were admitted to 10 different hospitals. Two of the patients had positive blood cultures, seven nursing home residents exhibiting diarrheal illness died, four of whom were culture confirmed (the other three were not cultured). All of the culture confirmed patients were exposed to Gideon municipal water.

Ten culture confirmed patients did not reside in Gideon but all traveled to Gideon frequently to either attend school (8), use a day care center in town (1), or work at the nursing home (1). The earliest onset of symptoms in a culture-confirmed case was on November 17 (this patient was last exposed to Gideon water on November 16). A CDC survey indicated that approximately 44 percent of the 1104 residents, or almost 600 people, were affected with diarrhea between November 11 and December 27, 1993 in Gideon, Missouri. Nonresidents who drank Gideon water during the outbreak period experienced an attack rate of 28 percent (Angulo et al., 1997).

On January 14, 1994 an EPA field team, in conjunction with the Centers for Disease Control and Prevention (CDC) and the State of Missouri initiated a field investigation which included a sanitary survey and microbiological analyses of samples collected on site. A system evaluation was conducted in which EPANET was used to develop various scenarios to explain possible contaminant transport in the Gideon system.

The investigation clearly implicated consumption of Gideon municipal water as the source of the outbreak. Speculation focused on the flushing program conducted on November 10. It was observed that the pump at well 5 was operating at full capacity during the flushing program (approximately 12 hours) which would indicate that the municipal tanks were discharging during this period.

During the EPA field visit, a large number of pigeons were observed roosting on the roof of the 378 m^3 (100,000 gal) municipal tank. Shortly after the outbreak, a tank inspector found holes at the top of the Cotton Compress tank, rust on the tank, and rust, sediment and bird feathers floating in the water. According to the inspector, the water in the tank looked black and was so turbid that he could not see the bottom. Another inspection, conducted after the EPA's field study, confirmed the disrepair of the Cotton Compress tank and also found the 378 m^3 (100,000 gal) municipal tank in such a state of disrepair that bird droppings could, in the opinion of the inspector, have entered the stored water. Bird feathers were in the vicinity or in the tank openings of both the Cotton Compress and the 378 m^3 (100,000 gal) municipal tank.

It was hypothesized that taste and odor problems might have resulted from a thermal inversion that may have taken place due to a sharp temperature drop prior to the day of the complaint. If stagnant or contaminated water were floating on the top of a tank, a thermal inversion could have caused this water to be mixed throughout the tank and to be discharged into the system resulting in taste and odor complaints (Fennel et al., 1974). Turbulence in the tank from the flushing program could have stirred up the tank sediments which were transported into the distribution system. It is likely that the bulk water and/or the sediments were contaminated with *Salmonella* serovar *typhimurium*.

It was initially speculated that the backflow valve between the Cotton Compress and the municipal system might have failed during the flushing program. After the outbreak, the valve was excavated and found to be working properly. The private tank was drained accidentally after the outbreak during an inspection and so it was impossible to sample water in the tank bowel. Sediment in the private tank contained *Salmonella* serovar *typhimurium* dulcitol negative organisms as did samples found in a hydrant sample and culture confirmed patients. The *Salmonella* found in a hydrant matched the serovar of the patient isolate when analyzed by the CDC laboratory comparing DNA fragments using pulse field gel electrophoresis. The isolate from the tank sediment, however, did not provide an exact match with the other two isolates. No *Salmonella* isolates were found elsewhere in the system.

The purpose of the system's evaluation was to study the effects of distribution system design and operations, demand, and hydraulic characteristics on the possible propagation of contaminants in the system. Given the evidence from the survey and the results from the valve inspection at the Cotton Compress, it was concluded that the most likely contamination source was bird droppings in the large municipal tank. Therefore, the analysis concentrated on propagation of water from the large municipal tank in conjunction with the flushing program. This did not rule out other possible sources of contamination, such as cross connections.

The systems layout, demand information, pump characteristic curves, tank geometry, flushing program, etc., and other information needed for the modeling effort were obtained from maps and demographic information and numerous discussions with consulting engineers and the city and the DNR officials. EPANET, a hydraulic/water quality program, was used to conduct the contaminant propagation study (Rossman, 1994).

Data from the simulation study, the microbiological surveillance data and the outbreak data were utilized to provide insight into the nature of both general contamination problems in the system and into the "outbreak" itself. The water movement patterns showed that the majority of the special samples which were coliform and fecal coliform positive occurred at points that lie within the zone of influence of the small and large tanks. During both the flushing program and for large parts of normal operation, these areas are predominately

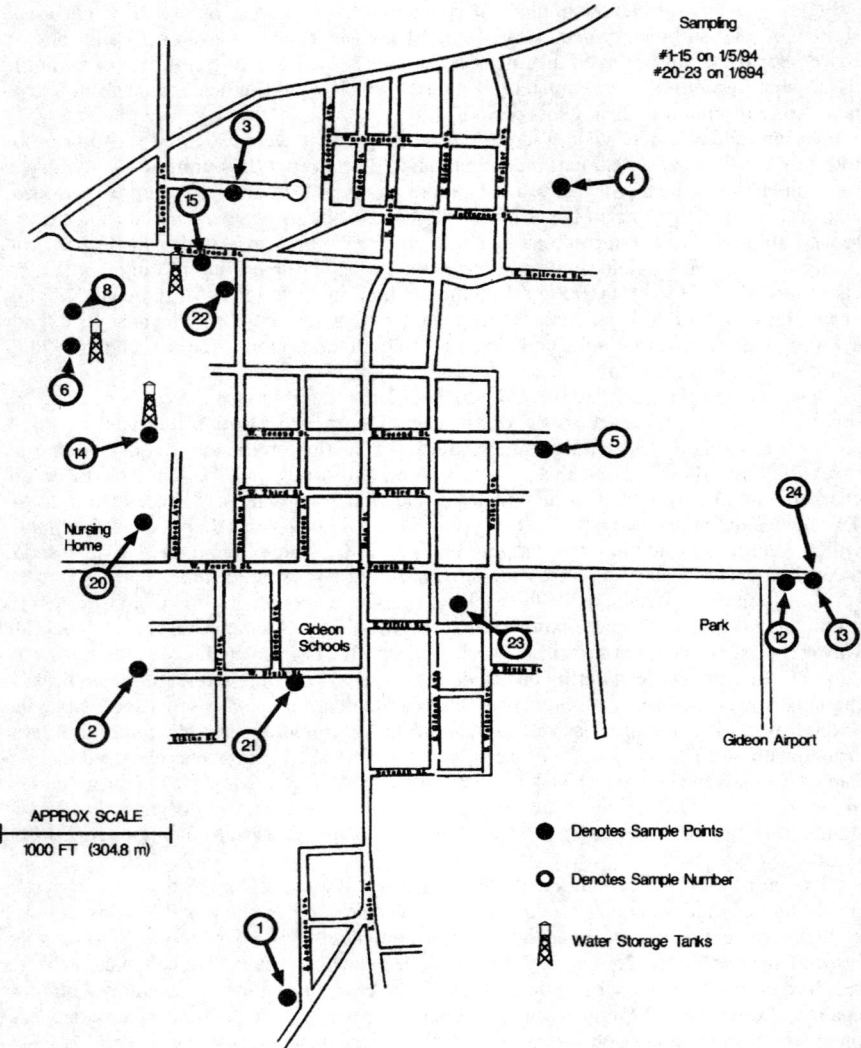

FIGURE 9.10 Street map and sampling points in Gideon.

served by tank water which might suggest that the tanks are the source of the fecal contamination since the tank yielded positive FC samples prior to chlorination.

Figure 9.10 is a street map for Gideon, showing the location of sampling points used in the filed study and some critical locations in the network. Data from the early cases, in combination with the water movement data, were utilized to infer the source of the outbreak. Using data supplied by the CDC and the water movement simulations, an overlay of the areas served by the small and large tanks during the first 6 h of the flushing period and the

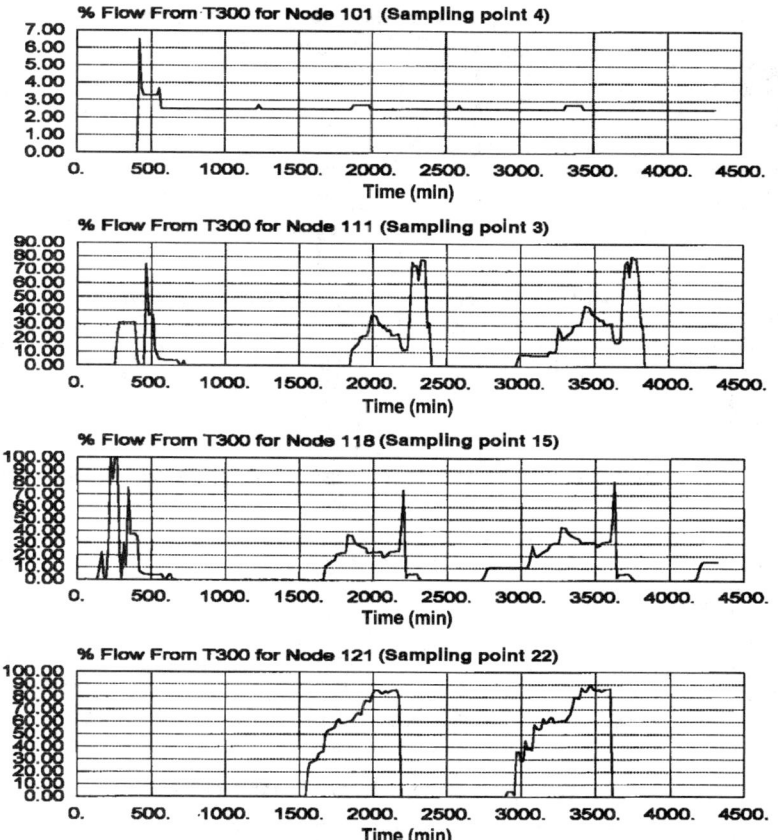

FIGURE 9.11 Percent of water from tank T300 at nodes 101, 111, 118, and 121 for 6-h simulation period.

earliest recorded cases was created, as shown in Figs. 9.11 and 9.12. As was seen in the earliest recorded cases, the positive *Salmonella* hydrant samples were found in the area that was primarily served by the large tank, but outside the small tank's area of influence, during the flushing period. One can conclude that during the first 6 h of the flushing period, the water which reached the residence and the Gideon School was almost totally from the large municipal tank. Therefore, it was logical to conclude that these locations should experience the first signs of the outbreak, which makes a strong circumstantial case for the large tank as the contamination source.

Based on the results of the DNR/DOH sampling program, it is likely that the contamination had been occurring over a period of time, which is consistent with the possibility of bird contamination. If the cause were a single event, the contaminant would most likely have been "pulled" through the system during the flushing program.

It is clear that some of the data utilized in this analysis could be very typical of that from a covert attack. Clearly, the response of the local public health authorities was reasonable and consistent with the data. However, note in Fig. 9.9, that the flushing event occurred on

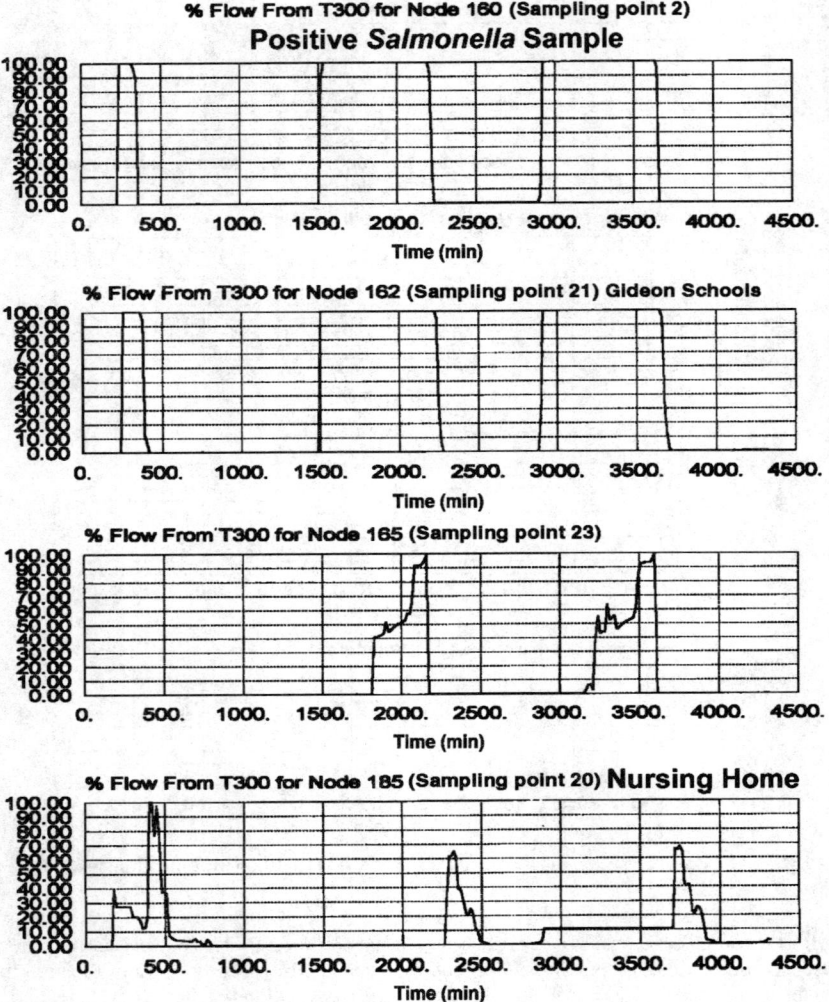

FIGURE 9.12 Percent of water from tank T300 at nodes 160, 162, 165, and 185 for 6-h simulation period.

November 10, the peak of the students absences was November 12, and the recognition of a public health emergency occurred on December 18, approximately a 4–5 week lag.

9.5 CURRENT TRENDS IN WATER QUALITY MODELING

Distribution system water quality modeling has evolved from the "bare bones" hydraulic models, available in the seventies and eighties, to much more sophisticated models. Current models include complex tank mixing models, fully integrated Geographic Information

Systems (GISs) and flexible user interface features. Several case studies are discussed in this section and demonstrate the extended capability of water quality models.

9.5.1 Case Studies

Two case studies that illustrate the application of the current generation of models are Cholet, France and Southington Connecticut. Cholet (population 60,000) is a municipality located in the west of France which derives water from the Moulin de Ribou and Verdon reservoirs (Heraud et al., 1997). It has two treatment plants, the largest of which treats water from the Moulin de Robou reservoir. Production is approximately 30,000 m^3/day and the distribution system consists of over 280 km of pipes and includes two tanks. "Piccolo" a hydraulic model, developed by the Research Center of the Lyonnaise des Eaux Groupe, was used to analyze flow in the network. Results from the study demonstrated that the main system was actually made up of two hydraulically independent subsystems. Continuous online chlorine residual monitor was incorporated as a part of the water quality modeling effort.

Southington, Connecticut has a distribution system with 299 km (186 mi) of pipe and nine wells which are capable of pumping more than 0.2965 m^3/day (4700 gal/min) (Aral and Masila, 1997; Aral et al., 1996). Three municipal reservoirs are also incorporated into the system. EPANET, in conjunction with a GIS system, was used to simulate four volatile organic chemical exposure scenarios that represented pumping conditions for 1970, 1974, 1979. The groundwater had been contaminated by volatile organic chemicals during the 1970s. The study team reached the following conclusions:

- VOC contamination exposure can exhibit significant spatial variation from one census block to another even when census blocks are adjacent to each other within a specified radius.
- Peak demand conditions may not yield the maximum exposure.
- Hydraulic and water quality modeling is very effective for quantifying the exposure of populations to past contamination.

9.5.2 Mixing in Storage Tanks

Storage tanks are the most visible components of a water distribution system but are least understood in terms of their impact on water quality. Their major function is to provide hydraulic reliability for fire fighting needs and for providing reliable service. However they may serve as vessels for complex chemical and biological changes that may cause the deterioration of water quality. Grayman and Clark (1993) conducted studies that highlighted the importance of the design, location, and operation of tanks on water quality. It was concluded that water quality degrades as a result of long residence time in tanks. Mau et al. (1995) developed compartment models to represent the different mixing conditions in tanks assuming steady-state conditions, which were extended by Clark et al. (1996) assuming nonsteady-state conditions.

The City of Azusa, California initiated a study of the effect of the Ed Heck Reservoir on its water quality (Grayman et al., 1996; Boulos et al., 1996). This tank had a capacity of 14,782.61 m^3 (4 mil gal) and was built to provide operational and emergency storage and to maintain contact time for the water leaving the Azusa water treatment plant. Unlike many system storage tanks it operated with simultaneous inflow and outflow. The study placed primary emphasis on providing an understanding of the hydraulic mixing and free chlorine residual concentration distribution in the reservoir. The reservoir was found to be completely mixed with two exceptions. Short circuiting existed between the inlet and outlet and a stagnant zone was found in the center core of the reservoir indicating the possibility of stratification or partitioning in the reservoir.

Storage tank models in the early versions of water quality distribution models such as EPANET were relatively simple. In general, they assumed complete mixing or used two-compartment models, even though the actual mixing regime might have been much more complex. A part of the AwwaRF/EPA study by Grayman et al. (2000) examined the following three approaches to describing the behavior of tanks—the use of physical models, a simplified systems model that emphasized the input and output of the reservoir or tank, and a computational fluid dynamics model based on mathematical equations. They found that each approach had some advantages.

9.6 SUMMARY AND CONCLUSIONS

The distribution network is the most vulnerable part of the water system and there are many opportunities to breach the network. A dedicated terrorist with a small pump could target a section of the system and deliberately inject a contaminant into the network from a home, an office building, warehouse, fire hydrant, or pumping station. Backflow prevention devices which can provide protection against this type of attack are typically installed only in industrial and commercial facilities or can be disengaged. They are not usually installed in residences and fire hydrants. Depending on the hydraulic design and demand conditions, much of the distribution system could be impacted, potentially affecting a large number of individuals and resulting in many casualties. Several studies have demonstrated this effect. It would be very difficult to track the spread of the contaminant and to find the source of the injection. If the agent is biological in nature, the injection event would be over before anyone is aware that an attack has occurred.

Once a contamination event has been verified and characterized, it will be necessary to determine the fate of the contaminant in the system. Not only will this require sampling locations to be established in the system but will also require an understanding of the hydraulic behavior of the distribution system. The final step will be to purge the contaminant from the system and confirm that the contaminant is no longer present.

In responding to a contamination threat it is important to effectively use monitoring to confirm that a threat exists and to understand movement of water in the system. Two case studies are presented to show that modeling techniques in conjunction with monitoring data could be used to identify the source of the threat, where contaminants may move in the system and to suggest remediation approaches.

REFERENCES

Angulo, F. J., S. Tippen, D. J. Sharp, B. J. Payne, C. Collier, J. E. Hill, T. J. Barrett, R. M. Clark, E. E. Geldriech, H. D. Donnell, and D. L. Swerdlow, "A Community Waterborne Outbreak of Salmonellosis and the Effectiveness of a Boil Water Order," *Am. J. Public Health* 87(4): 580–584, 1997.

Aral, M. M., and M. L. Maslia, "Exposure Assessment Using Simulation and GIS," *Proceedings of the 1997 CSCE/ASCE Environmental Engineering Conference*, held in Edmonton, Alberta, Canada, pp. 885–892, July 22–26, 1997.

Aral, M. M., M. L. Maslia, G. V. Ulirsch, and J. J. Reyes, "Estimating Exposure to Volatile Organic Compounds from Municipal Water-Supply Systems: Use of a Better Computational Model," *Arch. Environ. Health* 51(4): 300–309, July/August 1996.

Biswas, P., C. Lu, and R. M. Clark, "A Model for Chlorine Concentration Decay in Drinking Water Distribution Pipes," *Water Res.* 27(12): 1715–1724, 1993.

Boulos, P. F., T. Altman, P. A. Jarrige, and F. Collevati, "Discrete Simulation Approach for Network-Water-Quality Models," *J. Water Resour. Planning Management, ASCE* 121(1): 49–60, 1995.

Boulos, P. F., W. M. Grayman, R. W. Bowcock, J. W. Clapp, L. A. Rossman, R. M. Clark, R. A. Deininger, and A. K. Dhingra, "Hydraulic Mixing and Free Chlorine Residual in Reservoirs," *J. Am. Water Works Assoc.* 88(7): 48–59, July 1996.

Buchberger, S. G, J. T. Carter, Y. H. Lee, and T. G. Schade. Random Demands, Travel Times and Water Quality in Deadends. Report 90963F. *American Water Works Association Research Foundation*, Denver, CO, p. 470, 2003.

Clark R. M., and W. M. Grayman, *Modeling Water Quality in Drinking Water Distribution Systems*, American Water Works Association, 6666 West Qunicy Avenue, Denver, CO, 1998.

Clark, R. M., "Assessing the Etiology of a Waterborne Outbreak: Public Health Emergency or Covert Attack," in Jennifer Hatchett (ed.), *Proceedings of the First Water Security Summit*, Haested Press, Haested Methods, Waterbury, CT, pp. 170–179, 2001.

Clark, R. M., "Water Quality Modeling-Case Studies," in L. W. Mays (ed), *Water Distribution Systems Handbook*, McGraw-Hill, New York, pp. 13.1–13.49, 2000.

Clark, R. M., A. Vicory, and J. A. Goodrich, "The Great Ohio River Oil Spill of 1988: A Case Study," *J. Am. Water Works Assoc.* 82(3): 39–44, March 1990.

Clark, R. M., and R. A. Deininger, "Minimizing the Vulnerability of Water Supplies to Natural and Terrorist Threats," *Proceedings of the American Water Works Association's IMTech Conference*, held in Atlanta, GA, pp. 1–20, April 8–11, 2001.

Clark, R. M., and R. A. Deininger, "Protecting the Nation's Critical Infrastructure: The Vulnerability of U.S. Water Supply Systems," *J. Contin. Crisis Management* 8(2): 73–80, 2000.

Clark, R. M., E. E. Geldreich, K. R. Fox, E. W. Rice, C. H. Johnson, J. A. Goodrich, J. A. Barnick, and F. Abdesaken, "Tracking a *Salmonella* Serovar *Typhimurium* Outbreak in Gideon, Missouri: Role of Contaminant Propagation Modeling," *J. Water Supply Res. Technol. - Aqua.* 45(4): 171–183, 1996.

Clark, R. M., E. E. Geldreich, K. R. Fox, E. W. Rice, C. H. Johnson, J. A. Goodrich, J. A. Barnick, F. Abdesaken, J. E. Hill, and F. J. Angulo, "A Waterborne *Salmonella Typhimurium* Outbreak in Gideon, Missouri: Results, From A Field Investigation," *Int. J. Environ. Health Res.* 6(3): 187–193, September 1996.

Clark, R. M., F. Abdesaken, P. F. Boulos, and R. Mau, R., "Mixing in Distribution System Storage Tanks: Its effect on Water Quality," *J. Environ. Eng.* 122(9): 814–821, 1996.

Clark, R. M., G. Smalley, J. A. Goodrich, R. Tull, L. A. Rossman, J. T. Vasconcelos, and P. F. Boulos, "Managing Water Quality in Distribution Systems: Simulating TTHM and Chlorine Residual Propagation," *J. Water SRT - Aqua* 43(4): 182–191, 1994.

Clark, R. M., W. M. Grayman, J. A. Goodrich, R. A. Deininger, and A. F. Hess, "Field Testing Distribution Water Quality Models," *J. Am. Water Works Assoc.* 83(7): 1991.

Clark, R. M., W. M. Grayman, R. M. Males, and A. F. Hess, "Modeling Contaminant Propagation in Drinking Water Distribution Systems," *J. Environ. Eng., ASCE* 119(2): 349–364, 1993.

Fennel, H., D. B. James, and J. Morris, "Pollution of a Storage Reservoir by Roosting Gulls," *J. Soc. Water Treat. Exam.* 23: 5–24, 1974.

Grayman, W. M., and R. M. Clark, "Using Computers to Determine the Effect of Storage on Water Quality," *J. Am. Water Works Assoc.* 85(7): 67–77, July 1993.

Grayman, W. M., L. A. Rossman, C. Arnold, R. A. Deininger, C. Smith, J. F. Smith, and R. Schnipke. *Water Quality Modeling of Distribution System Storage Facilities*. American Water Works Association Research Foundation. Denver, CO, 2000.

Grayman, W. M., R. A. Deininger, A. Green, P. F. Boulos, R. W. Bowcock, C. C. Godwin, "Water Quality and Mixing Models for Tanks and Reservoirs," *J. Am. Water Works Assoc.* 88(7): 60–73, July 1996.

Heraud, J., L. Kiene, M. Detay, and Y. Levy, "Optimized Modelling of Chlorine Residual in a Drinking Water Distribution System with a Combination of On-Line Sensors," *J. Water SRT-Aqua*, 46(2): 59–70, 1997.

International Life Sciences Institute, Risk Science Institute, *Early Warning Monitoring to Detect Hazardous Events in Water Supplies*. ILSI PRESS, Washington DC, 1999.

Islam, M. R., "Modeling of Chlorine Concentration in Unsteady Flows in Pipe Networks," Ph.D. Thesis, Washington State University, Pullman, WA, May 1995.

Mau, R., P. Boulos, R. Clark, W. Grayman, R. Tekippe, and R. Trussell "Explicit Mathematical Models of Distribution System Storage Water Quality," *J. Hydraulic Eng.* 121(10): 699–709, October 1995.

PL107-188. Public Health Security and Bioterrorism Preparedness and Response Act of 2002.

President. White Paper. The Clinton Administration Policy on Critical Infrastructure Protection: Presidential Decision Directive 63, 1998, available at: http//www.ciao.gov/CIAO_Document_Library/paper598.htm

Rice, E., "Disinfection," *Controlling Disinfection By-Products and Microbial Contaminants in Drinking Water*, United States Environmental Protection Agency, Office of Research and Development, Washington DC, EPA/600/R-01/110, December 2001.

Rossman, L. A., *EPANET Users Manual*, Drinking Water Research Division, U.S. EPA, Cincinnati, OH, 1994.

Rossman, L. A., R. M. Clark, and W. M. Grayman, "Modeling Chlorine Residuals in Drinking Water Distribution Systems," *J. Environ. Eng.* 120(4): 803–820, July/August 1994.

U.S. EPA., "25 years of the Safe Drinking Water Act: History and trends," EPA 816-R-99-007, Office of Water, December 3, 1999.

Vasconcelos, J. J., L. A. Rossman, W. M. Grayman, P. F. Boulos, and R. Clark, "Kinetics of Chlorine Decay," *J. Am. Water Works Assoc.* 89(7): 54–65, July 1997.

Vasconcelos, J. J., P. F. Boulos, W. M. Grayman, L. Kiene, O. Wable, P. Biswas, A. Bhari, L. A. Rossman, R. M. Clark, and J. A. Goodrich, "Characterization and Modeling of Chlorine Decay in Distribution Systems," Report No. 90705, *AWWA Research Foundation*, 6666 West Quincy Avenue, Denver, CO, 1996.

CHAPTER 10
RECONSTRUCTING HISTORICAL CONTAMINATION EVENTS

Walter M. Grayman,* Robert M. Clark,[†] Benjamin L. Harding,[‡] Morris Maslia,[§] and Jeff Aramini[¶]

10.1 HISTORICAL CONTAMINATION EVENTS

A water supply system is inherently quite susceptible to contamination. Contaminants may be introduced accidentally, naturally, or on purpose at a diverse range of locations in the large spatial extent of the system covering the source water, the treatment process, and the distribution system. Figure 10.1 illustrates a range of potential contamination scenarios.

Groundwater and surface water sources may be contaminated near to the entry point into the water system or at significant distances upstream or upgradient of the entry point. For surface water sources, the contaminant may move quickly with the flow and reach the intake in relatively short time spans. With a groundwater source, contaminant movement may take several years or even decades to travel to a well and may result in long-term contamination of the well or may be a short-term, immediate, and intense event if the contamination occurs at the well site.

Within the water treatment/distribution system, contamination can occur in a wide range of locations. Cross-connections between sewage systems and water systems, intrusion into pipes, contamination of storage facilities, and accidental or purposeful entry of contaminants through hydrants or service connections are all plausible among the scenarios that can and have resulted in contamination of distribution systems.

Though many contamination events are relatively minor in consequence and may pass with few impacts and without detection, other events can and have been significant in their impacts. A thorough understanding of past contamination events and how and why they occurred and evolved is important for many reasons:

---

*W. M. Grayman Consulting Engineer, Cincinnati, Ohio
[†]Environmental Engineering and Public Health Consultant, Cincinnati, Ohio
[‡]Hydrosphere Resource Consultants, Inc., Boulder, Colorado
[§]Agency for Toxic Substances and Disease Registry, Atlanta, Georgia
[¶]Health Canada, Guelph, Ontario

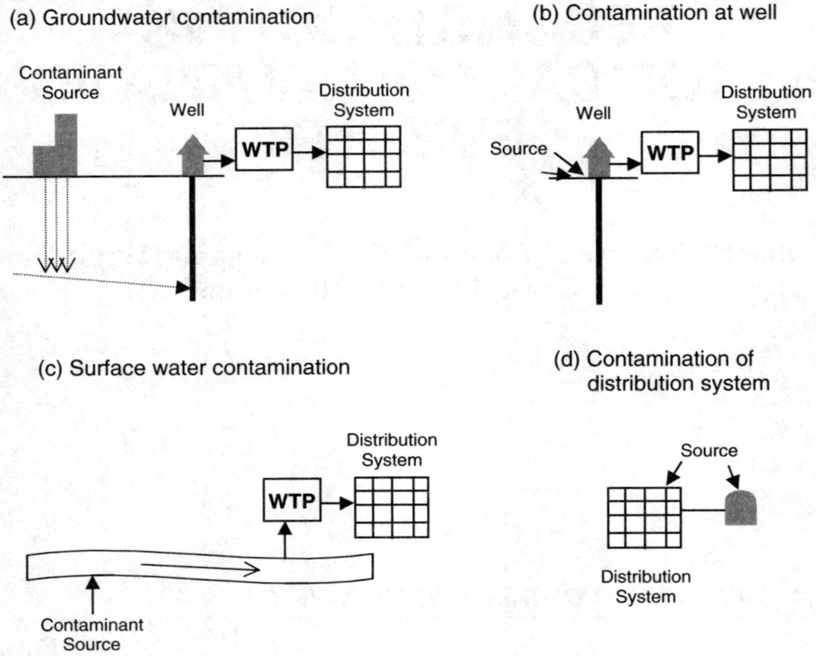

(a) Groundwater contamination

(b) Contamination at well

(c) Surface water contamination

(d) Contamination of distribution system

FIGURE 10.1 Contamination scenarios.

- Remediating and correcting the specific causes of the contamination event
- Identifying the full extent of damage and injury
- Identifying liability for the event
- Providing general knowledge that will reduce the likelihood of contamination events and the vulnerability of water systems to contamination

The process of understanding the contamination event is referred to here as historical reconstruction of the event. Reconstruction can vary from a qualitative review of the general events and causes that led up to the contamination to a highly quantitative assessment utilizing hydraulic and water quality models to trace the movement of the contaminant within the source water and through the distribution system.

The time lag between the contamination event and the occurrence of reconstruction historically can vary from a few hours or days after an event to many years or decades following the event. This time scale results in significantly different methodologies and objectives in reconstructing the event. For the purpose of assessment, events can be divided into three categories though there may be some overlap for a specific event:

1. *Active event identified.* In this case, the contamination event is detected while it is still active. Primary emphasis in this situation is the rapid assessment of the situation in order to locate, disable, and control the contamination, and to prevent further impacts on customers and the water system. Historical reconstruction may involve reviewing and assessing the operation of the system over the period immediately preceding the event.

2. *Recent past contamination event.* In many cases, a contamination event may result in waterborne diseases that are identified only after several days through increased activities in emergency rooms, increased sale of pharmaceuticals, increased absence at work or schools, or acute disease among the elderly or immunocompromised. Since the contamination may already have passed through the distribution system, primary emphasis is upon preventing further illness, and identifying and rectifying the underlying problem.

3. *Far past contamination event.* Contamination events that are or were not detected can lead to long-term contamination of a water system. This was most common in the period prior to 1980 when routine sampling of finished water was limited and constituents such as volatile organics and pesticides were later discovered in the drinking water. However, it is still a problem as the potential health effects of known chemicals at lower concentrations are identified. For these long-term situations, it is sometimes necessary to reconstruct the water system characteristics and operation for a period of many years, sometimes dating back over a half century. Most of these historical reconstructions have been associated with legal actions or government studies.

10.1.1 Types of Contamination Events

Contamination events may also be categorized based on the underlying time scale and resulting health impacts. Short-term, acute contamination events are generally caused by a contaminant being introduced into the water supply over a period of minutes, hours, or days. The contaminant may reach a limited part of the distribution system or may spread over a wide area and can result in waterborne illness and death that occurs within a short period (generally on the order of days) following exposure to the water. At the other end of the spectrum are long-term, chronic contamination events generally caused by contamination that enters the water supply over periods of months or years. The resulting long-term exposure can lead to a wide range of illnesses that appear years or even decades after the initial exposure. Examples of these two general categories of events are described below followed by four historical reconstruction case studies.

Short-Term Acute Contamination Events. Waterborne disease outbreaks undoubtedly date back to prehistoric times. Scientific assessment of such outbreaks, linked to municipal water supplies, was first associated with the benchmark methods used by John Snow in Britain in the 1850s. Through a meticulous case mapping study in London, Snow demonstrated that cholera occurred much more frequently among customers of one specific water company (Snow, 1855). Since that time, many suspected waterborne disease outbreaks have been confirmed by observing differing rates of illness among populations served by different water supplies (MacKenzie et al., 1994; Bowie et al., 1997). Unfortunately, in communities supplied with drinking water originating from multiple sources and undergoing extensive mixing within a distribution system, a simple examination of the spatial distribution of cases may be of little epidemiological value when attempting to identify the source of contamination (Moorehead et al., 1990).

Significant waterborne disease outbreaks in North America that have been the subject of historical reconstruction include an outbreak of *E. coli* serotype 0157:H7 in Cabool, Missouri in 1989 (Geldreich et al., 1992), a *Salmonella* outbreak in Gideon, Missouri in 1993 (Clark et al., 1996), widespread increase of diarrhea in Milwaukee, Wisconsin linked to the organism Cryptosporidium (Fox and Lytle, 1996), and an outbreak in Walkerton, Ontario in 2000 associated with waterborne *E. coli* O157:H7 and/or *Campylobacter* (Bruce Grey Owen Sound Health Unit, 2000). In the cases in Cabool, Gideon, and Walkerton, water distribution system models were developed shortly after the outbreaks to help assess the situation and to determine the likely causative factor. The Gideon and Walkerton cases are the subject of detailed case studies described later in this chapter.

Long-Term Chronic Contamination Events. The book, *A Civil Action* (Harr, 1995), details the investigation and legal activities associated with the contamination of the groundwater supply serving Woburn, Massachusetts by industrial chemicals and its subsequent health impacts in that city. The 2000 movie *Erin Brockovich* describes the settlement of a toxic court case in Hinckley, California brought by citizens who alleged significant health impacts from chromium contamination of the groundwater source by Pacific Gas & Electric Company's local gas compressor station. These were highly publicized, long-term contamination cases. However, other similar contamination events have resulted in more extensive, formal historical reconstructions (Grayman, 2001). In many cases, the availability of information on the reconstruction of events is limited by court orders or settlements associated with legal actions involving these cases.

During and following the well-publicized legal case in Woburn, Massachusetts, extensive study and research were performed to help understand the extent, cause, and dynamics of the contamination event. Groundwater models were applied to estimate how the pollutants may have moved from the alleged locations where they were introduced on the surface to the wells where they entered the water distribution system. Simplified, steady-state models of the water distribution system were applied to describe their likely movement through the water distribution system to identify the extent of customers that may have been exposed to the contaminated water (Murphy, 1986).

In recent cases, more sophisticated modeling capabilities have been employed. Long-term (multi-year) simulations of the movement of trichloroethylene (TCE) in the distribution system were simulated in Phoenix and Scottsdale (Harding and Walski, 2000). The Agency for Toxic Substances and Disease Registry (ATSDR) modeled the movement of water from well sources through the Dover Township, NJ distribution system over the period from 1962 to 1996 in conjunction with epidemiological studies investigating clusters of childhood leukemia and nervous system cancers (Maslia et al., 2001). Uncertainty was incorporated in the historical reconstruction of the movement of TCE and the perchlorate anion through the distribution system in Redlands, California (Harding and Grayman, 2002). The Dover Township and Redlands cases are the subject of detailed case studies described later in this chapter.

Models of contaminant transport in the vadose zone (Charbeneau and Daniel, 1993) and in the groundwater (Mercer and Waddell, 1993) are frequently used in reconstruction cases to trace the movement of contaminants from the point of contamination to wells. Recent studies have emphasized the importance of incorporating uncertainty in this modeling (Normani and Sykes, 2002; ASCE, 2003).

10.2 GIDEON, MISSOURI

10.2.1 Water Quality Models and Public Health Surveillance

A waterborne outbreak that occurred in Gideon Missouri provides an excellent example of "a short-term acute contamination event." It also illustrates how local and state environmental health specialists and water utility operators may be the first to observe and recognize an environmental problem potentially caused by a covert attack. For example, as will be discussed in the case study, unusual patterns of school absences and increased levels of morbidity and mortality in a local nursing home provided early indicators of the outbreak. These same patterns might occur in a covert attack. In addition, the case study illustrates how water quality models may be used to help identify the source of a contamination event and provide the basis for historical reconstruction of the event.

10.2.2 The Gideon Outbreak

Gideon, Missouri located in Southeastern Missouri is typical of many small midwestern towns. In 1990, the population of Gideon was 1104. The median income was $14,654 (25 percent of the population was below the poverty level) and the unemployment rate was 11.3 percent. Major employers in Gideon were a nursing home with 68 residents and a staff of 62 and the Gideon schools with 444 students (kindergarten through 12th grade). The topography was flat and the predominant crop in the area was cotton.

The Gideon municipal water system was originally constructed in the mid-1930s. Water was obtained from two adjacent 396 m (1300 ft) deep wells, which were not disinfected at the time of the outbreak. The distribution system consisted primarily of small diameter (5, 10, 15 cm (2, 4 and 6 in)) heavily tuberculated and corroded unlined steel and cast iron pipes. Under low flow or static conditions the water pressure was close to 3.5 kgf/cm^2 (50 psi) but under high flow or flushing conditions the pressure dropped dramatically. In the Cotton Compress yards, the major industry in the area, water was used for equipment washing, in rest rooms and for consumption. The entire Cotton Compress water system was isolated from the Gideon system by a back flow prevention valve shortly after the Cotton Compress tank was built.

On November 29, the Missouri Department of Health (DOH), became aware of two high school students from Gideon who were hospitalized with culture confirmed salmonellosis (Clark et al., 1996). Within 2 days, five additional patients living in Gideon were hospitalized with salmonellosis (one student, one child from a day care center, two nursing home residents, and one visitor to the nursing home). In December of 1993, six to nine cases of diarrhea were reported at a local nursing home in Gideon, Missouri, raising the possibility of a waterborne outbreak (Clark et al., 1996). The State Public Health Laboratories identified the isolates as dulcitol-negative *Salmonella* and since this is a reportable disease, samples were sent to CDC. The CDC laboratories identified the organism as serovar *typhimurium*. Interviews conducted by the DOH suggested that there were no food exposures common to a majority of the patients. All of the ill persons had consumed municipal water. Figure 10.2 shows the absentee statistics for the Gideon schools, which are the same type of data that one might see in a covert attack.

The Missouri Department of Natural Resources (DNR) was informed that the DOH suspected a water supply link to the outbreak. Water samples collected by the DNR on December 16 were positive for FC. On December 18, the city of Gideon, as required by the DNR, issued a "boil water" order. Signs were posted at city hall, in the grocery store, and two area radio stations announced the boil water order.

Several water samples collected by DNR on December 20 were also found to be FC positive. On December 23, a chlorinator was placed on line at the city well by DNR, and nine samples were collected by the DOH and DNR from various sites in the distribution system. None of the samples contained chlorine but one sample collected from a fire hydrant was positive for dulcitol-negative *Salmonella* serovar *typhimurium*. The Missouri DOH had informed the CDC about the outbreak in Gideon in early December and requested information about dulcitol negative *Salmonella* serovar *typhimurium*. On December 17, DOH informed CDC that contaminated municipal water was the suspected cause of the outbreak and (on December 22) invited CDC to participate in the investigation. A flyer explaining the boil order, jointly produced by DOH and DNR, was placed in the mailboxes of all the homes in Gideon on December 29 and the privately owned water tower was physically disconnected from the municipal system on December 30. DNR mandated that Gideon permanently chlorinate their water system.

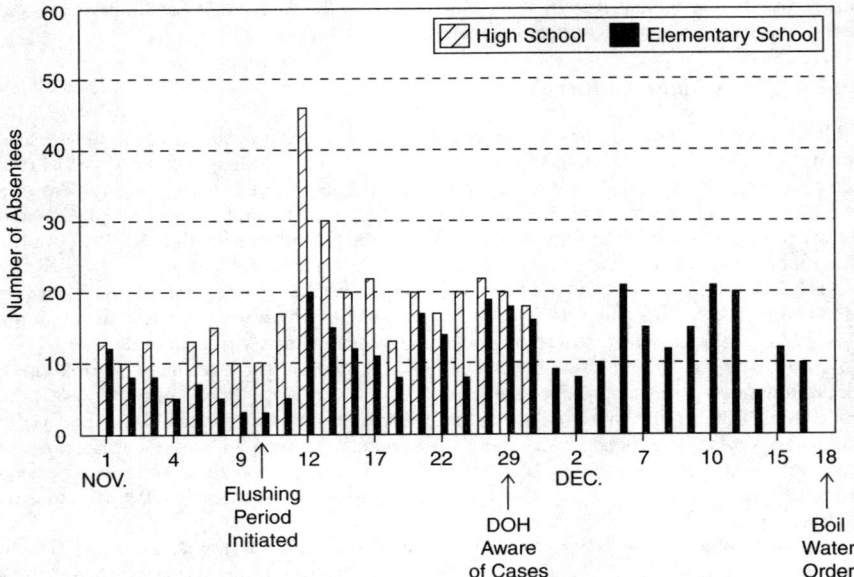

FIGURE 10.2 School absentees and sequence of events during Gideon outbreak.

By January 8, 1994 the DOH had identified 31 cases with laboratory confirmed salmonellosis and the state public health laboratories had identified 21 of these isolates as dulcitol-negative *Salmonella* serovar *typhimurium*. Fifteen of the 31 culture confirmed patients were hospitalized (including two patients hospitalized for other causes and who developed diarrhea while in the hospital). The patients were admitted to 10 different hospitals. Two of the patients had positive blood cultures, seven nursing home residents exhibiting diarrheal illness died, four of whom were culture confirmed (the other three were not cultured). All of the culture confirmed patients were exposed to Gideon municipal water.

Ten culture confirmed patients did not reside in Gideon but all traveled to Gideon frequently to either attend school (8), use a day care center in town (1), or work at the nursing home (1). The earliest onset of symptoms in a culture confirmed case was on November 17 (this patient was last exposed to Gideon water on November 16). A CDC survey indicated that approximately 44 percent of the 1104 residents, or almost 600 people, were affected with diarrhea between November 11 and December 27, 1993 in Gideon, Missouri. Nonresidents who drank Gideon water during the outbreak period experienced an attack rate of 28 percent (Angulo, et al., 1997).

By the end of December, Gideon municipal water was clearly implicated as the source of the outbreak. Speculation focused on a sequential flushing program conducted on November 10 involving all the 50 hydrants in the system. The program was started in the morning and continued throughout the entire day. Each hydrant was flushed for 15 min. at an approximate rate of 2.8 m^3/min (750 gal per min). It was observed that the pump at well 5 was operating at full capacity during the flushing program (approximately 12 h), which would indicate that the municipal tanks were discharging during this period. The flushing program was conducted in response to taste and odor complaints. Initially, speculation regarding the cause of the outbreak focused on the Cotton Compress tank, which was constructed in 1930 and appeared to be heavily rusted and in an obvious state of disrepair.

It was hypothesized that the taste and odor problems may have resulted from a thermal inversion that may have taken place due to a sharp temperature drop prior to the day of the complaint. If stagnant or contaminated water was floating on the top of a tank, a thermal inversion could have caused this water to be mixed throughout the tank and to be discharged into the system (Fennel et al., 1974). Turbulence in the tank from the flushing program could have stirred up the tank sediments, which might have been transported into the distribution system. It is likely that the bulk water and/or the sediments were contaminated with *Salmonella* serovar *typhimurium*. Figure 10.2, which shows the relationship in time between the flushing program and the increase in school absentees, clearly implicates the flushing activity as the cause of the problem.

On January 14, 1994 an EPA field team, in conjunction with the Centers for Disease Control and Prevention (CDC) and the State of Missouri initiated a field investigation which included a sanitary survey and microbiological analyses of samples collected on site. During the EPA field visit, a large number of pigeons were observed roosting on the roof of the 378 m^3 (100,000 gal) municipal tank. Shortly after the outbreak, a tank inspector found holes at the top of the Cotton Compress tank, rust on the tank, and rust, sediment, and bird feathers floating in the water. According to the inspector, the water in the tank looked black and was so turbid he could not see the bottom. Another inspection, conducted after EPA's field study, confirmed the disrepair of the Cotton Compress tank and also found the 378 m^3 (100,000 gal) municipal tank in such a state of disrepair that bird droppings could, in the opinion of the inspector, have entered the stored water. Bird feathers were in the vicinity or in the tank openings of both the Cotton Compress and the 378 m^3 (100,000 gal) municipal tank.

As indicated earlier, it was initially speculated that the backflow valve between the Cotton Compress and the municipal system might have failed during the flushing program. After the outbreak, the valve was excavated and found to be working properly. The private tank was drained accidentally after the outbreak during an inspection, so it was impossible to sample water in the tank bowel. However, sediment in the private tank contained *Salmonella* serovar *typhimurium* dulcitol-negative organisms as did samples found in a hydrant sample and culture confirmed patients. The *Salmonella* found in a hydrant matched the serovar of the patient isolate when analyzed by the CDC laboratory comparing DNA fragments using pulse field gel electrophoresis. The isolate from the tank sediment, however, did not provide an exact match with the other two isolates. No *Salmonella* isolates were found elsewhere in the system.

10.2.3 Systems Analysis

As part of the EPA investigation, a systems evaluation was conducted to study the effects of distribution system design and operations, demand, and hydraulic characteristics on the possible propagation of contaminants in the system. Given the evidence from the survey and the results from the valve inspection at the Cotton Compress, it was concluded that the most likely contamination source was bird droppings in the large municipal tank. Therefore, the analysis concentrated on propagation of water from the large municipal tank in conjunction with the flushing program. Contamination from other sources such as cross connections were not ruled out.

The systems layout, demand information, pump characteristic curves, tank geometry, flushing program, and other information needed for the modeling effort were obtained from maps and demographic information and numerous discussions with consulting engineers and city and DNR officials. EPANET, a hydraulic/water quality program, was used to conduct the contaminant propagation study (Rossman, 2000).

EPANET was calibrated by simulating flushing at specific hydrants, where head loss was available, assuming a discharge of 2.8 m^3/min (750 gpm) for 15 min. The "C" factors were adjusted until the head loss in the model matched head losses observed in the field.

Data from the simulation study, the microbiological surveillance data and the outbreak data were used to provide insight into the nature of both general contamination problems in the system and into the outbreak itself. The water movement patterns showed that the majority of the special samples that were coliform and fecal coliform positive occurred at points that lie within the zone of influence of the small and large tanks. For both, the flushing program and for large parts of normal operation, these areas are predominately served by tank water, which led to the conclusion that the tanks were the source of the fecal contamination since there were positive FC samples prior to chlorination.

Data from the early cases, in combination with the water movement data, were utilized to infer the source of the outbreak. Using data supplied by CDC and the water movement simulations, an overlay of the areas served by the small and large tanks during the first 6 h of the flushing period and the earliest recorded cases was created. The earliest recorded cases and the positive *Salmonella* hydrant sample were found in the area that was primarily served by the large tank, but outside the small tank's area of influence, during the flushing period. It was concluded that during the first 6 h of the flushing period, the water, which reached the residence and the Gideon School, was almost totally from the large tanks as shown in Fig. 10.3.

Therefore, it was concluded that these locations should experience the first signs of the outbreak. A comparison between the model's predictions and the outbreak data showed that this, in fact, is what happened making a strong circumstantial case for the large municipal tank as the contamination source.

It is likely that, based on the results of the DNR/DOH sampling program, the contamination had been occurring over a period of time, which is consistent with the possibility of bird contamination. Because of the flushing program, water from the contaminated tank was "pulled" through the system causing the outbreak. At the end of the study EPA provided input to DNR on the criteria necessary to lift the boil water order (Angulo et al., 1997).

10.2.4 Summary and Conclusions

It has become conventional wisdom that drinking water distribution systems are vulnerable to a contamination attack. However, detecting such an attack is likely to depend on the sensitivity of state and local public health surveillance systems. The first signs of such an attack may be an increase in the incidence of morbidity and mortality in a community. An increased number of school absentees, or sickness, or death in nursing homes may be the specific signs that such an attack has occurred.

Even once the occurrence of a public health emergency has been identified, taking specific action to deal with the emergency may not be easy. For example, in the Gideon example, within 2 days after the flushing period was initiated the increase in school absentees was dramatic. However, the Department of Health became aware of a potential waterborne outbreak only two-and-a-half weeks after the first increase in school absentees and 3 weeks after the flushing period. A boil water order and mandatory chlorination were initiated after that. In case of a covert attack, such a delay in response might be disastrous.

A tool that has genuine potential for identifying and mitigating a public health emergency is water quality modeling. In the Gideon case study, the application of water modeling along with other physical evidence was critical to the identification of the cause of the outbreak.

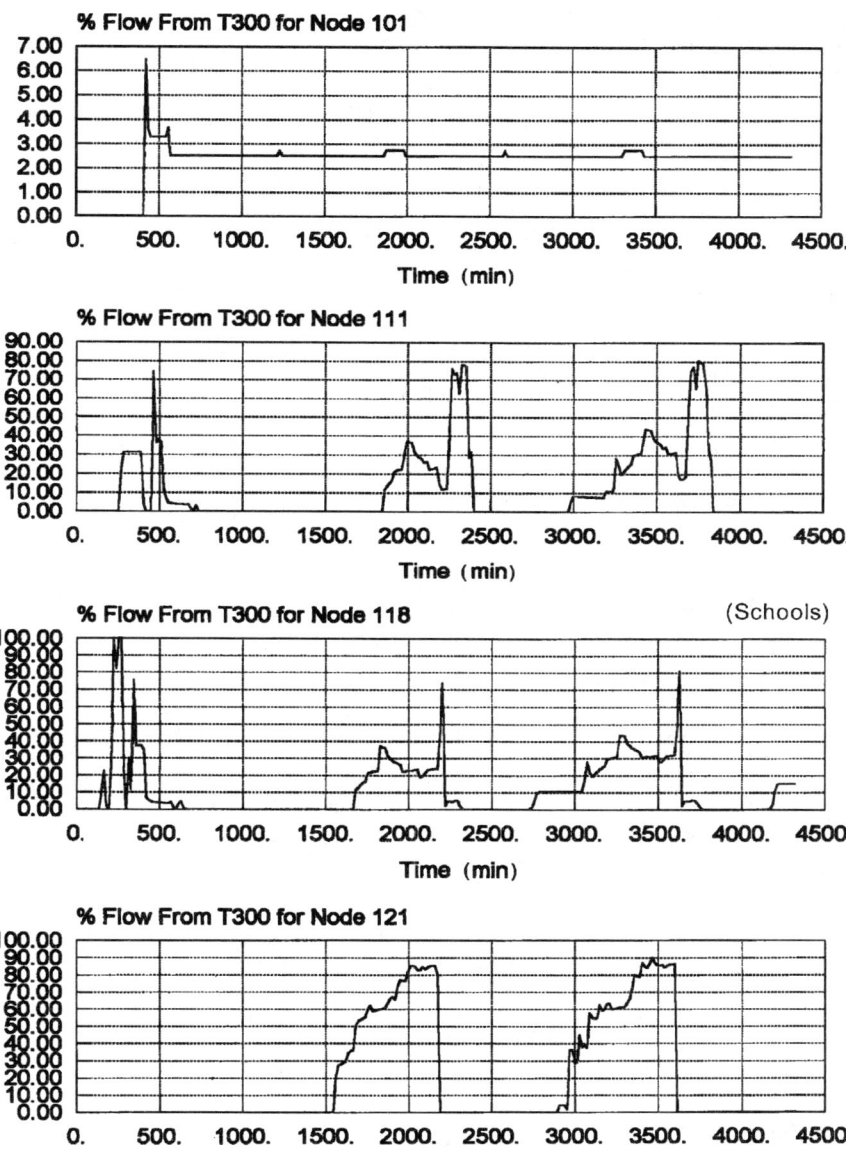

FIGURE 10.3 Percent of water from tank T300 at nodes 101, 111, 118, and 121 during 72-h simulation period.

10.3 WALKERTON, ONTARIO WATERBORNE OUTBREAK

In May of 2000, the first documented outbreak of *Eschericia coli* O157:H7/ *Campylobacter* spp. gastroenteritis associated with a municipal water supply in Canada took place in the small rural town of Walkerton, Ontario (Bruce Grey Owen Sound Health Unit, 2000). At the time of the outbreak, the town's drinking water originated from three drilled wells (Wells 5, 6, and 7), each contributing to a common distribution system. Using a cross-sectional study, it was demonstrated that during the outbreak, residents living in homes connected to the municipal water supply and consuming Walkerton water were 11.7 times more likely to have developed gastroenteritis than those not exposed to Walkerton water (Bruce Grey Owen Sound Health Unit, 2000). A hydrogeological assessment revealed that two of the three wells (Wells 5 and 6) were under the influence of surface water, and a microbiological investigation found *E. coli* O157:H7 and *Campylobacter* spp. on a cattle farm adjacent to Well 5 with identical molecular characteristics to those isolated from patients.

This section describes how a retrospective model of the Walkerton water distribution system, together with the temporal-spatial characteristics of gastroenteritis cases and controls, was used in descriptive and conditional logistic modeling exercises to help identify the well most likely responsible for the outbreak.

10.3.1 Methodology

The overall approach of this study was to investigate the temporal-spatial relationships between cases of gastroenteritis and hypothesized water distribution system contamination scenarios. Exposure was assessed by modeling the proportional concentrations of a hypothetical contaminant originating from each of Walkerton's three wells using a commercially available hydraulic analysis software package. As contaminant concentrations at points in a distribution system are related to both the age of the water (relative to the time of contamination) and the proportion of water originating from the contaminated source, this exposure measurement is an appropriate proxy for pathogen concentrations. Six *exposure scenarios* were chosen to assess predetermined hypothesized exposure events. Results of the exercise were assessed both qualitatively and quantitatively.

Modeling the Walkerton Distribution System. The objective of the water modeling exercise was to recreate the pattern of water flow throughout the town's distribution system immediately before and during the outbreak period. This was accomplished with the use of GIS (Arcview; ESRI, Redlands, CA) and a water distribution software package (WaterCAD; Haestad Methods, Waterbury, CT).

Modeling of the Walkerton water system involved inputting the following parameters into WaterCAD:

- Pipe diameter and length, location, age, and composition of all water pipes
- Size, storage capacity, and active volumes of the two stand pipes (water towers)
- Well pump specifications (including pump curves)
- Pipe friction.

WaterCAD models the system as a network of nodes, which can be source (well), storage (tanks), or demand (customer) locations, connected by links (pipes and pumps). Each metered commercial user was questioned about water use during the critical period, and individualized temporal water demand patterns were assigned to each *commercial node*. Residential users were assigned to the nearest *residential node*. As residential users in Walkerton were not metered, hourly demand was estimated using the daily volume of water supplied to the system after accounting for commercial users and fire events, and literature-based hourly demand patterns (Kindler and Russel, 1984; Obradovic and Lonsdale, 1998). Well pump controls were added to the model and pump *on* and *off* times were set in accordance with the historical pump records. Computerized data from the supervisory control and data acquisition (SCADA) system for the water supply containing 15-min pumpage rates for the three wells were utilized.

Flushing of the distribution system commenced on May 19 as a result of contamination suspicions. As a result, the water-flow model was not considered reliable after May 18 and all analyses in the present investigation only considered exposure dates up to midnight of May 18.

Exposure scenarios were created by adding a hypothetical inert contaminant to each well at predetermined times and concentrations. WaterCAD provides the ability to follow the movement and relative concentrations of contaminants through a distribution system. Considering the shape of the epidemic curve, the computerized well pumpage data, and the hypothesized importance of the May 12 to 14 rain event (Bruce Grey Owen Sound Health Unit, 2000), six contamination scenarios were chosen (Table 10.1). For each scenario,

TABLE 10.1 Exposure Scenarios Evaluated in the Walkerton Waterborne Outbreak

Scenario/exposure variable	Well	Contamination pattern	Hypothesis tested
[5a]	5	Contamination (100%)* from 12:00 a.m. May 9, and continuing.	Well 5 heavily contaminated when it started up on May 9.
[5b]	5	Partial contamination (25%)* from 12:00 a.m. May 9 to 12:00 a.m. May 12, then full contamination (100%).	Well 5 slightly contaminated when it started up on May 9; the contamination was augmented by heavy rainfall on the 12th and the 13th.
[5c]	5	Contamination (100%)* from 12:00 a.m. May 9 to 12:00 a.m. May 12.	Well 5 heavily contaminated when it started up on May 9 but the contamination only lasted for 3 days.
[6]	6	Contamination (100%)* from 12:00 a.m. May 9, and continuing.	Well 6 heavily contaminated when it started up on May 9.
[7a]	7	Contamination (100%)* from 12:00 a.m. May 8, and continuing.	Well 7 heavily contaminated the day before it shut off prior to the outbreak peak.
[7b]	7	Contamination (100%)* from 12:00 a.m. May 1, and continuing.	This scenario allowed the testing of whether or not Well 7 water was protective.

*Contamination (%) refers to relative level of contamination.

WaterCAD provided the hourly hypothetical contaminant concentration at each residential node. Diagrams and graphs were generated using S-Plus (Insightful; Seattle, WA), Excel (Microsoft; Redmond, WA), and Arcview to visualize the temporal-spatial pattern of contaminants among the exposure scenarios investigated.

Case and Control Selection. Cases were identified through hospital admission, emergency room visits, and telephone calls to the local health unit (Bruce Grey Owen Sound Health Unit, 2000). Analyses were limited to individuals living in homes supplied by the municipal water distribution system during the outbreak period. A *case* was defined as a person with diarrhoea or bloody diarrhoea commencing between May 11 and May 20, or having a stool specimen positive for *E. coli* O157 or *Campylobacter* spp., or being diagnosed with haemolytic uremic syndrome. Since the objective of the analysis was to identify the well(s) most likely responsible for the outbreak, only primary cases were used (i.e., the first person ill within a household). Furthermore, given the potential for enhanced person-to-person transmission of infectious agents within the two Walkerton multiunit seniors' residents, inhabitants of these two locations were excluded.

In an attempt to investigate the temporal-spatial nature of the cases, a number of descriptive analyses were performed. Together with graphing cases by onset date, age, and sex, case density rates were mapped overtop the Walkerton water distribution network. Case density rates were calculated as the number of primary cases divided by the number of households per 500 m^2. Residential addresses were geocoded for input into GIS using house/apartment address and a commercially available street network file (CanMap; DMTI Spatial; Markham, ON). Geocoordinates were verified using aerial photography and by a ground survey.

Controls used in the conditional logistic regression exercise were identified through a random telephone cross-sectional survey of Walkerton (Bruce Grey Owen Sound Health Unit, 2000). Walkerton residents whose homes were supplied with municipal water (except for individuals living in the two seniors' residences) and who had consumed municipal water from May 8 to 18 were eligible to be controls. Of these individuals, those who did not have diarrhoea and were not culture positive for *E. coli* O157:H7 or *Campylobacter* during the course of the outbreak were designated as *controls*.

Conditional Logistic Regression. Conditional logistic regression was used to quantitatively evaluate the relationship between exposure to potentially contaminated well water from each of Walkerton's three wells and the likelihood of becoming a case (Breslow and Day, 1980). Using a median incubation period of 2 to 5 days for both *E. coli* O157:H7 and *Campylobacter* (Chin, 2000), the mean source well contaminant concentrations supplied to homes 2 to 5 days prior to illness onset date were calculated for each case. This was done for each of the six exposure scenarios investigated (Table 10.1). Control exposure values were calculated in a similar manner, but unlike cases who only contributed one observation over the study period, controls contributed 10, one for each day. Given that both the contaminant concentrations and the likelihood of being a case varied greatly by day, it was necessary to match on the day (Rosenbaum and Rubin, 1985; Rothman and Greenland, 1998).

SAS (SAS Institute; Cary, NC) was used to run the conditional regression analysis. Potential correlation arising from multiple observations made on the controls was handled using the PHLEV macro (Thernau and Grambsch, 2000). This macro, which is an extension of the PHREG procedure in SAS, estimates a robust covariance matrix and provides an estimate of variance to adjust for correlation among observations (Lin and Wei, 1989).

The PHLEV procedure fits the following Cox's proportional hazards model:

$$\frac{h_i(t)}{h_j(t)} = \exp\{\beta_1(x_{1i} - x_{1j}) + \cdots + \beta_{k1}(x_{ki} - x_{kj})\} \quad (10.1)$$

TABLE 10.2 Correlation Among Walkerton Water Exposure Scenario Contaminant Concentrations

Scenario[*]		[5a]	[5b]	[5c]	[6]	[7a]	[7b]
	Well	5	5	5	6	7	7
[5a]	5	1	0.52	0.82	−0.12	−0.47	−0.78
[5b]	5	0.52	1	0.04	−0.08	−0.24	−0.39
[5c]	5	0.82	0.04	1	−0.08	−0.38	−0.65
[6]	6	−0.12	−0.08	−0.08	1	−0.24	−0.45
[7a]	7	−0.47	−0.24	−0.38	−0.24	1	0.61
[7b]	7	−0.78	−0.39	−0.65	−0.45	0.61	1

[*]Refer to Table 10.1 for variable explanations (null hypothesis: no correlation, p value <0.0001 for all estimates).

where $h_i(t)$ = hazard function associated with the ith individual at time t
β_k = kth covariate
x_i = level of exposure for the ith individual

Parameter estimates given by the PHLEV procedure correspond to the β's

A total of eight independent variables were available for modelling—age, gender, and the six exposure variables associated with the six contamination scenarios (Table 10.1). As was predicted, significant correlation was observed among the six exposure variables (Table 10.2). Given the nature of water flow in a distribution system, source well proportional-volume contributions at any point in the distribution system must sum to a total of 100 percent. Considering the inherent collinearity among exposure variables, these variables were modeled in three independent groups—Well 5 scenarios, Well 6 scenarios, and Well 7 scenarios. Variables that were found to be significant using an initial univariate screening p-value threshold of 10 percent were considered for further analysis (except for age which was forced into all multivariate models). Final models were chosen on the basis of the robust chi-square statistic and the Akaike information criterion (Akaike, 1974). Interpretations of the final models in the context of the overall study objectives were made by assessing the entire suite of final models.

The analysis described above was performed on the entire case-control data set, which consisted of cases with onset dates between May 11 to 20, and their matched controls. In addition, to test the possibility that the water exposure effect might be different for earlier and later cases, separate analyses were also conducted on cases with onset dates between May 11 to 16, and May 17 to 20, and their respective controls. It was hypothesized that given the inherent uncertainties in the water-flow model (potentially compounded over time), calculated exposure levels may be more accurate in the early stages of the outbreak.

10.3.2 Results

Descriptive Analyses. Figure 10.4 shows the average node contaminant concentrations resulting from Scenario [5b]. It appears that Well 5 disproportionally serviced the southern regions of the town during the period of interest. Except for variations in the absolute range of contaminants, Scenario [5a] (average node contaminant range—0.0 to 73.4 percent) and Scenario [5c] (average node contaminant range—0.0 to 16.1 percent) resulted in similar

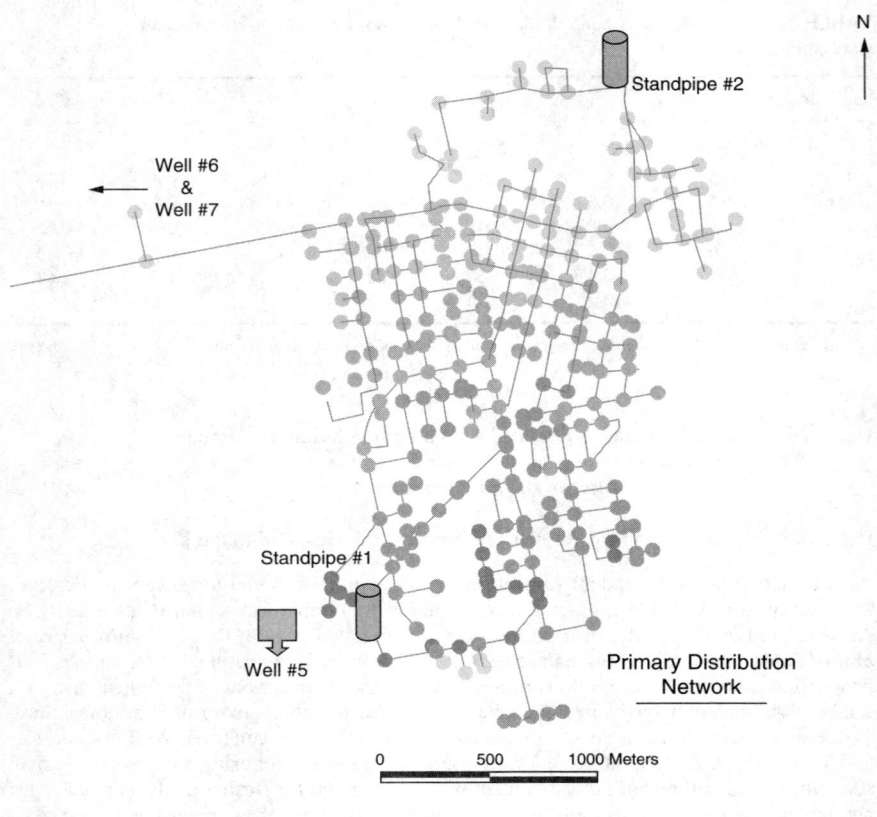

FIGURE 10.4 Walkerton Well 5: May 9 contamination augmented by heavy rainfall (Scenario [5b])—average node contaminant concentration (May 6–18).

distribution patterns and are thus not shown here. Figures 10.5 and 10.6 show the spatial pattern of average node contaminant concentrations resulting from Scenario [6] and Scenario [7b], respectively. Once again, except for a variation in the absolute range of average contaminant concentrations, Scenario [7a] (average contaminant range—0.0 to 84.2 percent) resulted in a similar spatial contaminant concentration pattern to that of Scenario [7b] and is thus not presented. As can be appreciated from Figs. 10.5 and 10.6, Well 6 and Well 7 appear to have disproportionally supplied the northern areas of the town.

A total of 378 persons met the case definition—211 females and 167 males. The median age of cases over the entire study period was 29.5 years (mean—30.9 years, range <1 to 93 years). A total of 367 persons met the control definition—204 females and 163 males.

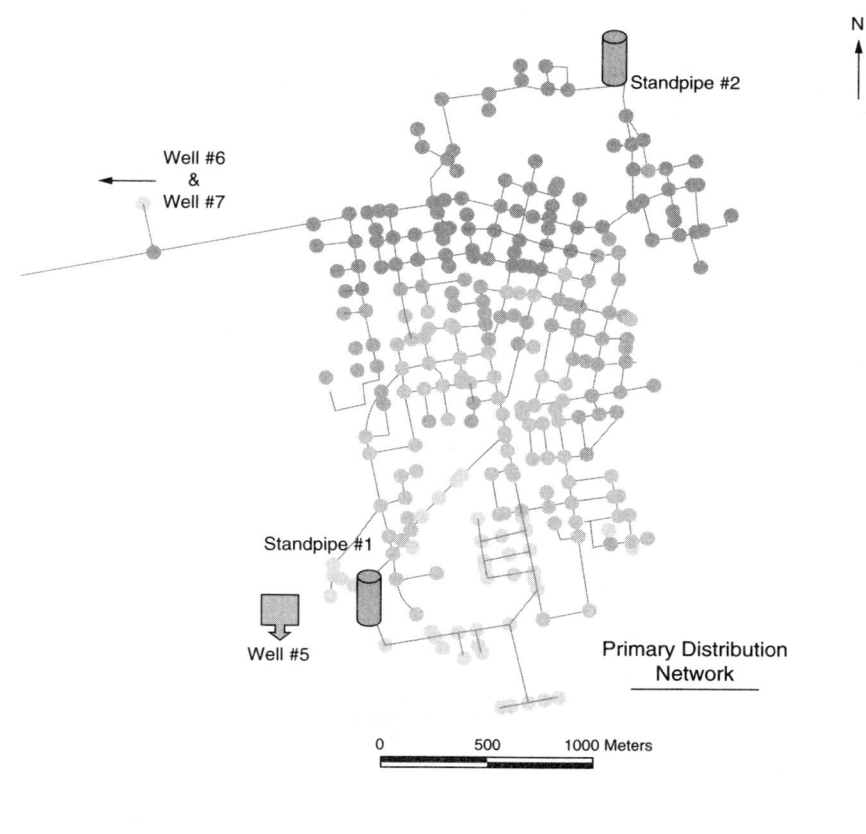

FIGURE 10.5 Walkerton Well 6: Contamination beginning May 9 (Scenario [6])—average node contaminant concentration (May 6–18).

The median age of controls was 42.0 years (mean—40.7 years, range <1 to 94 years). Figure 10.7 demonstrates the age and gender breakdown of cases by date of onset.

In an attempt to visualize the spatial distribution of cases in Walkerton, a map of the case density rates (primary cases/households per 500 m^2) was produced (Figure 10.8). In this diagram, circles of varying shades of grey (corresponding to the range of case density rates) appear in the centre of each 500 m^2 area of analysis. As is often the case with disease rates, rate stability (and thus overall significance) is dependent on the size of the denominator (Marshall, 1991; Heisterkamp et al., 1993). To allow a visualization of both the case rate and the number of residences in the denominator, the sizes of the circles have been scaled to reflect the number of denominator residences. Comparing the spatial distribution of cases (Fig. 10.8) with the spatial variations in exposures associated with the contamination

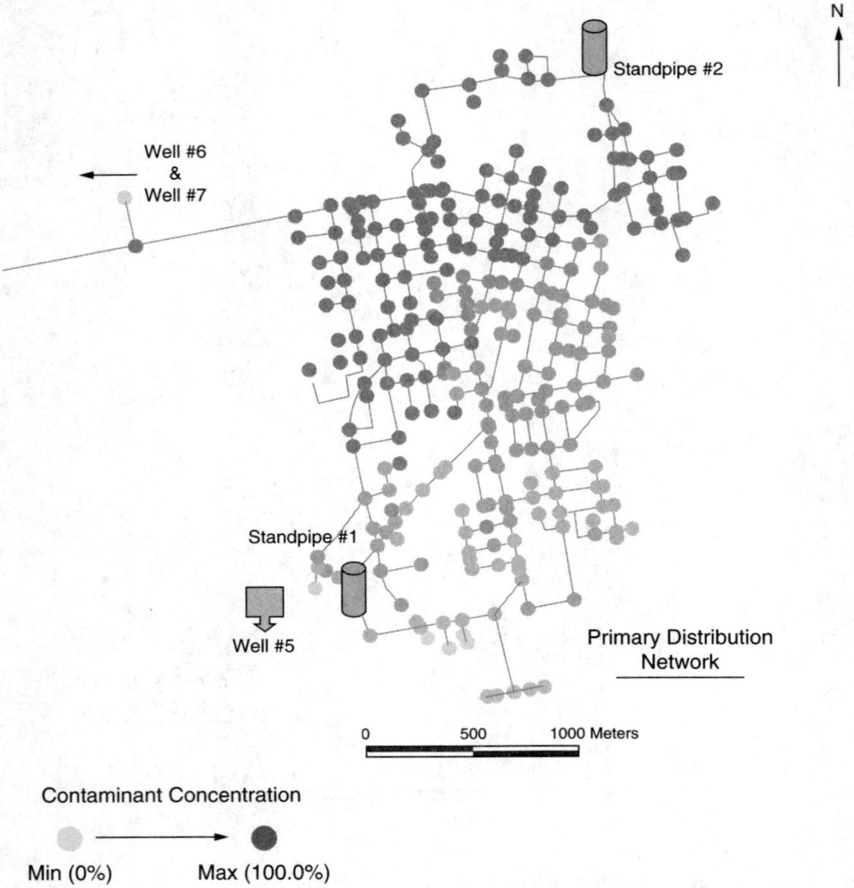

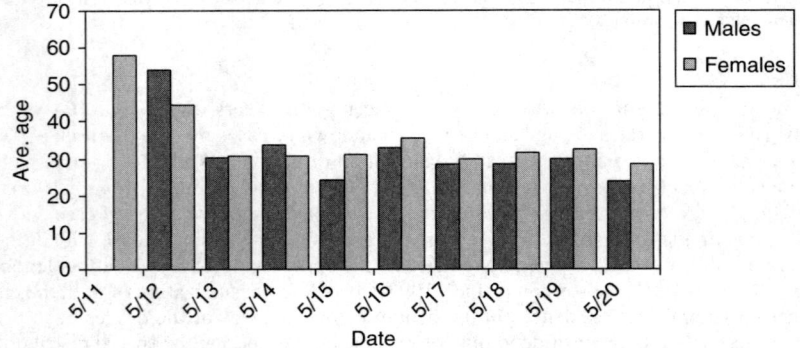

FIGURE 10.6 Walkerton Well 7: Contamination beginning May 1 (Scenario [7b])—average node contaminant concentration (May 6–18).

FIGURE 10.7 Age and gender distribution of Walkerton cases by date of onset.

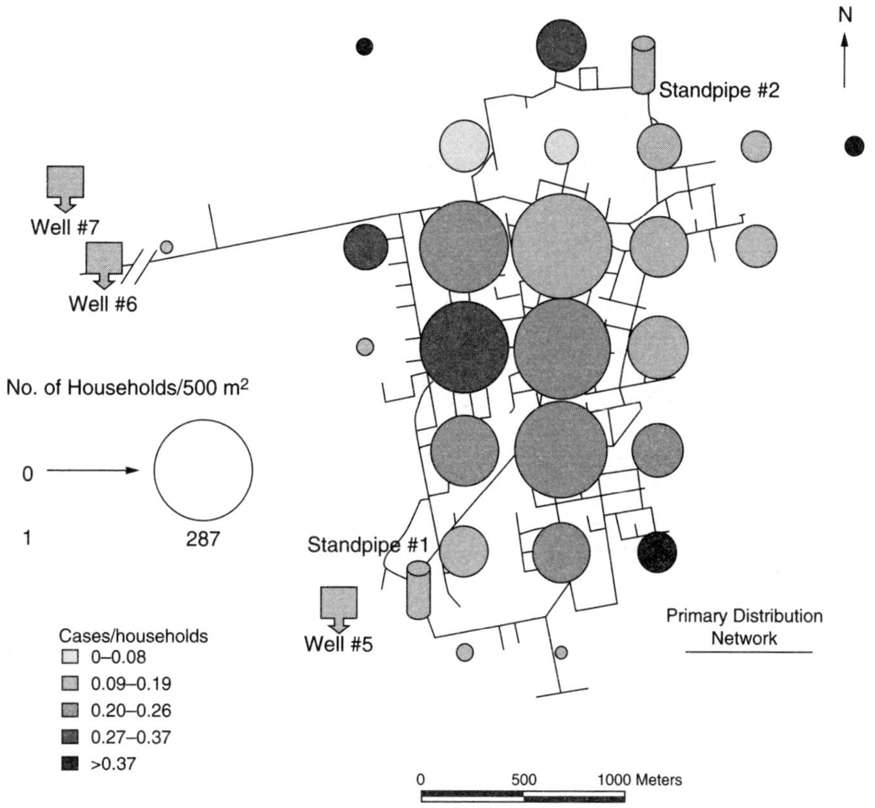

FIGURE 10.8 Walkerton outbreak primary case rates per 500 m² (May 11–20).

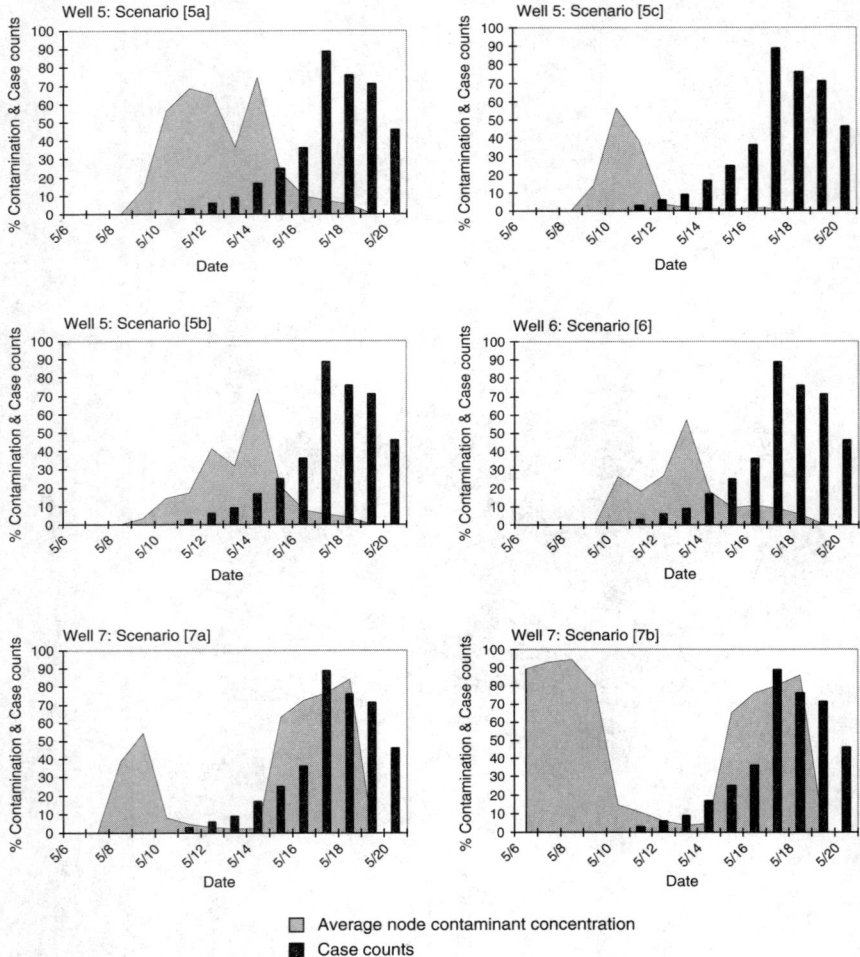

FIGURE 10.9 Average node contaminant concentrations associated with six exposure scenarios and epidemic curve, Walkerton waterborne outbreak (May 2000).

scenarios (Figs. 10.4 to 10.6), it appears that the case density rate pattern most closely resembled the contamination pattern associated with Well 5.

Figure 10.9 displays the average daily node contaminant concentrations associated with each exposure scenario along with the epidemic curve of the cases. It appears that the epidemic curve most closely resembled the shape of exposure levels associated with Scenario [5b].

The case and control average exposure contaminant concentrations by date of onset, associated with each of the six exposure scenarios, are shown in Fig. 10.10. In general, it appears that cases were associated with higher levels of Well 5 water than controls (Scenarios [5a], [5b], and [5c]), and controls were associated with higher levels of Well 6 and 7 water than cases (Scenarios [6], [7a], and [7b]).

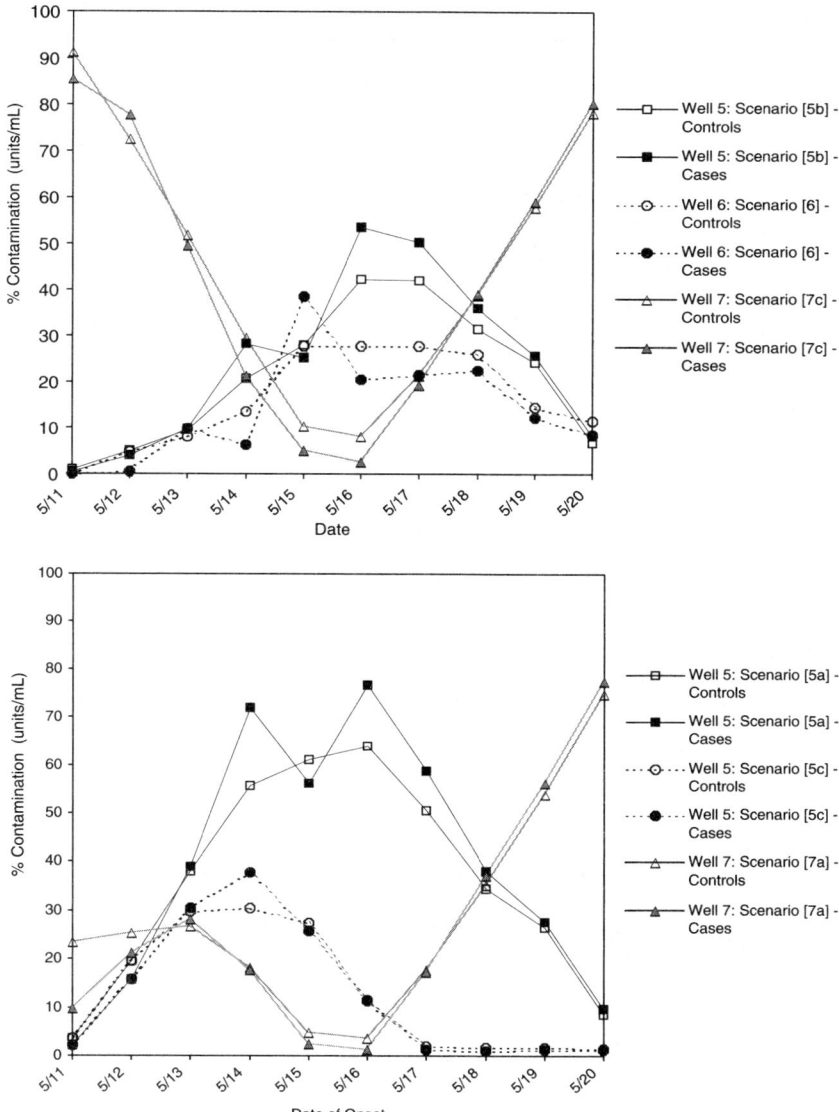

FIGURE 10.10 Case and control mean 2- to 5-day contaminant concentrations associated with the six exposure scenarios investigated, Walkerton waterborne outbreak (May 2000).

Conditional Logistic Regression. Univariate parameter estimates and p-values are presented in Table 10.3. The multivariate conditional logistic regression results are presented in Table 10.4. Seven final models emerged, three associated with the May 11 to 20 onset period (models A1, A2, A3), and two associated with both the May 11 to 16 (models B1, B2) and May 17 to 20 (models C1, C2) onset periods. For all subsets of data, gender was not a

TABLE 10.3 Walkerton Outbreak Conditional Logistic Regression Univariate Results

Variable	Scenarios*							
	Age	Gender	[5a]	[5b]	[5c]	[6]	[7a]	[7b]
Onset period A (May 11 to 20)								
Parameter estimate	−0.014	−0.004	0.010	0.015	−0.005	−0.014	0.004	−0.008
p value	0.000	0.980	0.002	0.000	0.518	0.004	0.488	0.082
Onset period B (May 11 to 16)								
Parameter estimate	−0.007	0.176	0.008	0.018	0.005	−0.003	−0.050	−0.032
p value	0.171	0.442	0.048	0.010	0.579	0.590	0.006	0.005
Onset period C (May 17 to 20)								
Parameter estimate	−0.017	−0.073	0.013	0.015	−0.179	−0.022	0.010	0.001
p value	0.000	0.645	0.004	0.001	0.001	0.000	0.150	0.881

*Refer to Table 10.1 for variable explanations.

TABLE 10.4 Walkerton Outbreak Conditional Logistic Regression Final Models

Model	Exposure variable*	Parameter estimate	Robust SE	p value	Variable range	Max OR (95% CI)
May 11 to 20						
A1	[5b]	0.0152	0.0041	0.0002	0.0–82.1%	[+] 3.47 (1.79–6.76)
	Age	−0.0138	0.0032	0.0000	0–94 years	[−] 3.66 (2.03–6.57)
A2	[6]	−0.0132	0.0047	0.0050	0.0–86.3%	[−] 3.11 (1.41–6.89)
	Age	−0.0137	0.0032	0.0000	0–94 years	[−] 3.45 (1.45–8.18)
A3	[7b]	−0.0085	0.0047	0.0724	0.0–99.9%	[−] 2.33 (0.93–5.88)
	Age	−0.0139	0.0032	0.0000	0–94 years	[−] 3.70 (2.06–6.63)
May 11 to 16						
B1	[5b]	0.0179	0.0070	0.0105	0.0–70.1%	[+] 3.50 (1.34–9.11)
	Age	−0.0066	0.0049	0.1753	0–94 years	[−] 1.81 (0.74–4.47)
B2	[7b]	−0.0323	0.0112	0.0039	0.0–99.9%	[−] 25.09 (2.81–224)
	Age	−0.0071	0.0049	0.1467	0–94 years	[−] 1.95 (0.79–4.79)
May 17 to 20						
C1	[5b]	0.0143	0.0046	0.0017	0.0–82.1%	[+] 3.24 (1.56–6.75)
	Age	−0.0167	0.0035	0.0000	0–94 years	[−] 4.81 (2.53–9.13)
C2	[6]	−0.0214	0.0057	0.0002	0.0–66.6%	[−] 4.16 (2.53–9.13)
	Age	−0.0167	0.0035	0.0000	0–94 years	[−] 4.80 (2.53–9.11)

*Refer to Table 10.1 for variable explanations.

significant predictor of disease. Also provided in Table 10.4 are the maximum and minimum values of each significant independent variable, together with their associated maximum odds ratios and 95 percent confidence levels.

Among the six well exposure variables, the one associated with Scenario [5b] (May 9 contamination of Well 5 augmented by heavy rainfall on May 13) was the only exposure variable identified as a significant risk factor in each data subset investigated (Table 10.4).

Analysis of Results. The results of this study provide strong evidence to support what is now considered the most likely series of events leading up to the infection of an estimated 1286 Walkerton residents with *E. coli* O157:H7 and/or *Campylobacter* (Bruce Grey Owen Sound Health Unit, 2000). The results of the temporal-spatial analyses presented here strongly support the theory that Well 5 was the primary, if not the only well involved in the Walkerton outbreak. Although hydraulic models have been used on occasions over the last decade to aid in the investigation of waterborne outbreaks (Geldreich et al., 1992; Clark et al., 1996; Clark and Grayman, 1998), many of the approaches taken in the present waterborne outbreak investigation have not been previously described.

In reviewing the results of the descriptive analyses, both the spatial and temporal results provide suggestive evidence supporting the Well 5 contamination hypothesis. Although the spatial distribution of cases (Fig. 10.8) did not match exactly any of the average contaminant concentration spatial patterns (Figs. 10.4 to 10.6), it appears to most closely resemble the pattern associated with Well 5 (Fig. 10.4), compared to Well 6 and 7 scenarios (Figs. 10.5 and 10.6). Temporally, the resemblance of the shape of the epidemic curve to the daily distribution system average contaminant levels associated with Scenario [5b] (Fig. 10.9) not only suggests the involvement of Well 5, but also the potential importance of the rainfall event.

The spatial and temporal observations described above are consistent with the comparison of case and control exposure levels associated with the scenarios evaluated (Fig. 10.10). In general, cases appeared to be associated with higher levels of Well 5 water (Scenarios [5a], [5b], and [5c]), and controls with higher levels of Well 6 and Well 7 water (Scenario [6] and Scenarios [7a] and [7b], respectively).

A somewhat unrelated observation with respect to the objectives of the present investigation, but an important observation nonetheless, was the average age of cases when examined by date of illness onset (Fig. 10.7). The average ages for individuals with dates of onset of May 11 and 12 were greater than the average age of those with later onset dates. Although the biological explanation for this observation is unknown, reference is often made to the greater sensitivity of elderly individuals to communicable enteric infections (Mandell et al., 2000). Potential implications of this observation with respect to surveillance programs aimed at the early detection of waterborne events using elderly populations as sentinels are intriguing.

The results of the conditional logistic regression exercise support the findings of the descriptive analyses. Among the six well exposure variables, the one associated with Scenario [5b] (May 9 contamination of Well 5 augmented by heavy rainfall on May 13) was the only exposure variable identified as a significant risk factor in each data subset investigated (Table 10.3). The significance of this variable throughout the entire study period provides strong supportive evidence for the Well 5 contamination hypothesis. Furthermore, the Scenario [5b] exposure variable appeared to be a more important predictor than the other two Well 5 exposure variables ([5a] and [5c]). This finding supports the hypothesis that the contamination event likely extended well into the study period, and that the heavy rains which occurred on May 12 and 13 likely influenced the degree of contamination. The importance of extreme rainfall in waterborne disease events has also been suggested by a recent review of past United States outbreaks (Curriero et al., 2001).

Over the entire study period (May 11 to May 20), individuals exposed to the maximum 3-day average Well 5 contaminant concentration associated with Scenario [5b] (82.1 percent), had a 3.5 times greater likelihood of becoming a case compared to individuals not exposed to any of the Well 5 contaminated water during the same exposure period.

It was difficult to assess the significance of one well exposure variable while controlling for another given the inherent colinearity among variables. Nevertheless, with the Well 6 and Well 7 exposure variables identified in the final models being important protective factors, it is unlikely that a second well also contributed to the contamination of the water supply. The analyses suggest that individuals residing in homes receiving greater contributions of water from Well 6 and/or Well 7 were less likely to become ill. In fact, in the early stages of the outbreak, it appears that individuals from homes supplied with 100 percent Well 7 water were 25 times less likely to succumb to gastroenteritis compared to residents whose homes received no Well 7 water during the same time period. Interestingly, whereas Well 7 appeared to provide the greatest protective effect during the early stages of the outbreak, Well 6 became protective towards the end of the study period.

Given the nature of the data used in this investigation, it is prudent not to over-interpret the absolute values of the regression parameters. Exposure misclassification associated with the water flow model itself, the generalization of the incubation period, and the assumption that all cases became infected after consuming tap water at home, were all likely contributors to some degree of error in the regression parameters. However, as misclassification bias was likely non-differential in nature, thus favoring the null hypothesis (Rothman and Greenland, 1998), it is probable that the true exposure effects were actually greater than those presented here.

10.3.3 Conclusions

The results of this study clearly support the hypothesis that Well 5 was the primary, if not the only, well involved in the Walkerton *E. coli/Campylobacter* waterborne outbreak. Moreover, the results suggest that the extreme rainfall event, which occurred just prior to the peak in the outbreak, may have played a significant role. These results are consistent with a hydrogeological assessment that found that Well 5 was under the influence of surface water, and the results of an environmental investigation that found *E. coli* O157:H7 and *Campylobacter* spp. on a cattle farm adjacent to Well 5 with identical molecular characteristics to those isolated from patients, together with the detection of *E. coli* O157:H7 in Well 5 water by Polymerase Chain Reaction (Bruce Grey Owen Sound Health Unit, 2000).

Approximately 150 years ago, Dr. John Snow identified drinking water from the Southwark and Vauxhall Company as the likely source of cholera for several regions of London, England. Fortunately for Snow, the various water companies supplying London at the time had independent distribution systems, providing Snow with distinct exposure groups for comparison. Today, it is not uncommon for cities and towns to have more than one primary source of drinking water, feeding into a single distribution system with multiple connections to strengthen the integrity of the network. Fortunately for epidemiologists today, analytical tools exist to model water flow through complicated distribution networks allowing for a quantifiable temporal-spatial exposure assessment.

10.4 DOVER TOWNSHIP (TOMS RIVER), NEW JERSEY

10.4.1 Introduction

In the spring of 1995, the Agency for Toxic Substances and Disease Registry (ATSDR) requested that the New Jersey Department of Health (now the New Jersey Department of

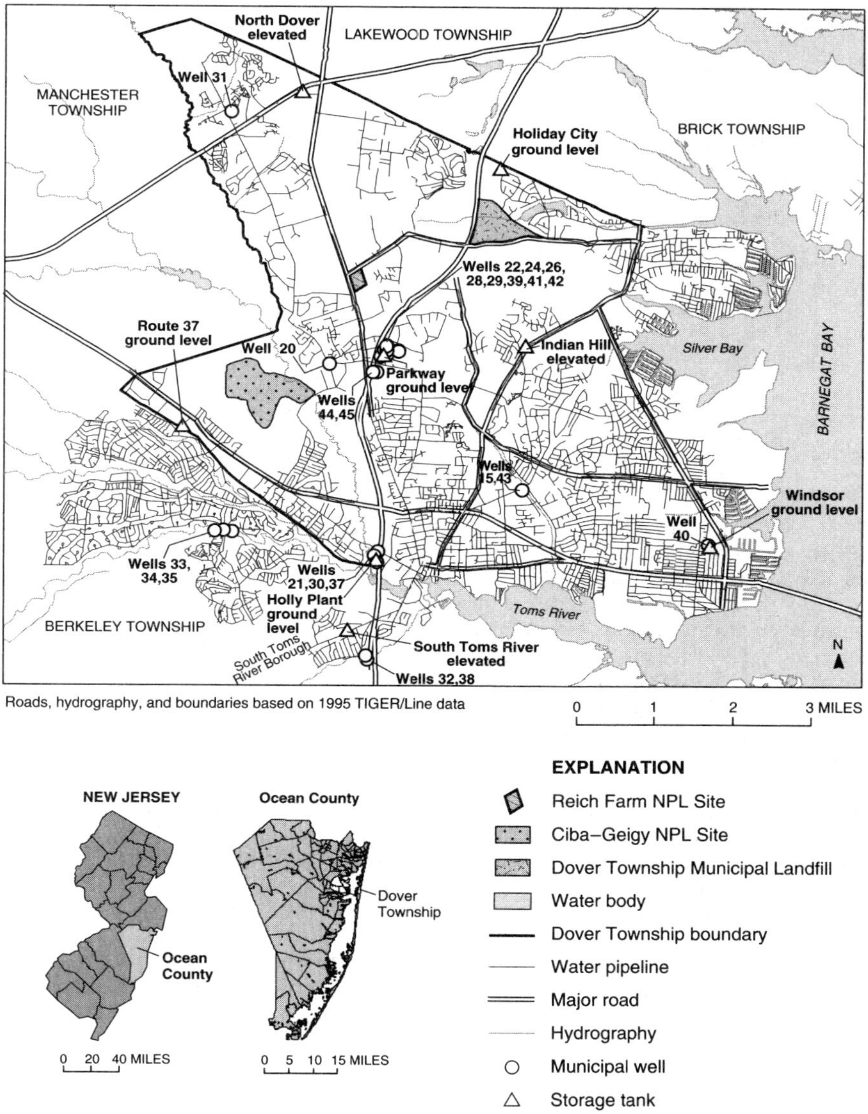

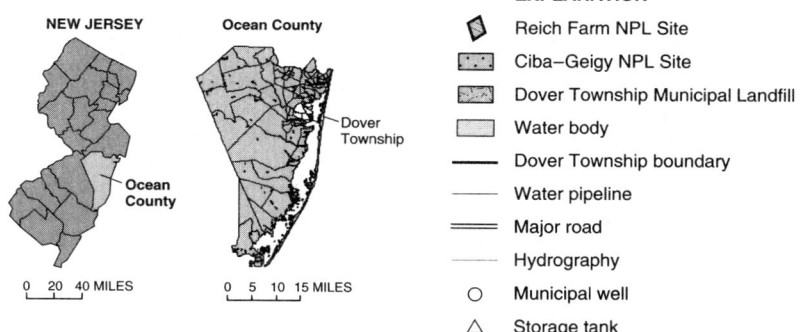

FIGURE 10.11 Investigation area, Dover Township, Ocean County, New Jersey (*modified from Maslia et al., 2001*).

Health and Senior Services [NJDHSS]) evaluate the childhood cancer incidence in the Toms River section of Dover Township, Ocean County, New Jersey (Fig. 10.11). In August 1995, the NJDHSS completed a preliminary evaluation of data from the New Jersey State Cancer Registry from 1979 through 1991 and concluded that childhood cancer incidence in Dover Township and the Toms River section was higher than expected for all malignant cancers combined, brain and central nervous system cancer (CNS), and leukemia (Berry, 1995).

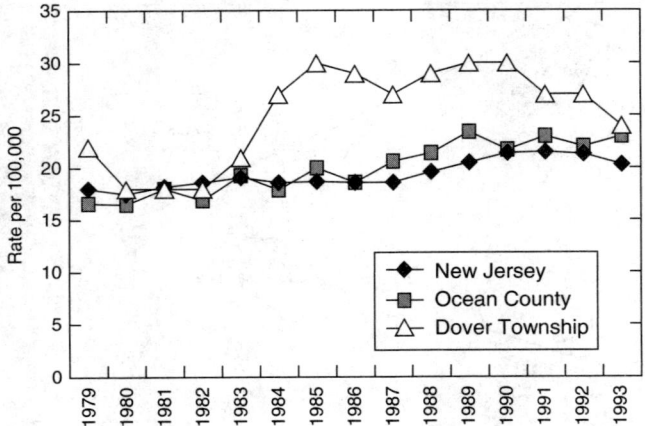

FIGURE 10.12 Time trend in childhood cancer rates, 1979–1995 (*from Berry and Haltmeier, 1997*).

In March 1996, the NJDHSS and the ATSDR developed a Public Health Response Plan (PHRP) describing actions the agencies would take to investigate childhood cancers and environmental concerns in Dover Township (NJDHSS and ATSDR, 1996). The PHRP included an updated childhood cancer incidence analysis through 1995 and evaluations of potential environmental exposure pathways relative to two National Priorities List (NPL or Superfund) sites in Dover Township (Fig. 10.11)—Ciba-Geigy (ATSDR, 2001b) and Reich Farm (ATSDR, 2001d). In addition, an environmental evaluation was conducted for the Dover Township Municipal Landfill (ATSDR, 2001c) and an extensive water-quality evaluation was conducted of the United Water Toms River (UWTR) community water supply (ATSDR, 2001a).

Childhood Cancer Incidence. As part of the PHRP, the NJDHSS expanded its preliminary cancer incidence analysis to 1995. This evaluation included all childhood (under 20 years of age) cancers combined and groupings of childhood cancer types for Ocean County, Dover Township, and the Toms River section of the Township. A time-trend analysis of childhood cancer found that in Dover Township childhood cancers were elevated from 1979 through 1995 (Fig. 10.12). (Standardized Incidence Ratio (SIR) of 1.3 and 95 percent Confidence interval (CI) of 1.1 to 1.7.) The elevations were most pronounced among female children (Fig. 10.13) under age five in Toms River for acute lymphocytic leukemia (SIR = 9.2, 95 percent CI = 2.5–23) and for brain and CNS cancers (SIR = 11.5, 95 percent CI = 2.3–34), with a peak in incidence from the mid-1980s through the early 1990s (Berry and Haltmeier, 1997).

Environmental Exposure Pathways. Environmental evaluations of the two Superfund sites in the community identified potential exposure pathways, including contamination of several public water-supply wells (ATSDR, 2001a,b,d). Testing of the public water supply revealed a previously undiscovered contaminant, styrene-acrylonitrile (SAN) trimer in groundwater from several wells in the Parkway well field (ATSDR, 2001a). This compound was one of the substances dumped at the Reich Farm Superfund site in 1971 (ATSDR, 2001d). The Parkway well field (Fig. 10.11) is a major source of potable water for the distribution system, obtaining a substantial amount of its water from the shallow, high yielding, Kirkwood-Cohansey aquifer (Maslia et al., 2000a). The Parkway well field,

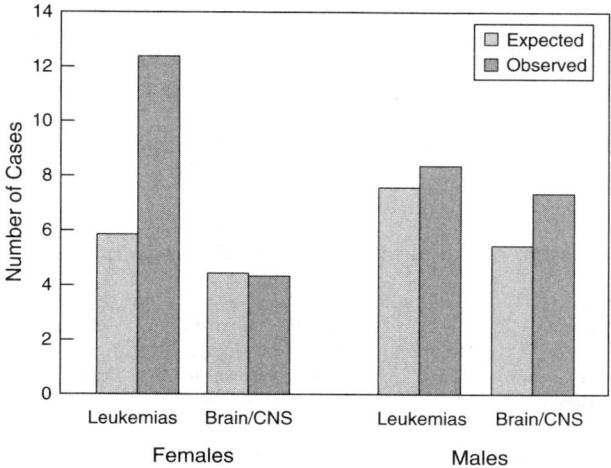

FIGURE 10.13 Childhood cancer incidence, ages 0–19 years, Dover Township, New Jersey, 1979–1995 (*from Berry and Haltmeier, 1997*).

which began production in 1971, is located about 1 mi south of the Reich Farm Superfund site.

Chemical wastes were illegally disposed of at Reich Farm in 1971, which lead to contamination of local groundwater, including the Kirkwood-Cohansey aquifer. (ATSDR, 2001d). The specific chemical compounds, historically contaminating the groundwater, were not well characterized. However, chlorinated solvents and the SAN trimer were found in the contaminant plume. The ATSDR and the NJDHSS concluded that the Reich Farm Superfund site was a public health hazard due to past exposures (ATSDR, 2001d).

The Ciba-Geigy Superfund site (Fig. 10.11), located in Dover Township, produced organic dyes and intermediate products, epoxy resins, and specialty chemicals, beginning in 1952 (ATSDR, 2001b). Disposal of liquid and solid wastes from the manufacturing process on company property resulted in groundwater and soil contamination. The groundwater in the vicinity of this Superfund site has been contaminated with a variety of volatile organic compounds (VOCs), in addition to metals and possibly other chemical compounds. Contaminated groundwater from public water supply at the Holly well field (Fig. 10.11) was identified in the mid-1960s (Toms River Chemical Corporation, 1966) and documented to be contaminated during the 1970s (ATSDR, 2001b). Based on evaluations conducted as part of the public health assessment process, the ATSDR and the NJDHSS concluded that the Ciba-Geigy Superfund site represented a public health hazard due to past exposures (ATSDR, 2001b).

Epidemiologic Study. In 1997, the NJDHSS and the ATSDR began designing a case-control epidemiologic study of childhood cancers that occurred in Dover Township. In a case-control study, a population is delineated and cases of diseases arising in that population over a specified time period are identified. The exposure experiences of the case group are compared to the exposure experiences of a sample group of nondiseased persons in the population from which the cases arose. The exposure experiences that are more common among the diseased cases may be considered possible risk factors for the disease (Rothman and Greenland, 1998). A case-control study design was selected because it was the best method for studying rare diseases (Schlesselman, 1982), such as childhood cancers. The study focused on two age groups in which elevated rates of cancer were previously found in Dover Township—children diagnosed before 20 years of age and children diagnosed before age five (NJDHSS, 2003). The study was designed to focus on specific hypotheses about certain environmental exposure pathways. One of the primary hypotheses that were tested was that study cases had higher exposure to public water supplies with documented contamination (the Parkway and Holly well fields) than did study controls.

10.4.2 Assessment of Contaminated Public Water Supply

To assist NJDHSS with the contaminated drinking water exposure assessment component of the epidemiologic study, ATSDR developed a water-distribution model using the EPANET 2 software (Rossman, 2000). The model was used to simulate historical characteristics of the water-distribution system serving Dover Township from 1962 through 1996. Given the paucity of historical contaminant-specific data during most of the period relevant to the epidemiologic study, modeling focused on estimating the percentage of water that a study subject might have received from each well and well field that supplied the water-distribution system. This modeling approach lead to the development of the "proportionate contribution" concept wherein at any given point in the distribution system, water may be derived from one or more sources in differing proportions (Maslia et al., 2000a,b).

Methods and Approach. Because of the lack of historical hydraulic and water-quality information, the water-distribution system was characterized using data gathered during an extensive field investigation in 1998. The field investigation consisted of two components: (1) determining spatial locations of distribution system facilities (wells, tanks, pump, and hydrants) and (2) equipping hydrants with continuous-recording digital data loggers and monitoring supply sources (wells, pumps, and tanks) to measure system responses during winter-demand (March) and summer-demand (August) periods. Twenty-five hydrants located throughout the distribution system were equipped with data loggers to simultaneously collect information on system response (Maslia et al., 2000a).

An "all-pipes" hydraulic network model was calibrated to present-day conditions (1998) using field investigation results. The reliability of the calibrated model was successfully demonstrated through a water-quality simulation of the transport of a naturally occurring conservative element (barium) and comparing results with data collected in March and April 1996 at 21 schools and six points of entry to the water-distribution system. Results of the field-data collection activities, model calibration, and reliability testing are described in Maslia et al. (2000a,b).

To describe the historical distribution-system networks specific to the Dover Township area, databases were developed from diverse sources of information. These data were applied to EPANET 2 and simulations were conducted for each month of the historical period—January 1962 through December 1996 (420 simulations). After completing the 420 monthly analyses, source-trace analysis simulations were conducted to determine the percentage of water contributed by each well or well field operating during each month for all study subject (cases and controls) locations. Results of these analyses—the percentage of water derived from the different sources that historically supplied the water-distribution system—were provided to health scientists for their analysis of the association between exposure to contaminated water and the occurrence of disease (childhood cancers) in developing exposure indexes to assess the environmental factors being considered by the epidemiologic study. An important fact to note is that throughout the investigation, persons involved with the water-distribution modeling effort were blinded to the status and location of epidemiologic study subjects.

Specific Data Needs. The simulation approach to the historical reconstruction of the water-distribution system in the Dover Township area required knowledge of the functional as well as the physical characteristics of the distribution system. Accordingly, six specific types of information were required: (1) pipeline and network configurations for the distribution system; (2) potable water-production data including information on the location, capacity, and time of operation of the groundwater production wells; (3) consumption or demand data at locations throughout the distribution system; (4) storage-tank capacities, elevations, and water-level data; (5) high-service and booster pump characteristic curves; and (6) system-operations information such as the on-and-off cycling schedule of wells and high-service and booster pumps, and the operational extremes of water levels in storage tanks.

Yearly historical network configuration maps were developed for the period 1962 through 1996 and are presented in Maslia et al. (2001). These data indicated that water-distribution system complexity increased significantly over the time span of the historical period. For example, the 1962 water-distribution system served nearly 4300 customers from a population of about 17,200 persons and was characterized for modeling by

- Approximately 2400 pipe segments ranging in diameter from 2 to 12 in and comprising a total service length of 77 mi
- Three groundwater extraction wells with a rated capacity of 1900 gal/min (gpm)
- One elevated storage tank and standpipe with a combined rated storage capacity of 0.45 million gallons (Mgal)
- Production of about 1.3 million gallons per day (MGD) during the peak-production month of May

By contrast, in 1996—the last year of the historical reconstruction period—the water-distribution system served nearly 44,000 customers from a population of about 89,300 persons and was characterized for modeling by

- More than 16,000 pipe segments ranging in diameter from 2 to 16 in and comprising a total service length of 482 mi
- Twenty groundwater extraction wells with a rated capacity of 16,550 gpm
- 12 high-service or booster pumps
- Three elevated and six ground-level storage tanks with a combined rated capacity of 7.35 Mgal
- Production of about 13.9 MGD during the peak-production month of June

Analysis of production data indicates that historical distribution systems could be characterized by three typical demand periods each year: (1) a low- or winter-demand period, generally represented by the month of February—designated as the minimum-demand month; (2) a peak- or summer-demand period, represented by one of the months of May, June, July, or August—designated as the maximum-demand month; and (3) an average-demand period, generally represented by the month of October—designated as the average-demand month.

Water-production data were gathered, aggregated, and analyzed for each well for every month of the historical period. These data were obtained from the water utility (Flegal, 1997), Board of Public Utilities, State of New Jersey, Annual Reports (1962–1996), and NJDHSS data searches. The production data were measured by using in-line flow meters at water-supply wells.

Monthly production data can be represented graphically as shown in a three-dimensional plot (Fig. 10.14). Referring to this plot, the x axis is the year (1962–1996), the y axis is the month (January–December), and the z axis is the total monthly production in million gallons.

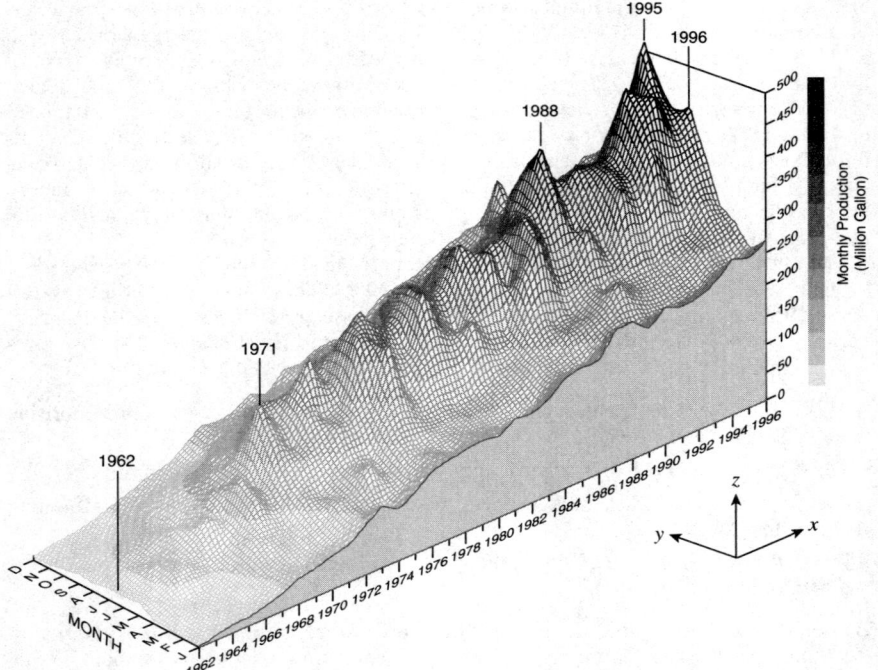

FIGURE 10.14 Three-dimensional representation of monthly production of water, Dover Township area, New Jersey *(from Maslia et al., 2001)*.

TABLE 10.5 "Master Operating Criteria" Used to Develop Operating Schedules for the Historical Water-Distribution System, Dover Township area, New Jersey

Parameter	Criteria
Pressure*	Minimum of 15 psi; maximum of 110 psi at pipeline locations, including network end points
Water level	Minimum of 3 ft above bottom elevation of tank; maximum equal to elevation of top of tank; ending water level should equal the starting water level
Hydraulic device online date	June 1 of year installed to meet maximum-demand conditions
On-and-off cycling: manual operation	Wells and high-service and booster pumps cannot be cycled on-and-off from 2200 to 0600 hours
On-and-off cycling: automatic operation	Wells and high-service and booster pumps can be cycled on-and-off at any hour
Operating hours	Wells should be operated continuously for the total number of production hours, based on production data†

*Generally, for residential demand, minimum recommended pressure is about 20 psi. However, for some locations in the Dover Township area (mostly in areas near the end of distribution lines) lower pressures were simulated.
†See Maslia et al. (2001) for production data (Appendix B) and hours of operation (Appendix D)
Source: From Maslia et al. (2001).

Maximum production is shown to occur in the months of May, June, July, or August. In addition, considerable production increases occurred in 1971, 1988, and 1995. These years are characterized on the plot by sharp peaks.

As noted previously, to simulate the distribution of water for each of the 420 months of the historical period, network configuration, demand, and operational information were required. Before 1978, operational data were unavailable requiring development of system-operation parameters—designated as "Master Operating Criteria." These are based on hydraulic engineering principles necessary to successfully operate distribution systems similar to the one serving the Dover Township area (Table 10.5). From 1978 onwards, for selected years, operators of the water utility provided information on the generalized operating practices for a typical "peak-demand" (summer) and "nonpeak demand" (fall) day. These guidelines were used in conjunction with the "Master Operating Criteria" to simulate a typical 24-h daily operation of the water-distribution system for each month of the historical period.

Examples of historical water-distribution system operating schedules for the maximum-demand months of May 1962 and June 1996 are shown in Tables 10.6 and 10.7, respectively. These tables indicate the hour-by-hour operation of wells and high-service and booster pumps during a typical day of the maximum-demand month for the given year. Note that in 1962 (Table 10.6), high-service and booster pumps were not part of the distribution system and, therefore, only groundwater wells were operated to supply demand by discharging water directly into the distribution system (wells 13, 14, and 15). In 1968, high-service and booster pumps were added to the distribution system. From that year onwards, some wells supplied storage tanks, then high-service and booster pumps were operated to meet distribution-system demands (wells 21 to 30, 40, and 42); other wells still discharged directly into the distribution system (refer to Tables 10.6 and 10.7 for details).

10.4.3 Simulation Methods

The application of simulation methods to the historical reconstruction analysis (application of EPANET 2) using the specific network data for the Dover Township area was accomplished in two steps. First, hydraulic modeling was conducted whereby average network conditions were simulated for every month of the historical period (420 simulations). These simulations were completed under balanced flow conditions that honored hydraulic engineering principles and that conformed to the "Master Operating Criteria" (Table 10.5). Second, using the results of the monthly network hydraulic simulations, water-quality simulations (source-trace analysis) were conducted for each water source (point of entry) of the network in order to determine the monthly proportionate contribution of source water at all locations in the Dover Township area serviced by the water-distribution system.

With respect to the scheduling of groundwater well operations, EPANET 2 utilizes "pattern factors" which correspond to the hourly operations of supply wells. These pattern factors along with the operational extremes of storage tank water levels were manually adjusted during each of the 420 monthly network simulations to achieve balanced flow conditions. This approach to simulation was designated as the *manual adjustment process*. Simulation results presented in the "Historical Reconstruction Simulation Results" section of this part of Chap. 10 were obtained using the *manual adjustment process*.

A second simulation approach was also utilized to achieve balanced flow conditions for each of the 420 monthly networks of the historical period. This approach to simulation was designated the *genetic algorithm* or *GA optimization* approach and required the development of an innovative methodology known as the progressive optimality genetic algorithm

TABLE 10.6 Water-Distribution System Operating Schedule, Dover Township area, New Jersey, May 1962

Well ID*	Hour of the day†																							
	0	1	2	3	4	5	6	7	8	9	10	11	12	13	14	15	16	17	18	19	20	21	22	23
													Groundwater well											
Holly (13)										■	■	■	■	■	■	■								
Holly (14)											■	■	■	■	■	■	■	■	■	■	■	■		
Brookside (15)	■	■	■	■	■	■	■	■	■	■	■	■	■	■	■	■	■	■	■	■	■	■	■	■

[May is maximum-demand month for 1962; hour of day in gray means well operating; ■ Groundwater well]

*Wells discharge directly into the distribution system.
†Hour of the day: 0 is midnight; 12 is noon, respectively.
Source: From Maslia et al. (2001).

TABLE 10.7 Water-Distribution System Operating Schedule, Dover Township area, New Jersey, June 1996.

Well ID[a]	Hour of the day[b]																							
	0	1	2	3	4	5	6	7	8	9	10	11	12	13	14	15	16	17	18	19	20	21	22	23
Groundwater well																								
Brookside (15)																								
Route 70 (31)																								
South Toms River (32)																								
Berkeley (33)																								
Berkeley (34)																								
Berkeley (35)																								
South Toms River (38)																								
Pump ID																								
High-service or booster pump																								
Holly pump 1[c]																								
Holly pump 2[c]																								
Holly pump 3[c]																								
Parkway pump 1[d]																								
Parkway pump 2[d]																								
Holiday City pump																								
St. Catherine's pump[e]																								
South Toms River pump 1																								
South Toms River pump 2																								
Windsor pump 1[f]																								
Windsor pump 2[f]																								
Windsor pump 3[f]																								

[June is maximum-demand month for 1996; hour of day in gray means well or pump operating; ▒ Groundwater well; ▓ High-Service or Booster pump]

[a] Wells discharge directly into the distribution system; Indian Head well (20) out of service.
[b] Hour of the day: 0 is midnight, 12 is noon, respectively.
[c] Holly pump 1, Holly pump 2, and Holly pump 3 supplied by Holly ground-level storage tanks and Holly well 30.
[d] Parkway pump 1 and Parkway pump 2 supplied by Parkway ground-level storage tank and Parkway wells 22, 24, 26, 28, 29, and 42.
[e] Also know as Route 37.
[f] Windsor pump 1, Windsor pump 2, and Windsor pump 3 supplied by Windsor ground-level storage tank and Windsor well 40.

Source: From Maslia et al. (2001).

(POGA) and is an automated objective simulation technique (Guan and Aral, 1999a,b; Aral et al., 2001a,b,c). The GA simulations utilized the balanced flow conditions obtained by the manual adjustment process as starting conditions, although, because of the robustness of the POGA approach, such a requirement is unnecessary. The GA technique was used to address the following questions:

- If a balanced flow operating condition was achieved using the manual adjustment process, was the resulting operating condition the only way the system could have successfully operated?
- Could alternative or additional operating conditions be defined such that system operations would also be satisfactory or even "optimal?"

Thus, the POGA methodology was used in conjunction with EPANET 2 to simulate alternative and possibly optimal water-distribution system operations and to assess the effects of variations in system operations on the results of the proportionate contribution simulations. Results achieved using the POGA methodology are presented in the "Sensitivity Analysis" section of this part of Chap. 10.

Hydraulic Modeling. To conduct the historical simulations, model parameter values input to EPANET 2 required variation that reflected the change in the historical data. For example, data documenting the installation year of network pipelines were available on an annual basis and thus model parameters describing the pipeline network were modified in the EPANET 2 simulations on an annual basis. Data documenting water production were available on a monthly basis (Fig. 10.14) and thus, EPANET 2 model parameters associated with production were varied for each month of the historical period simulations. For other model parameters, such as the on-and-off cycling of wells, data were not available throughout the entire historical period. Quantitative estimation and qualitative description methods were used to derive values required to conduct the EPANET 2 simulations. A summary of model parameters, data availability, and the time-unit variation required to conduct the historical reconstruction simulations using EPANET 2 is provided in Table 10.8.

Routinely, simulation of water-distribution systems, similar to the historical water-distribution system that serviced the Dover Township area, would require detailed descriptions of system operations, such as the on-and-off scheduling of high-service and booster pumps and groundwater wells for the entire period of simulation. In order to simplify these rigorous data requirements, a surrogate or alternative method was devised. Balanced flow conditions were maintained, and the measured volumes of monthly water production were used while avoiding the need for detailed system operations data, which were not available for most of the historical period.

For the Holly, Parkway, and Windsor treatment plants (Fig. 10.11),[*] the actual network consists of a groundwater well (or wells) pumping water and discharging the water into a storage tank. Then high-service or booster pumps discharge water from the storage tank into the distribution system based upon some predetermined operating schedule and demand requirements.[†] This physical or "real-world" representation is shown in Fig. 10.15A and was the method used to represent the distribution of water during the simulation of the present-day system (Maslia et al., 2000a,b). This method is referred to as the "Well-Storage

---

[*]The term treatment plant is used by the water utility to identify all distribution-system facilities associated with a particular point of entry such as wells, storage tanks, water treatment, and high-service or booster pumps.
[†]For purposes of modeling, water treatment was not included in the distribution system.

TABLE 10.8 Summary of Model Parameters, Data Availability, and Time-Unit Variation for Historical Reconstruction Analysis, Dover Township area, New Jersey.

Model parameters	Data availability	Time-unit variation for historical reconstruction analysis	Notes
Network pipeline data	1962–96*	Annual	Assumed operational date of January 1 for in-service year
Hydraulic device in-service date	1962–96*,†	Annual	Assumed operational date of June 1 for in-service year
Pipe roughness coeffcient	1998	No variation	Maslia et al. (2000a)
Pipe diameter values	1998	No variation	Maslia et al. (2000a)
Pump-characteristic data	1998	No variation	Maslia et al. (2000a)
System production data	1962–96†	Monthly	Figure 3; Maslia et al. (2001)
Point-demand (node) values	October 1997–April 1998	Monthly	Maslia et al. (2000a)
Pattern factors (system operations)‡: 1962–77	None	Hourly	Maslia et al. (2001)
Pattern factors (system operations)‡: 1977–87	Typical peak day (summer) and nonpeak day (fall) for selected years¶	Hourly	Maslia et al. (2001)
Pattern factors (system operations)³: 1988–96	Typical peak day (summer) and nonpeak day (fall) for selected years; 1996; and March and August 1998¶,§	Hourly	Maslia et al. (2000a, 2001)
Nodal concentration or percent contribution of water from specified source	March and April 1996 barium sample collection and transport simulation§	24-h average	Simulated 24-h average of percent contribution of water to model node from water source point of entry (well or well field)

*Data from Flegal (1997).
†Data from annual reports of the Board of Public Utilities, State of New Jersey (1962–96).
‡Model parameters include groundwater well on-and-off cycling schedules simulated by using pattern factors in EPANET 2 and starting water levels in storage tanks.
¶Data from Richard Ottens, Jr., Production Manager, United Water Toms River, Inc., written communication, 1998.
§Refer to Maslia et al. (2000a).
Source: From Maslia et al. (2001).

Tank-Pump" or WSTP simulation method and the corresponding distribution system is referred to as the WSTP system. Using this method (Fig. 10.15A) to calibrate the model to present-day conditions required the following information:

- Known operating schedules for groundwater well on-and-off cycling
- Observed storage tank water-level variations

A. Well-Storage Tank-Pump (WSTP) simulation method

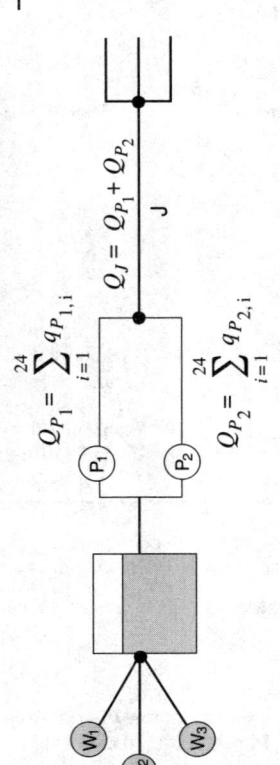

B. Supply-Node-Link (SNL) simulation method

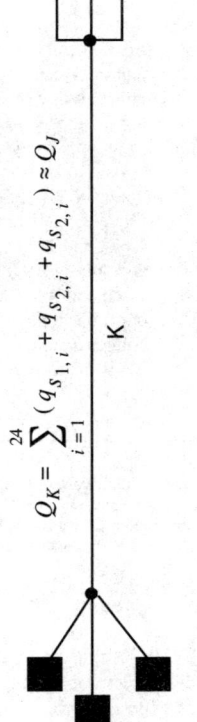

FIGURE 10.15 Distribution system representation of groundwater well, storage tank, and high-service and booster pump combination for (A) physical, "real-world" network, and (B) model network used for historical reconstruction analysis *(from Maslia et al., 2001)*.

- Realistic high-service and booster pump-characteristic curve
- Known operating schedules for the on-and-off cycling of high-service and booster pumps

The model parameter, that was required by the NJDHSS health scientists to compute exposure indexes for the epidemiologic study, was the proportionate contribution of water from wells and well fields to locations throughout the historical pipeline networks. Thus, the distribution of water delivered to the pipeline locations was the item of interest rather than the specific operation of the WSTP combination, which delivered the water. In order to simplify the simulation of the WSTP combination and, thus, reduce data requirements for simulation, a method of idealizing the WSTP combination was developed—designated the Supply-Node-Link or SNL simulation method. This surrogate simulation method eliminated the need for including the storage tank and high-service and booster pump combinations in the historical simulations (Fig. 10.15B). Thus, the Holly, Parkway, and Windsor Avenue treatment plants were represented in historical water-distribution system simulations using the SNL method.

To demonstrate that the surrogate SNL simulation method supplies the distribution system with an equivalent amount of water when compared to the "real-world" WSTP simulation method, both simulation methods were applied to the present-day (1998) water-distribution system for conditions existing in August 1998. Measured and simulated high-service pump flows—using the WSTP simulation method—are compared with simulated flows for the SNL method representing the Holly and Parkway treatment plants in Fig. 10.16. The results obtained using both the WSTP and the SNL methods produce nearly identical simulated flow. Total simulated supply to the distribution system from the Holly treatment plant over the 48-h period using the SNL method was 5.62 Mgal, which is nearly identical to the measured supply of 5.63 Mgal. For the Parkway treatment plant, simulated flow using the SNL method was 8.53 Mgal which is less than 3 percent different from the measured flow of 8.32 Mgal. Thus, results obtained using both the WSTP and the SNL methods produce nearly identical simulated flows, thereby confirming the appropriateness of representing the "real-world" WSTP distribution system (Fig. 10.15A) with the surrogate SNL distribution system (Fig. 10.15B) for historical reconstruction analyses.

Water-Quality Modeling (Source-Trace Analysis). To model the water quality of a distribution system, EPANET 2 uses flow information computed from the hydraulic network simulation as input to the water-quality model. The water-quality model uses the computed flows to solve the equation for conservation of mass for a substance within each link. Details of the specific mathematical formulation of the water-quality simulator and the solution technique are provided in the EPANET 2 Users Manual, as are the model input data requirements (Rossman, 2000).

Identifying the source of delivered water in a distribution system is necessary when trying to determine the exposure of water users to chemical or biological constituents. Males et al. (1985) developed a method using simultaneous equations to calculate the spatial distribution of variables such as percentage of flow, concentration, and travel times that could be associated with links and nodes, under steady-flow conditions. Grayman et al. (1988) developed a water-quality model that used flows previously generated by a hydraulic model and a numerical method to route contaminants—conservative and nonconservative—through a distribution system. This type of model has become known as a dynamic water-quality model. EPANET 2 is also a dynamic water-quality model, and has the ability to compute the percentage of water reaching any point in the distribution system over time from a specified location (source) in the network—the "proportionate contribution" of water from a specified source. To estimate the proportionate contribution of water, a source location is assigned a value of 100 percent. The resulting solution provided by the water-quality simulator in

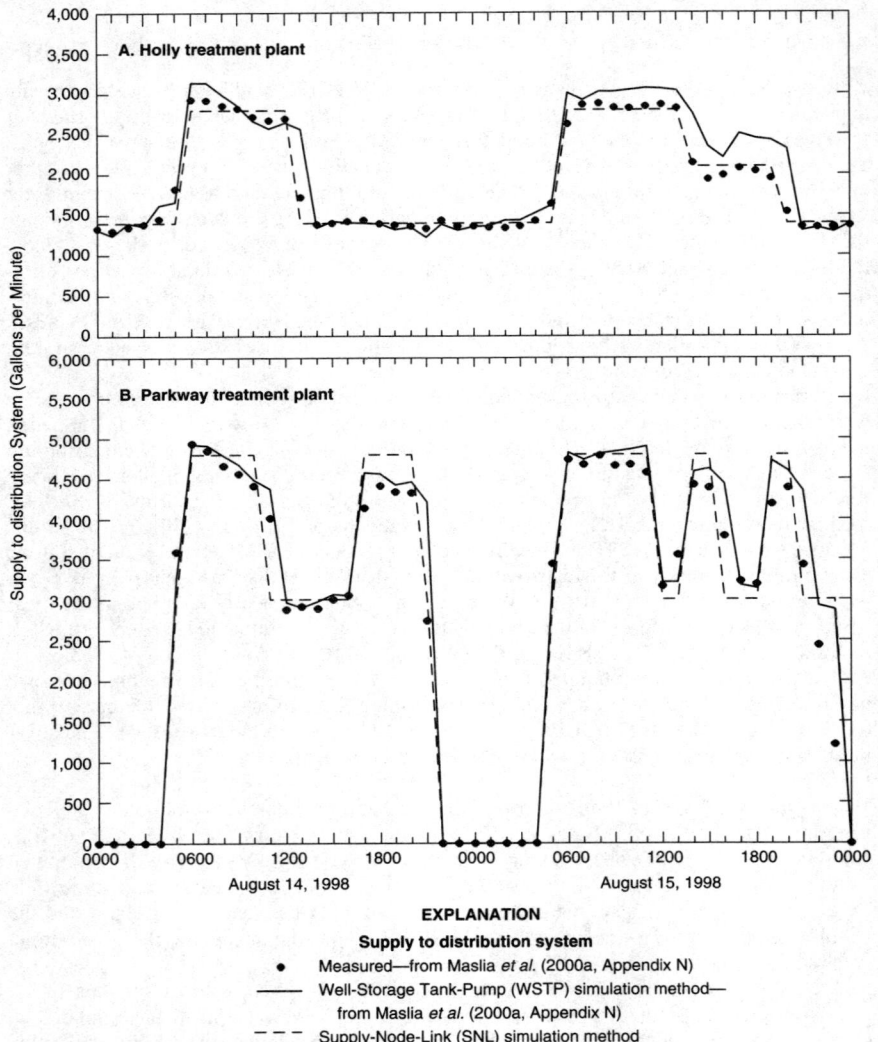

FIGURE 10.16 Measured and simulated flows using Well-Storage Tank-Pump (WSTP) and Supply-Node-Link (SNL) simulation methods, Dover Township area, New Jersey, August 1998 at (A) Holly treatment plant and (B) Parkway treatment plant *(from Maslia et al., 2001)*.

EPANET 2 then becomes the percentage of flow at any location in the distribution-system network (for example, a demand node) contributed by the source location of interest.

For the historical reconstruction analyses, a source-trace analysis was conducted for every month of the historical period. Source nodes were assigned a value of 100 percent in order to estimate the proportionate contribution of water to locations in the historical distribution-system networks. Initial conditions must be flushed out of the distribution system before retrieving the proportionate contribution results (Maslia et al., 2000a, p. 55). Accordingly, the monthly historical network models were run for simulation periods of approximately 1200 h to reach a state of stationary water-quality dynamics (dynamic equilibrium). Results of the source-trace analyses reported herein represent the last 24 h of the 1200 h of the simulation period. Hydraulic time steps of 1 h and water-quality time steps of 5 min were used. For some monthly simulations in the 1980s, the water-quality time steps were reduced to 1 min. These smaller water-quality time steps were necessary to ensure that the mass balance summed to 100 percent. Results of the source-trace analyses are presented and discussed in the next section.

10.4.4 Historical Reconstruction Simulation Results

Examples of historical reconstruction simulation results are shown in Figs. 10.17 and 10.18. In Fig. 10.17, the areal distribution of simulated proportionate contribution results for all model nodes (pipeline junctions) are shown for the maximum-demand month of July 1988, using the Parkway well field as the point of entry (source point). The simulated proportionate contribution results are divided into three intervals (1 to 10, 10 to 50, and 50 to 100 percent) and a gray shade is assigned to all nodes within each interval (results are not shown for proportionate contribution of less than 1 percent). Using this method to display results for the entire historical period requires a different map for each operating well or well field for each specific month and year. When the Parkway well field did not contribute water to a pipeline junction in July 1988, such as the pipelines serving the southern and southwestern areas of Dover Township, nodes representing pipeline junctions are not displayed (compare Figs. 10.17 and 10.11).

Simulated proportionate contribution results can also be viewed in terms of selected pipeline locations. Five geographically distinct pipeline locations are selected from the historical networks to represent the spatial distribution of proportionate contribution results. These locations are identified in Fig. 10.17 as locations A, B, C, D, and E. The percentage of water contributed by every well and well field to each selected pipeline location (A to E) for maximum-demand months for seven selected years (1962, 1965, 1971, 1978, 1988, 1995, and 1996), is shown in Fig. 10.18 using a stacked column graph. The proportionate contribution of water, in percent, from each operating well or well field for the time of interest is stacked one on top of the other within each column. Referring to Fig. 10.17, the proportionate contribution results indicate that the Parkway well field contributed in the range of 50 to 100 percent of the water to pipeline location C. Inspection of the July 1988 graph in Fig. 10.18 for the same pipeline location indicates simulated proportionate contribution of approximately 55 percent, which is in agreement with results shown in Fig. 10.17.

Analysis of the proportionate contribution of water from wells and well fields to selected network locations in the Dover Township area illustrates the increasing complexity and operational variability of the distribution system throughout the historical period. For simulation results shown in Fig. 10.18, the five geographically distinct pipeline locations (A to E) were selected from the historical networks to represent the spatial distribution of proportionate contribution results. (These locations are also identified in Fig. 10.17). Comparison of the May 1962 results with the June 1996 results indicates the increasing complexity of the network and distribution-system operations and how such operations influenced the proportionate contribution of water to specific locations. In May 1962, only two well fields (Holly and Brookside) provided water to any one location; whereas, in June 1996, as many as seven well fields provided water to the distribution system (for example, pipeline location E, for June 1996 in Fig. 10.18), with the Parkway well field providing approximately 75 percent of the potable water to pipeline location E.

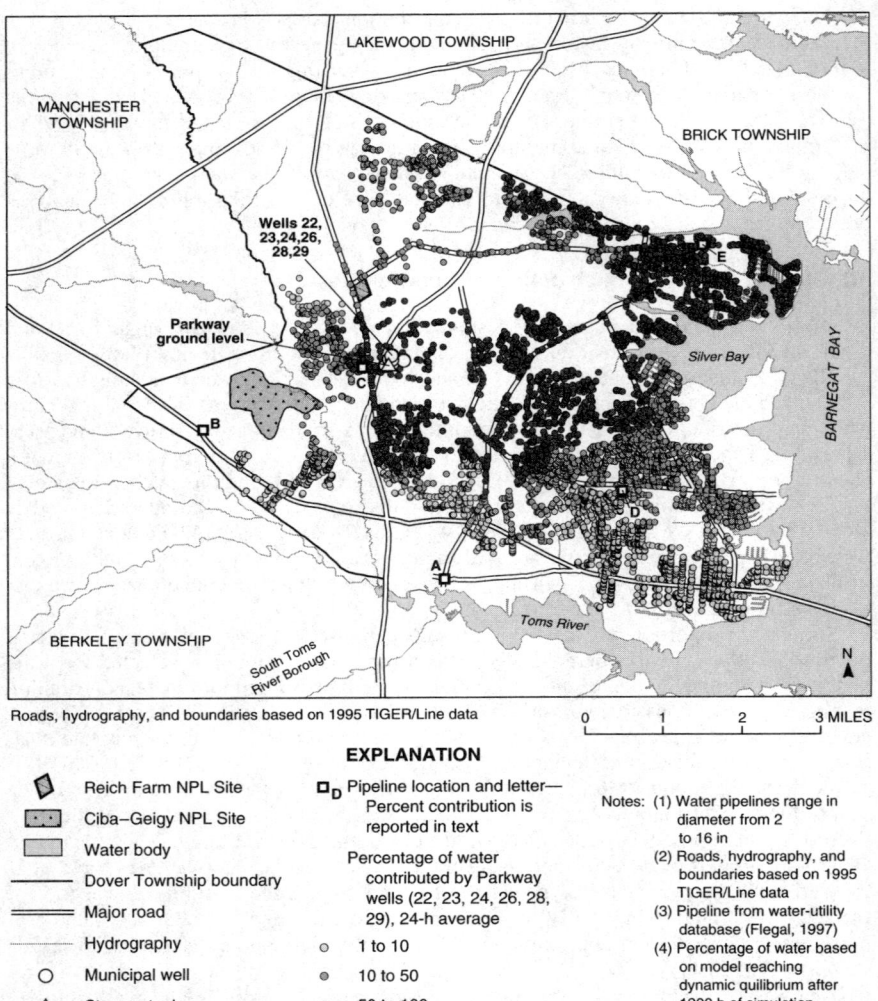

FIGURE 10.17 Areal distribution of simulated proportionate contribution of water from the Parkway wells (22, 23, 24, 26, 28, 29) to locations in the Dover Township area, New Jersey, July 1988 conditions (*from Maslia et al., 2001*).

The simulation results shown in Fig. 10.18 demonstrate that the contribution of water from wells and well fields varied by time and location. However, the results also show that certain wells provided the predominant amount of water to locations throughout the Dover Township area. Readers who are interested in the proportionate contribution of water from specific water sources, at specified times, during the historical period of 1962 through 1996, should refer to Maslia et al. (2001).

In Fig. 10.18, the sum of the proportionate contributions of water from all wells and well fields to any pipeline location should be 100 percent. Because of numerical approximation

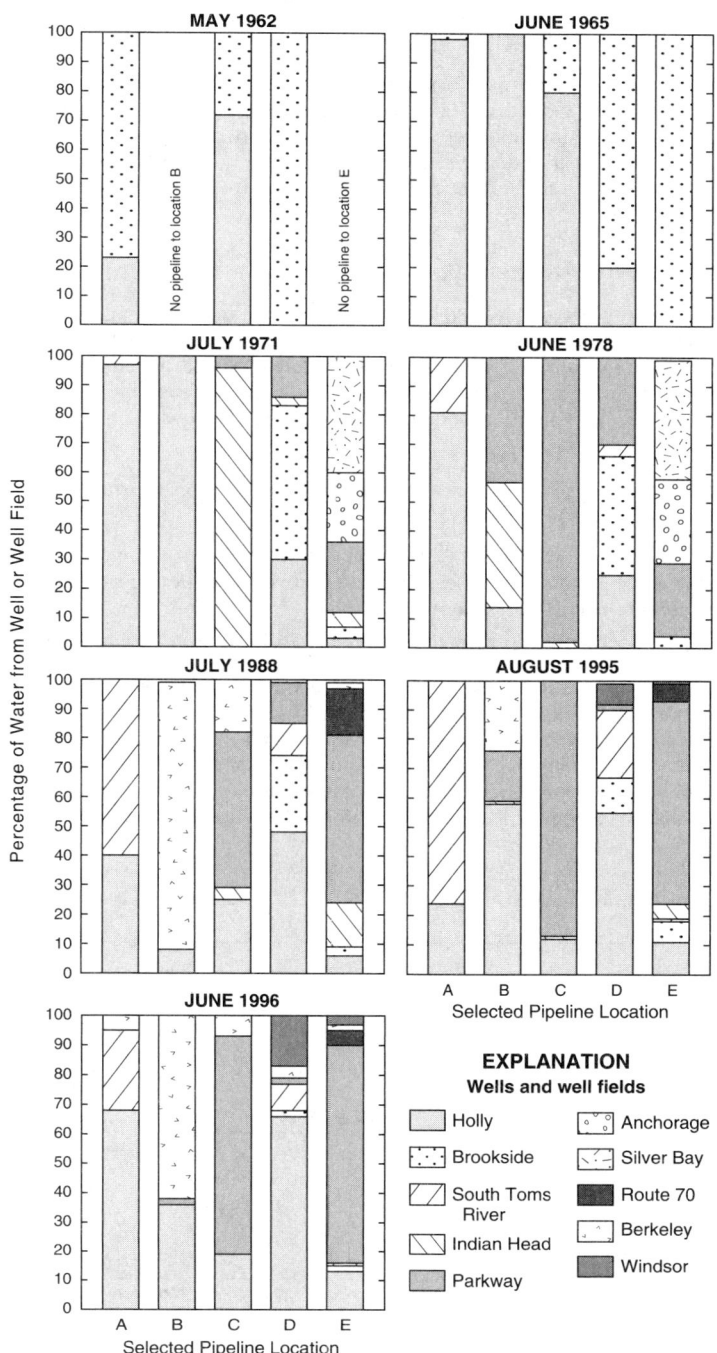

FIGURE 10.18 Simulated proportionate contribution of water from wells and well fields to selected locations, using the manual adjustment process, peak-demand conditions, Dover Township area, New Jersey (*from Maslia et al., 2001*). [Selected pipeline locations shown on Fig. 10.17]

and round off, however, the total contribution from all wells and well fields may sum to slightly less or slightly more than 100 percent at some locations. Such results are expected when using numerical simulation techniques. In the historical reconstruction analysis conducted for the water-distribution system serving the Dover Township area, the sum of the proportionate contribution results at any location ranged from 98 to 101 percent.

10.4.5 Sensitivity Analyses

To address issues of uncertainty and variability of system operations and to test the sensitivity of the proportionate contribution results to variations in model-parameter values, alternate operating conditions were investigated. Four types of operational and hydraulic constraints were varied during sensitivity analyses in order to determine the effects of constraint changes on the proportionate contribution results. The constraints subjected to variation were as follows:

- EPANET 2 pattern factors assigned to wells and supply nodes (operational variation in value and time of day)
- Minimum pressure requirements at model nodes
- Allowable storage tank water-level differences between the starting and ending time of a simulation (hour 0 and 24, respectively)
- Daily system operations represented by a typical 24-h day over a month-long period

For the first three constraints or constraint sets, the innovative POGA methodology (previously described in the section on "Hydraulic Modeling") was used to conduct the sensitivity analyses by synthesizing the on-and-off cycling patterns and pattern factors for groundwater wells and supply. Figure 10.19 shows an example of EPANET 2 pattern factors derived using the manual adjustment process and corresponding pattern factors from a sensitivity analysis using the POGA methodology. The pattern factors schedule pumping

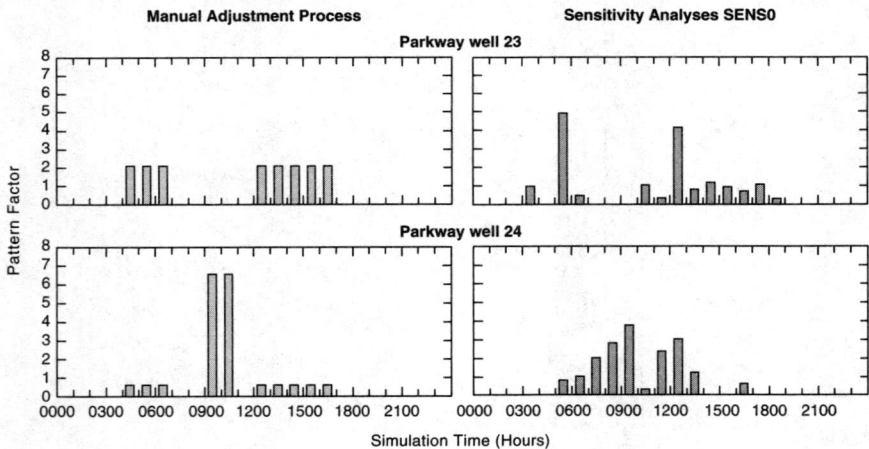

FIGURE 10.19 Pattern factors derived using the manual adjustment process and sensitivity analyses using the progressive optimality genetic algorithm (POGA) methodology, for Parkway wells 23 and 24, July 1988 conditions. See Fig. 10.17 for well locations (*from Maslia et al., 2001*).

at supply nodes representing Parkway wells 23 and 24 (Fig. 10.17) operating in July 1988. From Figure 10.19 we have the following pattern factor information for Parkway wells 23 and 24:

- *Parkway well 23.* From 0500 to 0600 h, the pattern factor derived using the manual adjustment process is about 2; whereas, the pattern factor derived using the POGA methodology is about 5. From 1200 to 1300 h, the pattern factor derived using the manual adjustment process is again about 2; whereas the pattern factor derived using the POGA methodology is about 4.
- *Parkway well 24* From 0900 to 1100 h, the pattern factors derived using the manual adjustment process are about 6.5; whereas, pattern factor derived using the POGA methodology is about 4 from 0900 to 1000 h and less than 0.5 from 1000 to 1100 h.

Although pattern factors for some hours of operations show marked differences (like those in Fig. 10.19), the simulated proportionate contributions of water simulated using pattern factors, derived from the application of the POGA methodology, show little difference throughout the Dover Township area when compared to corresponding proportionate contribution of water simulated using the manual adjustment process. This result is shown in Fig. 10.20 by using the stacked column graph format to compare results for July 1988 for the proportionate contribution of water from wells and well fields to five selected pipeline locations (A to E) derived using the manual adjustment process and the POGA methodology. Figure 10.20 indicates that proportionate contribution of water results at specific historical pipeline locations in the Dover Township area are nearly identical and therefore, the difference between the two methods of deriving historical system operations is insignificant.

Results of sensitivity analyses conducted using the historical reconstruction process indicated the following:

- There was a narrow range within which the historical water-distribution systems could have successfully operated and still satisfy hydraulic engineering principles and the "Master Operating Criteria"
- Daily operational variations over a month did not appreciably change the proportionate contribution of water from specific sources

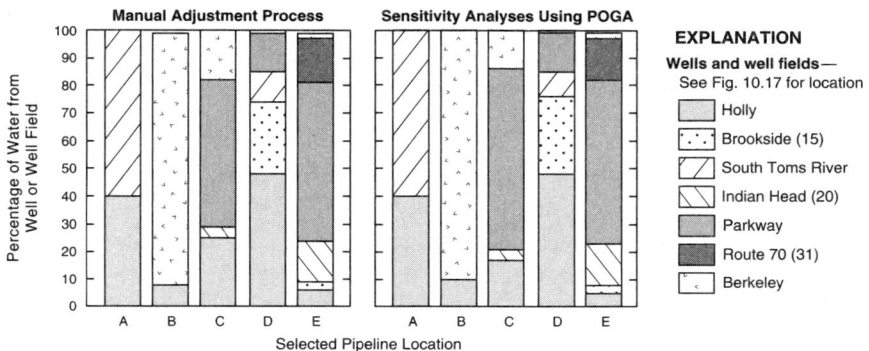

FIGURE 10.20 Simulated proportionate contribution of water derived from the manual adjustment process and the progressive optimality genetic algorithm (POGA) methodology for selected pipeline locations, Dover Township area, New Jersey, July 1988 conditions (*from Maslia et al., 2001*).

TABLE 10.9 Leukemia (age 0–19) and Prenatal Exposure to Parkway Well Field Water, Dover Township Area, New Jersey, 1982–1996.

Group	Contribution of water source	Cases/controls	Odds Ratio (95% confidence interval)
Males and females	Less than 10%	17/70	—
	10% to less than 50%	1/9	0.5 (0.1–4.3)
	50% or greater	4/8	2.2 (0.5–9.4)
Females only	Less than 10%	8/39	—
	10% to less than 50%	1/7	0.8 (0.1–8.8)
	50% or greater	4/5	5.0 (0.8–31)

Source: From NJDHSS, 2003.

Therefore, the reconstructed historical water-distribution systems were determined to be the most plausible and realistic scenarios under which the 1962–1996 historical water-distribution systems were operated.

10.4.6 Findings and Conclusions

Health scientists conducting the case-control epidemiologic study used the percentage of water derived from the different sources that supplied the water-distribution system (proportionate contribution) to derive exposure indexes for each study subject. Results from the case-control study showed that there was an association between prenatal exposure to contaminated community water and leukemia in female children (NJDHSS, 2003). For example, female leukemia cases were five times more likely to have been exposed during the prenatal period to a high percentage of Parkway well water than were control children (i.e., the *Odds Ratio* in Table 10.9). The slightly elevated Odds Ratio for males and females combined is accounted for by the female findings. There were no elevations in males alone. In this analysis *high* exposure means that 50 percent or more of the water during the prenatal period came from the Parkway well field, and *low* means less than 10 percent. When the amount of self-reported tap water consumption was factored into the exposure assessment, the Odds Ratio increased to 6.0 (NJDHSS, 2003). Exposure to Parkway well field water over other time intervals was also investigated because of uncertainty about when the well field became contaminated. From these analyses, the Odds Ratio was highest for the period 1984 to 1996 when case children were 15 times more likely to have been exposed to Parkway well field water than controls (NJDHSS, 2003). Two important observations should be made about these findings:

- From descriptive epidemiology discussed previously (Fig. 10.12), there was a peak in childhood cancer incidence during the mid-1980s to early 1990s.
- Upon examining the cancer data more closely, all female cases with high exposure to Parkway well water (greater than 50 percent proportionate contribution) were found to be born after 1983.

These results are significant because out of hundreds of cancer cluster investigations, only two—Woburn, Massachusetts and Dover Township, New Jersey—have shown an association between environmental exposures and childhood cancer (Costas et al., 2002; NJDHSS, 2003).

These findings would not have been possible without the results derived from the innovative water-distribution system modeling efforts. These efforts have led to developing new methods for evaluating the accuracy of modeling results and exposure classification techniques that are critical components of epidemiologic studies. Some of the innovations documented by the Dover Township historical reconstruction analysis are as follows:

- A new approach, "proportionate contribution analysis" was developed that utilized water-distribution system modeling source tracing to quantify exposure on a monthly basis for all locations historically served by the distribution system.
- Through the use of an innovative genetic algorithm approach (POGA), historical water-distribution system operating schedules were synthesized. Sensitivity analyses indicated operating system changes did not appreciably change the proportionate contribution of water to Dover Township locations.
- The association between exposure and disease would not have been possible without developing the integrated approach using environmental science, engineering evaluations, and epidemiologic analyses.

Historical reconstruction of environmental exposure is never an easy task. The procedures and results summarized herein (and the detailed analyses in Maslia et al. [2001]), represent one of the most comprehensive, best documented, and quality-controlled studies of its kind.

10.4.7 Acknowledgments

A study and investigation of this magnitude is not conducted alone or without input and assistance from many individuals and organizations. As such, the author would like to acknowledge colleagues at ATSDR, Multimedia Environmental Simulations Laboratory at the Georgia Institute of Technology (MEESL-GT), NJDHSS, U.S. Environmental Protection Agency, National Risk Management Research Laboratory (U.S. EPA-NRMRL), and the U.S. Geological Survey (USGS).

The author would like to specifically acknowledge Richard E. Gillig, Juan J. Reyes, Jason B. Sautner, and Robert C. Williams of ATSDR for assistance with the water-distribution modeling aspect of the study; Mustafa M. Aral of MESL-GT for assistance with and development of numerical simulation methodologies; Jerald A. Fagliano of NJDHSS for suggestions and advice from the epidemiologic perspective; Lewis A. Rossman of USEPA-NRMRL for assistance with requested modifications to the EPANET 2 water-distribution system model; and S. Jack Alhadeff and Thomas R. Dyar of the USGS Center for Spatial Analysis Technologies for assistance with preparing historical aerial photographs. Cartographic assistance in preparing illustrations for the original report (Maslia et al., 2001) was received from Carolyn A. Casteel, Caryl J. Wipperfurth, and Bonnie J. Turcott of the USGS.

10.5 REDLANDS, CALIFORNIA

10.5.1 Background Information

The city of Redlands California lies in the San Bernardino valley of California, approximately 60 mi east of central Los Angeles. In 1981, a routine analysis for chlorination byproducts revealed the presence of trichloroethylene (TCE) in a sample of water from the Redlands water system. Subsequent water quality analyses revealed that a number of wells supplying the city were contaminated with TCE. In 1997, the perchlorate anion (ClO_4) was also detected in several wells.

In 1996, the first of a series of lawsuits was filed in California State Court alleging that the source of these contaminants was a manufacturing facility located upgradient from the most seriously contaminated wells. One of these lawsuits claimed that plaintiffs were harmed by exposure to toxic chemicals that were improperly disposed of at the manufacturing site and found their way into groundwater that was subsequently extracted through the city's wells and delivered to water customers, including the plaintiffs. The plaintiffs' burden of proof requires them to establish, among other things, that they were actually exposed to contaminated water at their homes, places of work, or other locations, and that the amounts of contaminants that entered their bodies as the result of these exposures were sufficient to cause them harm. To establish this proof, plaintiffs, among other things, had to reconstruct the historical conditions in the water distribution system of the city of Redlands over a period from the mid-1950s to the late 1990s.

10.5.2 The Redlands Water System

Currently, the Redlands water distribution system consists of two water treatment plants, more than 30 wells, 17 reservoirs, 39 pump stations, and more than 350 mi of pipe organized into seven primary pressure zones. Figure 10.21 shows the distribution system in 1998.

Between 1955 and 1996 the Redlands Water System provided water to a population that grew from 22,000 to 65,000 people. Over this period, water was supplied from approximately 40 wells and two surface water sources with new sources being developed in response to growth and other sources being retired due to inefficiency or contamination. On an annual basis, a little more than half the water used by the city was supplied by wells, with this fraction ranging from slightly less than 30 percent in winter to about 60 percent in summer.

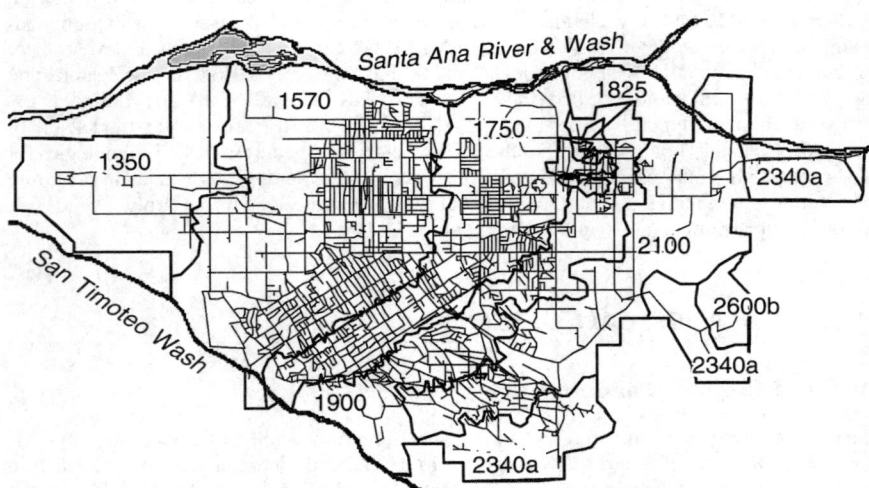

FIGURE 10.21 Redlands California water distribution system.

Operation of the Redlands system followed two distinct patterns. Base loads are served as much as possible by the Henry Tate WTP, which is located high on the eastern edge of the city. During the winter months, production from the Tate plant is sufficient to serve most of the city, so water flows generally out of higher zones in the east and into the lower zones in the west. As water use increases in the spring and summer the production from the Tate plant must be supplemented by supplies from the Horace Hinkley WTP, along the northern edge of the city, and wells. Under these circumstances operation of the system becomes more complex, with water being boosted up from lower zones in which relatively large wells are located into the central zones.

10.5.3 General Methodology

As part of litigation, several forensic reconstructions of water quality in the Redlands water distribution system were done. The reconstruction described here involved estimates of both human exposure to toxic contaminants, and the whole-body intakes of these chemicals. Exposures were expressed as time-series estimates of concentrations of chemicals in water at locations used by the plaintiffs in the litigation. Intakes were expressed as estimates of the mass or weight-normalized mass of contaminants that entered people's bodies. Exposure and intakes were estimated in a stochastic context using the Monte Carlo simulation techniques.

Water Quality Analysis. The water quality analysis employed the principle of superposition. Transfer coefficients were estimated that represented the response of a given node to a unit value of contamination at a given well. Numerically, the transfer coefficient represents the percentage of water at a given node that originates at a given source. Concentration estimates at nodes were calculated by summing the contribution to that node from each source according to Eq. (10.2):

$$C_n = \sum_s c_s t_{sn} \qquad (10.2)$$

where C_n = concentration at node n
c_s = concentration at source s
t_{sn} = transfer coefficient from sources to node n (s by n matrix)

Transfer coefficients were estimated using EPANET 2 (Rossman, 2000). To account for the variability inherent in the system, one-year-long time-series EPS analyses were used (Harding and Walski, 2000.) For exposure and intake analyses, transfer coefficients were averaged to a quarterly basis.

Because the reconstruction of exposures and intakes would be done stochastically on a quarterly (three-month) basis, the objective of the water quality modeling was to reproduce the statistics of the transport of contaminants in the water distribution system. Thus, it was not necessary to reproduce precisely the true value that occurred at any given moment in the system.

Estimates of Exposures and Intakes. Exposure to a contaminant is defined in terms of intensity and duration. In this work, the intensity of an exposure is measured by the concentration of a contaminant in the water at a location where an individual was present. The duration of the exposure is the period of time that the individual was at a given location where contaminated water was present.

The intake of a chemical is the amount of the chemical that enters the human body as a result of activities during an exposure. Examples of activities include, among other things, drinking, showering, bathing, swimming, or using a swamp cooler. These activities in the presence of contaminated water can lead to intakes through three pathways—ingestion, inhalation, or dermal absorption. The fundamental intake equation is shown in Eq. (10.3) (U.S. EPA, 1992):

$$I = C \times ED \times IR \tag{10.3}$$

where I is the whole-body intake, C is the concentration, ED is the exposure duration, and IR is the intake rate.

IR is a function of individual characteristics and the nature of the activity that leads to the exposure. Estimation of IR may require only a few or many variables. C is calculated using Eq. (10.2).

Stochastic Analysis. Estimates of exposures and intakes were expressed as credibility intervals, which were calculated using the Monte Carlo simulation techniques. The Monte Carlo simulation is a widely used stochastic technique that incorporates uncertainty in the analysis by representing selected processes probabilistically. In this approach, most of the variables used in calculating C and IR are expressed as probability distributions. In the case of Eq. (10.2), both c_s and t_{sn} are expressed as probability distributions. Equations (10.2) and (10.3) are calculated many times (in this case, 10,000 times). In each calculation, values of variables are generated randomly from their probability distributions. The result is a population of values of I. This population is sampled to determine the upper and lower limits of the credibility interval.

10.5.4 Development of Data

Development of the data used in a historical reconstruction is always a challenge. This is particularly true of the configuration of the pipe network, as most municipalities do not maintain copies of obsolete maps of their distribution system. Additionally, operating rules for pumps, valves and tanks are often not completely documented. In the western United States, water production records are often reasonably complete due to state water rights or permit administration requirements. Reconstruction of conditions in the city of Redlands system followed this pattern.

Facilities. Development of a historical reconstruction of the Redlands pipe network and facilities began with a model developed as part of a 1998 water system facilities master plan. This model was then "deconstructed" to represent earlier epochs (discrete set of years over which the distribution system was relatively unchanged) based on several sources of information, including old maps, previous planning studies, construction-related documents, regulatory reports, state highway construction information, interviews with former employees and aerial photographs. Aerial photographs were generally used to define the timing of development of new service areas although they sometimes helped define the timing of changes to the interior of the pipe network caused by highway development. Pump curves were assumed to remain roughly unchanged over time. Network changes were defined in several epochs.

Operating Rules. Operating rules were developed from those currently in use. Some written information was available about historical operating rules and additional insights were gained by talking to a former manager of the system. With a few exceptions, the layout of the central pressure zones remained unchanged over the study period. On this basis, operating rules relative to those zones were assumed to be similar to those in use today. The peripheral pressure zones were developed recently enough that the operating rules for these zones were also assumed to be similar to those in use today. Notable changes to the hydraulic structure of the system occurred when an intermediate zone was split, when the Hinkley WTP was brought on line, and when blending operations were put into place to deal with nitrate levels in some wells. When inferring operating rules where significant changes in the system had occurred, the principle was to use the lowest cost source that could provide acceptable system pressures.

Water Supply. Essentially complete monthly records of water production at wells and WTPs were available. However, these data were not always available on a daily basis, particularly early in the study period. Additionally, the daily data that were available contained anomalies, such as transcription errors and missing values. In order to provide a homogeneous water supply data set, daily data for total system supplies and supplies from the two WTPs were interpolated from monthly data using a cubic spline algorithm. Daily supplies from wells were generated by the use of a dispatch algorithm that simulated a reasonable pattern of well dispatch and that preserved monthly production volumes. While this method did not produce exact agreement with available historical records, it did produce patterns that had similar statistics.

Water Use. The water quality analysis required a daily time-series of water use and a diurnal pattern at each model node where a water use was represented. These data were developed from system-wide water supply data and land use data.

System daily water use was set equal to system daily water production, under the assumption that operators generally succeeded in matching supply with demand over periods of a day or so. The daily time-series of system-wide water use was spatially distributed to model nodes on the basis of a GIS analysis of land use. A system-wide spatial database of land use classification was obtained for the current period. This database was adjusted to reflect historical conditions at several earlier periods through reference to historical aerial photographs and documents.

Typically, over the study period, a particular parcel underwent one transition from a low-intensity land use to its current, higher-intensity use. The most common transition was from agricultural or vacant land to residential development. In a fewer number of cases land would be redeveloped from one built-up use to another.

To compare the land use data with the aerial photographs, a vector representation of the current land use database was projected over raster representations of historical aerial photographs and "heads-up" digitizing techniques were used to identify the land use transition(s) for a parcel. The time of transition for a parcel was estimated based on the timing of the aerial photograph and other ancillary information that might be available. The result was a spatial database with attributes that represented the history of land use for each parcel.

Each land use classification was assigned an average daily water use intensity, in units of flow rate per unit area. More intensive uses, such as high-density residential developments, use water at a higher rate per unit area than lower intensity uses. In addition, each classification was assigned one of several diurnal patterns of water use. Water use intensity and diurnal patterns were estimated based on previous field studies in areas with similar water use characteristics.

The service area for each water-use node was delineated by the use of a proximal polygon technique in GIS software. The daily average water use for each node's service area was established based on the land uses within that service area. The daily average water use at a node is scaled up or down in proportion to total system use to obtain daily water use for each day of the analysis:

$$Q_{id} = \frac{Q_{\mathrm{avg}_i}}{\sum_{j=1}^{n} Q_{\mathrm{avg}_j}} \times S_d \qquad (10.4)$$

where Q_{id} = water use at node i on day d
S_d = total system water use on day d
Q_{avg} = average daily water use at node i and j
n = number of nodes in the system

The average daily nodal water use is calculated based on the area of each type of land use in the node service area and the water use intensity on each land use:

$$Q_{avg_i} = \sum_{k=1}^{L} f_k A_{ik} \qquad (10.5)$$

where L = number of different land uses within the node service area
A_{ik} = area of land use k within the service area of node i
f_k = water use intensity for land use k in units of flow/area.

The calculations of water use were done using GIS and relational DBMS software.

Four diurnal patterns had previously been determined on a 2-h time step for developed land uses, and a fifth was developed for irrigated open space. These five diurnal patterns were associated with each land use in the land use database. The pattern of diurnal water use at a node was determined as the weighted composite of the patterns associated with each of the land uses within the node service area.

$$p_{ij} = \frac{\sum_{k=1}^{L} \overline{P}_{jk} f_k A_{ik}}{Q_{avg_i}} \qquad (10.6)$$

where p_{ij} = fraction of daily water use at node i occurring during diurnal time step j
$\overline{P}_{jk}$ = fraction of daily water use for land use k occurring during period j
f_k = areal water use intensity for land use k
A_{ik} = area of land use k within the service area of node i
L = number of land uses within the service area of the node.

Nodal water use for each period was calculated as follows:

$$Q_{idj} = p_{ij} \times Q_{id} \qquad (10.7)$$

where Q_{idj} = water use at node i during the jth period of day d
p_{ij} = fraction of daily water use at node i occurring during period j
Q_{id} = average daily water use at node i on day d.

10.5.5 Modeling Procedure

Water Quality Modeling. Transfer coefficients were calculated using a modified version of EPANET 2. Modifications to EPANET allowed it to read a new time-series file, to run arbitrarily-long time-series analyses using data in the time-series file, to statistically aggregate water quality and hydraulic outputs over periods specified by the user, and to write out those results at the end of the aggregation period. Individual runs were made for each contaminated source for each year of the study period. Each run was for 365 days, or 366 days in a leap year. Water quality conditions at the beginning of an annual run were initialized based on the ending conditions of the previous-year's run for that source. Considerable code was developed to automate the process of preparing data for a run, generating a run of the model, and processing data after a run. Because of the considerable amount of time required for model runs, the model control codes automatically distributed runs over as many as 10 computers running on a network. Sensitivity analysis model runs were made in addition to the "production" runs. In these runs, model data and parameters were perturbed to quantify the sensitivity of results to those perturbations.

Monte Carlo Simulation. Probability distributions for the wellhead concentrations of contaminants were provided by hydrogeologists and groundwater scientists based on groundwater fate and transport modeling. These probability distributions reflected uncertainty in the source term, transport through the vadose zone, and transport in groundwater. A probability distribution reflecting uncertainty in the water distribution fate and transport modeling was developed based on the results of sensitivity analyses and the modelers' professional judgment. This probability distribution was applied to the transfer coefficient t_s in Eq. (10.2). Probability distributions were provided by medical experts for the variables and parameters used in the calculation of the intake rate in Eq. (10.3). The form of the intake rate function differs with different activities and pathways. In addition, the probability distributions of the variables and parameters used in the function vary from individual to individual and over time. As a result, a considerable number of probability distributions were required to define the variables and parameters necessary to calculate the intake rate.

The stochastic model was implemented using Monte Carlo techniques. This simulation combines a long time-series structure with other dimensions including the identity of individuals, locations of activities, multiple contaminants, multiple activities, and multiple physiological pathways. Because of the multidimensional nature of the analysis, readily-available commercial Monte Carlo packages were not suitable for this analysis and a code was written to implement the Monte Carlo model.

The model used three principal databases. The first of these were data representing the parameters of probability distributions for wellhead concentrations of contaminants. These were provided on an annual basis. The second database contained point values for the transfer coefficients on a quarterly basis. The third database contained the exposure duration and parameters for probability distributions for variables and parameters used in the calculation of the intake rate.

In execution, the Monte Carlo code made an analysis for each quarter of each year in the study period. For a given quarter the code traversed the nodes in the water quality model at which, during that quarter, individuals spent time and undertook activities that led to intakes. This set of nodes represented a subset of the nodes in the model. For each of these nodes, the code retrieves the transfer coefficients for each source contributing to the node along with the parameters of the probability distributions defining the wellhead concentrations. Using, in addition, the probability distribution for uncertainty in the transfer coefficients, the code constructs a population of nodal concentrations for each contaminant.

These populations of concentrations are passed on to routines that calculate populations of intakes for different pathways for each individual for each quarter according to Eq. (10.3), using, in turn, the code that calculates the intake rate for different activities and pathways. Populations of cumulative lifetime values of intakes are also calculated for each individual. The intake populations were sampled to establish estimates of the lower 2.5 percentile and upper 97.5 percentile credibility limits.

10.5.6 Results

The maps in Fig. 10.22 display the spatial distribution of contaminated water for the years 1970, 1975, and 1980. The irregular polygons represent the node service areas. The thematic schema of these maps is based on the sum of transfer coefficients, for contaminated wells, at the nodes in the water distribution model. This representation indicates, for each node

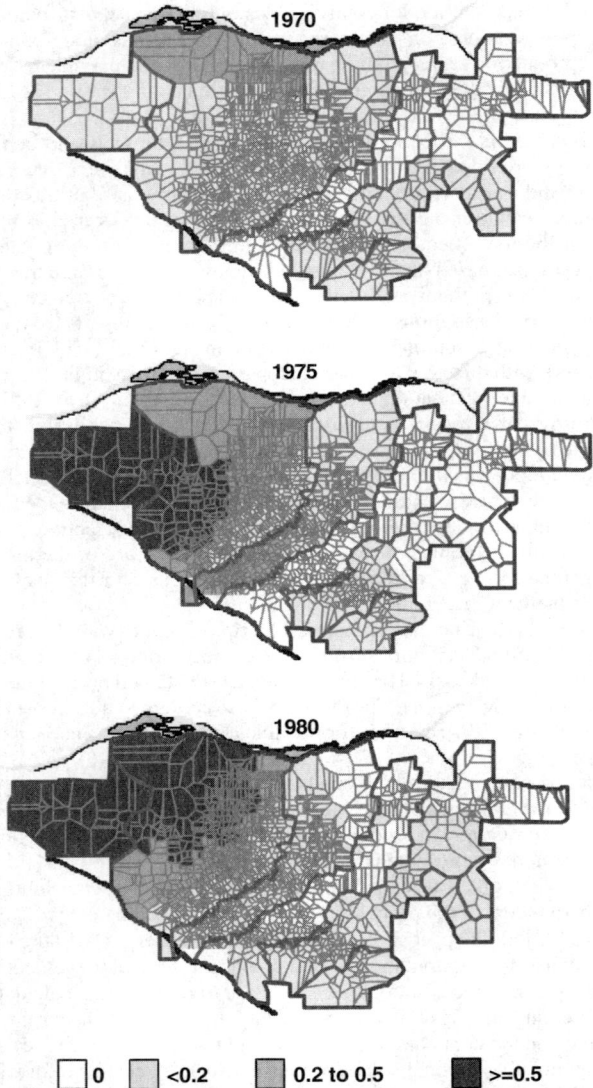

FIGURE 10.22 Modeled spatial distribution of contaminants, Redlands California.

service area, the percentage of water supplied at the node that originated at one or more contaminated well. A node service area that received its entire water supply from one or more contaminated well would show a value of 1.0, while a node that received no contaminated water would show a value of zero. Values are averaged over the four quarters of the year.

RECONSTRUCTING HISTORICAL CONTAMINATION 10.51

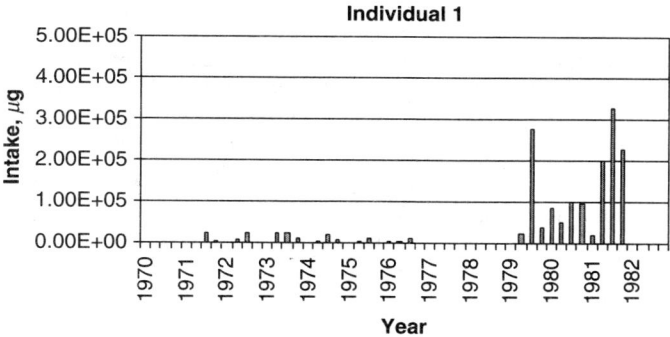

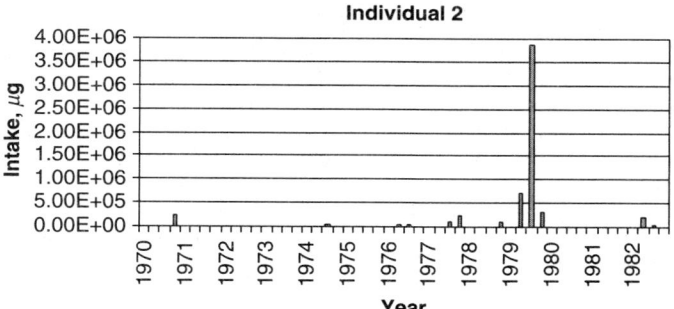

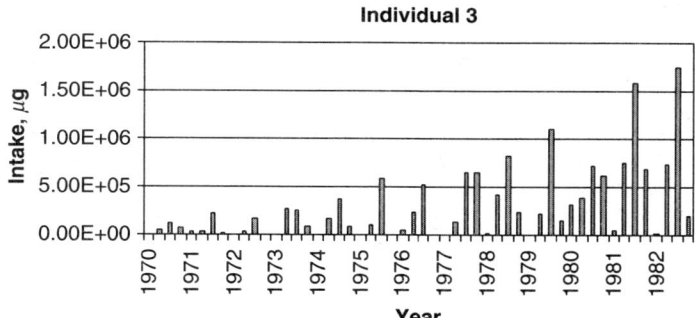

FIGURE 10.23 Selected individual intakes of perchlorate, Redlands California.

The chart in Fig. 10.23 displays the time-series of the 97.5 credibility limit for annual total intake for three selected individuals for tricholoroethylene (TCE). The variation in the intakes for the individual number 1 are largely driven by changes in water quality. This person resided close to a well that model results indicate initially produced water with lower levels of contamination, but which, in about 1976 began producing water with much higher concentrations of TCE. This individual's living situation was relatively stable, as was the operation of the well. The increase in intakes is largely due to the increased concentration of TCE in the well water. (During 1977 and 1978 this person lived outside the area.)

Individuals 2 and 3 both resided in an area substantially supplied by a large contaminated well field. Individual 2 lived somewhat farther from the well field than did individual 3. Both attended school at the same location that was less influenced by the contaminated well field. This is one factor in the seasonal pattern of intakes for individual 3. In the third quarter of 1979, however, conditions in the system were such that concentrations of TCE at both residences were similar. The intakes for individual 2 were considerably higher than those for individual 3 because individual 2 reported that he/she spent roughly twice as much time showering and used a higher flow rate in the shower. TCE can be introduced into the body during showering by inhalation and dermal absorption.

10.6 SUMMARY

Water supply systems are inherently susceptible to accidental, natural or purposeful contamination. Short-term, acute contamination events are generally caused by a contaminant being introduced into the water supply over a period of minutes, hours or days. The contaminant may reach a limited part of the distribution system or may spread over a wide area and can result in waterborne illness and death that occurs within a short period. Long-term, chronic contamination events are generally caused by contamination that enters the water supply over periods of months or years. The resulting long-term exposure can lead to a wide range of illnesses that appear years or even decades after the initial exposure.

Though many contamination events are relatively minor in consequence and may pass with few impacts and without detection, other events can and have been significant in their impacts. A thorough understanding of past contamination events and how and why they occurred and evolved is important for many reasons:

- Remediating and correcting the specific causes of the contamination event
- Identifying the full extent of damage and injury
- Identifying liability for the event
- Providing general knowledge that will reduce the likelihood of contamination events and the vulnerability of water systems to contamination

There are several statistical and modeling methods that can be used to assist in retrospective (historical) evaluations of past events. In many cases, models of water distribution systems have been applied to study the likely movement and spread of the contaminant from the source through the distribution system. Developing and applying such models is challenging because of the need to reconstruct some data based on limited information. These challenges and the degree of uncertainty tend to increase as the length of time between the event and the model construction increase. Various methods for incorporating uncertainty such as Monte Carlo simulation and sensitivity analysis have been applied in order to provide a more realistic estimate of the impacts of the contamination event.

REFERENCES

Akaike, H., "A New Look at Statistical Model Identification," *IEEE Transac. Auto. Control* 19: 716–722, 1974.

Angulo, F. J., S. Tippen, D. J. Sharp, B. J. Payne, C. Collier, J. E. Hill, T. J. Barrett, R. M. Clark, E. E. Geldriech, H. D. Donnell, and D. L. Swerdlow, "A Community Waterborne Outbreak of

Salmonellosis and the Effectiveness of a Boil Water Order," *Am. J. Public Health* 87(4): 580–584, 1997.

Aral, M. M., J. Guan, and M. L. Maslia, "Identification of Contaminant Source Location and Release History in Aquifers," *J. Hydrol. Eng.* 6(3): 225–234, 2001a.

Aral, M. M., J. Guan, M. L. Maslia, and J. B. Sautner, "Reconstruction of Hydraulic Management of a Water-Distribution System using Genetic Algorithms," Multimedia Environmental Simulations Laboratory Report No. MESL-01-01, School of Civil and Environmental Engineering, Georgia Institute of Technology, Atlanta, October, 2001b.

Aral, M. M., J. Guan, M. L. Maslia, J. B. Sautner, R. C. Williams, and J. J. Reyes, "Reconstruction of Hydraulic Management of a Water-Distribution System Using Genetic Algorithms," *Proceedings of the World Water and Environmental Resources Congress 2001*, Environmental and Water Resources Institute of the American Society of Civil Engineers, Orlando, FL, May 20–24, 2001c [CD-ROM].

ASCE, *Groundwater Quality Modeling and Management Under Uncertainty*, in S. Mishra (ed.), Reston, VA, 2003.

ATSDR, *Health Consultation, Drinking Water Quality Analyses, March 1996 to June 1999, United Water Toms River, Dover Township, Ocean County, New Jersey*, Agency for Toxic Substances and Disease Registry, Atlanta, GA, 2001a.

ATSDR, *Public Health Assessment for Ceiba-Geigy Corporation, Dover Township, Ocean County, New Jersey, EPA facility ID: NJD001502517*, Agency for Toxic Substances and Disease Registry, Atlanta, GA, 2001b.

ATSDR, *Public Health Assessment for Dover Township Municipal Landfill (a/k/a Dover Township Landfill), EPA facility ID: NUJD980771570 and Silverton Private Well Contamination (a/k/a Silverton wells), EPA facility ID NJD981877780, Dover Township, Ocean County, New Jersey*, Agency for Toxic Substances and Disease Registry, Atlanta, GA, 2001c.

ATSDR, *Public Health Assessment for Reich Farm, Dover Township, Ocean County, New Jersey, EPA facility ID: NJD980529713*, Agency for Toxic Substances and Disease Registry, Atlanta, GA, 2001d.

ATSDR, *Summary of Findings, Historical Reconstruction of the Water-Distribution System Serving the Dover Township Area, New Jersey: January 1962–December 1996, September 2001*, Agency for Toxic Substances and Disease Registry, Atlanta, GA, 2001e.

Berry, M., and P. Haltmeier, "Childhood Cancer Incidence Health Consultation: A Review and Analysis of Cancer Registry Data, 1979–1995 for Dover Township (Ocean County), New Jersey, final technical report," New Jersey Department of Health and Senior Services, Division of Environmental and Occupational Health Services, Consumer and Environmental Health Services, Trenton, NJ, 1997.

Berry, M., Letter (unpublished) to Steve Jones, ATSDR Region 2 from Michael Berry, Environmental Health Service, New Jersey Department of Health, August 31, 1995.

Board of Public Utilities, State of New Jersey, *Annual Report of United Water Toms River (a/k/a Toms River Water Company), Year Ended December 31*, Trenton, NJ, 1962–1996.

Bowie, W. R., A. S. King, D. H. Werker, J. L. Isaac-Renton, A. Bell, S. B. Eng, and S. A. Marion, "Outbreak of Toxoplasmosis Associated with Municipal Drinking Water," *Lancet* 350: 173–177, 1997.

Breslow, N. E., and N. E. Day, *Statistical Methods in Cancer Research*, Vol 1, IARC Scientific Publications, London, UK, 350 pp., 1980.

Bruce Grey Owen Sound Health Unit, The Investigative Report of the Walkerton Outbreak of Waterborne Gastroenteritis, 2000, available at: http://www.publichealthgreybruce.on.ca/_private/Report/SPReport.htm

Charbeneau, R. J., and D. E. Daniel, "Contaminant Transport in Unsaturated Flow," in D. R. Maidment (ed.), *Handbook of Hydrology*, Chap. 15, McGraw-Hill, New York, 1993.

Chin, J. (ed.), *Control of Communicable Diseases Manual*, 17th ed. American Public Health Association, Washington DC, 624 pp., 2000.

Clark, R. M., "Assessing the Etiology of a Waterborne Outbreak: Public Health Emergency or Covert Attack," *Proceedings of the Water Security Summit*, Heastad Methods, Waterbury, CT, pp. 170–179, 2001.

Clark, R. M., and W. M. Grayman, *Modeling Water Quality in Drinking Water Distribution Systems*, American Water Works Association, Denver, CO, 223 pp., 1998.

Clark, R. M., E. E. Geldreich, K. R. Fox, E. W. Rice, C. H. Johnson, J. A. Goodrich, J. A. Barnick, and F. Abdesaken, "Tracking a *Salmonella* Serovar *Typhimurium* Outbreak in Gideon, Missouri: Role of Contaminant Propagation Modeling," *J. Water Supply Res. Technol. - Aqua* 45(4): 171–183, 1996.

Costas, K., R. S. Knorr, and S. K. Condon, "A Case-Control Study of Childhood Leukemia in Woburn, Massachusetts: The Relationship Between Leukemia Incidence and Exposure to Public Drinking Water," *Sci. Total Environ.* 300: 23–35, 2002.

Curriero, F. C., J. A. Patz, J. B. Rose, and S. Lele, "The association between extreme precipitation and waterborne disease outbreaks in the United States, 1948–1994," *Am. J. Public Health* 91: 1194–1199, 2001.

Fennel, H., D. B. James, and J. Morris, "Pollution of a Storage Reservoir by Roosting Gulls," *J. Soc. Water Treat. Exam.* 23: 5–24, 1974.

Flegal, G. J., *Written Communication to Morris L. Maslia*, United Water Toms River, Toms River, NJ, February 25, 1997.

Fox, K. M., and D. A. Lytle, "Milwaukee's Crypto Outbreak: Investigation and Recommendations," *J. Am. Water Works Assoc.* 88(9): 87–94, 1996.

Geldreich, E. E., K. R. Fox, J. A. Goodrich, E. W. Rice, R. M. Clark, and D. L. Swerdlow, "Searching for a Water Supply Connection in the Cabool, Missouri Disease Outbreak of *Eschericia coli* O157:H7," *Water Res.* 26: 1127–1137, 1992.

Grayman, W. M., "Reconstructing Impacts of Industrial Contamination of Groundwater on Drinking Water," *Proceedings of the Water Security Summit*, Haestad Methods, Waterbury, CT, pp. 252–258, 2001.

Grayman, W. M., R. M. Clark, and R. M. Males, "Modeling distribution-system water quality: dynamic approach," *ASCE J. Water Resour. Planning Management* 114(3): 295–312, 1988.

Guan, J., and M. M. Aral, "Optimal Remediation with Well Locations and Pumping Rates Selected as Continuous Decision Variables," *J. Hydrol.* 221: 20–42, 1999b.

Guan, J., and M. M. Aral, "Progressive Genetic Algorithm for Solution Optimization Problems with Nonlinear Equality and Inequality Constraints," *Appl. Math. Modeling* 23: 329–343, 1999a.

Harding, B. L., and T. M. Walski, "Long Time-series Simulation of Water Quality in Distribution Systems," *J. Water Resour. Planning Management, ASCE* 126(4): 199–209, 2000.

Harding, B. L., and W. M. Grayman, "Historical Reconstruction of Contamination in a Distribution System Incorporating Uncertainty. Linking Exposures and Health: Innovations and Interactions," *12th Annual Conference of the International Society of Exposure Analysis, 14th Conference of the International Society for Environmental Epidemiology.*, Vancouver, BC, 2000.

Harr, J. A., *"Civil Action,"* Random House, New York, 1995.

Heisterkamp, S. H., G. Doornbos, and H. Gankema. "Disease Mapping Using Empirical Bayes and Bayes Methods on Mortality Statistics in Netherlands," *Stat. Med.* 12: 1895–1913, 1993.

Kindler J., and C. S. Russel (eds.) *Modeling Water Demands*, Academic Press, London, UK, 248 pp., 1984.

Lin, D. Y., and L. J. Wei, "The Robust Inference for the Cox Proportional Hazards Model," *J. Am. Stat. Assoc.* 84: 1074–1078, 1989.

MacKenzie, W. R., N. J. Hoxie, M. E. Proctor, M. S. Gradus, K. A. Blair, D. E. Peterson, J. J. Mazmierczak, D. G. Addiss, K. R. Fox, J. B. Rose, and J. P. Davis, "A Massive Outbreak in Milwaukee of Cryptosporidium Infection Transmitted Through the Public Water Supply," *N Engl. J. Med.* 331: 161–167, 1994.

Males, R. M., R. M. Clark, P. J. Wehrman, and W. E. Gates, "Algorithm for Mixing Problems in Water Systems," *ASCE J. Hydraulic Eng.* 111(2): 206–219, 1985.

Mandell, G. L., J. E. Bennett, and R. D. Dolin (eds.) *Principles and Practices of Infectious Diseases*, 5th ed., Churchill Livingston, Philadelphia, PA, 3261 pp., 2000.

Marshall, R. J., "Mapping Disease and Mortality Rates Using Empirical Bayes Estimators," *Appl. Stat.* 40: 283–294, 1991.

Maslia, M. L., J. B. Sautner, and M. M. Aral, *Analysis of the 1998 Water-Distribution System Serving the Dover Township area, New Jersey: Field-Data Collection Activities and Water-Distribution System Modeling*, Agency for Toxic Substances and Disease Registry, Atlanta, GA, 2000a.

Maslia, M. L., J. B. Sautner, M. M. Aral, J. J. Reyes, J. E. Abraham, and R. C. Williams, "Using Water-Distribution System Modeling to Assist Epidemiologic Investigations." *Journal of Water Resources Planning and Management, ASCE*, 126(4): 180–198, 2000b.

Maslia, M. L., J. B. Sautner, M. M. Aral, R. E. Gillig, J. J. Reyes, and Williams, R. C, *Historical Reconstruction of the Water-Distribution System Serving the Dover Township Area, New Jersey: January 1962–December 1996*, Agency for Toxic Substances and Disease Registry, Atlanta, GA, 2001.

Mercer, J. W., and R. K. Waddell, "Contaminant Transport in Groundwater," in D. R. Maidment (ed.), *Handbook of Hydrology*, Chap. 16, McGraw-Hill, 1993.

Moorehead, W. P., R. Guasparini, C. A. Donovan, R. G. Mathias, R. Cottle, and G. Baytalan, "Giardiasis Outbreak from a Chlorinated Community Water Supply," *Canad. J. Public Health* 81: 358–362, 1990.

NJDHSS and ATSDR, *Dover Township Childhood Cancer Investigation Public Health Response Plan*, New Jersey Department of Health and Senior Services, Trenton, NJ, and Agency for Toxic Substances and Disease Registry, Atlanta, GA, 1996.

NJDHSS, *Case-Control Study of Childhood Cancers in Dover Township (Ocean County), New Jersey*, New Jersey Department of Health and Senior Services, Division of Epidemiology, Environmental and Occupational Health, Trenton, NJ, 2003.

Normani, S. D., and J. F. Sykes, "Uncertainty in Water Distribution System Contaminant Concentrations Resulting from Contaminated Groundwater. Linking Exposures and Health: Innovations and Interactions," *12th Annual Conference of the International Society of Exposure Analysis, 14th Conference of the International Society for Environmental Epidemiology*, Vancouver, BC, 2002.

Obradovic, D., and P. Lonsdale, *Public Water Supply: Data, Models and Operational Management*, Routlege, New York, 462 pp., 1998.

Rosenbaum, P. R., and D. B. Rubin, "The Bias Due to Incomplete Matching," *Biometrics* 41: 103–116, 1985.

Rossman, L. A., *EPANET 2 Users Manual*, National Risk Management Research Laboratory, USPA, Cincinnati, OH, 2000.

Rothman, K. J., and S. Greenland, *Modern Epidemiology*, Lippincott-Raven, Philadelphia, 1998.

Schlesselman, J. J., *Case-Control Studies; Design, Conduct, Analysis*, Oxford University Press, New York, 1982.

Snow, J., *On the Mode of Communication of Cholera*, John Churchill, London, UK, 191 pp., 1855.

Thernau, T. M., and P. M. Grambsch. *Modeling Survival Data: Extending the Cox Model*, Springer, New York, 350 pp., 2000.

Toms River Chemical Corporation, *Memorandum to R.K. Sponagel from P. Wehner Concerning the Water Supply Situation*, Toms River, NJ, March 21, 1966.

U.S. EPA, "Guidelines for Exposure Assessment," FRL-4129-5. U.S. Environmental Protection Agency, Washington DC, May 29, 1992.

CHAPTER 11
SOURCE WATER EARLY WARNING SYSTEMS

Walter M. Grayman,[*] Rolf A. Deininger,[†] Richard M. Males,[‡] and Richard W. Gullick[§]

11.1 INTRODUCTION: AN OVERVIEW OF EARLY WARNING SYSTEMS

Public water systems using surface water sources are vulnerable to a variety of disruptions in water quality as a result of accidental, intentional, or natural contamination of supplies. Surface water sources, particularly rivers, can be subjected to a variety of contaminants which can change rapidly both in nature and concentration. The frequency, magnitude, and type and location of contamination are highly variable and stochastic, and thus cannot be predicted in a deterministic manner.

Protecting the water that is used as a supply source for a potable water treatment plant is of utmost importance. This is the first stage of the multiple-barrier approach to water treatment and supply, which includes selection and maintenance of high-quality source water, its treatment and disinfection, maintenance of distribution system water quality, and water quality monitoring.

Many utilities practice some degree of monitoring of their source water. These data, however, are often limited in the number of parameters measured, and are collected at a frequency that is not conducive to detecting sudden changes in water quality (e.g., weekly, monthly, or quarterly sampling). Early warning systems are designed to address these limitations and to improve on the information available, and to utilize these and other information in an integrated approach for identifying and responding to contamination before it enters the drinking water supply.

Source water protection programs and early warning monitoring systems can help to ensure the highest quality intake water and to reduce treatment costs. They can also lead to reduced potential for regulatory violations (maximum contaminant levels) or compromises in treatment process integrity. Furthermore, monitoring systems can sometimes detect unauthorized waste discharges.

---

[*]W.M. Grayman Consulting Engineer, Cincinnati, Ohio.
[†] The University of Michigan, Ann Arbor, Michigan
[‡]RMM Technical Services, Inc., Cincinnati, Ohio
[§]American Water, Voorhees, New Jersey

11.2 WATER SUPPLY SYSTEMS SECURITY

Though the principles of early warning systems apply to water quality changes in any part of the water field, this chapter focuses on monitoring for impacts on source waters and does not directly address threats to the water supply infrastructure or treated water in the distribution system. Many experts consider these latter two possibilities to be of greater risk than deliberate source water contamination since the large volume of most surface water sources would provide a large amount of dilution, and the water would still be treated and disinfected (Brosnan, 1999).

11.2 FRAMEWORK FOR ASSESSING EARLY WARNING SYSTEMS

Early warning systems are an integral part of the operation of a water system. Thus, the basic problem may be viewed as one of designing and operating the overall system (including monitors, raw water intakes, treatment, and storage) in order to minimize the risks associated with degraded drinking water quality under various cost and technology constraints (Grayman et al., 2001).

Though early warning systems are frequently equated to monitoring instrumentation used to detect contaminants in the water, in reality an effective early warning system must include a wide range of components that work together to prevent contaminants from entering a water supply. These components include the following:

- A mechanism for detecting the likely presence of a contaminant in the source water
- A means of confirming the presence of the contamination, determining the nature of the contamination event and predicting when the contamination will affect the source water at the intake sites and the intensity (concentration) of the contamination at the intake
- An institutional framework generally composed of a centralized unit that coordinates the efforts associated with managing the contamination event
- Communication linkages for transferring information related to the contamination
- Various mechanisms for responding to the presence of contamination in the source water in order to mitigate its impact on water users

These components are described below and presented in greater detail later in this chapter. The elements of an early warning system are displayed schematically in Fig. 11.1.

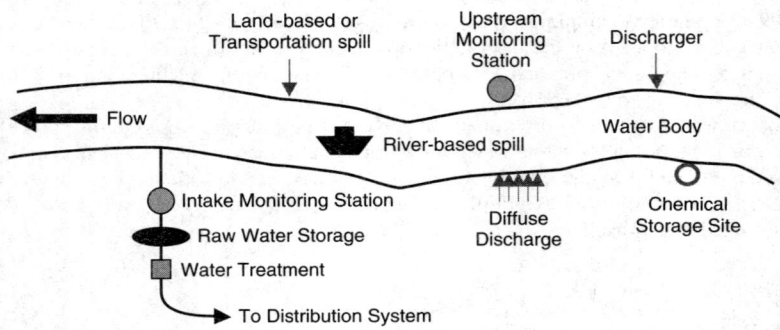

FIGURE 11.1 Elements of an early warning system.

11.2.1 Detection Mechanisms

The initial step in an early warning system is the determination of the likelihood of an unusual contamination event present in the water body that is serving as a raw water source. There are three primary mechanisms for making this initial detection—water quality monitors, self reporting by the entity that has caused the contamination event, or sighting and reporting by the public or by private or public organizations. In an early warning system, the primary objective of the detection phase is to raise a warning, in a very timely manner, that the source water has been contaminated. This is equivalent to "raising a red flag" or yelling, "help." Though further information on the type and extent of the contamination is obviously useful, initially the aim is to merely identify that a problem has occurred. A more complete characterization of the nature of the contamination must then follow.

Monitors. Monitors used in early warning systems can be classified as conventional instruments and biomonitors. Conventional monitors employ chemical and physical methods to measure concentrations of specific compounds or values of indicator parameters. Biomonitors measure the stresses placed on living organisms by contaminants or other changes in the environment. A wide range of organisms including fish, mussels, clams, *daphnia* (water fleas), and bacteria have been used as the basis of biomonitors. Biomonitors do not directly identify the type of contaminant but sometimes indicate the presence of some form of contaminant. For both categories of monitors, depending upon the sensitivity of the monitor, the particular type of monitor, and the monitoring frequency, the instrument may or may not respond to the presence of a contaminant. Two types of potential errors are false positives (monitor declares the presence of a contaminant when there is none present) or false negatives (inability to detect a contaminant that is present).

Public Reporting. Public reporting of a contamination event provides a second mechanism that is used in early warning systems. This mechanism is most effective with larger contamination events—events that result in fishkills, events involving contaminants that are readily detectable by sight or odor, and events in more heavily populated areas. In most cases, such reports are made to the police or fire emergency phone numbers. The effectiveness of this method is dependent upon a population that has been sensitized to reporting such events. In some countries such as Japan, this mechanism is the most common early warning method. In addition to reporting by the public, reporting by other agencies and groups can be an effective early warning mechanism. For some watersheds, water utilities or governmental agencies employ people to patrol these sensitive areas to identify contamination events.

Self-Reporting. The most effective mechanism for identifying the presence of a contaminant should be self-reporting by the discharger of the contaminant. If the discharger is aware of the contamination event then this method can result in a report that is the most timely and exact in terms of location, type, and extent of the contaminant. In most developed countries, laws require self-reporting of any significant spill. However, the compliance with such laws varies significantly around the world. In some countries, dischargers will not report spills in order to avoid fines if they feel they can get away with it. In the United States, it can be surmised that a combination of significant fines if a spill is not reported along with poor public relations if the discharger is caught not reporting a spill has led to a very significant increase in self-reporting of spills.

11.2.2 Confirmation and Characterization of Contamination Events

Following the detection of a contamination event by one of the methods described above, the next step in the early warning process is confirmation of the event and development of further data on the event needed for an effective response. Confirmation involves additional sampling and testing to ascertain that the initial detection was correct. Incorrect detection may be due to false positives associated with monitoring instrumentation or incorrect public reports. Some advanced monitoring stations automatically take samples at fixed intervals and store these samples for a fixed period (e.g., 24 h). Other stations are designed to take samples automatically when a monitor detects an unusual event. In either case, the samples can then be analyzed using standard tests to confirm and characterize the nature of the contaminant.

Characterization of a contamination event involves the following actions:

- Determine the specific contaminant(s) involved.
- Determine the spatial and temporal variation in concentration in the source water.
- Identify the likely source of the contaminant (if unknown).
- Assess the dynamic behavior of the contaminant in the water body (mixing and decay behavior).
- Predict the movement of the contaminant within the water body so that the time that the leading and trailing edges of the contaminant plume reach water intakes and also the likely concentration at the intake can be predicted.
- Determine impacts on the waterway itself (e.g., fish kills).

Characterization of the contaminant is generally accomplished through collection of samples in the field, field and laboratory monitoring, in-stream tracking of the event, and use of mathematical models to predict the movement of the event in the water body. Depending upon the extent and severity of the event, the amount of fieldwork and monitoring can vary between a few samples and a massive field effort to assess and track the event over many days and weeks (DeMarco et al., 1994). More advanced characterization includes the use of mathematical models to assess and predict the movement of a spill in a water body.

11.2.3 Institutional Issues

In many spill situations, multiple agencies and institutions may be involved in the detection, coordination, and mitigation of a spill event. Since rapid response is generally important in dealing with a spill, preplanned institutional responsibilities, protocols, and arrangements are needed.

Experience around the world has shown that effective spill response requires a lead organization to serve as the overall coordinator during emergency events. A lead agency can be either a regional agency that coordinates the event over an area that can include several water intakes or a water utility that maintains one or many intakes and treatment plants that may be impacted by a spill. A wide range of organizations serve as the coordinating agency—international agencies (Rhine River), state agencies (Louisiana), Federal-state commissions (Ohio River), water utilities (U.K.), a group of water agencies (Japan), and private organizations (St. Clair River). Interest in early warning monitoring networks has increased in recent years, and such systems are currently being developed for the Upper Mississippi, Schuylkill, Delaware, Allegheny, Monongahela, and Susquehanna rivers using a variety of institutional arrangements (Gullick, 2003).

Other agencies that may be involved in spill situations in the United States include the National Response Center, the Coast Guard, state and federal environmental agencies, the U.S. Army Corps of Engineers, local health and environmental agencies, and water utilities. Effective interaction among the agencies is a key to successful operation during a spill situation.

11.2.4 Communication Linkages

Communication linkages provide the mechanisms for information transfer during spill events. These can include communications between field monitors and centralized coordinating centers, between industries and coordinating centers, between coordination centers and water utilities, and inter-agency communications. Actual linkages may vary depending upon the specific institutional arrangements and requirements. For example, water utilities may function as the central coordinating agency during spill events, thus eliminating the intermediate communication steps. In other cases, governmental reporting rules may require additional steps and communication linkages during spill events. In all communication linkages, delays may occur due to technology, institutional requirements, human error or judgement, or reluctance to report a spill.

11.2.5 Responses to an Early Warning

There are several responses that can be taken to mitigate the impacts of a spill event: (1) closure of water intakes and use of alternate sources, (2) cleanup of the spill prior to impacting water intakes, (3) removal of the contaminant through normal treatment processes, (4) advanced treatment options, (5) enhanced temporary chemical treatment at the treatment plant, and (6) public notification.

Water Intake Closure. Closure of water intakes provides the most absolute barrier to a spill impacting a drinking water supply. The two limitations to intake closure are (1) in order to be effective, the spill must be detected prior to or immediately after reaching the intakes, and (2) to optimize effectiveness, the intakes must be closed for the entire period that the contaminant levels are greater than acceptable levels at the intake. Closure of an intake requires an alternate water source either in the form of finished water from another treatment plant or an alternative raw water source to the treatment plant served by the closed water intake, or sufficient in-system storage to satisfy demand. If the water intake can only be closed for a limited time period (e.g., a few hours), then this places a premium on an accurate prediction of the concentration at the intake so that the intake can be closed during the period of highest concentration.

In some water systems, there are alternative water intakes that draw water from the same source but at different locations. In the case of intakes in reservoirs or lakes where the water body may be stratified, the intakes may be located at different depths in order to avoid drawing contaminated water. For rivers, alternative intakes may be located in different lateral positions (i.e., right bank, left bank, center) so that a contamination event that does not affect the entire cross-section can be avoided.

Closure of water intakes can be facilitated by the presence of raw water storage basins that can serve as emergency water supplies in case of intake closure. Typically, such supplies can provide water for a period of a few days. The availability of such supplies means that water intakes can be closed more routinely, and in fact, allows for closure upon initial detection of a nearby spill without having to wait for complete characterization of the spill. Raw water storage is more common in Europe than in the United States.

Spill Cleanup. Ideally, it would be best if the contaminant could be removed from the water body prior to its impact on water intakes. Unfortunately, this alternative is limited in most cases. The most common case where spill cleanup or control is a viable solution is oil spills because oil has a density less than water and under some circumstances will float on the top of a water body.

Cleanup generally involves containment of the spill and removal of the oil products. The three principles of mechanical protection include containment to hold the oil in place during recovery, deflection of the oil away from sensitive areas or towards containment sites, and exclusion of the oil to prevent impacts upon a sensitive area such as a water intake (Michel et al., 1994). Oil booms are the most common mechanism for controlling the movement of oil spills. At some surface water intakes, oil booms are permanently installed to divert surface oil away from the intakes.

Effects of Normal Treatment Processes on Contaminant Removal. The effectiveness of treatment processes on removal of a particular contaminant depends upon many factors—contaminant type, contaminant concentration, and other water quality considerations such as pH and temperature. Various advanced treatment options such as granular activated carbon are available to provide additional barriers to contaminants reaching the water user. In Europe, groundwater injection or infiltration is frequently used as such a barrier. Typically, water is drawn from the raw water source followed by limited treatment. This water is injected and infiltrated into a shallow groundwater table and extracted after a period of several weeks. Further treatment is then provided. Such systems also serve effectively as raw water storage facilities, thus allowing relatively long periods during which water intakes can be closed.

Emergency (Enhanced) Water Treatment. For some contaminants, there are temporary treatment procedures available that can reduce the concentrations leaving the treatment plant. If an early warning system detects a contaminant or a group of contaminants and the normal treatment process will not remove the contaminant(s), then several options may exist.

Coagulation/Sedimentation. In a treatment plant that uses coagulation/sedimentation, an increase in the coagulant dose will increase the removal rate of the contaminants. For example, the rating of this process for removing synthetic organic chemicals (SOC) is poor to good. Operating this process at the highest feasible coagulation dose will remove a higher percentage of the contaminant(s).

Powdered Activated Carbon. Powdered activated carbon (PAC) is primarily used to control tastes and odors but with higher dosage rates may also be effective in removing some organics associated with spill events. PAC is most effective in removing SOC, pesticides/herbicides, and dissolved organic carbon (AWWA, 1999). It is less effective but can also be used to remove some volatile organic chemicals (VOC). The estimated efficiency of adding PAC varies over a wide range and is dependent upon many factors. The appropriate dosage is best determined by jar tests. For spills, dosage rates of over 100 mg/L are sometimes required. A utility that has perceived threats from upstream discharges by industries and/or spills might do a series of tests to predetermine required carbon addition for a number of compounds. Even if the contaminant has not been tested previously, at least the procedure would have been worked out and can be applied under emergency conditions.

The literature has tables that predict the efficiency of powdered activated carbon for specific compounds in clean water (see AWWA, 1999, for a review of literature on this topic). These values are just guidance values since in a spill a combination of compounds are frequently found, and the river water itself is a complex matrix of many substances. Finally, it should be noted that some kind of feeding equipment must be available to mix the carbon evenly into the raw water. Additionally, sufficient on-site chemical storage is required to satisfy the demands of a contaminant event.

Disinfection. An increase in the disinfection dose is always possible. If the treatment system has an ozonation system online, then increasing the ozone dosage will destroy more of the contaminants. Higher chlorine dosages are effective against bacterial contaminants but when phenols are present, addition of chlorine produces chlorophenols that greatly affect the taste and odor of the water and make matters worse. Strong oxidants such as potassium permanganate or hydrogen peroxide also oxidize organics, but are not normally stored onsite.

Public Notification. Notification of the public is generally considered to be the last resort by most water utilities. Such notification can be in the form of a boil water order or an order not to use water for consumption purposes. Under emergency conditions, the public media is the most common notification mechanism. Use of targeted automated telephone calls is emerging as an effective notification mechanism in the emergency response field.

11.3 EARLY WARNING MONITORING METHODS

While technology exists to monitor for a wide range of compounds in the environment, the requirements for monitors as part of an early warning system are more limiting. Ideally, an early warning system monitoring system must cover a broad spectrum of compounds, must provide results in a timely manner, and ideally operate with minimal human intervention. On the other hand, accuracy is not generally as important when compared to laboratory instrumentation used for regulatory compliance. There are tradeoffs between costs and the range and type of monitors used. Selection of the specific methods for monitoring the parameters of concern should be based on a variety of factors, including sensitivity, speed, desired frequency of analysis, available means of data development and retrieval, labor and maintenance requirements, initial and operating costs, availability of space, and environmental requirements.

Water quality monitors include physical, chemical, radioactive, and microbiological analyses, as well as bio monitoring systems that use living organisms as broad spectrum indicators of potential changes in water quality. References for rapid or on-line monitoring techniques for the water industry include AwwaRF and CRS PROAQUA., 2002; Frey et al., 2001; Grayman et al., 2001; Gullick, 2001; Gullick et al., 2003; Dippenaar et al., 2000; Pollack et al., 1999; and Reinhard and Debreaux, 1999. A brief overview of select methods for early warning is provided below.

Some of the more common physical and chemical monitoring methods used in early warning systems include simple probes (e.g., turbidity, pH, temperature, conductivity, dissolved oxygen, chlorophyll); relatively simple batch tests (e.g., immunoassays for herbicides), and more advanced monitoring for chemicals (e.g., fluorescence for oils; chromatography for oil and petroleum constituents, volatile organic chemicals and phenols). Some of the primary surrogates used include turbidity, dissolved oxygen, odor, conductivity, and general measures of organic carbon content (e.g., oxidant demand, total organic carbon). However, some of the parameters that are easily and inexpensively monitored via on-line probes (e.g., temperature, conductivity, pH) provide little information in detecting many spill events (e.g., oil spills). The more advanced monitors are more expensive and require more maintenance and expertise, but are better for detecting many spill events.

11.3.1 Physical Analyses

Most physical monitoring methods are relatively rapid (e.g., turbidity, conductivity, temperature, odor), and many can generate continuous real-time on-line data.

Continuous online turbidity measurements are regularly used in treatment process control applications, and more expensive on-line particle counters are also sometimes used. Large increases in turbidity are frequently correlated with adverse changes in microbial water quality, since both turbidity and microbial concentrations often increase substantially in surface waters during and subsequent to storm events due to surface runoff. LeChevallier et al. (1998) cites an example where high *Cryptosporidium* loadings at one intake can typically be avoided by shutting that intake and using water from onsite storage when turbidity rises above a certain level (e.g., >15 NTU).

The presence of unusual odors can be a useful indicator for certain contamination events, including those resulting from algal byproducts (e.g., geosmin and methylisoborneol), phenols, petroleum products, and assorted volatile organics. One means for detecting odors is the so-called "smell bell," where the water is heated to help release (volatilize) the odors into a glass container, which is then smelled by trained operators. However, it requires trained personnel with good noses, and thus is seldom performed more often than once per shift or once per day, thus limiting its use in early warning systems. Based on recent research in the food and chemical industry, electronic odor sensing technologies ("electronic noses") may be available in the future for use in the analysis of water (Grayman et al., 2001).

11.3.2 Chemical Analyses

Many standard chemical analyses can be used for early warning monitoring, and several methods have been adapted for automated on-line applications and remote data access. A summary of the relative costs and pros and cons for different early warning monitoring technologies, for select chemical constituents, is presented in Table 11.1.

On-line Analytical Probes. On-line analytical probes are relatively inexpensive, easy to use, can provide continuous or nearly continuous monitoring with remote access to data, and are available from a variety of manufacturers. Many inorganic ions can be quantified via ion-selective electrodes (ISE), including pH, elemental anions (e.g., chloride, bromide, fluoride, and iodide), ammonium, nitrite/nitrate, cyanide, certain metals (e.g., lead, cadmium, copper, aluminum, and manganese), and several other inorganic pollutants (Table 11.2). Probes are also available for turbidity, chlorophyll, and dissolved oxygen. Some manufacturers combine a variety of electrodes into one convenient and efficient multiparameter instrument. Since probes can foul in many raw water environments, some models use self-cleaning systems to reduce maintenance requirements.

Dissolved Oxygen. The dissolved oxygen (DO) concentration has a significant effect on the survival of aquatic life, and for early warning applications it is typically measured with a simple on-line probe. A decrease in DO can indicate the presence of organic compounds from sewage or surface water runoff. In addition, diurnal fluctuations in DO can be indicative of the presence of algae, and thus DO is sometimes used in conjunction with chlorophyll and turbidity measurements to monitor for algal blooms.

Nitrate and Ammonia. Nitrate and ammonia/ammonium may be measured with an ion specific electrode. More sensitive but more expensive instruments for on-line colorimetric and UV analyses are also available. Both parameters may be indicative of agricultural pollution (fertilizers). Ammonia may come from sewage and animal waste discharges.

Metals. Ion specific electrodes are available for certain metals, including lead, cadmium, copper, aluminum, and manganese. Anodic stripping voltametry/polarography is an alternative for rapid analysis (<1 to 10 min) of low levels (ng/L range) of certain metals, and is used online at various monitoring stations in Europe. The instruments ($10,000 to $17,000) can detect four to six metals simultaneously, but the method is restricted to amalgam forming metals (e.g., Cd, Cr, Cu, Pb, Zn), and is subject to matrix interferences. Colorimetric methods are relatively inexpensive, typically apply to a single metal, and are subject to more interferences than more sophisticated methods. Atomic absorption spectrometry (AAS) and

TABLE 11.1 Methods for Detecting a Variety of Chemical and Radioactive Constituents

	Approach								
	High end			Medium end			Low end		
Constituents	$100,000	Pros	Cons	$10,000	Pros	Cons	$1000	Pros	Cons
Ions (salts)				IC	Fast, broad, sensitivity		Ion probe	Sensitivity	Selective
Metals	ICPMS	Fast, broad ID, sensitivity	Staff, lab	AAS Polarography	Fast, sensitive Fast, fairly selective	Staff, lab Selective	Ion probe	Sensitivity	Selective
Polar organics	LCMS	Broad ID	Staff, lab	LC	Broad ID	Staff, lab	UV		Lack of sensitivity
				TOC	Broad ID	Lack of sensitivity			
Nonpolar organics	GCMS	Broad ID	Staff, lab	LC	Broad ID	Staff, lab			
Volatiles, oil, hydrocarbons	GCMS	Broad ID	Staff, lab	P&T-GC GC Fluorescence (oil, HC)	Broad ID Broad ID Broad ID	Staff, lab Staff, lab Interferences	Smell bell	Fast	Human testers
Specific compounds	GCMS, LCMS	Broad ID	Staff, lab				Immunoassay (pesticides)	Fast, specific	Staff

(*Continued*)

TABLE 11.1 Methods for Detecting a Variety of Chemical and Radioactive Constituents (*Continued*)

Constituents	Approach	High end			Medium end			Low end		
		$100,000	Pros	Cons	$10,000	Pros	Cons	$1000	Pros	Cons
Biotoxics					Biomonitors	Continuous, fast	Lack of specific ID			
Radiation					Tritium	Fast, specific	Not available online			
					Gamma-Detector	Fast, broad ID, available online	Lack of specific ID			
					Beta or alpha Detector	Fast	Lack of specific ID, lab, evaporation step; not available online			

Notes: Broad ID = can monitor for many compounds simultaneously, Selective = monitors for a single compound, AAS = atomic absorption spectrometry (furnace or flame), GC = gas chromatography, HC = hydrocarbons, IC = ion chromatography, ICPMS = inductively coupled plasma mass spectroscopy, ID = identification, LC = liquid chromatography, MS = mass spectrometry, P&T = purge and trap, UV = ultraviolet, Biomonitors = fish, daphnids, mussels, algal fluorescence, and luminescent bacteria.

Source: Modified from Brosnan (1999).

TABLE 11.2 Ion Specific Electrodes Used in Monitoring Raw Water

Ion	Type	Range (ppm)	Interferences
Ammonium	PVC membrane	0.1–18,000	Potassium
Bromide	Solid state	0.4–80,000	S, I, CN
Cadmium	Solid state	0.01–11,000	Ag, Hg, Cu, Pb, Fe
Calcium	PVC membrane	0.2–40,000	Pb, Hg, Cu, Ni
Chloride	Solid state	1.8–33,000	S, I, CN, Br, OH, NH_3
Copper	Solid state	0.0006–6350	Ag, Hg, Cl, Br, Fe, Cd
Cyanide	Solid state	0.1–260	S, I, B, Cl
Fluoride	Solid state	0.02 to saturation	OH
Iodide	Solid state	0.006–127,000	S, CN, Br, Cl, NH_3
Lead	Solid state	0.2–20,700	Ag, Hg, Cu, Cd, Fe
Nitrate	PVC membrane	0.5–62,000	I, CN, BF_4
pH	PVC membrane	1–14	
Surfactant	PVC membrane	1–12,000	
Hardness	PVC membrane	0.4–40,000	Cu, Zn, Ni, Fe

plasma emission spectroscopy (ICP/MS) instruments are expensive and typically available only in commercial laboratories. One promising new technology uses fluorescent molecules that react to specific metals in the presence of ultraviolet light, and has been applied to analysis of zinc, mercury, and cadmium (Bronson et al., 2001). Other developing methods for a variety of heavy metals include enzyme sensors and biosensors using genetically engineered microorganisms (Rogers and Gerlach, 1999).

General Organic Chemical Parameters. Total organic carbon (TOC) and ultraviolet light absorption at 254 nm (UV-254) are general measures of organic content that can be performed in minutes and online. Though TOC is generally more sensitive and thus used more often for early warning, its natural variability in source waters is often greater than the concentrations of specific organics of concern. Simpler bench-scale test kits for organic carbon are also available.

Oxidant Demand and Oxidant Residual. Oxidant demand can be a general indicator of organic carbon content and ammonia in the source water. Since many utilities practice pre-oxidation (e.g., chlorine, chlorine dioxide, ozone, or permanganate) and use on-line monitors to measure downstream oxidant residual, the oxidant demand can be calculated if the oxidant dose and flow rates are known. Oxidant residual is not, of course, applicable to raw waters, but can be a useful warning measure for changes in distribution system water quality if residual disinfection is used by the utility.

Oil and Petroleum. The primary techniques for on-line monitoring of floating oil use light scattering, and for dissolved oil use fluorescence, though each method has several limitations (He et al., 2001). Common chemical and physical interferences (e.g., particles, detergents, and floating debris) can cause false alarms to be inconveniently frequent and make it difficult to track an oil spill during rain events that increase turbidity. Most commercial oil-in-water monitors use light scattering techniques, and thus are primarily useful only for major spills (e.g., for a 0.33-mm layer or greater of floating product). Fluorometry can be used for dissolved gasoline, diesel, jet fuel, and oil components (e.g., BTEX—benzene, toluene, ethyl benzene, and xylenes), as well as chlorophyll from algae. Continuous fluorescence oil detectors ($12,000 to $24,000) are very sensitive (low µg/L range in fairly clean water) and are used in several monitoring programs worldwide, though turbidity and humic substances can interfere. While manual solvent extraction methods are labor intensive, some European monitoring stations use an automated system for extraction and spectrophotometric analysis of total dissolved hydrocarbons (between 0.2 and 10 mg/L). There is a need for improved on-line monitors for low concentrations of oil. The introduction of genetically engineered microorganisms as biosensors for BTEX (Rogers and Gerlach, 1999) may prove useful in the future.

Organic Chemicals. Manual and on-line gas chromatographs (GCs; $30,000 to $50,000) are used in several early warning systems worldwide to monitor for volatiles or other organic chemicals (including fuel oil components). Liquid chromatography is more expensive ($50,000 to $100,000) and generally requires an operator. Mass spectrometry (MS) is even more expensive and would be used primarily during the event confirmation step to provide accurate identification of organics in select samples.

Pesticides. Pesticide (herbicides and insecticides) contamination of surface waters is often seasonal as it primarily results from nonpoint source rainfall runoff from agricultural areas during periods of high pesticide application. The inexpensive batch ELISA (enzyme-linked immunosorbent assay) procedure is often used for the popular herbicide atrazine, takes approximately 40 min, and compares reasonably well to GC/MS results for concentrations on the order of 3 µg/L (the USEPA standard) (Lydy et al., 1996).

Radioactivity. Early warning for radioactivity in surface waters can be applicable for facilities downstream from a nuclear power plant or other potential large source. Both gross radioactivity and specific radioactive substances may be measured. Tritium (hydrogen-3) may be an especially good indicator for nuclear power waste since it behaves as a conservative tracer in water and would reach an intake prior to other radioactive constituents that have larger retardation factors. Monitoring stations on the Rhine River measure for total alpha, total beta, tritium, cesium-137, and strontium-90 activity.

11.3.3 Microbiological Analyses

Conventional methods of microbial analysis require a relatively long time period (hours or days) for isolation and reproduction (amplification) of the microbial species, and many tests are specific only to a single species or class of organism. As such, these analyses are not often used for early warning applications. However, numerous significant recent advances in microbial monitoring and related technology offer increased sensitivity, specificity and/or more rapid analysis, including DNA microchip arrays, rapid DNA probes, immunologic techniques, cytometry, laser scanning, laser fingerprinting, optical technologies, and luminescence (e.g., Rose and Grimes, 2001; Grayman et al., 2001; Foran and Brosnan, 2000; Quist, 1999; Rogers and Gerlach, 1999). Most of these methods are still being developed or were only recently introduced. However, their use is likely to increase in the future. Relatively rapid existing methods for microbes are summarized in Standard Methods (SM) Part 9211 (APHA et al., 1998) and Venter (2000).

Nucleic acid-based systems measure the genome of the organisms, thus giving a high degree of specificity, but sample processing typically takes at least 2 to 4 h. Several different kits are available for these tests. Rapid DNA probes are species-specific and use a robot-assisted micro-plate analysis of amplified samples of DNA (Quist, 1999). DNA microchip arrays are a new developing technology that can detect and identify multiple microorganisms within 4 h. Laser scanning cytometry can be used to rapidly detect any organism for which there is a specific antibody, but the instruments are expensive.

Immunoassays use target-specific fluorescent antibodies that bind with an antigen of the target species. Test kits for a variety of pathogens are available that are relatively rapid, inexpensive, sensitive and simple to use (www.aoac.org/testkits/microbiologykits.htm).

Commercial methods for measuring bacterial counts within 8 to 24 h are readily available (e.g., Colilert, Colifast, and Colisure). Recent advances have reduced the potential analysis time for bacteria (e.g., total coliforms, *E. coli*, or heterotrophic plate counts, HPC) down to the range of 4 to 8 h or less. As an example, a new modification of SM 9211C.1 using adenosine triphosphate bioluminescence allows quantification of HPC within minutes (Lee and Deininger, 1999).

The conventional tests for protozoan parasites such as *Giardia* and *Cryptosporidium* (e.g., U.S. EPA 1622 and 1623) require extensive training and are too time-consuming for early warning monitoring applications. Commercial instruments are available that can provide for screening of protozoan parasites in aqueous samples, but the tests still take a few hours due to sample preparation requirements.

By detecting algae blooms at their earliest stages, the algae can be treated in the reservoir before they grow out of control, thus reducing taste and odor problems and saving on treatment costs. Several commercial continuous monitors are available that rely on an on-line fluorescence detector to measure chlorophyll a, the principle photosynthetic pigment in all algae. Some probes combine these measurements with those for water clarity (turbidity) and oxygen to provide early warning of algal blooms, and cost around $5000. A more expensive and sophisticated system was used in Los Angeles to detect algae in supply reservoirs, and resulted in substantial cost savings for treatment chemicals (Morrow et al., 2000).

11.3.4 Biological Monitoring (Bioalarms)

The sheer magnitude of the number of pollutants of concern, and the inability to monitor many of them continuously or at all, has lead to the use of online biomonitors that measure changes in the behavior or properties of living organisms resulting from stresses placed on them by the presence of toxic materials. Conceptually biomonitors are analogous to the canaries used by miners to detect the presence of toxic gases. Though biomonitors do not provide information on the specific contaminant or cause of the effect, they warn that there is something unusual in the water that is affecting the organisms, thus warranting further investigation such as specific chemical analyses (Penders and Stoks, 1999). Some biomonitors respond rapidly to elevated concentration levels of a wide range of toxic compounds, and some can also be used to assess low-level chronic contamination by persistent, bioaccumulative toxins (e.g., from xenoestrogens, biocides, pharmaceuticals, and pesticides).

Biomonitors include use of fish, mussels, and *daphnia* (water fleas); delayed algal fluorescence; and luminescent bacteria response. Modern biomonitors typically measure changes in movement or physiological responses by an organism as it reacts or tries to avoid exposure to toxic chemicals in the water. Because different species respond to different chemicals to varying degrees, it is often recommended to simultaneously use different types of biomonitors, including some from different trophic levels (Penders and Stoks, 1999; LAWA, 1998). Examples of modern biomonitors include fish monitors that digitally track fish as they attempt to avoid contaminated water entering a tank, and *daphnia* monitors that use digital cameras that follow the behavior of individual *daphnium* and compare this movement to normal behavior. The simpler bacterial tests using luminescent bacteria are promising means to determine the toxicity of the river water. Likewise, the delayed fluorescence of algae can be relatively easily measured. A report of German field experiences rated the dynamic *daphnia* test as the first priority for developing a bioalarm station, followed in order by fluorescent algae, bacteria tests, and mussel monitors (LAWA, 1998). Fish monitors were not recommended primarily because the sensitivity was problematic and not reproducible (e.g., there were problems with both false alarms and the systems not responding to pollution events) (LAWA, 1998).

There are very few biomonitors in the United States but they are widely used in Europe (LAWA, 1998) and in Asia (most notably in Japan). On the Rhine River, *Daphnia* monitors are most widely used, while some also use fish, mussels, algae, and bacteria to test the water with organisms from different trophic levels. In the United States, the U.S. EPA research laboratories in Cincinnati are investigating the effectiveness of biomonitors at different trophic levels (VonderHaar et al., 2002), and *daphnia* toximeters were used for assessing source water quality during the 2002 Winter Olympics in Salt Lake City, Utah (Yates et al., 2002).

Purchase costs for commercial systems typically range from about $10,000 to near $50,000 and up. The manual batch bacteria tests can be the least expensive in terms of capital costs. The algae, *daphnia*, and mussel tests are fairly comparable in expense (around $20,000 to $40,000). The cost of fish monitor units (Stoks, 1998; LAWA, 1998) can vary considerably with commercial units being quite expensive, while "home-made" avoidance units can be constructed quite inexpensively. With the exception of the luminescent bacteria test, operating costs for these methods are fairly low, and are comprised primarily of replacement organisms and electricity.

False positive results can result from interferences from a variety of environmental factors other than contaminants (e.g., temperature changes or low oxygen). Information on the sensitivity and minimum detection limits of on-line biomonitors is relatively limited, and there is a relative lack of sensitivity for some chemicals of interest. Other drawbacks include a high cost for more sophisticated biomonitors, and maintenance requirements for the living systems. The interpretation of the signals from biological monitors is also an important consideration, and as this improves it is likely that the value of biomonitors will increase.

11.3.5 Emerging Monitoring Methods and Research and Development Needs

Emerging electronic noses and rapid bacterial methods are identified as areas in which developments are taking place which are likely to increase their future use in early warning systems. Numerous ongoing AwwaRF and Water Environment Research Foundation research projects investigate rapid and on-line monitoring technologies. Generally speaking, however, many of the advances in monitoring technologies occur from research in other scientific fields (e.g., the food and beverage industry, analytical chemistry, the sensor industry, and the military), including biosensor and biochip technology, fiber optics, genetically-engineered organisms, immunoassays, microelectronics, and others. Several U.S. government organizations conduct research on rapid and/or on-line monitoring systems for a variety of contaminants, including the U.S. EPA (e.g., Panguluri, 1999; Rogers and Gerlach, 1999), and the U.S. Army's Joint Service Agent Water Monitor program (Brosnan, 1999).

11.4 SURFACE WATER MODELS IN EARLY WARNING SYSTEMS

11.4.1 Introduction

Surface water modeling encompasses a wide range of methods that can be used to predict the hydraulics and water quality in an aquatic environment under a set of conditions. The technical basis for such models extends back for well over a century and computerized methods can be traced back to the 1950s. There are several general reviews of models published in recent years (Ambrose et al., 1996, McCutcheon et al., 1993, Huber, 1993, DeVries and Hromadka, 1993).

In the context of early warning systems, the specific type of modeling of interest involves the modeling of transient water quality conditions associated with spills or other contamination events and their potential impact on drinking water intakes. This class of models is frequently referred to as "spill models."

11.4.2. Characteristics of Spill Models

Spill models are a class of models that are used to trace the movement and fate of transient contaminants in receiving waters. In the context of early warning systems for drinking water supplies, the particular area of interest is fresh water bodies that can serve as such sources. Though groundwater is frequently used as a source of water and, indeed, is very susceptible to contamination, the time scale and mechanisms associated with movements of contaminants in groundwater differs greatly from surface water processes.

Spill models are generally used in real-time or near real-time situations. The presence of a contaminant in a water body is determined through monitoring, self-reporting or other means and the goal is to estimate how the contaminant will affect drinking water intakes. In this situation, time is generally of the essence so that the water utility may take some appropriate action (e.g., close their intake, modify treatment processes). In many cases, there are tradeoffs between the accuracy of the model predictions and the time it takes to make the predictions. This requirement generally limits activities such as extensive real-time field data collection, elaborate laboratory analysis, sophisticated calibration procedures, and significant intervention by an expert modeler. Thus the general requirements for a spill model as part of an early warning system are that it must provide a sufficient level of accuracy, in a timely manner, with outputs in a readily usable format.

There are three basic components of any spill model—a flow module, a water quality transport module, and a fate module. The flow module describes the movement of the water; the water quality transport module describes the processes by which the contaminant concentration changes due to the hydrodynamic forces; and the fate module describes the impacts of physical, chemical, and biological processes on the form and concentration of the contaminant. These modules may be represented in separate models that are interconnected through input-output or may be integrated into a single model that represents the entire process.

11.4.3 Incorporation of a Model in an Early Warning System

In developing an early warning system modeling capability, there is a range of alternatives in terms of incorporating a model, which includes the following:

- Development of a special purpose model.
- Modification of an existing model.
- Use of an existing model without modification.

Each of these approaches can be effective and the selection of an approach depends upon the particular circumstances. The first two approaches require skills in computer programming. The second approach (model modification) assumes the availability of source code that can be modified. Source code is available for many public domain models. Many models require extensive and involved input files. If the third approach is selected, then mechanisms for simplifying the input process, in order to facilitate use of the model under emergency situations, are frequently required. Such an approach was used in developing an early warning model for the Ohio River (Grayman et al., 2001).

Tables 11.3 and 11.4 provide a summary of several general-purpose models that can and/or have been used as the basis of an early warning system. These models include flow models, water quality models, and fate models. They include riverine, lake, and estuarine models. Most of these models are public domain models available from the U.S. EPA, the USGS and the U.S. Army Corps of Engineers. Three of the models are commercial software products.

11.4.4 Review of Selected Early Warning System Models

Unlike the general-purpose models described above, there are several models that have been developed and implemented specifically for use as spill models. These models fall into two categories—(1) models that utilize existing general purpose models surrounded by an application-specific shell to expedite the model for use in spill simulation and (2) models that have been explicitly developed or adapted for use in simulation of spills. Each of these approaches generally lead to a product that simulates transient spill conditions and that facilitates use under emergency spill situations. In this section, several such models are presented.

Riverine Spill Modeling System (RSMS). The Riverine Spill Modeling System (RSMS) is a general-purpose tool used to assist in rapid assessment of the impact of a spill on a river system (Grayman et al., 2001). It is designed to predict the transport of a constituent as a result of a spill of known quantity and duration, at a known point on a river or a first-order tributary of that river.

The RSMS makes use of the BLTM (Branched Lagrangian Transport Model), a model developed by the U.S. Geological Survey, as the basic computational engine for defining transport in the river (Jobson and Schoellhamer, 1987). The RSMS surrounds the BLTM "computation kernel" with a simple input and output framework, making it much easier to use in an emergency response environment.

Required input includes the following:

- A data file describing the flow and area in the river during the duration of the analysis.
- A data file describing the river mile structure of the river and its tributaries.
- Information on the spill:
  - Location (mainstem and tributary river miles)
  - Time of spill
  - Duration (hours)
  - Quantity (pounds)
  - Parameters describing the decay and dispersion associated with the spill
  - Various parameters controlling the details of the model run.

Output includes a variety of types of graphical displays of the spill impact, and the ability to export various types of data to a Microsoft Excel spreadsheet. The model is implemented as a Microsoft Visual Basic 5 application, making use of a variety of other programs to perform the needed calculations. For the RSMS to be used successfully in a short-term emergency spill response situation, mechanisms must be in place for the creation of the needed flow file in the desired format corresponding to the hydraulic situations at the time of the spill. As well, the corresponding river mile structure file must exist.

TABLE 11.3 General Summary of Hydraulic and Water Quality Models

Model	Agency	Comments	Websites
WASP5	EPA	Multicompartment water quality model	http://www.epa.gov/ATHENS/research/modeling/wasp.html
CE-QUAL-RIV1	WES	One-dimensional flow and water quality model	http://www.wes.army.mil/el/elmodels/riveinfo.html
CE-QUAL-W2	WES	Two-dimensional flow and water quality model	http://www.wes.army.mil/el/elmodels/w2info.html
CE-QUAL-ICM	WES	one-, two-, or three-dimensional water quality model	http://www.wes.army.mil/el/elmodels/estuinfo.html
DYNHYD5	EPA	One-dimensional tidal flow model	http://www.epa.gov/ATHENS/research/modeling/wasp.html
BLTM	USGS	One-dimensional water quality model for branching/looping rivers & estuaries	http://water.usgs.gov/software/bltm.html
BRANCH	USGS	One-dimensional flow model	http://water.usgs.gov/software/branch.html
CH3D-WES	WES	Three-dimensional hydrodynamic model	http://chl.wes.army.mil/software/ch3d/
RMA2	WES	Two-dimensional vertically averaged hydrodynamic model	http://smig.usgs.gov/cgi-bin/SMIC/model_home_pages/model_home?selection=rma2
RMA4	WES	Two-dimensional water quality model	http://chl.wes.army.mil/software/tabs/docs/rma4_6_05_01.pdf
MIKE 11	DHI	One-dimensional hydraulic and water quality model	www.dhi.dk
MIKE 21	DHI	Two-dimensional hydraulic and water quality model	www.dhi.dk
MIKE 3	DHI	Three-dimensional hydraulic and water quality model	www.dhi.dk

Notes: EPA = U.S. Environmental Protection Agency, USGS = U.S. Geological Survey, WES = Corps of Engineers Waterways Experiment Station, DHI = Danish Hydraulics Institute.

TABLE 11.4 Summary of Model Characteristics

Model	WQ	Hyd	1	2xy	2xz	3	River	Lake	Estuary
WASP5	X		X	X	X	X	X	X	X
CE-QUAL-RIV1	X	X	X				X		
CE-QUAL-W2	X	X	X		X		X	X	X
CE-QUAL-ICM	X	X				X	X	X	X
DYNHYD5		X	X	X			X		
BLTM	X		X				X		
BRANCH		X	X				X		
CH3D-WES	X					X	X	X	X
RMA2		X		X			X	X	X
RMA4	X			X			X	X	X
MIKE 11	X	X	X				X		
MIKE 21	X	X		X				X	X
MIKE 3	X	X				X	X	X	X

Notes: $2xy$ is a two-dimensional model with variation in the lateral and longitudinal directions; $2xz$ is a two-dimensional model with variations in the longitudinal and vertical directions.

A modified version of RSMS was implemented for the entire 1000 mi (1600 km) Ohio River mainstem from Pittsburgh, Pennsylvania until it joins the Mississippi River near Cairo, Illinois mainstem of the Ohio River for the Ohio River Valley Water Sanitation Commission (ORSANCO). The modifications in the RSMS code were made for consistency with an earlier spill model used as part of ORSANCO's early warning system (Grayman et al., 1993). The model uses flow and stage information generated by the FLOWSED hydraulics model (Johnson, 1982). FLOWSED is run daily by the U.S. Army Corps of Engineers as part of the reservoir operations group to generate a 5-day forecast for the Ohio River and lower portions of major tributaries. The resulting hydraulics files are downloaded on a daily basis by ORSANCO in preparation for an application of RSMS in case of a spill event.

RSMS generates output results in several alternative formats. Typical output displays include a plot of concentration vs. river mile at a selected time (Fig. 11.2) and a plot of time of leading edge, peak and trailing edge vs. river mile (Fig. 11.3). The latter plot is especially useful to operators of downstream water intakes in determining when and for how long to close an intake during a spill event.

Rhine Alarm-Model. The Rhine River is a major navigational river and a source of drinking water for several European countries. Over its length of 1320 km (820 mi) the river also traverses major industrial and agricultural areas. As a result, it is susceptible to transient contaminant events associated with spills that can significantly impact its water quality. A fire at the Sandoz Company in Basel, Switzerland in 1986 led to a significant chemical spill to the Rhine River that resulted in the development of a regional early warning system (Deininger, 1987). An analysis of the incident also showed the inadequacy of existing prediction methods for such events and led to the development of the Rhine Alarm-Model (van Mazijk, 1996).

Rather than using existing models, the hydraulic and water quality modules were developed specifically for use in the Rhine Alarm-Model. The primary assumptions of the model are as follows:

- Only longitudinal dispersion occurs.
- Directly after the spill, the contaminant is completely mixed over the river cross-section.

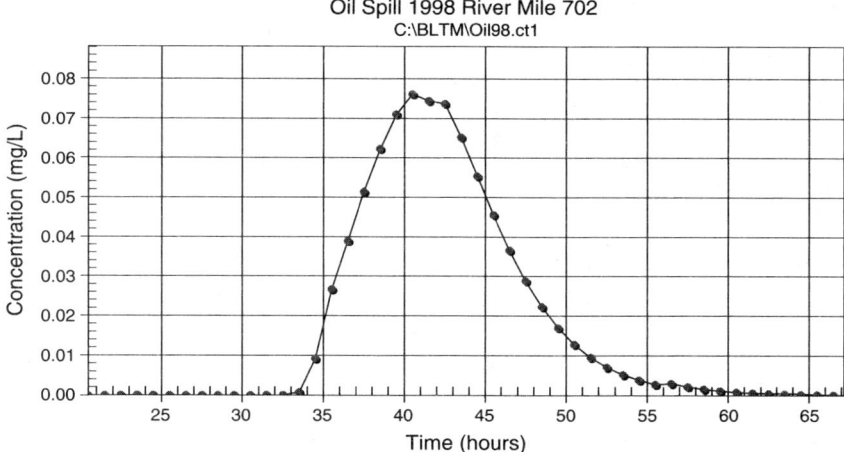

FIGURE 11.2 Example RSMS output of concentration vs. time.

- Flow conditions are steady.
- The pollutant is conservative or decomposes linearly with time.

The river is represented as a series of branches and nodes in the model. In each branch, the flow and velocity is constant and related to water level through a stage-discharge relationship. The classical advection-dispersion transport model is modified to represent dead zones along the bottom and sides of the river where the flow is nearly stagnant. Due to temporary

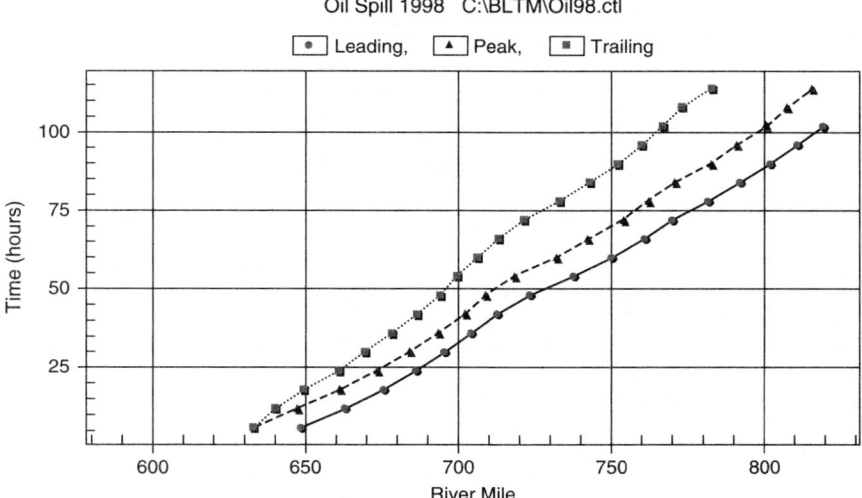

FIGURE 11.3 RSMS example output of time of leading edge, peak, and trailing edge.

entrapment of portions of the pollutant in these dead zones, the transport velocity of the pollutant plume is reduced. A lag coefficient for representing the difference between mean flow velocity and actual transport velocity of the pollutant due to dead zones and incomplete lateral mixing is introduced in the model.

The lag coefficient and the longitudinal dispersion coefficients were calibrated and verified by means of tracer experiments in Switzerland, France, Germany, and the Netherlands (Spreafico and van Mazijk, 1993). The model is implemented at major monitoring stations along the Rhine River so that spill computations can be made following the detection or reporting of spills.

R-TOT Model. Over a million people receive their drinking water from the lower Mississippi River between Baton Rouge and New Orleans, Louisiana. This stretch of river is also lined with numerous major chemical industries and extensive barge traffic. As a result, the potential for spills is quite significant. The implications in terms of drinking water are exacerbated by the limited storage at many of the water utilities leading to limited periods during which intakes can be closed and emphasizing the need for timely and accurate warnings of the passage of spill events at water intakes. This situation led to the development of the Louisiana Early Organic Contaminant Detection System (EWOCDS) on the Mississippi River (Romanowsky, 1985) and the R-TOT (river time-of-travel) model (Waldon, 1998).

The use of existing general-purpose advection-dispersion models was rejected by the model developers because of unacceptable prediction errors, primarily in their inability to represent the skewed concentration profiles common on the Mississippi River. Alternatively, a modified plug flow (MPF) model was developed to represent time-of-travel in the river. The MPF model calculates travel time for the spill peak and leading and trailing edges as though each were transported by plug flow at its own characteristic velocity. The model represents the dispersive transport within an active area of the channel and a slack water area (or dead zone). The velocity of the leading edge is associated with the velocity of water and solute within the active flow area. Travel times associated with the peak and the trailing edge take into account the interchange between the active area and the dead zone. Empirical equations were developed relating velocities to stream discharge. Parameters for these relationships were established through the use of rhodamine-WT dye tracer studies on the river. Results of the dye studies were also used to develop a characteristic shape of the concentration curve so that peak concentrations are calculated as a function of contaminant mass, stream discharge, and the leading and trailing edge travel times.

The R-TOT model has been in use on the Mississippi River for several years and has recently been revised using Microsoft Excel worksheets and macros in order to improve the user interface and simplify support. In addition to its use as a near real-time model, the model has also been applied in spill contingency planning and in identifying the possible stream reaches where spills of unknown origin may have originated.

REMM Model. The Riverine Emergency Management Model (REMM) is a computer program with associated river, chemical, and geographic information data files that computes the time of travel, and optionally, the fate of a chemical spill on a river system for various flow conditions (U.S. Army Corps of Engineers 1997a, 1997b). Its primary purpose is to give emergency planners the capability to make a reasonable determination of the travel time and the fate of chemical spills at locations downstream from a given location. The St. Paul District of the US Army Corps of Engineers developed this program for the State of Minnesota. To date, 330 mi of the Mississippi River main stem from Grand Rapids to Minneapolis, Minnesota have been modeled.

REMM is an interactive program used to compute travel times and chemical parameters at locations along a river system. Program design philosophy is to keep things simple, reliable, and easy to use. The program does no backwater analysis, flow routing, or unsteady flow computations. Rather, river stage-discharge-velocity relationships are coded at selected river locations. The following assumptions are implicit in the analytical techniques that are a part of the travel time algorithms used in the program:

- Flow is steady and gradually varied.
- Flow is one-dimensional. Velocity components in directions other than the direction of flow are not accounted for.
- The travel time between any two points can be expressed by the formula

$$\text{Travel time} = \text{Distance} / \text{Velocity}$$

Water quality is modeled through transport and fate routines. Instantaneous and complete mixing of the pollutant in the water column is assumed. Transport is represented by the classical advection-dispersion model. Fate routines represent volatilization, hydrolysis, sorption, and evaporation. A library of chemical properties is supplied with the model and used in the fate routines (Mills, 1985). The evaporation algorithm used for modeling petroleum products is based on Shen (1991) with parameter values from Stolzenbach (1977), and Fingas and Syndor (1980).

RiverSpill. RiverSpill is a Geographic Information System (GIS)-based software tool with integrated database capability that is used to track and model the flow and concentration of contaminants in the surface source waters of a public water supply (Samuels et al., 2002). The system contains a stream flow and transport model and uses the following national (U.S.) databases:

- U.S. EPA Reach File database
- U.S. EPA Public Water Supplies database
- USGS Real-Time Stream Flow Gauge database

The integrated system calculates, locates, and maps the population at risk from the introduction of contaminants to the public water supply. The transport model calculates the time of travel based on real time stream flow measurements, decay, and dispersion of a contaminant introduced into surface water. Additionally, the model can perform upstream tracing to identify potential sources and area of contamination.

Oil Spill Models. Though most earlier oil spill models addressed the ocean environment, in the past decade there have been increased efforts in modeling inland oil spills on rivers. Yapa and Shen (1994) have provided a comprehensive review of models that are specifically designed for river oil spills. They identify the following hydraulic, physical, chemical, and environmental processes that affect the fate and transport of an oil spill in a river:

- Advection due to wind and current
- Spreading of surface oil due to turbulent diffusion and mechanical spreading
- Emulsification and spreading of oil over the depth of the river
- Changes in mass and physical/chemical properties due to weathering
- Interaction of oil with the river shore lines
- Attachment of oil droplets to suspended particulates
- Photochemical reactions and microbial biodegradation

Riverine oil spill models are summarized in Table 11.5.

TABLE 11.5 Summary of River Oil Spill Models

	RiverSpill	WPMB	NRDAM	ROSS	ROSS2	ROSS3
Flow computation	Empirical one-dimensional (secondary currents computed at bends)	Time dependent two-dimensional depth averaged	Mean flow data (supplied)	Time dependent quasi-two-dimensional one-dimensional network coupled with stream tube	Time dependent quasi-two-dimensional one-dimensional network coupled with stream tube	Similar to ROSS2 but on non rectangular bathymetry based
Transport computations	(Surface layer) one-dimensional, provisions to compute some two-dimensional effects	two-dimensional surface layer only	three-dimensional surface layer & suspension, vertical uniform velocity	two-dimensional surface layer only	two-dimensional surface layer & suspension over the depth	Same as ROSS2 but grid is non-rectangular & size varies with water level
Advection and turbulent diffusion	Yes	Yes	Yes	Yes	Yes	Yes
Mechanical spreading	Yes but always circular	Yes	Yes	Yes	Yes	Yes
Variable shoreline deposition	No	Yes	Yes	Yes	Yes	Yes
Evaporation	No	No	Yes	Yes	Yes	Yes
Dissolution	No	No	Yes	Yes	Yes	Yes
Vertical mixing	No	No	Yes	Yes	Yes	Yes
GIS data	No	No	Yes	No	No	No
Simulation in the presence of ice	No	No	Yes	Yes	Yes	Yes
Biological effects	No	No	Yes	No	No	No
Grid sizes used	N/A	1000 m	100's to 1000's	500 ft/1000 ft	200 ft/500 ft/1000 ft	Variable
Reference	Tsahalis (1979)	Fingas and Sydor (1980)	Reed et al. (1991)	Shen and Yapa (1988)	Shen et al. (1993)	Yapa et al. (1994)
Application	Lower Mississippi River	Montreal harbor area	St. Marys, St. Clair, Detroit, Niagara, St. Lawrence	St. Marys, St. Clair, Detroit	Ohio, Monongahela, Allegheny, St. Lawrence	St. Clair River - Lake system

Source: Based on Yapa and Shen (1994).

11.5 EARLY WARNING SYSTEMS: CASE STUDIES

11.5.1 Advanced Early Warning Systems Around the World

Early warning systems may be classified in terms of their institutional structure, degree of sophistication, and physical extent. Institutional structures can be classified as regional or local. Regional systems can involve several water utilities and water intakes, multiple monitoring stations, and some form of centralized communications system. Localized systems typically are operated by a single entity and may contain a few or only a single monitoring station. The degree of sophistication generally depends on the number of monitoring instruments and the degree of automation involved.

Advanced early warning systems are defined here as those systems that are extensive in size and/or scope, generally employ significant on-line state-of-the-art monitoring equipment, and utilize monitoring, modeling, and communications in an integrated system for providing warning of contaminants in the source water. Based on this definition, there are only a relatively few advanced early warning systems around the world. The characteristics for several advanced early warning systems around the world are summarized in Table 11.6. Additional details on selected systems are described below.

11.5.2 ORSANCO Ohio River Early Warning System

The Ohio River Valley Water Sanitation Commission (ORSANCO) operates an early warning system serving the Ohio River mainstem and lower reaches of major tributaries (Grayman et al., 2000). Primary detection of spills and contaminants is provided by an organics detection system (ODS) composed of 15 gas chromatographs and reports to the U.S. Coast Guard's National Response Center. The early warning system was developed in response to major spill events in 1976 (carbon tetrachloride) and 1988 (oil). ORSANCO serves as the primary coordinating agency during events performing mathematical modeling of the movement of the spill in the river, collecting field data, and disseminating information to downstream water utilities.

11.5.3 Rhine River

The River Rhine and its tributaries flow through nine countries, the largest of which is Germany. Roughly 50 million people live in the catchment area and use it for water supply and navigation. One of the largest accidents on the river happened in 1986 when burning warehouses filled with pesticides were doused with water that ran into the Rhine. Thousands of fish died. This accident, plus the discharge from industry and cities placed a heavy burden on the river. There are now roughly 30 monitoring stations in the basin and probably the largest number of biomonitors using fish, *daphnia*, mussels, algae, and bacteria (Wilken et al., 2000).

11.5.4 River Dee Early Warning System

The River Dee is a major water source in northern Wales and England. As a result of a phenol spill in 1984, an early warning system composed of three water-quality-monitoring stations and gauging stations was established (Grayman et al., 2001). The stations are run by Hyder Laboratories and Sciences and financed by the water companies that draw water from the river. The stations operate unattended (except for routine maintenance several times per week) and

TABLE 11.6 Summary of Advanced Early Warning Systems Around the World

River	Country	Administration	Monitoring program	Comments	Websites
Ohio River	U.S.	ORSANCO	Organics Detection System (15 GCs)	Federal-state commission working with water utilities	www.orsanco.org
Mississippi River	U.S.	Louisiana DEQ	8 GCs for organics detection	Cooperative effort between the State, water utilities, and industries	www.deq.state.la.us/surveillance/ewocds/index.htm
Rhine River	Germany, Holland, Switzerland	International Commission for the Protection of the Rhine	9 international stations +20 national monitoring stations	Multinational early warning system. Extensive use of biomonitors	www.iksr.org
River Trent	U.K.	Severn Trent Water	One station at intake	Provides real time warnings and historical database	
River Dee	U.K.	Hyder Lab. and Sciences	3 stations	Cooperative effort among 3 water companies & government	
River Tyne	U.K.	Northumbrian Water Group	2 stations	Wide range of advanced monitors	
Llobregat River	Spain	Grupas Aguas de Barcelona	10 stations	Extensive network of automated monitors	

River	Location	Organization	Monitoring	Comments	Website
River Seine	France	SEDIF	Automatic monitoring stations & samplers serving 3 plants	Combines sophisticated treatment, monitors and early warning system	
North Saskatchewan River	Canada	EPCOR	2 stations located at intakes	Includes on-line monitors for chemical dosing decisions	
St. Clair River	Canada	ORTECH Environmental Inc.	1 monitoring station	Effective system in industrialized area since 1987	
Yodo River	Japan	Yodo River Water Quality Consultative Committee	Monitors at intakes	Cooperative effort among 10 water companies Unique monitoring systems	
River Han (and other rivers)	Korea	National Institute of Environmental Research	20 stations on four rivers	Combination of standard & advanced instruments and biomonitors	www.nier.go.kr
Danube River	Parts of 17 European countries	International Commission for the Protection of the Danube River	Mostly conventional monitors	Primarily a network for sharing spill information 11 nation commission	www.icpdr.org
Moselle River	France & Germany	International Commission for the Protection of the Mosel and the Saar	Several advanced monitoring stations with chemical and biomonitors	Primarily agricultural area with good water quality	www.iksms-cipms.org
Elbe River	Germany and Czech Republic	International Commission for the Protection of the Elbe	17 monitoring stations	Significant improvement in water quality since the reunification of Germany	www.arge-elbe.de www.bafg.de/html/ikse/ikse.htm

information is provided through telephone lines. The monitors vary among the stations based on the likely vulnerability at each location. On-line monitors include formaldehyde, ammonia, phenol, VOCs, herbicides, and standard water quality parameters.

11.5.5 Paris Early Warning System

The Isle de France water board (SEDIF) supplies water to two-thirds of the Paris area for a total population of about 4 million people. More than 95 percent of the raw water originates from surface water sources (the Seine River and tributaries) and is processed in three large treatment plants. In the past twenty years, there have been more than 400 incidents of significant pollution of the rivers. About 40 percent of the incidents required a change in the treatment processes, about 7 percent led to a reduced water production, and about 3 percent required a shutdown of the treatment plant. In response, early warning systems have been installed on the rivers (Grimaud, 1990). The early warning system is composed of a series of monitoring stations composed of automatic analyzers and self-emptying samplers. Each of the stations analyze for some or all of the following parameters—TOC, hydrocarbons, ammonia, nitrate, cyanide, heavy metals, toxicity, and the traditional parameters DO, temperature, pH, and conductivity. The self-emptying sampler systems are designed to temporarily store water samples so that in case of an incident the samples can be analyzed and the characteristics of the occurrence can be determined. In summary, the water treatment plants in the Paris area combine sophisticated treatment trains and comprehensive early warning systems composed of monitors, automatic samplers, forecast models, and inventories of toxic chemicals stored upstream of their intakes.

11.5.6 Yodo River Early Warning System, Japan

Water utilities in the vicinity of Osaka are members of the Yodo River Water Quality Consultative Committee. The three major member water utilities of the committee coordinate the monitoring and early warning activities in this river. Immediate notification is provided to water utilities when a contaminant is found. Oil spills form the majority of the reported contaminant events though others include pesticides, phenol and a wide range of chemicals. Monitoring facilities include several unique biomonitoring stations, advanced TOC and UV monitors, gas chromatographs, standard water quality monitors, and odor detector units used by treatment plant operators (Grayman et al., 2001).

11.5.7 River Trent (UK)

Severn Trent Water, one of the largest water companies in the United Kingdom, traditionally utilizes the Severn River as its primary water source. However, droughts in the 1970s and then the 1990s spurred action resulting in the use of the lower quality Trent River water as an alternative source. The River Trent drains the domestic and industrial areas of Birmingham, Nottingham, and Leicester. It contains the treated sewage from 4 million people plus industrial discharges and urban, highway, and agricultural runoff.

In order to monitor the water quality in the river, an early warning system utilizing an advanced monitoring station was installed near the intake to a new water treatment plant (Drage et al.,1998). The monitoring station generates alarms based on water quality and telemeters this data to Severn Trent facilities. Two types of alarms are issued—a high alarm results in the addition of powdered activated carbon (PAC) to the water and a high-high alarm results in immediately shutting the water intake gates. The monitoring station is operated by

Severn Trent Laboratories (STL) that is located 40 mi (64 km) away in Coventry. STL remotely monitors the results via personal computer and modem and visits the station three times per week to check the instruments and to perform routine maintenance. The primary components of the on-line monitoring station include liquid chromatography (for detecting organic substances); purge & trap gas chromatograph (volatile organics); conventional multiple parameter monitor for pH, D.O., conductivity, temperature, turbidity; ion selective electrode (ammonia and nitrate); TOC monitor, a nitrification inhibition toxicity monitor; ion chromatography (for bromide); and a surface oil detector using infra red reflectance. An auto sampler automatically takes samples for later analysis if needed. Technicians visiting the station use a "smell bell" to help identify any abnormalities (musty, earthy, fishy, etc.) in the source water.

11.5.8 St. Clair River (Canada)

The St. Clair River is a 64-km (40 mi)-long river that forms the border between the United States and Canada and connects Lake Huron and Lake St. Clair. On the Canadian side there is a large petrochemical that is one of the most intensive and highly integrated industrial complexes in Canada. In the 1950s, there were quite a number of spills and discharges of petrochemicals to the river, and it was decided that a continuous water quality monitoring station was necessary (Kuley and Frais, 1997). The resulting station is owned by the Lambton Industrial Society (LIS), an environmental co-operative created in 1952 to address both water and air pollution problems in the area with operating funds for LIS contributed by the area industries. The actual operation of the station is contracted out to ORTECH Environmental Inc., a Canadian environmental research organization. In the mid-1980s, a permanent, remote, unattended monitoring system that provides data on 20 volatile organic compounds (VOCs) was installed. This station has been on line continuously since May 1, 1987. Hourly water quality reports are available to LIS members in real time by modem connection to a data acquisition and central computing system.

11.6 RISK-BASED ANALYSIS OF EARLY WARNING SYSTEMS

11.6.1 Introduction

Contamination events in rivers are best described as highly stochastic. It is not known with any significant level of certainty when the next spill will occur, where it will occur, and what will be the characteristics of that spill (quantity, chemical type, duration). There is even uncertainty in what the hydrologic state of the river will be at the time of the spill, thus affecting the amount of dilution and velocity.

In most situations, spills are considered to be relatively rare occurrences. However, the consequences of a spill or other contaminant event (such as non-point sources or combined sewer overflows), especially if the contamination is not identified, can be very high. Resources can be spent in various ways to lessen the potential impacts associated with spills. Specifically, actions can be taken to (1) decrease the probability of spills occurring; (2) increase the probability of identifying a spill if it occurs; and (3) reducing the impacts of spills through treatment, increased raw or treated water storage, intake management, or development of alternative water sources. The problem is further complicated due to the possibility of false positive identification of spills through monitoring equipment, and the existence of other methods of determining the occurrence of spills, such as through self-reporting.

Therefore, in order to effectively evaluate, operate, or design an early warning system, it is necessary to consider the stochastic nature of the problem. Waldon et al. (1989) demonstrated the use of simple probabilistic techniques in the analysis and design of the early warning system for the Lower Mississippi River. In that study, the probability of the detection of an instantaneous spill as it passes one or more monitors was analyzed and based on these probabilities alternative monitor locations and sampling frequencies were considered. It was found that a daily sampling frequency was inadequate for their river situation, and as a result, a more frequent sampling rate was implemented. They concluded that a simple probabilistic approach was proven to be a useful tool and that future studies might include a Monte Carlo simulation in order to provide a more detailed analysis. Grayman and Males (2002) carried this further in the development and application of a Monte Carlo simulation based tool called Spill Risk for evaluating alternative early warning systems in a riverine setting.

11.6.2 Spill Risk Model

The Spill Risk model is a general purpose, data-driven model, useful for strategic planning on early warning systems. It is implemented as an event-driven Monte Carlo simulation model. Spill events are randomly generated and are routed through a simplified transport model. Monitors determine the presence of the spill, and notify the treatment plant, which can respond with intake closures, or enhanced treatment. Different response policies are available for action at the treatment plant, depending upon the distance to the spill, and the desire to act before or after confirmation of the spill. The overall metric for system behavior is the population exposure over time to treated water with concentrations above maximum contaminant levels (MCL) or other user-specified acceptable levels. The model also allows for specification of values that relate to strengthened institutions, such as probability of self- or public reporting of spills.

Monte Carlo Simulation. Monte Carlo simulation is a well-known technique for analyzing complex physical systems where probabilistic behavior is important. This technique is widely used in modeling probabilistic systems in the water resources area and many scientific fields. Events are represented as probabilistic occurrences (i.e., probability distribution of streamflows, probabilities of spills of different substances, magnitudes and duration, etc.). The relationships of these events (e.g., how a river responds to a spill) are embodied in the model, which is then run many times, with varying inputs based on the probabilities of the events. The results are recorded for each simulation run and are summarized statistically. In this fashion, the interacting probabilities result in statistics for the total system. A wide range of situations can be examined along with alternative designs and operations. The result is both an expected value and a distribution of results.

Model Formulation. In the model, the river is represented as a single one-dimensional reach, with seasonal hydrology. Multiple possible dischargers (referred to as spill generators) exist along the reach. Spill generators can be at fixed locations, representing known plant sites, and nonfixed (probabilistic) locations (e.g., barges or other vessels traveling along the waterway, or spills on adjacent highways). A simplified advective-dispersion transport model routes contaminants from point of origin down the reach.

Monitors are located at known locations and are characterized by sampling frequency and detection limits for various constituents. Different monitors can sample for different constituents, at different frequencies. One or more treatment plants are located along the reach. Removal rates are specified for each treatment plant, by constituent, under normal and "enhanced" treatment. Treatment plants are also characterized by the length of time that their intakes can be closed (to let a spill pass by).

Population exposure to treated water with concentrations above the maximum contaminant level (MCL) or some other designated critical concentration is taken as the metric for overall system behavior and is calculated by the following equation where C_t is the concentration of the contaminant at the treatment plant at time t:

$$\text{Population exposure} = \text{population} \times \sum_{\text{Leading Edge Time}}^{\text{Trailing Edge Time}} \Delta t (C_t - \text{MCL}) / \text{MCL} \quad (11.1)$$

The calculations are performed for each year of the simulation, and for the user-defined number of iterations. Statistics are generated for each iteration and for the scenario as a whole. Results are stored in a database for comparison between scenarios, and text format files are generated. The Spill Risk model includes a large number of parameters. Some of the parameters are readily derivable from historical records. Other parameters may require more subjective decisions or consensus among a panel of local experts. For those parameters that have the greatest inherent uncertainty, sensitivity analyses can be used by varying the parameters over a reasonable range.

Application to the Ohio River. The model was applied to a 322-km (200-mi) industrialized stretch of the Ohio River to study the effectiveness of the present early warning system on three water intakes. The Ohio River Valley Water Sanitation Commission (ORSANCO) operates an early warning system composed of 15 gas chromatographs on the Ohio River mainstem and tributaries. The effectiveness of the early warning system was found to vary significantly among the three intakes based on the following factors—the proximity of the intake to upstream spill generators; the location of monitors relative to the intakes; the sampling frequency and minimum detection limit for the instruments; the length of time that the water intakes could be closed; and the type of normal treatment routinely applied at the water treatment plant. An effectiveness index, defined as [Impact reduction/Impacts associated with no actions], was used to rate the effectiveness of the early warning systems. Values for impact reduction and impacts associated with no actions were calculated by the model for each intake based on a simulation of 3000 iterations of 1-year duration each. Based on the application of the Spill Risk model, the effectiveness was found to vary between 0.94 and 0.51 for the three intakes.

11.7 EARLY WARNING SYSTEM DESIGN AND OPERATION

Early warning systems should be viewed as an integral part of the operation of a water system. Therefore, an overall context for decision making relative to early warning systems may be viewed as one of designing and operating the overall system (including monitors, raw water intakes, treatment, and storage) in order to minimize the risks associated with degraded drinking water quality, under various cost and technology constraints.

A two-step risk-based process is outlined for making decisions related to early warning systems as illustrated in Fig. 11.4. This process recognizes the fact that there will always be some level of risk that an unacceptable level of contamination will be present in the water delivered to customers. In the proposed process, the level of risk is explicitly determined and is evaluated in terms of its acceptability. If the risk level is not acceptable then alternatives should be evaluated in terms of their cost and ability to lower risk levels.

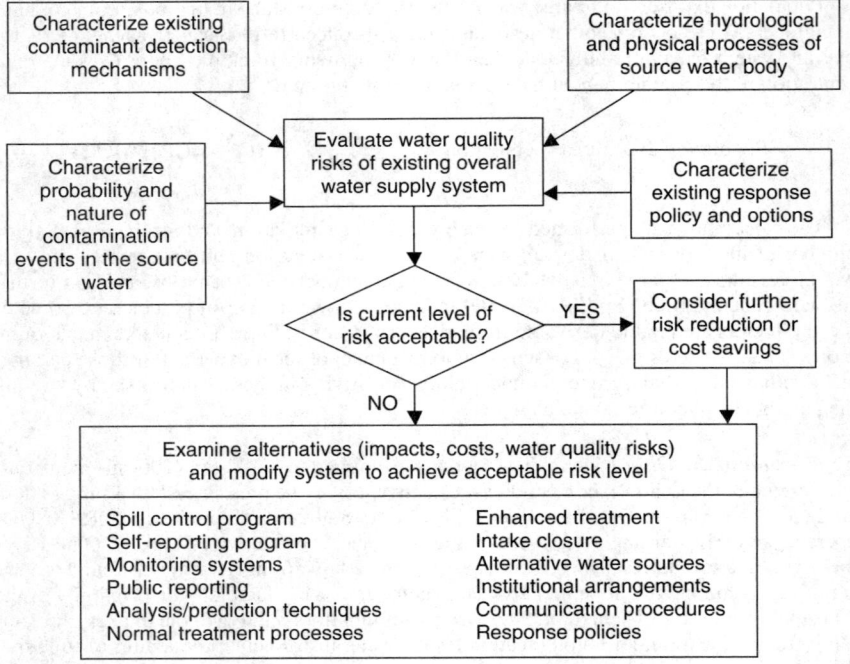

FIGURE 11.4 Risk-based decision-making process for early warning systems.

ACKNOWLEDGMENTS

The information in this chapter draw heavily on the work performed by the authors of this chapter on a recently completed project of the American Water Works Association Research Foundation (AwwaRF) titled, "Design of Early Warning and Predictive Source-Water Monitoring Systems." The Foundation financial support and the foresight in sponsoring this important research topic is gratefully acknowledged. Comments and views presented may not necessarily reflect the views of AwwaRF officers, directors, affiliates or agents.

REFERENCES

Ambrose, R. B., T. O. Barnwell, S. C. McCutcheon, and J. R. Williams, "Computer Models for Water-Quality Analysis," in L. W. Mays (ed.), *Water Resources Handbook*. McGraw-Hill, New York, 1996.

AwwaRF, and CRS PROAQUA, *Online Monitoring for Drinking Water Utilities*, E. Hargesheimer, O. Conio, and J. Popovicova, (eds.), in AWWA Research Foundation and American Water Works Association, Denver, CO, 425 pp., 2002.

APHA-AWWA-WEF (American Public Health Association, American Water Works Association, and the Water Environment Federation), *Standard Methods for the Examination of Water and Wastewater*, 20th ed., American Public Health Association, Washington DC, 1998.

AWWA (American Water Works Association), *Water Quality & Treatment A Handbook of Community Water Supplies*, 5th ed., McGraw-Hill, New York, 1999.

Bronson, R. T., J. S. Bradshaw, P. B. Savage, S. Fuangswasdi, S. Chul Lee, K. E. Krakowiak, and R. M. Izatt, "Bis-8-hydroxyquinoline-Armed Diazatrithia-15-crown-5 and Diazatrithia-16-crown-5 Ligand Possible Fluorophoric Metal Ion Sensors," *J. Org. Chem.* 66(14): 4752–4758, 2001.

Brosnan, T. M. (ed.), *Early Warning Monitoring to Detect Hazardous Events in Water Supplies—An ILSI Risk Science Institute Workshop Report*. International Life Sciences Institute, Washington DC, 37 pp., 1999, available at: http://www.ilsi.org/file/EWM.pdf

Deininger, R. A. "The survival of Father Rhine." *Jour. AWWA*, 79(7): 78–83, 1987.

DeMarco, J., D. H. Metz, D. J. Hartman, R. C. Pohlman, C. A. Shrive, and P. M. Hall, "Coping With Accidental Spills in the Ohio River Watershed," in *Proceedings of the AWWA Water Quality Technology Conference*, Denver, CO, American Water Works Association, 1994.

DeVries, J. J., and T. V. Hromadka, "Computer Models for Surface Water," in D.R. Maidment (ed.), *Handbook of Hydrology*, McGraw-Hill, New York, 1993.

Dippenaar, A. J., et al., "State of the Art Regarding On-Line Control and Optimisation of Water Systems (International Report)," *Water Supply (Rev. J. IWA)* 18(1/2): 245–289, 2000.

Drage, B. E., J. E. Upton, and M. Purvis, "On-Line Monitoring of Micropollutants in the River Trent (U.K.) with Respect to Drinking Water Abstraction," *Water Sci. Technol.* 38(11): 123–130, 1998.

Fingas, M., and M. Sydor, *Development of an Oil Spill Model for the St. Lawrence River*, Environment Canada Technical Bulletin No. 116. Inland Waters Directorate, Water Planning and Management Branch, Ottawa, Canada, 1980.

Foran, J. A., and T. M. Brosnan, "Early Warning Systems for Hazardous Biological Agents in Potable Water," *Environ. Health Perspec.* 108(10): 993–995, 2000.

Frey, M., L. Sullivan, and E. Lomaquahu, "Practical Application of Online Instruments," *Proceedings of the AWWA Annual Conference and Exposition*, Washington DC, AWWA, Denver, CO, 2000.

Grayman, W. M., S. R. Kshirsagar, and R. M. Males, *A Geographic Information System for the Ohio River Basin*, U.S. EPA, Cincinnati, OH, 1993.

Grayman, W. M., R. A. Deininger, and R. M. Males, *Design of Early Warning and Predictive Source-Water Monitoring Systems*, AWWA Research Foundation, Denver, CO, 292 pp., 2001.

Grayman, W. M., R. A. Deininger, and R. M. Males, "River Basin Early Warning Systems for Source Water Contamination," *Proceedings of the IWA 9th International Conference on Watershed and River Basin Planning*, IWA, London, U.K, 2002.

Grayman, W. M, and R. M. Males, "Risk-Based Modeling of Early Warning Systems for Pollution Accidents," *Water Sci. Technol.* 46(3): 41–49, 2002.

Grayman, W. M., A. H. Vicory, and R. M. Males, "Early Warning System for Chemical Spills on the Ohio River," in R. A. Deininger, P. Literathy, and J. Bartram (eds.), *Security of Public Water Supplies*, NATO Science Series, 2. Environment, vol. 66, Kluwer Academic Publishers, Dordrecht, 2000.

Grimaud, A., T. Vandevelde, and J. P. Morvan, "Automatic Stations for the Monitoring of Pollutants in Rivers," in *Proceedings of the AWWA Annual Conference*, AWWA, Denver, CO, 1990.

Gullick, R. W., *Monitoring Systems for Early Warning of Source Water Contamination*, American Water Works Service Company, Voorhees, NJ, 219 pp., 2001.

Gullick, R. W., "Developing Interjurisdictional Early Warning Monitoring Networks in the U.S.," in *Proceedings of the AWWA Source Water Protection Symposium*, Albuquerque, NM. AWWA, Denver, CO, 2003.

Gullick, R. W., W. M. Grayman, R. A. Deininger, and R. M. Males, "Design of Early Warning Monitoring Systems for Source Waters," *J. AWWA*, 95(11): 58–72, 2003.

He, L.-M., L. L. Kear-Padilla, S. H. Lieberman, and J. M. Andrews, "New Generation of Online Oil-in-Water Monitor," Preprints of Extended Abstracts (*Division of Environmental Chemistry*), 221st ACS National Meeting, San Diego, CA, 41(1): 372–376, 2001.

Huber, W. C. "Contaminant Transport in Surface Water," in D. R. Maidment (ed.), *Handbook of Hydrology*, McGraw-Hill, New York, 1993.

Jobson, H. E., and D. H. Schoellhamer, *Users Manual for a Branched Lagrangian Transport Model*, Water-Resources Investigations Report 87-4163, U.S. Geological Survey, Reston, Virginia, 1987.

Johnson, B. H., *Development of a Numerical Modeling Capability for the Computation of Unsteady Flow on the Ohio River and its Major Tributaries*, Waterways Experiment Station, U.S. Army Corps of Engineers, Vicksburg, MS, 1982.

Kuley, E. J., and W. Frais, "Monitoring of Volatile Organic Compounds in the St. Clair River by the Lambton Industrial Society," in *Proceedings of the AWMA 1997 Annual Meeting*, Toronto, Ontario, 1997.

LAWA (Working Group of the Federal States on Water Problems), "Recommendation on the Deployment of Continuous Biomonitors for the Monitoring of Surface Waters," Compiled by the LAWA 'Biomonitoring' Committee. Translated from the original Oct. 1995 German text, updated in July 1998, 1998.

LeChevallier, M. W., W. D. Norton, M. Abbaszadegan, and T. M. Atherholt, *Variation in Giardia and Cryptosporidium Levels in the Delaware River*, American Water Works Service Company, Voorhees, NJ, 124 pp., 1998.

Lee, J. Y., and R. A. Deininger, "A Rapid Method for Detecting Bacteria in Drinking Water," *J. Rapid Meth. Automation Microbiol.* 7: 135–145, 1999.

Lydy, M. J., D. S. Carter, and C. G. Crawford, "Comparison of Gas Chromatography/Mass Spectrometry and Immunoassay Techniques on Concentrations of Atrazine in Storm Runoff," *Arch. Environ. Contam. Toxicol.* 31: 378–385, 1996.

McCutcheon, S. C., J. L. Martin, and T. O. Barnwell, "Water Quality," in D. R. Maidment (ed.), *Handbook of Hydrology*, McGraw Hill, New York, 1993.

Mills, W. B., *Water Quality Assessment: A Screening Procedure for Toxic and Conventional Pollutants in Surface and Groundwater*, Parts 1 and 2, PB86-122496 and PB86-122504, U.S. EPA, Athens, Georgia, 1985.

Morrow, J. H., B. N. White, M. Chimiente, and S. Hubler, "A Bio-Optical Approach to Reservoir Monitoring in Los Angeles, California," *Issues Adv. Limnol.* 55: 179–191, 2000.

Panguluri, S., R. C. Haught, M. C. Meckes, and M. Dosani, "Remote Water Quality Monitoring of Drinking Water Treatment Systems," in *Proceedings of the 1999 AWWA Water Quality Technology Conference*, Tampa, FL, AWWA, Denver, CO, 1999.

Penders, E. J. M., and P. G. Stoks, "Biological Early Warning Systems in Drinking Water Protection," Poster presentation at the ILSI Workshop on Early Warning Monitoring to Detect Hazardous Events in Water Supplies, Reston, Virginia (May 17–18, 1999), 1999.

Pollack, A. J., A. S. C. Chen, R. C. Haught, and J. A. Goodrich, *Options for Remote Monitoring and Control of Small Drinking Water Facilities*, Battelle Press, Columbus, OH, 301 pp., 1999.

Quist, G. M., "Water Quality Systems to Offer Instantaneous Microbial Detection," *Water Cond. Purific.* 41(3): 68–73, 1999.

Reed, M., D. French, and S. Feng, "Natural Resource Damage Assessment Model," in *Proceedings of the ASCE National Conference on Hydraulic Engineering*, ASCE, New York, 1991.

Reinhard, M., and J. Debreaux, "New and Emerging Analytical Techniques for Detecting Organic Contaminants in Drinking Water," in *Identifying Future Drinking Water Contaminants*, National Research Council, National Academy Press, Washington DC, pp. 120–134, 1999.

Rogers, K. R., and C. L. Gerlach, "Update on Environmental Biosensors," *Environ. Sci. Technol.* 33(23): 500A–506A, 1999.

Romanowsky, P., *Progress in the Development of Louisiana's Lower Mississippi River Water Quality Management Program*, Louisiana DEQ, Baton Rouge, 1985.

Rose, J. B., and D. J. Grimes, *Reevaluation of Microbial Water Quality: Powerful New Tools for Detection and Risk Assessment*, American Academy of Microbiology, Washington, DC, 18 pp., 2001.

Samuels, W. B., R. Bahadur, D. E. Amstutz, J. Pickus, and W. Grayman, "RiverSpill: A GIS-Based Real Time Transport Model for Source Water Protection," in *Proceedings of the Watershed 2000 Specialty Conference*, WEF, Alexandria, Virginia 2002.

Shen, H. T., *A Mathematical Model for Oil Slick Transport and Mixing in Rivers*, Department of Civil and Environmental Engineering, Report No. 91-1, Clarkson University, Potsdam, NY, 1991.

Shen, H. T., and P. D. Yapa, "Oil Slick Transport in Rivers," *J. HY, ASCE*, 114(5): 529–543, 1998.

Shen, H. T., P. D. Yapa, D. S. Wang, and X. Q. Yang, *A Mathematical Model for Simulating Fate and Transport of Oil Spills in Rivers (ROSS2)*, CRREL Special Rep. 93-21. U.S. Army Corps of Engineers, Hanover, NH, 1993.

Spreafico, M., and A. van Mazijk, *Alarm Model Rhine: A Model for the Operational Prediction of the Transport of Pollutants in the Rhine River*. IRC/CHR Committee of Experts, Report No. I-12 of the CHR, Lelystad (in German), 1993.

Stoks, P. G., "From River Water to Drinking Water: On-line Systems in Water Quality Control," in *Proceedings of the 1998 IWSA Symposium on On-line Monitoring and Control of Water Supply*, Amsterdam, 1998.

Stolzenbach, K. D., *A Review and Evaluation of Basic Techniques for Predicting the Behavior of Surface Oil Slicks*. Ralph M. Parsons Laboratory for Water Resources and Hydrodynamics. Report No. 222. National Oceanic and Atmospheric Administration, MIT Cambridge, MA, 1997.

Tsahalis, D. T., "Contingency Planning for Oil Spills: Riverspill—A River Simulation Model," in *Proceedings of the Oil Spill Conference*, U.S. Coast Guard, EPA and API, Washington DC, 1979.

U.S. Army Corps of Engineers, *River Emergency Management Model. User Manual Version 3.0.*, 1997a, available at: http://www.mvp-wc.usace.army.mil/org/remm/userman.pdf.

U.S. Army Corps of Engineers, *River Emergency Management Model. Technical Manual Version 3.0.*, 1997b, available at: http://www.mvp-wc.usace.army.mil/org/remm/techman.pdf.

Van Mazijk, A., *One-Dimensional Approach of Transport Phenomena of Dissolved Matter in Rivers*. Communications on Hydraulic and Geotechnical Engineering Report No. 96-3, Faculty of Civil Engineering, Delft University of Technology, Netherlands, 1996.

Venter, S. N., "Rapid Microbiological Monitoring Methods: The Status Quo," in *The Blue Pages*, International Water Association, London, UK, 2000, available at: http://www.iawq.org.uk

VonderHaar, S. S., D. Macke, R. Sinha, E. R. Krishnan, and R. C. Haught, "Drinking Water Early Warning Detection and Monitoring Technology Evaluation and Demonstration," in *Proceedings of the 2002 National Water Quality Monitoring Conference*, National Water Quality Monitoring Council, USGS, Reston, VA, 2002.

Waldon, M. G., K. K. O'Hara, and D. E. Everett, *Detection Probabilities and System Analysis of the Mississippi River Early Warning Organic Compound Detection System*, Louisiana Department of Environmental Quality, Baton Rouge, Louisiana, 1989.

Waldon, M. G., "Time-of-Travel in the Lower Mississippi River: Model Development, Calibration, and Application," *Water Environ. Res.* 70(6): 1132–1141, 1998.

Wilken, R. D., T. Knepper, and K. Haberer, "Early Warning Systems on the Rhine and Elbe in Germany," in R. A. Deininger, P. Literathy, and J. Bartram (eds.), *Security of Public Water Supplies*, NATO Science Series, 2. Environment, Vol. 66, Kluwer Academic Publishers, Dordrecht, 2000.

Yapa, P. D., and H. T. Shen, "Modelling River Oil Spills: A Review," *J. Hydraul. Res.* 32(5): 765–782, 1994.

Yapa, P. D., H. T. Shen, and K. Angammana, "Modeling Oil Spills in a River-Lake System," *J. Marine Syst*, 1994(4): 453–471, 1994.

Yates, D. G, D. O. Pitcher, and M. Beal, "Implementing Advanced Early Warning Systems to Safeguard Public Drinking Water," in *Proceedings of the AWWA Annual Conference and Exhibition*, New Orleans, LA, AWWA, Denver, CO, 2002.

CHAPTER 12
SECURITY HARDWARE AND SURVEILLANCE SYSTEMS FOR WATER SUPPLY SYSTEMS

Ron Booth,[*] Andy Bowman,[†] Forrest Gist,[‡]
and James R. Ringold[§]

12.1 INTRUSION DETECTION: PERIMETER, AREA, AND INTERIOR

12.1.1 Introduction

Intrusion detection is a critical element of an overall balanced security program. As the name suggests, the purpose of an intrusion detection system is to detect an intruder approaching a site, facility, or area as early as possible. In many cases, visual observation of an intruder may not be possible or may be unreliable due to darkness, lack of personnel, or visual shielding from landscaping or terrain. An intrusion detection system may provide the first and only indication that someone or something is trying to enter the facility premises.

12.1.2 Issues to Consider

Before beginning the design or evaluation of an intrusion detection system, it is important to have an understanding of some key issues;

- *What is the facility, space or area to be protected?*
  - What is the area or region to be protected?
  - How large is the area?
  - Does the area occupy a level surface?

---
[*]SER Security Programs, CH2M Hill Atlanta, Georgia
[†]SiteSecure, Inc., Sanford, Florida
[‡]Industrial Design and Construction CH2M Hill Portland, Oregon
[§]Protection Group, Inc., Dunedin, Florida

- Is the area enclosed?
- What is the facility, space or area to be protected?
- Is the area outdoors?
- What humidity, temperature conditions, wind conditions exist?
- Are small animals or children living nearby the protected space?

- *Who is the Perceived Threat?*
  - From whom are we protecting the space against?
  - Is the anticipated adversary an outsider (one who does not have permission to access the site), an insider (someone with regular, authorized access to the area) or an outsider collaborating with an insider?
  - What tactics, motivation, skills, knowledge, tools or weapons might the adversary use? (Protecting a facility from a skilled, trained terrorist with knowledge of the facility requires a different tactic than protecting against a high school vandal.)

- *What are known vulnerabilities of the area or space? What are key assets or targets at the space?*
  - What are the known vulnerabilities or soft targets?
  - What key assets exist at the facility? (Treated water, pumps, generators, chemical storage, SCADA workstations, etc.)

- *How will system monitoring take place?*
  - How will alarm signals be transmitted back to a monitoring system?
  - What monitoring system is in place to receive the alarms—a SCADA system or a separate intrusion detection system?
  - Who will monitor the alarms, and is this on a continuous basis, or as alarms come in?

- *What power and communication methods exist?*
  - What electrical power is available for security hardware, if any?
  - Will hardwired systems be used or are wireless communication methods being considered?
  - Will all wiring be protected within conduit?

12.1.3 Alarm Activation Time

For an intrusion detection alarm to be processed and subsequent action taken, the following must occur—The sensor must be activated, an alarm signal must be initiated, the alarm reported back to the control system, and the alarm assessed for further response.

To be effective, all of these steps must occur and the alarm condition occurs before the adversary can complete his task. Refer to Fig. 12.1 depicting a linear beam (photoelectric beam).

As such, it is important that the system detection occurs as early (quickly) as possible after intrusion. The goal is that the intrusion detection system provides early warning of intrusion, requires low maintenance, and has high-performance characteristics.

12.1.4 Performance Characteristics

How do we compare the performance of intrusion detection systems? We look at the following elements; the Probability of Detection, the False Alarm Rate, and the Probability of Defeat.

SECURITY HARDWARE AND SURVEILLANCE SYSTEMS

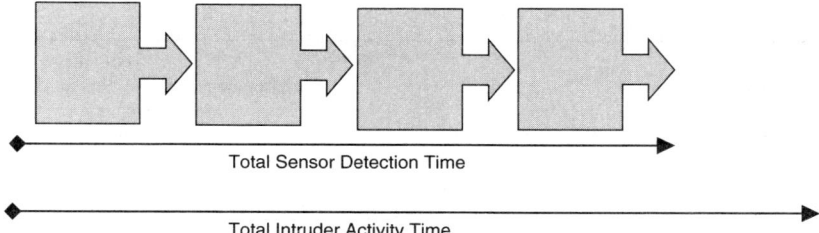

FIGURE 12.1 Linear beam (photoelectric beam).

Probability of Detection.   The probability of detection (Pd) is the probability that the system will detect an intruder. Under perfect conditions, this would be (1.0) or 100 percent detection. However, because no sensor is perfect, this number will be less that 1.0. Factors affecting probability of detection include the intended target, sensor design, installation settings, weather, climate, etc. The higher the sensitivity of the system, the higher the probability of detection. All things considered equal, the higher the probability of detection, the better the sensor.

False Alarm Rate.   The false alarm rate (FAR) is a measure of the frequency of invalid alarms or alarms that are not caused due to intrusion, but instead by other means (rain, animals, wind-blown debris, vibration, etc.) The lower the false alarm rate, the better the sensor.

Vulnerability to Defeat.   The vulnerability to defeat is a measure of the elements that can be used to defeat the sensor. There are two general ways to defeat an intrusion detection system—bypassing and spoofing.

Bypassing intrusion detection systems involves going around the detection area or detection space (jumping over a fence-mounted detection system).

Spoofing intrusion detection systems involve cloaking the intruder in some manner, allowing them to pass through the normal detection zone without generating an alarm (rubbing grease on the lens of a passive infrared detection system, causing the system to be masked).

The more difficult it is to defeat a sensor, the better will it perform.

12.1.5 Design Concepts and Goals

When designing any perimeter intrusion detection system, the goal is to achieve the best possible performance. These concepts apply:

- No gaps in coverage—a continuous line of detection around the perimeter area or interior space should exist.
- Suitability for physical and environmental conditions—the sensor must be appropriate for the area being monitored (temperature, humidity, rain, fog, wind, pollution).
- Layers of protection—a fundamental security concept is that multiple, layered detection systems are much more effective single systems. If one system is bypassed or defeated, the remaining systems are still in place to detect the intruder.

TABLE 12.1 Intrusion Detection Technologies

Sensor technology	Classification	Adverse environment	Potential defeat methods
Buried-line pressure/ seismic sensor	Exterior, buried-line	Hail, frozen ground	Bridging
Buried-line magnetic field sensor	Exterior, buried-line	Lightning, buried power lines	Bridging, Nonmagnetic materials
Buried-line ported coaxial cable	Exterior, buried-line	Runoff, standing water	Bridging, stilts
Buried-line fiber-optic cable	Exterior, buried-line	Hail, frozen ground	Bridging
Electro-mechanical vibration sensing	Exterior, fence-mounted	Ice coating, wind	Bridge, trench
Coaxial strain sensitive cable	Exterior, fence-mounted	Ice coating	Bridge, trench
Fiber-optic strain sensitive cable	Exterior, fence-mounted	Ice coating	Bridge, trench
Taut wire system	Exterior, fence-mounted		Bridge, trench
Active infrared sensor	Exterior, freestanding		Bridge, trench
Passive infrared sensor	Exterior, freestanding	Body temperature outdoor conditions, floodlights	Tunnel
Microwave	Exterior, freestanding	Surface snow	Trench
Dual technology (PIR + microwave)	Exterior, freestanding		
Glass break sensor	Interior	Noisy environments	Muffle breakage
Balanced magnetic door switch	Interior	Loose fitting doors,	Tailgate authorized user
Linear beam (photoelectric)	Interior		Tailgate authorized user
Video motion detection	Exterior, interior	Complex background	

12.1.6 Intrusion Detection Sensor Categories

Refer to Table 12.1, which provides a summary of the intrusion detection technologies, described within this section, including typical defeat methods.

Exterior or Interior.   Intrusion detection sensors may be classified as exterior or interior sensors. Exterior intrusion sensors are those that sense an intrusion crossing an outdoor perimeter boundary or area. Exterior sensors will have a lower probability of detection and higher false alarm rate due to the less predictable nature of their external environment. Interior intrusion sensors sense intrusion within an interior area, space or opening, or from a person moving or touching an object (doors, hatches, etc.).

Passive or Active.   Is the sensor a passive device, receiving energy (heat, vibration, and sound) from the environment? Or is the sensor actively transmitting energy (microwaves, photoelectric beams, infrared beams) into the environment and analyzing the received signal? Passive devices are more cost effective than active devices and require less power to operate.

Covert or Visible.   Is the sensor hidden from view (covertly installed) or is it visible (installed within view)?

Volumetric or Line/Boundary Detection. Is the sensor designed to monitor a volume of space, and will an alarm be generated if an intruder enters the space? Or is the sensor monitoring a line or boundary, and an alarm is generated if an intruder crosses the boundary?

Line of Sight or Terrain Following. Is the sensor detecting within a straight line of sight, or does the sensor follow the terrain to adjust for hills, valleys, depressions, etc.?

Exterior Sensor Types. Several types of exterior intrusion detection sensors exist, and they may be classified according to their type, method of use, style, and mode of application. The following exterior systems are most applicable to water system applications:

Buried Line Sensors. Buried line sensors include; pressure/seismic sensors, magnetic field sensors, buried ported coaxial cable, and buried fiber-optic cable sensor systems. Each of these systems relies on sensing the presence of an intruder by means of a buried cable system within the ground.

Pressure/Seismic Sensors. Pressure/seismic sensors use a buried cable system that senses minute vibrations or pressure changes caused when an intruder walks across or crosses the sensing boundary. The system may be tuned electronically to filter out wind-blown debris or small animal movements. (Manufacturer—Aritech)

Magnetic Field Sensors. Magnetic field sensors use a buried cable system that creates a balanced magnetic field at the surface of the monitored ground area. When an intruder passes over the sensing area, the magnetic field is disrupted in proportion to the mass of the intruder. The system can be tuned. (Manufacturers—Sensor Solutions, Honeywell)

Ported-Coaxial Buried Cable System. A buried ported-coax cable system uses a pair of buried coaxial cables, separated by a distance of 6 to 10 ft, which run parallel to the boundary of the area protected. These cables emit an electronic field that is disturbed when an intruder walks along or crosses the sensing area. (Manufacturer—Senstar Stellar)

Fiber-Optic Buried Cable System. Fiber-optic buried cable systems utilize fiber-optic cable buried below the ground surface. The cables are placed in a serpentine fashion parallel to the boundary of the area protected. A light source transmits a beam down the fiber optic cable. The amount of light refracted within the cable is measured and monitored continuously by a control system. When an intruder walks along the area containing the fiber, the soil shifts and the refracted light measurement changes. By adjusting the system, it can be tuned to filter out small movements for such items as small animals, rain, or wind-blown debris. (Manufacturers—Fiber SenSys, Magal)

Fence-Mounted Cabling Sensors. Several types of fence-mounted perimeter intrusion detection systems exist. These include electro-mechanical vibration sensing, coaxial strain sensitive cable, fiber-optic strain sensitive cable, and taut-wire systems. With all fence-mounted systems, it is critical that the fence construction be of high quality, with no loose fabric, flexing or sagging material, and having solid foundations for posts and gates. Otherwise, nuisance alarms may occur.

Electro-Mechanical Vibration Sensing. This system uses a series of electro-mechanical switches connected in series to a cable that is threaded within the fabric of the fence material. When sufficient vibration occurs due to climbing or cutting of the fence, the switch(es) open causing an alarm condition to occur.

Electro-mechanical intrusion detection systems are an economical fence-mounted sensor that is easy to install. They offer a high probability of detection if installed correctly on a properly constructed fence. Some designs do not allow for tuning of the system to accommodate wind vibrations and weather conditions. As with other fence-mounted sensors, tunneling below the fence and jumping across or bridging across the fence without contacting the fence material are possible methods of defeating the system. (Manufacturer—Senstar Stellar)

Coaxial Strain Sensitive Cable. This system uses a coaxial cable woven through the fabric of the fence. The coaxial cable transmits a dielectric field. As the cable moves due to strain on the fence fabric caused by climbing or cutting, the electric field changes are detected within the cable and an alarm condition occurs.

Coaxial strain sensing systems are readily available, and are highly tunable to adjust for field conditions due to weather and climate characteristics. Some coaxial cable systems are susceptible to electro-magnetic interference (EMI) and radio frequency interference (RFI). (Manufacturers—Senstar Stellar, Fiber SenSys, Southwest Microwave)

Fiber-Optic Strain Sensitive Cable. Similar to the Coaxial Strain Sensitive Cable system, the fiber-optic system uses a fiber optic cable (rather than a coaxial cable) woven through the fence fabric. Strain on the fence fabric causes microbending of the fiber cable, which is monitored by the control panel and generates an alarm condition.

Fiber-optic strain sensing systems are relatively newer detection systems but have a strong following. The systems are readily available, and are highly tunable to adjust for field conditions due to weather and climate characteristics. The system is impervious to lightning, electromagnetic interference, radio frequency interference or other electronic signals, and can be used over long distances.

Possible defeat measures include tunneling, jumping or bridging across the fence system. Careful climbing at corner posts may not generate sufficient vibration to generate alarm condition. (Manufacturers—Senstar Stellar, Fiber SenSys)

Taut-Wire Systems. A taut-wire fence detection system consists of a series of microswitches connected to tensioned wire installed on the fenceposts. Inside the microswitch is a center rod suspended within a conductive outer-rod. Upon movement of the taut wire, the microswitch activates, causing an alarm condition.

The taut-wire system is the most costly fence-mounted system available. It has a very high probability of detection and nearly zero false alarm rate. The system is in use at several of the world's highest security installations, due to its superb performance. Regular maintenance includes periodic retensioning of the taut-wires to ensure correct operation. (Manufacturer—Magal Systems)

Fence-Mounted Electric Field Sensors. A fence-mounted electric field sensor system uses an arrangement similar to an electrified cattle fence. A series of field wires and sensing wires are strung between fence posts, terminating in a control box, which contains the electric field generator and signal processor. When an intruder approaches or climbs the fence, the electric field strength received by the sensing wires is altered. The system can be tuned to filter out nuisance alarms due to small animals, etc.

Freestanding Exterior Sensors

Active Infrared sensors. Active infrared sensors emit infrared beams, which are received and measured by a separate receiving unit, placed up to 200 m distant. The infrared beams are usually interwoven into a criss-cross pattern, creating an invisible light curtain directly in front of the area being monitored. Intrusion is detected by a rapid, sustained reduction in the received infrared energy at the receiver channels.

When an intruder approaches the area monitored, the beam pattern is interrupted and an alarm condition is generated. With several beams in use, the system can discriminate against birds, wind-blown debris or small animals which disrupt only one or two beams at a time, as opposed to a human intruder, disrupting five or six beams. The system can be defeated by jumping or bridging over the sensor area without crossing the beam pattern. (Manufacturer—Takex America, Inc., Horton)

Passive Infrared Sensors. Passive infrared sensors operate by detecting the temperature differential between the background environment and an intruder. (A typical human emits an average of 50-60 continuous watts of energy). When a human or animal enters the area, its heat signature is noted and a temperature change is detected. When a temperature change is detected within the sensor's field-of-view detection pattern, it results in an alarm condition being generated.

Passive infrared sensors must be adjusted to aim at the area of coverage. Masking units can be applied to the sensor, allowing small animals to pass undetected if desired. Passive infrared units are best applied for detecting fast motion across the sensing pattern (moving perpendicular to the face of the sensor), and are not effective at detection of slow movements directly towards or away from the sensor. Care must be taken with outdoor units, not to aim at areas normally having very large vehicle traffic (which can be heat sources) or unwanted heat radiation sources (floodlamps, headlights, etc.), otherwise nuisance alarms may occur. (Manufacturer—DSC)

Microwave. Microwave sensors come in two styles—bistatic and monostatic. Bistatic microwave sensors use a transmitter and receiver pair. Monostatic microwave sensors use a single sensing unit, which incorporates both transmitting and receiving functions. With both types (bistatic and monostatic), the sensors operate by radiating a controlled pattern of microwave energy into the protected area. The transmitted microwave signal is received and a base level "no intrusion" signal level is established. Motion by an intruder causes the received signal to be altered, causing an alarm condition to be generated. One issue of note—microwave signals pass through concrete, steel, etc., and must be applied with care if roadways or adjacent buildings are near the area of coverage, otherwise nuisance alarms may occur. Many monostatic microwave sensors feature a cut-off circuit, which allows the sensor to be tuned to only cover within a selected region, helping to reduce nuisance alarms. (Manufacturers—MS Sedco, Southwest Microwave, Senstar Stellar, Crow Electronic Engineering, Inc.)

Dual Technology (PIR and Microwave). Dual technology sensors use a combination of passive infrared and microwave technology. In doing so, the sensor manufacturer hopes to minimize false alarms by utilizing the best features of each detection method. With most dual technology sensors, the individual sensor outputs may be optionally AND or OR combined. (AND requiring both PIR and microwave sensor units to receive an alarm before an output alarm condition is generated. OR requiring either PIR or microwave sensor units to receive an alarm to generate an output alarm condition). (Manufacturers—Protech Protection Technologies Inc. and Safeguards Technology Inc., Salco, Sentrol)

Interior Sensor Types. Many types of interior intrusion detection systems are in use today, including volumetric sensors, boundary penetration sensors, and proximity sensors. This section will focus on the sensors most applicable to water system facilities.

Interior Volumetric Sensors. Volumetric sensors monitor an internal area to detect the presence of an intruder. There are several types of volumetric sensors, including microwave, ultrasonic, passive infrared (PIR), and dual-technology (microwave and PIR).

Ultrasonic and Microwave Sensors. These operate by transmitting an ultrasonic or microwave signal within the protected area and receiving the reflected signal back. The reflected signal frequency will be identical to the transmitted frequency if there is no intruder present. Movement from an intruder causes changes in the frequency of the signal (due to the Doppler effect), causing an alarm condition. Microwave motion detectors use a higher frequency than ultrasonic, allowing it to detect motion through most interior walls.

Passive Infrared. Passive infrared (PIR) sensors are the most economical volumetric intrusion alarm available, because they do not transmit energy. They merely receive ambient energy from the environment. An interior PIR sensor detects sudden changes in the heat signature of the background environment. Radiant heat emitted from an intruder is sensed upon the intruder passing within the monitored area. Interior PIR units will typically filter out a small motion detected within one sensing quadrant but not another, as with small animal or rodent movement. Although coverage areas vary with model type, typical units serve areas of coverage approximately 50 × 50 square-feet. (Manufacturer—GE Interlogix, Bosch Security).

Dual Technology Sensors. Dual technology sensors are among the most prevalent volumetric intrusion detection sensors available, second only to PIR sensors. These sensors utilize both microwave sensor and PIR sensor circuitry within the housing. An alarm condition is generated if either the microwave or PIR sensor generates an alarm condition. The alarm settings may be adjusted to require both the microwave and the PIR unit to detect an intruder presence before an alarm condition is generated. (Manufacturer—Bosch Security, Protech)

Interior Boundary Penetration Sensors. Boundary penetration sensors detect the presence of an intruder across an interior boundary, such as a door, window, hatch, etc. The most typical boundary penetration sensors are glass break, door switch, and linear beam sensors.

Glass Break Sensor. There are three basic types of glass break sensors—acoustic sensors (listens for acoustic sound wave, matching broken glass frequency), shock sensors (feels the shock wave when glass is broken), and dual-technology sensors (senses acoustic and shock vibrations). Because a glass break sensor does not sense motion or intrusion from entering a door, hatch, etc., it is recommended to be used in conjunction with other methods such as volumetric, and not as the solitary method of detection. Glass break sensors are not recommended to be placed directly on the glass surface. (Manufacturer—GE Interlogix, Bosch Security, Ademco)

Door Switch. Door switches include contact switches, magnetic switches, and balanced magnetic switch. These switches may be used in a variety of applications, from monitoring doors to monitoring hatches, vaults, panel enclosure, etc. They are the workhorse of the security intrusion detection field. By far, the most effective is the balanced magnetic switch. This switch has internal circuitry that resists tampering or defeat from strong magnetic fields. By comparison, standard magnetic switches have been defeated by applying a strong magnet to the exterior of the door, (bypassing an alarm) and forcing the door open. (Manufacturer—GE Interlogix, Bosch Security)

Refer to Fig. 12.2, depicting a typical interior door monitoring installation example.

Linear Beam (Photoelectric Beam). Linear beam sensors (also referred to as a photoelectric beam or photoelectric eye) consist of a transmitter that emits a beam of light

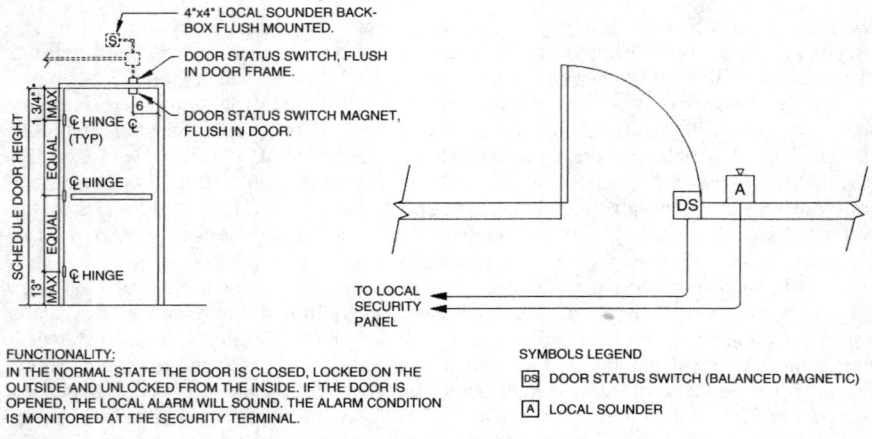

FIGURE 12.2 Interior door monitoring example.

(invisible to the human eye), and a receiver that receives the beam of light. If the beam of light is interrupted or broken by motion from an intruder, an alarm is triggered. Linear beam detectors can be surface mounted or recessed and require a straight line-of-sight between the transmitter and the receiver. (Manufacturer—GE Interlogix)

12.1.7 Video Motion Detection

Digital video motion detection is a technology that has seen significant advances recently. The system works on the principle of analyzing the video streams of closed-circuit television cameras, and comparing those video streams to a still image in the unit's memory. (The camera units must be fixed position, not pan/tilt units for the system to work). If the number of adjacent changed image pixels is greater than a threshold amount, the control circuitry determines that motion has occurred, and an alarm condition is generated. The units are particularly good in indoor locations with solid backgrounds such as a solid-painted wall or corridor, but experience more difficulty and require special tuning to work outdoors. With most units, certain regions are capable of being masked out, in order to be unaffected by adjacent vehicle motion on roads, wind movement at trees, and other motion events. Small digital video motion detection units can be purchased to serve individual camera units. Medium sized units can serve groups of 4 cameras and larger systems are expandable to serve hundreds of camera units.

12.1.8 Wireless Sensors

Recently, several companies have produced wireless intrusion detection sensors. In particular; door switches, PIR, and dual technology (microwave + PIR) sensors have been developed in a wide array of styles and configurations that use wireless technology to communicate an alarm condition. All of these sensors require power to operate, but in some cases internal batteries are sufficient to provide power (the unit generating a fault condition signal when battery power is low). These sensors can provide dramatic cost savings over hardwired sensors, particularly in difficult locations requiring asphalt trenching or significant conduit expense.

12.1.9 General Recommendations—Water Supply System Facilities

A few general recommendations for applying intrusion detection systems to Water Supply System Facilities are provided below. Not every recommendation will apply to each unique facility. For example, at some facilities it is not achievable to have perimeter fencing or perimeter detection because they are located within a park setting. However, this should provide some ideas that could apply to your particular facility.

Water Reservoirs and Elevated Tanks

Perimeter Detection. It assumes property boundary is not located inside a public use area such as a park. Provide continuous intrusion detection at the property perimeter. This could be accomplished by using fence-mounted intrusion detection or exterior PIR, microwave or dual technology (PIR + microwave) sensors. Connect the sensors to a constantly attended system, such as the SCADA system or a dedicated intrusion alarm system. A perimeter fencing system utilizing two parallel fences with a 10- to 20-yard stand off area between the fences is strongly recommended when using a perimeter detection system. The outer fence provides an animal barrier and identifies the perimeter of the site. The inner fence is outfitted with the intrusion detection system. This greatly reduces nuisance alarms from debris or animals. Occasional nuisance alarms must be tolerated with a perimeter detection system. If the perimeter detection system is calibrated in a manner to eliminate nuisance alarms the system will not be sensitive enough to detect an intruder.

Monitor Ladder. It assumes ladders have lockable ladder guard. Monitor the access ladder for unauthorized access via exterior motion sensing using PIR or dual technology sensor. Alternatively, mount a contact switch at the ladder guard, which is activated when the ladder guard is unlocked and opened. Wireless systems could be used if advantageous, reducing installation time and costs.

Monitor Vaults. It assumes vaults are locked. Monitor all vaults for unauthorized access, using door contact switches mounted on the interior of the protected space. Wireless systems could reduce costs.

Monitor Hatches. It assumes hatches are locked. Monitor all hatches for unauthorized access using door contact switches mounted on the interior of the protected space. For hatches that directly access treated drinking water, consider providing vibration/shock sensor that detects attempts to drill through hatch. Wireless systems could reduce costs.

Pump Stations

Perimeter Detection. Provide continuous intrusion detection at the property perimeter, using fence-mounted intrusion detection or exterior PIR, microwave or dual technology sensors.

Entrance Detection. Monitor all doors for open status using door sensing switches. Suggest balanced magnetic switches be used.

Volumetric Detection. Provide interior volumetric detection within the building. At minimum, orient sensor to apply coverage to all interior critical equipment (pumps, electrical motor control centers, PLC equipment, generator, fuel storage). Recommend that PIR or dual technology intrusion sensing be used.

Water Treatment Stations

Perimeter Detection. Provide continuous intrusion detection at the property perimeter, using fence-mounted intrusion detection or exterior PIR, microwave or dual technology sensors.

Entrance Detection. Monitor all doors for open status using door-sensing switches. Suggest balanced magnetic switches be used. Note—It is recommended that volumetric detection be used in conjunction with entrance detection (While a low-level adversary may use doors for access, a highly skilled adversary may not use the doors or windows, but instead may attempt to tunnel through the ceiling or walls.)

Volumetric Detection. Provide interior volumetric detection within the building at critical equipment areas (pump gallery, electrical motor control centers, PLC equipment areas, generator, fuel storage, chemical storage). Recommend that PIR or dual technology intrusion sensing be used. Some areas may benefit from video motion detection.

Monitor Clearwells. Provide clearwell intrusion monitoring at hatch locations, using door contact switches.

Monitor SCADA Control Rooms. It assumes SCADA system is in a lockable segregated room. Monitor door using door contact switch (balanced magnetic switch recommended.) Consider providing card access control. Consider video surveillance, coupled with video motion detection.

Monitor Chemical Storage and Dispensing Areas. It assumes storage areas are within lockable, segregated area. Monitor door using door contact switch (balanced magnetic switch recommended). Consider providing card access control. Consider video surveillance, coupled with video motion detection.

Raw Water Intake Stations

Perimeter Detection. Provide continuous intrusion detection at the property perimeter, using fence-mounted intrusion detection or exterior PIR, microwave or dual technology sensors. In remote wooded locations, care must be taken to choose a system that is not affected by animals, otherwise nuisance alarming will occur.

Entrance Detection. Monitor all doors for open status using door-sensing switches. Suggest balanced magnetic switches be used.

Volumetric Detection. Provide interior volumetric detection within the building at critical equipment areas (pump gallery, electrical motor control centers, PLC equipment areas, generator, fuel storage). Recommend that PIR or dual technology intrusion sensing be used. Some areas may benefit from video motion detection.

12.2 CLOSED CIRCUIT TELEVISION (CCTV): DETECTION, ASSESSMENT, CONTROL, AND ARCHIVE

12.2.1 Introduction

Closed circuit television (CCTV) has been used for security purposes for decades, but recent evolution in the digital hardware has made CCTV smarter, more reliable, more efficient, and more effective for premise security in all types of applications. CCTV is used to serve several basic security functions.

First, video serves as a surveillance tool. Cameras can be strategically located to provide a vantage point that guards or authorities would not have under practical circumstances. Cameras are often located on traffic stanchions, signposts, high masts, and tall building structures. The video from these cameras is then fed to a convenient location for observation by authorities.

Second, CCTV cameras act as a deterrent to would-be wrong doers. The simple presence of surveillance cameras often thwarts shop lifters, thieves, and assailants and vandals from achieving their goals. The realization that they may not finish their task without prosecution is enough to make many criminals second-guess their commitment to their goals.

Third, CCTV has been used for decades as a means to gather evidence in order to aid authorities in criminal investigations. Convenience stores, banks, hotels, and large retail establishments have been recording daily operations in hopes of catching images of criminals in the middle of their acts. In many instances, CCTV recordings are the only substantial evidence that prosecutors have to help link a particular perpetrator to crime. In the past, analog tape (VHS & Beta) has been the only acceptable media that would withstand scientific authentication against false incrimination due to falsification. But, as will be explained later, digital video recording has advanced to mainstream acceptance for use in prosecution.

Intrusion detection is the fourth major use for CCTV. In more recent years, CCTV has been used more frequently as a means to detect unwanted facility intrusions. With the advent of digital video processing techniques, live video is analyzed as it is being recorded for changes in the usually static image scene. Major changes in the pixels that make up the camera's view are detected by the signal processor and relayed in the form of an audio or video alarm to call attention to the camera's view. This process has made large CCTV monitoring facilities more effective with fewer operators required to perform the same monitoring function. More explanation of this extremely useful application will be provided later.

Lastly, access control authentication is another useful application for CCTV. Cameras are frequently located at facility entrance gates, main entry doors, and controlled access points.

A security guard or facility administrator typically compares a user's identification credential to the person's face on a CCTV monitor, and releases the door or gate mechanism. This provides an additional measure of authentication, as keys and access cards are easily lost or stolen. Later, more intelligent video applications will be discussed that explain how CCTV can be used with many electronic access control software applications to automatically compare a person's digitally stored facial image against the image on a CCTV camera.

As you can see, CCTV is an essential component of any facility security system. The details that follow will provide additional explanation of the topics in the introduction.

12.2.2 CCTV System Components

Cameras are the first component of an effective CCTV security system. There are thousands of different types of cameras used in surveillance systems, but all of them can be categorized into one of three categories—analog, digital, or IP cameras. Analog cameras are the traditional CCTV tool, and transmit a video image over a coaxial copper cable. These cameras are still very much mainstream, but are giving up sales volume to their digital counterparts, which offer enhanced clarity, resolution and image stability. Digital cameras also have the advantage of a variety of transmission media types including coaxial cable, S-Video, fiber optic, and IEEE 1394 "FireWire"™. However, copper coaxial cable and fiber optic cable are the predominant transmission media for digital cameras in most surveillance applications. With the enhanced offering of digital cameras in the marketplace, many manufacturers have also begun to offer IP (Internet Protocol) cameras, or those that are capable of transmitting over a LAN, WAN, or virtual network "out-of-the-box."

These cameras include a processor on board that compresses the video image into a common or proprietary format that is viewable with a web browser or special decoding software. IP cameras are typically "ethernet ready" and can be connected directly to a 10/100 TCP/IP network. Any workstation that has a compatible browser or decoding software utility is capable of viewing an IP camera connected to the network, provided the IP address of the camera is known. IP are very useful for small business applications, but have not found mainstream acceptance in large enterprise surveillance systems. Most IP cameras have a limited number of features as compared to their digital and analog surveillance counterparts, and are not offered in as many mounting styles, enclosure configurations and all-purpose applications.

The largest camera manufacturers offer all the three camera types mentioned above, but the analog and digital varieties have the widest variation in offerings. One manufacturer may offer a hundred combinations of features throughout their product line. However, all cameras are either fixed or movable (PTZ—pan/tilt/zoom). Both styles are offered in indoor and outdoor models with nitrogen pressurized domed enclosures for hazardous environments. The movable cameras are ideal for applications where a target or suspect needs to be followed with the video surveillance. They can also be programmed with an automated "guard tour" or controlled manually with a "joystick" by a surveillance operator. The highest quality these cameras offer is over 400 total lines of resolution, color video in lighted conditions, and automatic switchover to black and white video in low light conditions. Fixed cameras have the widest choice of lens options, but PTZ cameras have built-in zoom capabilities that offer as much as 25 times magnification optically with additional digital zoom capabilities for far-off targets.

One camera type that has not reached mainstream usage is the infrared camera or thermal imager. One reason may be the cost. A fixed IR camera in a weatherproof housing typically costs five to six times as much as a high-quality PTZ camera. This particular technology is very useful in outdoor, low-light areas with highly variable weather conditions and is used primarily for intrusion detection applications. The camera is not useful for assessment applications, however, as the infrared image is a grey-scaled image and does not provide sufficient clarity to distinguish facial or other physiological features between subjects.

The second major component of a CCTV system is the Switcher or Multiplexer. It is impractical to view surveillance video the same way we watch television at home. So, surveillance cameras are typically tiled on a video monitor so that multiple images can be viewed on the same screen. The multiplexer (or MUX) is the device that allows the user to accomplish this task. Typically, a monitor has a one-to-one relationship with the multiplexer, and the size of the MUX determines the number of cameras that can be viewed on screen. The cameras can be rearranged in several different formats on screen to make viewing of a particular camera larger, but the grid is usually limited to a 2×2, 3×3, 4×4, or 5×5 arrangement. In larger systems, a multiplexer is surrendered for a switcher, which incorporates the functions of the MUX with other key functions. The switch allows any of the multiplexed camera views to be switched to a "call-up" monitor, where the image is usually maximized on screen for enhanced viewing. This can also be automated upon alarm conditions from external devices to switch the associated camera to the "call-up" monitor. This is especially useful when there are an excessive number of cameras and a limited number of video operators. The keyboard that interfaces with the switcher typically incorporates a numeric keypad for programming and a joystick for camera control. This unit is similar to a gaming device except that the knob on the top of the joystick rotates left and right. This feature provides the zoom control for the PTZ cameras connected to the switch. In advanced level systems, this keyboard can also interface with digital recording devices to call-up recorded video for review on a monitor. It is noteworthy to mention that not all CCTV systems are set-up to provide live viewing of video as it is being recorded. In these instances, the CCTV cameras are input directly to the recording device, and the keyboard is used only to program and control the functions of any PTZ cameras connected to the video system.

The third major component of a CCTV system is the transport media. There are three types of transport media used for CCTV transmission between the viewing or recording apparatus (also referred to as the "Head-end") and the cameras—copper wire, fiber optic cable, and wireless hardware. The most predominant in small applications, especially indoor, is copper wire transmission. Most cameras and head-end equipment are built and configured to accept wired transmissions in the form of coaxial cable. Coaxial cable is not useful for long runs, where signal loss degrades the quality of the video, and in areas with high interference, or noise. Problems also arise when outdoor cameras are connected to the head-end equipment with coaxial cable, as power surges from lightning strikes cause frequent damage. There are even transmission products available that will transmit video and control data (for PTZ cameras) over an unshielded twisted pair cable similar to phone wire. The video is converted from a coaxial connection at the camera, using a baluns transformer, to the two-wire-twisted media and converted back to a coaxial connection at the head-end with a similar receiver. This media allows long-range transmissions over CAT5e rated cable with amplified transmitters to a range of approximately 1 mi. The manufacturers of such devices also claim that the twisted format of the wire also provides excellent resistance to interference from "noisy" devices, such as fluorescent light fixtures. Copper transport media is also used for IP cameras transmitted over CAT5 Ethernet cable in typical LAN/WAN wiring configurations. Without amplification, these transmission lengths are usually limited, however, and are prone to carrying power surges back to head-end equipment.

Fiber optic cable, on the other hand, is immune to power surges and offers extremely long transmission lengths. Fiber optic cable is a mainstream transmission medium for digital networking and has been used widely in the security industry in recent years as transport media for video and data. Several major manufacturers offer hundreds of products to facilitate the transport of closed circuit video from the field cameras to the head-end equipment. These products transmit video and data over glass fibers using light pulses, and with lasers can transmit such data for several miles before a receiver or repeater is required.

And because glass is non-conductive, an electric field cannot be introduced into the transport stream by interference from external devices. This makes fiber optic cable the ideal media for transmitting video between buildings, outdoor structures, and especially across long distances. Some devices can multiplex, or simultaneously transmit, as many as 32 video cameras across a single pair of fiber optic cable strands. And most camera manufacturers offer an optional fiber optic module inside the PTZ camera base to send video and receive control commands over a single fiber optic strand. The fiber optic transceivers at the head-end convert the video back to a coaxial cable and the control data to a two-wire configuration for connection to the switcher or recording equipment.

The last and the least common media type is wireless transmission. Wireless applications are more commonly used in outdoor installations where cabling is impractical or impossible. With some exceptions, the wireless media is similar to the hardware used for wireless WAN applications. Manufacturers of wireless video packages convert commercial quality CCTV cameras into IP ready cameras, which transmit video and data over the wireless WAN hardware. These cameras transmit back to the head-end, where an omni-directional receiver collects the data from the wireless cameras and converts it back to copper or fiber media for connection to switching or recording hardware. In some applications, the video remains in an IP format, and is viewed using a compatible browser or decoding software. This transmission method is growing more competitive with the fiber optic media, especially for new systems that do not have existing fiber optic cable to utilize. In some cases, the wireless configuration is more cost effective to implement than installing new fiber media and the associated transmission hardware.

The recording hardware and the application software are the last two major components of any CCTV system. These two items go hand-in-hand, but the software will be discussed in a later chapter dedicated to CCTV monitoring, control, and special applications.

The traditional appliance used for video recording is the analog tape recorder, or video cassette recorder (VCR). Only a couple of decades ago, it was offered in VHS as well as Sony's Beta format, but VHS and S-VHS are the only two mainstream VCRs available to the security industry. Many installations require the use of VCRs rather than digital recording media because they remain, in some jurisdictions, as the only reliably authenticated media type. Video cassette recorders in VHS or SVHS format from most major manufacturers come in two types: Real time and time-lapse recording. In real time, the recorder is constantly recording the video input stream, as long as media is present in the recorder. In time-lapse recording, the VCR records a fraction of the frames that compile real-time video, which is commonly considered 30 frames of video per second. Top of the line time-lapse recorders can record up to 720 h of video on a typical 120-min video cassette. Some recorders can also record on demand as stimulated by an external device, usually an alarm panel. This eliminates recording of unwanted video in some applications. Real time recording is usually done with a multiplexed view on the input channel. This allows as many as 16 separate video inputs to be recorded with a single VCR and is accomplished by using a secondary output from the multiplexer or a looping output from the multiplexed video monitor. Unfortunately, VCRs have serious limitations that have resulted in their near-obsolescence in most markets. Analog recording is just too cumbersome and has too many flaws, despite the advances in the technology of VCRs over the years. VCRs require the recording media to be changed out at regular intervals, have poor video quality when tapes are re-used, and make searching recorded video a near impossible task in short order. All of these problems have been overcome by digital recording.

Just as the video cassette has been replaced by the digital video disc and personal video recorders in the residential market, the VCR has been replaced with digital video recorders (DVRs) in the surveillance industry. The revolutionary accomplishments in the digital video marketplace have all but eliminated VCRs for all but a few select customers.

A digital video recorder, for lack of a better description, is little more than a specialized computer with a high-capacity hard drive. The DVR captures the video stream, digitizes it

SECURITY HARDWARE AND SURVEILLANCE SYSTEMS   **12.15**

into a digital format, and compresses the video using a compression algorithm so that it can be stored on a hard drive. Actually, it usually includes several large capacity drives to provide as much as 900 GB of storage per recorder in the best DVRs available. Since DVRs were designed to replace VCRs, they too are equipment rack mountable, and have a similar form factor. But that is where the similarities end. VCRs only have one or two video inputs, and are capable of recording only a single channel at a time. Digital recorders have as many as 32 separate video inputs, and are capable of recording all 32 video images simultaneously. Most DVRs are also capable of quadraplex operation, which allows a user to review recorded video, scan the DVR for alarm video and write video images to a portable video media (typically a CD-ROM or DVD) all while the DVR continues to record video from all of its inputs. VCRs are only capable of a single task at a given moment. This feature alone speaks volumes of the sophistication of digital recording, and as technology improves so will the recording capacity of the DVRs. The only major resistance that digital recording manufacturers have experienced is the issue of authentication. With the digital editing software available to the motion picture industry, digital video images can be altered almost seamlessly to the inexperienced observer. For this reason, manufacturers have created sophisticated authentication algorithms that are embedded in the recorded video stream. Most manufacturers have a software application that interfaces with the DVR utilizing an authentication tool. When a video clip is selected for authentication, the tool executes a function that utilizes the information in the pixels of the image. If the video stream is altered, the authentication formula spits-out a result different than that of the unaltered image using the previously defined algorithm. In this instance, the manufacturer can prove if a video clip has been altered, and this has been reliably demonstrated in courts of law throughout the country over the last several years. This invaluable feature has all but sealed the fate of traditional analog recording for even the most stubborn of customers.

Perhaps the most limiting factor in digital recording is the capacity of the hard drive in the recorder. Most recorders are designed to stand alone in a local or wide area video network, and are connected to the video observer through TCP/IP connectivity. So, most recorders are limited to their on-board storage capacity. Extended storage capacity is available on the recorder if it is configured to record at a slower frame rate, like the time-lapse VCR. However, if near full motion speed is desired, then most recorders can only store about a week's worth of video when all of the recorders' video inputs are used. If long-term storage is required or is desirable, the video must be transferred out of the recorder to another media. Several options are available depending on the needs of the user. If alarm video is the only long-term concern, then most DVRs support the export of video onto CD-ROM or DVD in a universally viewable format. This typically is a manual operation done by the surveillance operator. If automatic storage is preferred, then the DVR can be connected to an external tape library for video storage onto AIT-III (Advanced Intelligent Tape, developed by Sony). Each tape is capable of 100GB of storage, and the libraries are offered in robotic "jukebox" configurations that automatically load and unload the media from the tape drive. Tape libraries are commercially available with a capacity of as many as 600 tapes per unit for 60 TB of total storage. They can be configured to record over the tapes in the library, or new tapes can be loaded and the old tapes removed for archiving. Similarly, long-term storage can also be accomplished by connecting the DVRs to a Storage Area Network (SAN) or an external RAID (Redundant Arrays of Independent Disks) storage device. A typical RAID device will offer 8 to 14 hot-swappable redundant hard drives in a modular chassis. The total storage capacity of these units is measured in terabytes of data, and does not require media change out like tape libraries. They also have no moving parts, so they also tend to be more reliable. However, they do not provide any permanent means for archiving data, like the AIT data tape storage devices. Huge technological advancements are made for data storage applications every year, and span beyond the surveillance industry.

Digital video is just another form of data, and can be stored and transmitted like any other data format.

12.2.3 Special Considerations

Closed circuit television is an integral part of any comprehensive security system. It is, in some cases, the only means to properly assess a potential situation in progress or after the fact. CCTV hardware has proven invaluable in identifying the hijackers that were responsible for the September 11, 2001 terrorist attacks. And, with the advent of digital recording and digital video networks, it remains the most promising means for securing the nation's infrastructure from future terrorist attacks as well as domestic criminal activities. The most promising use of CCTV will continue to be the software applications that have been and are being developed to process digital video data to create a more intelligent and responsive surveillance system. Software and hardware applications that currently exist use digital video from CCTV systems to detect intrusions into secured areas. Enhancements in this technology have recently made digital video motion detection (DVMD) an extremely useful technology in even outdoor locations, which were previously too "noisy" to analyze because of inanimate motion in a typical video scene. Other applications also utilize facial character recognition to compare the facial features of potential criminals against a known database of criminal's facial images. This technology is currently in use in the United States, and will most likely become more prevalent in the nation's critical infrastructures.

Recent advancements in capturing, transmitting, storing and analyzing digital video has made CCTV detection, assessment control, and storage a more important design consideration than ever for any CCTV design and installation.

12.3 ACCESS CONTROL: PERIMETER AREA AND INTERIOR CONTROLLED ACCESS

12.3.1 Introduction

Access control systems have been around for centuries—Linus Yale invented the cylinder pin tumbler lock in the 1800s. Since then, an incredible number and variety of access control systems have been introduced.

The objective of an access control system is to permit only authorized personnel (including police and emergency response personnel) to enter and exit a restricted area. In some installations, an access control system is combined with a contraband detection system, so that unauthorized items or materials may not enter or leave the facility (an airport for example).

Electronic access control systems control entry into a perimeter, area or interior space by means of an electronic controller and associated components including locks, readers, sensors, and other equipment. Electronic access control specifies who can go where and when. An alarm should be generated by the electronic access control system when an entering individual is not properly identified. Upon an alarm receipt, the response team (security guard) should assess the problem, determine if the alarm is a nuisance alarm or is valid, and initiate the proper response to the situation.

This section describes access control devices applicable to Water Supply systems.

12.3.2 Layered Security Systems

Layered security systems are essential—the "security in depth" principle. In designing or evaluating an access control system it is important to establish "who should have access to

SECURITY HARDWARE AND SURVEILLANCE SYSTEMS

sensitive areas" and "what level of access—high, medium, or low level is needed?" This information is important in order to determine the level of security needed and what access control equipment should be used in order to meet the necessary requirements.

A typical water supply system may include four or five security access control levels:

- *Level 1*: Public zone. The area outside the perimeter fence accessible to the public.
- *Level 2*: Clearzone. The area between the fence and the locked building exterior.
- *Level 3*: Building lobby area. The area within the building lobby, prior to accessing the interior building circulation areas. May include a waiting area for visitors.
- *Level 4*: Internal Circulation Areas. Interior areas within the building.
- *Level 5* (If needed): High Value Areas. High value areas internal to the building, such as chemical storage areas, SCADA rooms, etc.

When choosing the proper access control system a balance must be struck between false rejection (authorized persons not allowed access) and false acceptance (unauthorized persons allowed access.) For example, biometric systems have a low false rejection rate, however the identification process can be lengthy, making their use impractical for high traffic areas.

12.3.3 Access Control Attributes

Access control systems are based upon something you possess, something you know, a physical attribute, or a combination of any of the three.

Possession Basis. In this case, the access control system relies on something in the possession of the authorized person. This can be a key, an identification card, or an access card. A disadvantage of this system is that the object can be lost or stolen and still permit entry.

Knowledge Basis. A knowledge based access control system relies on restricted information that the authorized person possesses. This includes lock combinations, personal identification numbers (PINs), or passwords. A disadvantage of this system is that the information can be copied from, coerced out of, or coaxed from an authorized person.

Physical Attribute. Electronic access control systems that rely on physical attributes of the authorized user as their basis are called biometric devices. These include fingerprint scanning, palm scanning, iris mapping, and voice recognition. These systems are typically more difficult to deceive. A disadvantage of these systems is the delay involved in the enrollment process required to add a user into the system memory, and the issue that some users voice concerns about having their features scanned by the system on a regular basis.

Combination. The most secure basis of all access control systems utilizes a combination of the above. In particular, access control systems using a combination of access cards (possession) and entry of a keypad PIN (knowledge) have proven to be highly reliable.

12.3.4 Locking Systems

The major types of entry control systems in use are key locks, keypads, and electrified locking systems. Each has its distinct advantages and disadvantages.

Key Locks. Key Locks are the most common access control system in use.
  Advantages. Key locks have the advantage of being simple, require no external power, and require the least cost investment.

Disadvantages. Keys can be lost or duplicated, keys are typically not tracked, keys do not keep a record of who has entered the building or area. Key locks require material expense to change out the lock if a key is lost.

With a key-based system, it is critical that a key control program be in place. Attributes of a good key control system include the following:

- Keys are assigned only to selected personnel and are recorded into a logbook.
- Keys are stamped with "Do Not Duplicate."
- Keys are required to be automatically turned in when employees are terminated or when reassigned to a different position.
- Locks are re-keyed on a periodic basis, and when keys have been lost.
- Key use is segmented so that one key does not fit all locks.
- Key distribution is segmented so that not all employees have access to every facility.
- Keys should be kept in a lockbox, when not in use.

Mechanical or Electrical Keypads. Keypads are the next step in flexibility and security. Keypads rely on a mechanical (or electronic) locking system to allow access. The user typically pushes an access code, usually four to six digits in length, in order to gain access. With a mechanical unit, programming is done manually, entering the valid code(s) by hand once the unit is in program mode. With electronic units the programming is usually done using a laptop that is plugged in or uses wireless infrared communication.

Advantages. Keypads can allow multiple codes to be used, they allow quick PIN code change if a user is terminated or his/her PIN code has been compromised. Some systems have a PIN retry lockout feature that disables the keypad of an entry reader for a specified amount of time after a specified number of improper PIN entries. This feature protects the system from intruders who tamper with a keypad-controlled access point by slowing down the process of trying all possible code combinations.

Disadvantages. Keypad codes can be copied by watching an authorized user (some have protective shrouding to limit this), they are more expensive than key locks, and do not typically track who has entered the area (assuming that one common PIN code is used by all). Typically keypads are not interconnected to a common integrated control system, which means that keypad changes must be made to all doors individually, rather than simultaneously.

Electrified Locking Systems. Electrified locking systems offer the most sophisticated, flexible, and secure locking systems. A variety of types are available. The advantage of an electrified locking system is that it offers the maximum flexibility for changing locking conditions. Typically, electrified locking systems are part of an integrated access control system, which enables user privileges to be changed across all doors simultaneously, should a user be terminated. However, with that flexibility comes cost—electrified locking systems have a higher first cost than key locks or keypad locks. Types of electrified locking systems include electric strikes, magnetic locks, and electrified mortise lock sets.

Electric Strikes. Electric strikes are a solenoid-operated device that is located in the strike (within the doorjamb) that will allow a locked latchbolt to be pushed through the strike when electrical power is applied to the strike. Typically electric strikes rely on 24-V ac or dc power sources. Electric strikes are probably the most commonly used electrified lock type. They are somewhat susceptible to prying efforts, so they are most frequently used on interior door applications, rather than perimeter door applications.

Magnetic Lock. A magnetic lock consists of a large coil of wire mounted to a doorframe. When current is passed through the coil, it creates a strong magnetic field. A large metal strike plate is also secured to the door and is held tightly against the coil of wire, due to the presence of the strong magnetic field. The door can be released (or "unlocked") by interrupting the flow of current through the coil, thereby removing the strong magnetic field. Magnetic locks are inherently fail-safe (meaning they unlock upon power failure). Magnetic locks are frequently used in high-security applications such as prisons, due to their exceptional holding strength (often 1200 lb. or more). Note—Magnetic locks with a holding force of 600 lb. or less should not be used for security applications.

Electrified Mortise Lockset. An electrified mortise lockset is a device that fits into a mortised cutout in the edge of a door. Mortise locksets are considered to be very secure and resist prying or forcing attempts on the door and frame assembly. Electrified mortise sets are frequently used on exterior doors and double door applications.

Fail-Safe Locking Systems, Fail-Secure Locking Systems. A term that is used with electrified locking systems is fail-safe and fail-secure. A fail-safe lock configuration will become unlocked with power removed. A fail-secure lock configuration remains locked when power is removed. A fail-secure lock must allow for emergency egress in the event of power failure. Typically, emergency exit pushbuttons are provided for this purpose.

12.3.5 Card Reader Systems

Card reader access control systems provide the most reliable, flexible method of controlling access to a facility. Card reader systems come in many configurations, from standalone systems, controlling only one door, to systems that are scaleable to provide enterprise-wide control for an entire corporation, spanning multiple continents. Newer card reader systems offer sophisticated database intelligence that allows integration with payroll, IT, and human resources databases. (If an employee is terminated, his/her access privileges are revoked within the access control system instantaneously, without delay.) Other access control systems offer seamless integration with video surveillance systems, such that access control alarms and video surveillance images may be displayed on common PC workstations.

Typical System Operation for a Card Reader System. The card reader system typically consists of a computer workstation—which displays alarm conditions and allows programming of the system. Local control panels control the doors, card reader units, and access cards. A printer unit may be used to print each event and alarm condition. Under normal operation, the system grants access at doors with card readers by comparing the time and location of any attempted entry with information stored in the memory. Access is granted only when the security card used has a valid entry code at the card reader for a designated time frame.

Refer to Fig. 12.3, depicting a card reader access control system (shown integrated with video surveillance).

Card Reader Capabilities and Functions. Some of the significant advantages that card reader systems have are the capability for event tracking and programmable software functions:

Event Tracking—Event Log. A log or list of security events that are recorded by the access control system and indicate the actions performed and monitored by the system. Each event log entry contains the time, date and any other information specific to the event.

Two-Man Rule Software. Software programming that is optional on many card reader systems. Prevents an individual cardholder from entering a selected empty security area unless accompanied by at least one other person. Once two cardholders are logged into the area, other cardholders can come and go individually, as long as at least two people are in the area. Conversely, when exiting, the last two occupants of the security area must card out together, so that there are never less than two cardholders in the area.

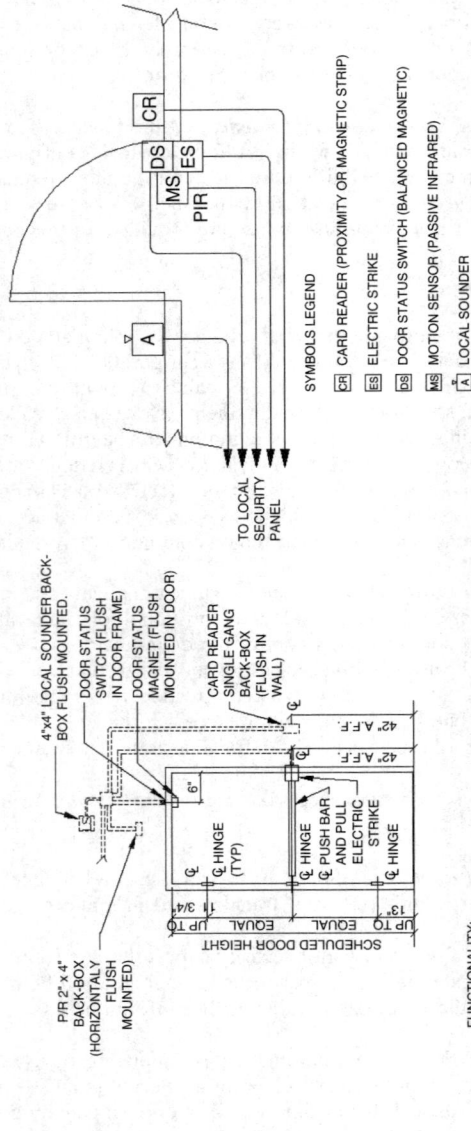

FIGURE 12.3 Typical single door card reader installation detail.

Antipassback Software. Software feature that prevents users from giving their cards to someone else to use. This feature is sometimes available with keypads. To prevent the same PIN from being used by many people, a time element can be programmed in—the PIN won't work again until that time expires. Some antipassback systems require that if a card is used to enter an area it must be used to exit that area before it can be used to gain access to a different or unrelated area. This feature also helps eliminate *piggy-backing* or tailgating by unauthorized persons. Refer to Fig 12.4, depicting a typical card reader installation detail.

12.3.6 Guard Tour System

Most access control systems feature integrated guard tour systems, which require designated security staff to conduct a security tour of a facility at specified frequencies and durations. Integrated guard tour systems feature access control sensors which are typically passive, matching the style of the card reader in use. The use of an integrated guard tour system within the access control system reduces hardware costs, programming effort, maintenance effort and training over a separate guard tour system.

Recommendations for an Effective System

- Capability for unlimited number of tours.
- Optional printed or on-screen reports.
- Documentation of hit, missed, duplicated or unidentified control points, and comprehensive incident reporting.
- Wireless technology shall be utilized wherever practical, such that conduit and wiring requirements are minimized for new tour points added.

12.3.7 Card Access Technologies

The most commonly used card access systems in use today rely on one of four technologies-magnetic stripe, Wiegand, proximity, or smart card.

Magnetic Stripe. As the name implies, the access card contains the coded information within a magnetic stripe embedded on the surface of the card, similar to credit card encoding.

Weigand. A card access technology relying on a series of wires embedded in a vinyl card. To initialize the Wiegand card, it is passed through a special reader to communicate a distinguishing pattern of ones and zeros to the access control system to identify a particular cardholder.

Proximity. A card access technology relying on a radio frequency link between the reader and the card (proximity reader and proximity access card). Encoded information is passed between the access card and reader, usually supplying a unique pattern enabling identification of the cardholder. The majority of new systems in use today rely on proximity technology.

Smart Card. A small card with a magnetic stripe and microchip for encoding data. A card reader reads the encoded data on the card, usually for access control or transactions (purchasing items, for example). Smart cards are used most frequently in Europe. Their use in the United States is not widespread, but is increasing.

12.3.8 Site Perimeter Access Control

Perimeter access control is one of the fundamental concepts in an overall balanced security system. In providing access control at the perimeter entrance of a facility, the purpose is to identify and turn away an unauthorized intruder approaching a site, facility, or area as early

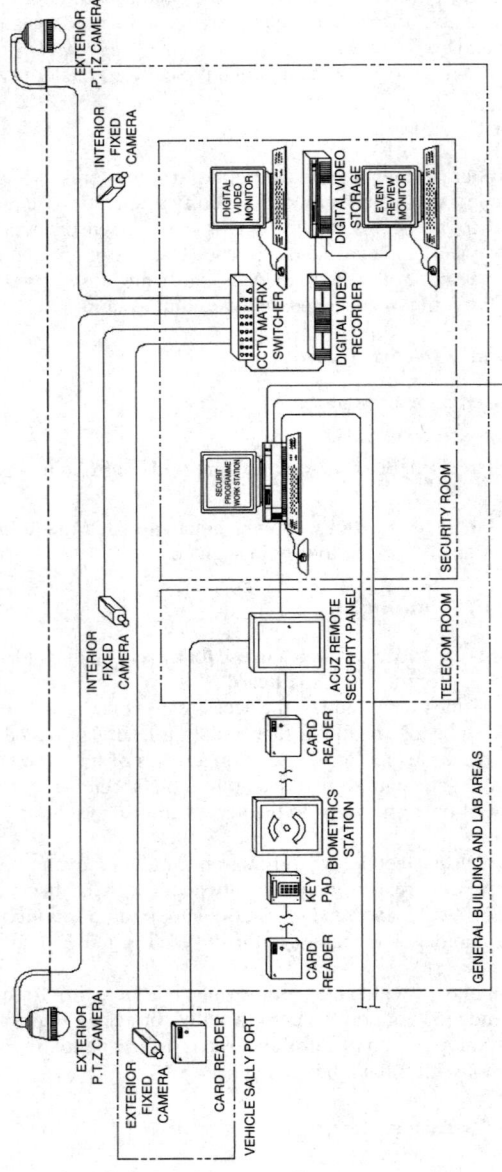

FIGURE 12.4 Typical access card system block diagram (with video surveillance).

as possible. A perimeter access control system may provide the first opportunity to restrict entry to an unauthorized individual trying to enter the facility premises.

Recommendations for Effective Site Access Control

- Establish a perimeter fence surrounding the facility at an appropriate standoff distance. This establishes a perimeter boundary and restricts access to the area surrounding the facility. Add appropriate signage to the fence.
- Install vehicle access barriers (including fencing, trees, berms, and bollards) which prevent vehicles from driving into the site at points other than official entrances.
- Establish a "No Stopping" zone along the roadway serving the facility, with appropriate signage. Security should monitor and patrol the roadway and have stopped or parked vehicles towed.
- Designate a clear-zone area from perimeter fence to building exterior. Within the clear-zone region, well-kept landscaping that eliminates hiding places around the perimeter of the facility will be maintained.
- Provide a vehicle checkpoint area for establishing the identity of all incoming vehicles attempting to access the perimeter. In high security applications, the security checkpoint could consist of a guardhouse adjacent to a vehicle sally port, detaining vehicles until the driver identity and vehicle contents are confirmed.
- Consider locating a separate visitor parking outside the secure clear-zone area.
- If a separate visitor parking is provided, install personnel gates/turnstiles for visitors entering the fenced perimeter clear-zone. Turnstiles include intercom station, CCTV video surveillance from the Security Office, and a card reader for staff use.

Clear-Zone Areas. An important concept in perimeter access control is a clear-zone. A clear-zone is an area surrounding the perimeter of a facility that is clear of shrubs and trees and features well-maintained landscaping that eliminates hiding places by an adversary. Clear-zones enhance visual observation by security personnel, and create a demarcation zone which makes unauthorized persons more noticeable. Clear-zone distances vary, but clear-zone areas ranging from 50 to 100 ft from perimeter fence to building exterior are common. Frequently, lighting is enhanced within clear-zone areas, making it easier for employees and passersby to observe and identify intruders. Within the clear-zone space surrounding the critical buildings, motion detection is sometimes installed, with instant-on high visibility lighting (3 to 5 ft-candles) which illuminates if personnel approach the building.

Standoff Distance. Another important concept in perimeter access control is standoff distance. Standoff distance is the distance between the outside perimeter (the public areas) to critical facilities or buildings inside the perimeter (the restricted access areas). Standoff distances are determined by evaluating the threat and potential consequences of unauthorized vehicle entry. For reference, Department of Defense documents have identified that 500 lb. of explosive will cause moderate to minor structural damage to buildings from 200 ft distant. An effective standoff distance will significantly minimize damage and injury caused by a vehicle bomb explosion.

Options for obtaining the optimum standoff distance can include one or more of the following:

- Restricting large vehicle traffic adjacent to the facility
- Redirecting traffic or adjusting approaching roadways so that vehicles pass further away from the facility

- Erecting vehicle barriers to prevent unauthorized entry
- Constructing a search area for vehicles entering the area
- Relocating truck deliveries away from the protected facility
- Strengthening vulnerable building elements (reinforcing windows, walls, etc.)
- Relocating the facility or asset to a safer location

Vehicle Checkpoints. A vehicle checkpoint area for detaining vehicles for identification is recommended in a perimeter access control system. The purpose is to screen all vehicles or pedestrians prior to accessing the property. In all cases, prior to granting entry to a visitor, the guard should collect the information from the visitor—the visitor's name, his/her company, the name of the staff member he/she wishes to visit, and the purpose of his/her visit.

In a simple system, a vehicle checkpoint can consist of a gate, with an intercom and video surveillance system. When a vehicle approaches, the driver requests permission to enter the facility using the intercom. After security staff have visually identified the visitor, access may be granted or denied from within the facility. Adding an exterior card reader on a pedestal outside the gate could serve to grant access to employees.

In more elaborate security installations, a guardhouse facility may be located at the entrance to a facility. A security officer, who screens all vehicles entering the site, staffs the guardhouse. Vehicles that are not permitted to enter the site are turned back, and must exit the site.

High security applications utilize vehicle sally ports to detain and screen incoming vehicles. A vehicle sally port consists of interlocking gates within a fenced area. Incoming drivers pass through the first gate, stopping at the second gate. Once both gates are closed and the vehicle is captured within the sally port, a security guard may confirm the identity of the driver and, if necessary, search the vehicle to confirm the contents. Once the vehicle and driver are approved, the second gate opens and the vehicle may drive onto the facility.

Delivery Access Control. Deliveries present a difficult security challenge for facilities. Particularly for water systems, with chemical deliveries a regular and necessary element, additional access control policies may be warranted.

- Consider adding provision for a CCTV video surveillance system. Deploy cameras to capture the vehicle license plate and driver facial features. During high-risk periods or for specific reasons, security staff could be trained and equipped to physically inspect vehicles entering and/or leaving the area.
- Ensure training of security personnel to keep detailed logs of deliveries and pickups, including driver information and destination.
- Adopt a procedure that requires faxed or electronically transmitted copies of delivery bills-of-lading information and driver identification are sent to security prior to the truck arriving on-site.
- Implement a procedure for ensuring that a driver who regularly picks up or delivers hazardous materials such as hazardous chemicals, is previously identified, given proper identification badges and trained properly in the facility security requirements.

Vehicle Access Control Barriers. In order to maintain access control for vehicles entering the perimeter boundary of a facility, gates and vehicle barriers are used. Several types and configurations are possible, including passive devices such as bollards, jersey barriers, berms, fencing, etc. Active vehicle barrier devices include horizontal sliding gates, vertical barrier arm gates, hydraulic pop-up barriers or retractable bollard assemblies. Whatever the configuration, the vehicle barrier or gate should be hardened enough to prevent a vehicle from crashing through it at high speed. Vehicle barrier calculations should be performed using computer modeling to simulate the performance against the anticipated vehicle weight and speed.

Some things to consider when selecting vehicle access barriers:

- Do not install barriers that require installation below ground, if there is a high water table. Freezing conditions or water collection may incapacitate the system.
- Do not install barriers at entrance and exit gates without also reinforcing the remaining accessible areas of the perimeter. (Use jersey barriers, bollards or aircraft reinforcing cable installed at fencing to reinforce the remaining perimeter areas accessible by vehicle.)
- Avoid extensive protection of a large facility perimeter. It will generally be more cost-effective to protect individual buildings or zones within the perimeter than trying to reinforce a very large perimeter area.
- Do not neglect to install barriers on the exit side, as well as the entrance.
- Avoid long, straight paths to a crash-resistant barrier. Where this cannot be avoided, provide a passive-type barrier maze to slow the vehicle.

12.3.9 Building Access Control Concepts

Once an effective access control system has been established at the site perimeter, the next step is to examine the building access control philosophy. An important concept for consideration is to limit the number of building entrances used by staff. This helps establish tighter control on building security while minimizing the number of allowable entrance locations.

Recommendations for Effective Building Access Control

- Keep all exterior doors locked at all times.
- Establish a primary entrance door and add access control, visitor intercom, and video surveillance equipment.
- Identify critical exterior circulation doors and add access control to those doors. These doors shall be designated as access-controlled doors, and only accessible by employees. (Access control methods could consist of either adding key locks, keypads, or card readers)
- Designate remaining doors not having exterior access control as exit-only. Remove exterior door hardware from exit-only doors. Ensure that on the interior side of the doors, appropriate exiting hardware remains, allowing free access under emergency egress conditions.
- Establish a secure lobby area, with hardened doors capable of being activated by security to go to "locked-down" mode.
- Provide a waiting area for visitors. Visitors should be required to sign-in, be assigned a color-coded visitor badge and escorted at all times while in the facility.
- Provide CCTV camera surveillance of incoming personnel into the secure lobby area using at least two cameras—one records body size and clothing characteristics and the other records close-up of facial characteristics. All CCTV surveillance will be recorded digitally.
- Establish a secure area to house CCTV monitors, security computer equipment, radios, etc. Restrict access to this room.

12.3.10 Interior Access Control Concepts

Now that access control has been established at the site perimeter and the building exterior, the interior access control may be examined. Certain building interior spaces may warrant additional access control, in order to secure those areas from inadvertent access, insider access, or unauthorized access. These may include chlorine storage and distribution areas, SCADA equipment and workstation areas, high-value equipment areas and laboratory areas.

Recommendations for Effective Access Control Scheme

- Provide identification badges for employees. These employee-issued identification badges should include a photograph, wearer's name, an expiration date, and color coding to represent clearance level to restricted areas.
- Visitors are issued a temporary visitor badge, which is uniquely colored to indicate visitor status. All visitors must be escorted.
- Consider adding layered access control to high-value areas within the facility (SCADA rooms, etc.).
- Segment access control such that only employees requiring access to high-value areas are permitted access, rather than all employees having access to all areas.
- Consider requiring an additional element beyond card access for high-value areas (for example, consider requiring card access and PIN for entry).

As an example, refer to Table 12.2, which depicts how segmented access zones might be applicable to the treatment plant of a water supply system.

TABLE 12.2 Water System—Layered Access Control (Example)

Security access zone	Affected areas	Access control process	Other complementary security measures
Zone 0	Public Areas outside perimeter fence	N/A - Public Zone	No trespassing signage Guard house with security officer checking vehicles High visibility site lighting Perimeter CCTV surveillance
Zone 1	Building lobby area	Visual inspection by security officer (or staff member) Badge display Inspection of parcels, packages	CCTV surveillance of incoming personnel (body size and facial features) Hardened blast-resistant exterior doors, with electronic mortise locks Interlocked exterior and interior lobby doors Door switch devices
Zone 2	Interior circulation corridor General mechanical spaces	Card access	Interior motion detection
Zone 3	Chlorine storage areas SCADA workstation areas Laboratory areas Security Equipment room	Card Access + PIN	CCTV surveillance of high value areas

12.4 MONITOR/CONTROL SYSTEMS: AREA DETECTION, VIDEO SURVEILLANCE, AUTHORIZED A/C

12.4.1 Introduction

Intelligent video surveillance and access control systems have drastically reduced the need for human intervention during monitoring operations of CCTV and access control systems (ACS). The manufacturers of such equipment have effectively empowered these security sub systems to make basic decisions on their own. These systems can be integrated together to work in concert with each other to provide immediate access to critical alarm and video information. The following is a detailed description of the predominant features that most high-level CCTV and access control systems possess.

12.4.2 Digital Video Management

A digital video recording system would be as cumbersome to use as a video cassette recorder if it were not for the video management software. There are scores of DVR manufacturers producing hardware for the surveillance industry, but the ones that sell the most units are the manufacturers that have developed the most user-friendly Graphical User Interface (GUI).

Camera Control/Recording Setup. These applications are almost always a Windows compatible software that provides a means to access the video stored on the recorder in an easily navigable environment. The applications are intuitive and require relatively little training to become proficient in their functions. These software applications provide a variety of utilities including a camera control/program module, a live and recorded video viewer, a recorded video search engine, and an alarm video management tool. The camera control module allows the user to control the pan, tilt and zoom functions of movable cameras, connected to the recording network, through the video management software. It also allows the user to program the recording schedules, recording speeds and other detailed features of the digital video recorder(s).

Live Viewing Module. The live/recorded video viewing module gives the user the ability to view video during or after the recording has taken place on the DVR. The video is typically viewed on a MediaPlayer type controller, and the video can often be multiplexed for 2 × 2 or 3 × 3 viewing of live video from multiple camera inputs. However, this function is dependent on the speed of the processor in the operator's workstation and the amount of RAM on board.

Searching Recorded Video. Of all of the management software capabilities, the search engine tool is perhaps the most important of the applications suite. The tool empowers the user to electronically search all of the video on the recorder for a given camera to find a specific event. The best applications allow the user to define the search area on a given video clip using a drafting style sizable box. The box is overlayed on top of the video image to define the area of interest where an event may have occurred. Then, the approximate time period or event date is entered into the search field and the search engine is started. The software then scans hundreds, even thousands, of hours of video to find video clips where activity has occurred within the area of interest. The user then plays the individual video clips to determine which one has the information of interest. This operation alone would save days of searching using an analog VCR, and truly makes managing the stored video a simple task.

Alarm Review Utility. Another useful application of video management software is the alarm viewer utility. This utility is generally an alarm notification tool that displays the live or recorded camera image as an alarm event is taking place. This module works with alarm inputs on the DVR to trigger on-demand recording or archiving of recorded video in response to an external alarm source, such as an intrusion alarm or access control violation. It will display the real-time video feed of the alarm, and stores the information as a video clip with up to 30 s of video prior to the alarm event. This allows the user to review the video prior to the alarm event to gain an accurate status of the alarm sequence.

Web Browser Application. Many of the best video management software packages also offer a web browser application that combines a number of the above referenced functions into a "watered-down" utility that can be operated from outside of the video network. The utility assigns access rights to a particular user that will allow them to dial into the digital management network from an off site workstation to review live, recorded, or alarm video stored on the network DVRs. This application is invaluable for system administrators and supervisors who may be empowered to make decisions based on recorded video events but do not have immediate access to the network site.

12.4.3 Intelligent Video Applications

Since the evolution of digitally networked video, software and firmware manufacturers have leaped at the opportunity to enhance the effectiveness of CCTV surveillance. These application engineers realized early that video data could be analyzed much like any other digital data. So, they began creating applications to reduce human intervention during CCTV surveillance, to streamline the audit processes and to create automatic system responses to alarm events. The following are examples of how intelligent video solutions have improved the effectiveness of digital video CCTV.

Access Control Authentication. An application that is currently being used in conjunction with access control systems is the video authentication feature. In a properly integrated system, the video camera associated with a particular access door or control point is "popped-up" on screen as a cardholder presents his or her card for access authorization. As a preprogrammed response in the ACS, the stored photographic image from the cardholder's badge is retrieved from the ACS database to allow the access control administrator to grant or deny entry. This is an extremely useful feature for high security control points and is streamlined by integrating the ACS and CCTV software on a shared workstation.

Facial Character Recognition. Another useful software application is facial character recognition. This application retrieves facial data from a person's image on a video screen and compares it to a known database of criminals or terrorists on a watch list. When and if a match is ever determined, an alarm notification box is activated on the CCTV workstation for the operator to acknowledge. The alarm window provides the comparison of the two facial images as well as a statistical probability value that the captured video image is the person from the watch list database. This application reduces the amount of manpower required to monitor a live CCTV system and drastically improves the probability that a known criminal can be found in a crowd of people by constantly comparing the images to a potentially large database of facial images. This task would be all but impossible for even a dozen CCTV operators at a live workstation.

Object/Target Tracking. In recent years, several manufacturers have released products capable of monitoring the quantity, speed and direction of objects in a fixed field of view.

These applications are frequently being used to count people entering a room, cars entering a parking lot and more commonly, vehicles in line at intersections. These applications initiate programmed responses to external applications, and in many cases, eliminate the need for human decision-making processes or intervention. These applications will prove useful in surveillance situations where an alarm should be initiated by vehicles or people entering restricted areas, such as cars or trucks in standing areas outside of airport terminals. These applications can also initiate an alarm when an object is moved or removed from a normally static field of view. For instance, if a cash register is removed from a camera's normal field of view, the CCTV system would initiate an alarm-requiring acknowledgement by the CCTV operator. A particular camera manufacturer uses a similar firmware module to enhance the functions of its PTZ camera line. The module provides the camera with the capability to track a human or object within the viewing constraints of the camera. This feature keeps the CCTV operator from having to follow a subject using the joystick controls of the system keyboard, which can often be a cumbersome task for a novice operator.

Behavior Tracking. A variation of the same object tracing software is also available as a behavior module. It is frequently used in retail applications where a particular type of human motion is a common indicator of a robbery. For instance, if a convenience store customer were to raise a straightened arm toward a store clerk in a camera's field of view for more than a few seconds, then an alarm output could be generated to a monitoring agency. This type of behavior is just a simple example of how human behavior is analyzed to become more intelligent video.

Digital Video Motion Detection (DVMD). DVMD is another intelligent video application that has similar characteristics to the previously described utilities. The reliability of other outdoor perimeter detection systems is such that DVMD represents a promising technology. DVMD processors collect data from a fixed camera image, and analyze it to look for changes in the "normal" scene. Significant advancements have been made to DVMD in recent years, which have dramatically improved the efficiency of the newest processors. These processors have achieved increased probability of detection while minimizing the false alarm rates in outdoor applications. Their detection algorithms have a number of variables that can be adjusted to increase the sensitivity of the detection device. The typical parameters are object size, object speed, and aspect ratio of the object to the background scene. These factors can all be adjusted to eliminate unwanted alarm conditions from blowing leaves, small animals, moving curtains, shadows, light reflections, birds, and insects. Unfortunately, DVMD remains a relatively unproven technology in outdoor applications. But there are several enhancements that can help alleviate the factors that frequently cause false alarms outdoors. For instance, using an Infrared camera instead of a color or black and white camera will all but eliminate disturbances from weather influences. A thermal imager is also unaffected by smoke, fog, and darkness, so lighting conditions do not need to be improved to enhance detection with an IR camera. So, in this application, the IR camera is definitely worth the additional expense.

12.4.4 Access Control Applications

Access control hardware and software is as prevalent as digital video recording equipment. There are literally scores of manufacturers who build systems that are installed in facilities all over the world. However, the most popular systems are those that have reliable hardware, scalability, flexible connectivity, a wide variety of options, and a user friendly

Graphical User Interface (GUI). As with video management systems, an ACS must be easy to use and easy to understand in order get maximum use of its capabilities.

Graphical Map Tool. Aside from the card readers themselves, the GUI is the next most critical component of an ACS. The GUI is the heart of the control system software, and is how the entire system is programmed and operated. The GUI usually includes a graphical map tool that allows the system administrator to create maps of the facility. Icons can then be placed on the maps to represent the controlled access points and other key inputs into the control system. These maps often include door position switches for overhead doors, perimeter gates, and fire exits as well as card reader locations and CCTV camera locations. These maps provide an easily understandable format for communicating alarm information, door status, and camera viewing vantage points. Without these maps, an ACS operator would have to have an intimate knowledge of the location of each access control point in order to interpret information from a database format.

Video Badging. The GUI usually includes a video badging utility also. These embedded software applications allow the ACS user/administrator to design, enroll, photograph, modify, produce, and even laminate photoidentification information on the access control cards. Having this utility within the ACS software makes the badging process much smoother since the information is entered into the database as the card is created and printed. It eliminates the need for a separate program and secondary step to enter the cardholder's information into the ACS database.

CCTV Integration. The GUI software for most major ACS manufacturers comes equipped with a CCTV interface built right in. The software typically includes a limited number of specific utility drivers that allow the software to communicate with camera switching equipment from the top CCTV hardware manufacturers. These drivers allow the access control software to call up live video during access control violations using the camera closest to the access control violation. This eliminates the need for the security administrator to switch from the ACS workstation to the CCTV keyboard to manually select a camera on the switcher. This step happens automatically with amazing precision. CCTV interface is an important consideration when selecting an access control system manufacturer for your specific projects.

Email Alarm Notification. Access control system software comes loaded with dozens of extra features that would require pages to properly describe. However, there are a couple of features that a novice ACS operator would appreciate. Most GUI applications include an email notification module that will alert an on-duty officer though an email ready pager or a personal digital assistant. This tool is extremely valuable for small installations where a full time ACS operator is not required on a daily basis. This allows the operator to perform other tasks while giving due diligence to the access control system.

Threat Level Adjustments. Another valuable utility is the threat level adjustment feature. This feature allows the ACS to be quickly upgraded to a preprogrammed higher security protocol in the event of a security breach or incident. A higher security environment may require occupants to present an access card and enter a PIN at control points equipped with a keypad card reader. Or, it may require third party authentication by the administrator at high security control points. The system can be upgraded to the next security level at any time by the system administrator through the GUI or, as with some advanced level systems, through any card reader keypad connected to the system. Since the advent of the Homeland Security threat level scale, access control system manufacturers have specifically marketed the merits of this feature. However, many systems have supported multiple preprogrammed security protocols long before September 11, 2001. It should come as no surprise to an outside observer that software applications dominate the most recent security system innovations. These digital appliances enhance the effectiveness of CCTV as well as electronic access control systems and even supplant security officers in many instances. Perhaps, as the technology is proven with each successive installation, these applications will achieve mainstream acceptance as tools of the trade.

12.5 SPECIAL APPLICATION: SECURITY SYSTEM COMPONENTS

12.5.1 Introduction

Even the best of comprehensive security system designs will, after installation, show areas where additional security monitoring will be required. These security components can be temporary, due to some factor such as site construction as an example, or it can be permanent, because comprehensive system testing shows a potential vulnerability.

12.5.2 Temporary Devices

Some examples of temporary devices are as follows:

Temporary Closed Circuit Television Monitoring. Temporary camera locations can be invaluable in covering areas that may block existing camera views or to cover areas or structures that may need to be monitored temporarily. Some applications may also warrant covert temporary camera installations in which a hidden camera is installed to monitor certain areas. This can be useful, for example, to mitigate employee theft or pilferage issues.

a. The most cost effective way to transmit video signals from cameras is via coaxial cable. There are several things that can be done to simplify the installation. These include the following:

  i. Use 12 or 24 VDC cameras with the power supply for the cameras located at the "head end" where the signals from the cameras are terminated.
  ii. Coax cable that provides the two conductors for the camera power in the same cable as the camera signal is available. By using this cable, only one cable is needed between the camera and the monitoring point. This enables a cost effective, rapid, and simplified installation.
  iii. Tremendous advances have been made in sending video images over a simple twisted pair of copper wire. Relative long lengths of a simple twisted pair of wires can transmit a very high quality video image. This technique utilizes a simple device at each end that converts the signal from a 75-Ω coax signal to a twisted pair signal. Many of these devices do not even require power other than the 1 V video signal. In addition, conventional telephone cables almost always contain extra pairs of wires that can be used for this type of video transmission. The video will not interfere with the telephone communications.

b. In many temporary installations, it is not practical or feasible to route video transmission cabling. In this case, wireless video transmission is a solution. There are a wide variety of wireless systems, some that require an FCC License, and some that transmit on frequencies that are exempt from licensing requirements. However, these exempt frequencies are used for other devices such are wireless telephones and are subject to interference. Some of the available transmission frequencies and their capabilities are as follows:

  i. 900 MHz.   This was the original frequency that was available. It is also the busiest frequency and will be subject to more interference. Also, the lower frequency needs a larger hole in the environment for the signal to pass. The signal is widest at a point midway between the receiver and transmitter. If this dimension is less than several feet in diameter, the quality of the video images will be degraded.
  ii. 2.4 GHz.   This is the frequency that newer wireless telephones use. This frequency has better penetration abilities because of the smaller diameter of the signal sine wave.
  iii. Higher frequencies are in the near microwave-to-microwave frequency area, and most require licensing. The higher the frequency the smaller the area required for the signal to pass.
  iv. Line of sight is obviously best for a clear signal transmission. But the signal will penetrate walls and other obstructions. However, anything that will reflect a visual image will also reflect the signal. Examples are a mirror, foil backing on insulation, metal siding, etc. The one object that absolutely blocks the signal is the human body because of the high water content. Other items with a saturation of water, wet wallboard for example, will also disrupt the signal. When setting up a wireless video system, a transmitter or receiver location adjustment of only a few feet may greatly improve the signal quality.
  v. Infrared can be used to transmit video, which has an advantage of being harder for others to intercept. However, it is no longer as popular as it used to be with the advent of radio systems.
  vi. For longer ranges, microwave can be used. FCC Licenses are required. However, line-of-site ranges of several miles are available.

c. If the area being monitored is wired with a computer network, net IP address cameras are an ideal solution. In this scheme, each network camera has its own IP address. Special software can be used to receive the images that are transmitted over the network.

12.5.3 Camera Limitations

With network cameras the following limitations apply:

  i. The bandwidth required by cameras on the network is considerable because of the large amount of information contained in real-time video. A T100 or faster network should be used and the number of cameras restricted.
  ii. To ease the burden on the network, it is desirable that the video signal is time-compressed, in which fewer than the 30 images per second, that makes up real-time video, can be transmitted. Ten images per second can still provide a lot of information, and reduce the required bandwidth by 66 percent, but will not provide a smooth real-time video image.

Temporary motion detection devices may also be required to cover areas blocked by temporary structures. There are several types of temporary motion detection systems that are available:

SECURITY HARDWARE AND SURVEILLANCE SYSTEMS 12.33

a. Microwave systems normally utilize a receiver and transmitter. However, passive microwave that reflects off of a reflective surface may be used. The advantage of the high-end systems is a range of up to 300 ft and good immunity to "invalid alarms."
b. In outdoor environments with many moving objects, such as trees swaying in the wind, etc., dual technology motion sensors are a good choice. These sensors utilize passive microwave and passive infrared combined into one unit. A signal from both technologies must be received for the detector to generate an alarm.
c. Active infrared systems that utilize a receiver and transmitter (in the same way as microwave systems) can be utilized. These systems use low-voltage devices and are easy to install. However, the range of these systems is lower than microwave systems.
d. Temporary closed circuit television cameras can also be used as motion detection devices. Digital multiplexers can be used to process 4, 8, or 16 camera signals and combine (multiplex) these signals into a single output, which can then be recorded. In addition to this capability, multiplexers can also be programmed to process movement in a video image and changes in the minute pixel images (the darkness or brightness of the image). When this change is detected, an alarm output can be activated. Most of these video movement processors can be set to ignore a portion of the video image frame. So, as an example, if there is a tree in the viewing area it can be ignored so that movement of the tree will not activate an alarm.

For high-security applications, biometric access control units can be provided. A biometric access control device is an access control device that matches a unique physical characteristic (hand, eye, fingerprint, etc.) of an individual to enable or deny access. Typical biometric devices include fingerprint reading, hand geometry reading, eye reading, voice recognition, signature recognition, and facial recognition.

a. Fingerprint reading devices utilize a fingerprint or palm print reading technology. The user places his/her finger or palm on a special reading device. This is the second most prevalent of technologies for biometric devices, capturing approximately 35 percent of the biometric market in 2002. Fingerprint data have a relatively small data size of 300 to 800 bytes.
b. Hand geometry reading devices are based on the profile of the hand. The user places his/her palm onto a special reading device that measures the length, width, thickness, and surface area of the hand. This is the leading biometric device technology, capturing approximately 46 percent of the biometric market in 2002. Typically, it takes less than 1 s to verify access.
c. Eye reading devices consist of two styles—iris mapping or retinal scanning. Iris mapping is typically preferred by users as a less invasive technology. Approximately 15 percent of biometric technology sold in 2002 used eye-reading technologies.
d. Voice recognition devices and signature recognition devices are not frequently used, as duplication of false entry can be allowed.
e. Facial recognition devices are in use in military and experimental applications, but have not been cost effective for widespread use.

Another high-security device that can be used in certain applications is an automated high-security-revolving portal. These are similar to a revolving door, except that they are integrated with access control, and optionally with integral body mass and metal detection sensing. If access is granted, the door automatically revolves, otherwise it backs up if access is denied. Some portals include integral CCTV camera coverage at portal door entrance, door exit, and a pinhole camera in the revolving door. Where warranted, these portals may be combined with biometric access control devices to ensure the highest degree of security.

12.6 DESIGN AND IMPLEMENTATION: GENERAL DESIGN AND IMPLEMENTATION CONSIDERATIONS

Below is a list of general recommended practices for the design and installation of access control and closed circuit television systems.

12.6.1 Closed Circuit Television (CCTV) System

General Recommendations

- Provide optimum view of facility with minimum number of cameras.
- Provide video coverage of all site entry points (i.e. vehicle gates, pedestrian gates, etc.)
- Provide video coverage of all exterior doors receiving card readers.
- Provide video coverage of all critical hatches and manholes.
- Provide a means for providing video coverage during primary power outages.
- Provide video coverage under all ambient lighting conditions (low light and sunlight).

Digital Video Recorder (DVR). Provide a secondary power source, 4-h duration capable, to supply power to the DVR during a primary power outage. Possible secondary power sources include

- Generator (NFPA 110 compliant)
- Battery backup
- Uninterrupted power supply (UPS)
- Provide variable frame-rate recording capability. (In other words, the DVR should be able to change to a higher-resolution rate of recording upon receipt of a relay contact or RS-232 signal, for evidentiary recording.)
- Provide capability of digitally tagging alarm events within the video footage for later review.
- Provide capability of remote pan, tilt, and zoom (PTZ) camera control via integral RS-232, RS-422 or similar communications connection.
- In case of primary power loss, if a secondary power source is not available the DVR should be capable of automatically rebooting and beginning immediate resumption of recording upon reconnection of power.
- Provide capability of backing up stored video information to CD-R or DVD-R media.

Cameras (exterior)

- Provide mechanical and electrical protection against tampering and vandalism to camera and associated cabling.
- Mount cameras to withstand wind load for area.
- Equip exterior cameras with sun-shield and heater/blower housing to minimize sun glare and fogging.
- Equip fixed position exterior cameras with varifocal, autoiris lenses.
- Provide cameras with automatic day/night switching capability. (Color to black and white.)

- Provide IR illuminators on cameras where minimum required lighting levels are not adequate to support proper operation and additional lighting is not desirable or practical. (Residential areas)
- Provide one or more preset camera position and focus locations that can be remotely called up via the RS-232, RS-422, or similar communications connection. (PTZ cameras only)

Cameras (interior)

- Provide mechanical and electrical protection for camera and associated cabling.
- Equip cameras with varifocal, autoiris lenses.
- Training and Startup
- Require the system integrator to provide minimum 8-h training for maintenance and operation to utility staff.

12.6.2 Access Control System (ACS)

General Recommendations

- Provide the capability for monitoring the status of the ACS at remote sites with off-site reporting via leased line or DSL. Provide capability now, implement in the future.
- Consider providing a method of monitoring of site perimeter fenceline. (IR beam detection, fiber-sense cable, microwave detection, etc.)
- Integrate access control system with CCTV system via RS-232 communication connections or dry-contact relay connections. In other words, when card reader is activated, camera pans automatically to preset position to capture the image.
- Provide magnetic door position sensors on doors, the areas where access is controlled by the Access Control System (card reader control.) In other words, add door monitoring to all exterior doors or roll-up doors in a building having an exterior card reader.

Card Reader

- Provide card reader on the nonsecure side of the site perimeter fence at the main entrance gate. Upon presentation of a valid card the ACS will de-energize a dry contact relay output to the SCADA (telemetry) system for notification of authorized occupancy, "Disarm" the ACS, and actuate the gate controller.
- Provide card reader on the secured side of the site perimeter fence at the main entrance gate. Upon presentation of a valid card the ACS will reenergize a dry contact relay output to the SCADA (telemetry), "Arm" the ACS, and actuate the gate controller.

Door Status Switch

- Provide magnetic gate position sensors on site perimeter gates.
- Provide magnetic door position sensors on doors, the areas where access is controlled by the Access Control System (card reader control.)

Training and Startup

- Require the system integrator to provide minimum 8 h training for maintenance and operation to utility staff.

12.6.3 Considerations for Central Security Areas

When the basic decision on alarm, surveillance, and access control systems is contemplated several considerations must be made (e.g. whether or not they must be monitored, adjusted for optimum reliable operation, set to give notification when activated, and/or kept in an archival history of activities and images stored):

The location of the "head end" computers and controllers should be located in as secure a location as possible. If the head end is monitored 24/7 and the monitoring of systems is the sole task of the monitoring personnel, the head end can be located in a secure inner part of a building.

The following elements should be included if possible:

a. A windowless, unmarked location offers higher security.
b. Separate HVAC system should be provided for this area if possible, sized to meet the heat load of the equipment within the room.
c. A dedicated power source with UPS protection should be utilized. A diesel backup generator should be considered for larger installations. As a minimum, 12 h of backup is recommended.
d. A mantrap entrance to the facility offers protection against unauthorized forced entry, or tailgating by unauthorized personnel.
e. Any windows including interior windows, should be bullet resistive.
f. Walls, including walls adjacent to windows should be bullet resistive.
g. Telephone lines into this area should not go through the normal telephone panel (if the telephone panel is located outside this area) but should be wired directly to the security monitoring area by a different route into the building.

Integrated systems that combine the monitoring of CCTV, access control, intrusion alarms, and fire alarms are becoming more common as the software and the power of the computers it runs on becomes more capable. This capability has been enhanced because of digital input from the systems that can be processed by current database management systems, such as Microsoft access, utilizing Sequel Server databases.

a. The era of dozens of monitors displaying video camera images is over. There is nothing to be gained by viewing scenes where nothing is happening. The advent of video motion detection combined with the digital matrix switching and digital hard drive storage of those images allows only those video images where there is movement to be brought to the attention of the operators. One monitor per digital switcher, and a monitor for study of a specific image of interest may be all that is required. It has been proven that a person can view a bank of monitors effectively for only a few minutes; to view them comprehensively for an extended period is not possible.
b. Graphic interfaces that can automatically display maps showing the location of an event and possibly a video image of the event are much easier for an operator to digest and evaluate than a printed recap of the event. However, this printed recap of the event is important as a record of the event. The printed record should indicate the time and date of the event as the operator that was monitoring the system at the time. These hardcopy printouts should be archived, for future reference.
c. The rapid evolution of CCTV image storage has rapidly transitioned from analog videotapes to digital storage. Digital storage methods are more reliable, scaleable, are easy to backup, and do not require daily human intervention, such as is necessary to change the videotape(s) and which is normally required on a daily basis. Digital storage systems come in many configurations but generally most utilize multiple hard drives, like those used in network servers, for short term storage of 30 days or less. Long-term digital

storage can be accomplished via compact disc (CD), digital video disc (DVD), or digital tape backup. Another advantage of digital storage is a drastic reduction time in recalling the images of a stored event.
- **d.** Digital storage also makes the display of employee photo ID along with the other personal information very easy. This simplifies confirming that the person attempting access with credentials is actually who he claims he is. This capability should be part of any system being considered.
- **e.** A graphic interface map (possibly with touch-screen capabilities) on the display with all the CCTV, access control, intrusion, and fire alarm devices depicted, allows rapid and correct acknowledgement of several types including

   - **i.** Acknowledgement of intrusion alarms by clicking on the device creating the alarm.
   - **ii.** Summary information and acknowledgement of fire alarm conditions by clicking on the map area displaying the alarm condition. Note—This does not preclude automatic fire alarm notification to the proper authorities, or other code-required elements necessary for the fire alarm system as required by the Authority Having Jurisdiction.
   - **iii.** Point-and-click interfaces with cameras, door, or other monitoring devices. This will allow the monitoring personnel to simply click on a camera icon to display the camera or click on a controlled door to manually grant access.

The computers on which these systems are operating should be equipped with RAID drive hardware where the stored data will not be lost in the event of a hard drive failure. In larger more elaborate systems, a redundant computer should be utilized so that a computer failure will not shut down the system. As with all computer systems, regular backups of critical information, with offsite secure storage are essential. This computer system should also be protected by placing surge arrestors on all incoming wires to prevent a surge from damaging the "head end." Fiber optic is the ideal data transmission medium for inputting data because it is immune to power surges, offers higher transmission speeds, and in many cases is cost competitive with copper cabling such as coax or twisted-pair.

All security control hardware should be located in a user-friendly environment. Comfortable and durable seating should be provided, which is very important to an operator sitting for several hours at a time. Lighting should not be harsh and it should be designed to not produce glare in the video displays. Carpet should be utilized if possible to deaden sound. Walls should be painted in pleasing tones, and wall hangings should also control the ambient sound of the room. Headsets should be considered to free both of the operator's hands for the required control manipulations. Each employee should have a small locker for headset storage and as a place to secure personal possessions while working.

Obviously, this is only a thumbnail sketch of considerations for designing a functional and secure "head end" control center that is a pleasant and therefore effective place to work. Try to consider all aspects of the environment in the conceptual stage and the result will be much more secure, efficient, and effective.

CHAPTER 13
OPTIMAL LOCATION OF ISOLATION VALVES: A RELIABILITY APPROACH

Sukru Ozger and Larry W. Mays
*Department of Civil and Environmental Engineering,
Arizona State University, Tempe, Arizona*

13.1 INTRODUCTION

Distribution systems of pipelines, pipes, pumps, storage tanks, and the appurtenances such as various types of valves, meters, etc., offer great opportunities for terrorism because they are extensive, relatively unprotected and accessible, and often isolated. The physical destruction of a water distribution system's assets or the disruption of water supply may be more likely than contamination. Loss of water pressure compromises firefighting capabilities and could lead to possible bacterial build-up in the system. Potential for creating a water hammer effect by opening and closing major control valves and or turning pumps on and off too quickly could result in simultaneous main breaks.

Control valves are used to regulate flow or pressure at different points of the system by creating headloss or pressure differential between upstream and downstream sections. The mission of isolation valves is to isolate a portion of the system whenever system repair, inspection or maintenance is required at that segment. The two most common types of isolation valves are gate valves and butterfly valves. Unlike the control valves, little attention has been paid to the layout and operation of isolation valves. Yet, from a reliability point of view, isolation valves are of great interest since they determine the extent of isolation should a portion of the system need to be isolated from the rest of the system. As an extreme example, a system with no valves would have to be shut down completely at the source for any maintenance. On the other hand, if there is sufficient valving in the system the outage or isolation can be limited to a small portion of the system.

The objective of this chapter is to describe a methodology for determining the optimal location of isolation valves based upon incorporating reliability aspects of a network. Under normal operating circumstances isolation valves are used to isolate a portion of the system that needs repair, inspection, or maintenance. During an emergency event that results from a terrorist activity, the isolation of a portion of a system may also be an important aspect of continuing the operation of a water distribution system.

Recent incidences of gas crisis in Phoenix metropolitan area, Arizona, and the biggest blackout in the history of United States in the Eastern portion of the country revealed that utilities that provided tens of millions of people were not equipped with emergency plans. The same incidences, although the causes were announced accidental, gave a clue as to how similar cases might arise in water supply systems as a result of terrorist attacks.

Arizona has no gasoline refineries and all of its supply must be imported from other states. There are two major pipelines carrying gasoline into the area. The western pipeline, called The Los Angeles line, supplies about 70 percent of the Valley's gasoline. The eastern pipeline brings fuel from Texas by way of Tucson. The gas crunch in Phoenix began when the 60,000-barrel-a-day pipeline supplying roughly 30 percent of the Valley's gasoline from Tucson ruptured, spilling about 10,000 gal on July 30. The line was shut down by the operator on August 8 because of concerns that there could be more problems. While the eastern pipeline was still closed, the western line also failed to deliver gasoline on August 21 due to a gravel truck in the Los Angeles area falling onto the pipeline. While the underground line suffered no damage, the line was closed for about seven hours since federal regulations required an inspection before reopening. Finally, the Phoenix gas crisis was quickly mitigated when petroleum began moving through the rerouted eastern pipeline on August 25, 2003. The incident was interesting because it showed what could happen if a similar crisis arose in a water supply system. It was also interesting that the shortage of gas caused panic among the people in the area and majority of the drivers topped up their tanks, although under normal circumstances they would not. In other words, while supply capacity went down by about 30 percent, demand from the area increased significantly making the problem harder to cope with.

The biggest blackout in the history of the United States happened in the summer of 2003 affecting several states and millions of people living on the Eastern portion of the country. The interesting consequence of the blackout, that lasted for several days, from a water supply point of view was that the cities of Cleveland and Detroit were mostly relying on pumping for water delivery and suffered severely from the loss of pumping capacity during the blackout.

13.2 DEMAND-DRIVEN ANALYSIS VS. PRESSURE-DRIVEN ANALYSIS

Hydraulics of a water distribution network can be approached from two different perspectives. The difference between them comes from the level of primacy given to nodal demands vs. nodal pressures. The first approach assumes that consumer demands are always satisfied regardless of the pressures throughout the system and formulates the constitutive equations accordingly to solve for the unknown nodal heads. This approach is called demand-driven analysis and is used by almost all the traditional network hydraulic solvers. Examples are EPANET by Rossman (1994) and KYPIPE by Wood (1980).

A water distribution network can be designed or operated using demand-driven analysis under normal operating conditions. This is usually done by adjusting such independent decision variables as simple or rule-based controls for pump operations, valve settings, reservoir/tank levels, etc. If a proper design is made for a given design demand loading, then nodal pressures throughout the system must be well above the minimum pressure required. On the other hand, if the same network were analyzed with a higher demand loading, it would not be surprising to see warning messages such as "Negative pressures at 6:00 h." from the EPANET model while nodal demands are fully satisfied at the same time period. Almost all demand-driven models possess a partial recognition of this weakness with similar warning messages when negative pressures are calculated from a network hydraulic analysis.

The major drawback of the demand-driven approach is that it fails to measure a partially failed network performance resulting from abnormalities in physical and nonphysical system components (e.g., pipe breaks, fire-fighting demands, etc.). In such cases, if demand-driven analysis is used, it may produce very unrealistic results, e.g., negative pressures are calculated at some nodes in the system. It is not uncommon that a hydraulic analysis using a demand-driven approach yields large negative pressures at one or more junctions of a distribution network.

The second approach to a network hydraulic analysis is called head-driven (occasionally called pressure-driven or pressure-dependent) analysis. Here, the primacy is given to pressures. A node is supplied its full demand only if a minimum required supply pressure is satisfied at that node. If the minimum pressure requirement cannot be met, then only a fraction of the nodal demand can be satisfied. That fraction is determined by a predefined relationship between nodal head and nodal outflow.

As mentioned above, demand-driven models work well under normal operating conditions. Major applications of head-driven analysis, on the other hand, are for analysis of network performance under partially failed conditions. Since a network reliability assessment requires evaluation of partially failed conditions of a network, pressure-dependent analysis is much more suitable for reliability assessment of water distribution networks.

One of the major problems with the pressure-driven approach is that it requires extensive field data collection to determine the relationship between nodal heads and nodal flows. Perhaps, that relationship is unique for each junction of a network since it depends on such physical characteristics as type of service connection and the development type that the junction serves (e.g., residential, high-rise, schools, hospitals, etc.). Calibration of models using field data is also part of the large effort needed to be able to use pressure-dependent analysis.

The second major drawback of head-driven analysis is that it lacks robust methods for the computational solution of the constitutive equations (Tanyimboh, Tabesh, and Burrows, 2001). Almost all the traditional network hydraulic models are built upon the demand-driven approach. Therefore, it is very desirable that reliability analysis techniques based on interpretations or transformations of existing demand-driven analysis from a pressure-driven approach be developed.

13.2.1 Hydraulic Equations for a Network Simulation Problem

Two sets of equations are needed to solve a network simulation problem. First, conservation of flow must be satisfied at each network junction. The second set of equations describes a nonlinear relation between flow and headloss in each pipe, such as the Hazen-Williams or Darcy-Weisbach equation. These equations form a coupled set of nonlinear equations whenever the network contains loops or more than one fixed-head source (Rossman, 2000a). The fact that most distribution systems are looped and/or have multiple sources requires the aid of a computer to solve the nonlinear set of equations using an iterative procedure. The basic equations in a network model are described for a network with N junction nodes and NF fixed grade nodes (tanks and reservoirs) by Rossman (2000b).

Let the flow-headloss relation in a pipe between nodes i and j be given as:

$$H_i - H_j = h_{ij} = rQ_{ij}^n + mQ_{ij}^2 \tag{13.1}$$

where H = nodal head
h = headloss
r = resistance coefficient
Q = flow rate
n = flow exponent
m = minor loss coefficient

The value of the resistance coefficient and the corresponding flow exponent depend on which friction headloss formula is being used. Most network models use one of the following flow-headloss relationships: (1) Hazen-Williams, (2) Darcy-Weisbach, and (3) Chezy-Manning.

In addition to the first set of equations described above, conservation of flow around all nodes must be satisfied which forms the second set of equations

$$\sum_j Q_{ij} - D_i = 0 \quad \text{for } i = 1, \ldots, N \tag{13.2}$$

where D_i = flow demand (or supply) at node i
  j = set of nodes directly connected to node i
  Q_{ij} = flow in pipe ij

By convention, flow is positive if it is into the subject node and negative, otherwise.

As previously mentioned the conventional network hydraulic solvers such as EPANET by Rossman (1994) and KYPIPE by Wood (1980) are demand-driven. Demand-driven analysis seeks a solution for all heads H_i and flows Q_{ij} that satisfy the two sets of equations, Eqs. (13.1) and (13.2), for a set of known heads at the fixed grade nodes. The basic assumption in the demand-driven analysis is that nodes are assigned demands that are assumed fully satisfied. For example, variables used in the solution algorithm of EPANET are divided into unknown and known variables (Rossman 1994):

The unknown variables are

y_s = height of water stored at tank node s
Q_s = flow into storage tank node s
Q_{ij} = flow in pipe connecting nodes i and j
H_i = hydraulic grade line elevation at node i (elevation plus pressure head)

and the known variables (constants) are

A_s = tanks cross-section area
E_s = elevation of node s
D_i = demand (or supply) at node i

The major assumption of demand-driven analysis—demands are fully satisfied—works well under normal operating conditions. On the other hand, it is not uncommon that under abnormal conditions a solution arises where pressures are negative or unacceptably low, since demands are satisfied regardless of validity of calculated pressures at network nodes (Bouchart and Goulter, 2000). A few examples of such abnormal conditions may occur when demand loading on the network exceeds the design demand loading—fire-fighting demands for which the network was not designed and closure of a portion of a network because of contamination or pipe main breaks. In reality, there begins a shortfall in the volume of water actually delivered to consumers as the pressures in the network drop below some threshold value. Naturally and ideally, a lower bound for this threshold value is zero gauge pressure.

Water distribution networks are almost always skeletonized to some degree, depending on the purpose of modeling, forcing demands from secondary networks to be lumped into nearby junctions. Because of various headlosses associated with secondary networks and different elevations of outlets in these networks, it is very difficult, if not impossible, to define a relationship between pressure at a junction and the flow available from that junction into the secondary network it serves. A tedious way of doing that is field data collection and calibration. A typical value for threshold pressure is below the minimum of 20 m in network studies (Bouchart et al., 2000). Tanyimboh et al. (1999) states that pressures of 15 to 25 m are the minimum acceptable standards with the use of network models. All this implies that demand-driven algorithms cannot cope with situations where pressures are less than satisfactory at some demand nodes (Tanyimboh, Tabesh, Burrows, 2001) and thus fail to depict deficient network performances.

Service Pressures. Criteria for assessing acceptable service pressures in a distribution network may vary from system to system. Therefore, there are no universally acceptable pressure ranges. Chin (2000) lists the following considerations for assessing the adequacy of service pressures:

1. Flow is adequate for residential areas when pressures are above 240 kPa (35 psi);
2. The pressure required at the street level for excellent flow to a 3-story building is about 290 kPa (42 psi);
3. Adequate flow to a 20-story building would require about 830 kPa (120 psi) of service pressure at the street level, which is not desirable because of the associated leakage and waste. Thus, very tall buildings are usually served with their own pumping system;
4. In general, pressures in the range of 410–520 kPa (60–75 psi) would be adequate for the following:

   - Supplying normal water delivery to buildings up to 10 stories
   - Providing adequate sprinkler service in buildings of four to five stories
   - Supplying water for fire protection
   - Handling fluctuations in pressure caused by clogged pipes and excessive length of service pipes

Chase (2000) states that pressures in most cases should be maintained above 207 kPa (30 psi) and below 689 kPa (100 psi) during normal operations. Pressures above 689 kPa tend to increase water waste through undetected leaks and may cause damage to residential and commercial plumbing systems or pipe breaks. The minimum pressure requirement may be lowered to 138 kPa (20 psi) during emergency conditions such as a fire. A 138-kPa pressure is sufficient to supply the suction side of pumps on a fire-pumper truck. According to Walski (2000), the threshold pressure value below which a failure to satisfy the full demands occurs is usually assumed to be below the minimum of 20 m (~196 kPa or 28.4 psi).

The Office of Water Services in England specifies a minimum acceptable static pressure of 7 m (~68.5 kPa or 10 psi) below which customers may be entitled to compensation for less than satisfactory service. On the other hand, it is not uncommon that minimum acceptable pressure is specified as high as 25 m (~245 kPa or 35 psi) to allow for possible increases in demand (Tanyimboh, Burd, Burrows, and Tabesh, 1999). In general, nodal pressures of 15 to 25 m will guarantee satisfactory service at all related stop taps in a distribution system. Pressures that are lower may cause a shortfall in supply and failure to supply the full demands. Tanyimboh et al. (1999) also suggests that in the absence of field test data, a good approximation to critical pressures can be obtained by the equivalent static pressure at the height above ground level of rooftop water tanks in typical 2-story residential areas.

Water using devices in residential houses may also require certain minimum pressures to operate. For example, most dishwasher manufacturers specify minimum working pressures anywhere from 20 to 40 psi.

Another factor that should be considered while estimating the critical pressure is the degree of network skeletonization. Typically, smaller pipes are excluded from hydraulic analysis in water distribution network models. The result is that total demands from the consumers located along those smaller lines are lumped into the nearby junctions located on larger pipes. Thus, when choosing the critical pressure for a junction, such characteristics as outlet elevations and headlosses of the secondary network that the junction serves must also be taken into account.

13.2.2 Head Driven Analysis

Demand-driven analysis explained in the previous section assumes that consumer demands in a distribution network are fully satisfied regardless of the validity of calculated pressures. It can be argued that many modern water using devices such as toilets, washing machines, dishwashers are essentially pressure-independent provided a minimum threshold pressure is available, which supports the assumption of primacy of demands. Most commercial dishwashers require a minimum water pressure of 20 to 40 psi.

The philosophical basis of head-driven analysis, the counter approach to demand-driven analysis, is explained by Rossman (2001) as follows:

> "... a distribution system can be thought of as a completely closed, pressurized system with a set of variable sized orifices that connect to various devices that are open to the atmosphere (e.g., sinks, toilets, water heaters, washing machines, pipe leaks, etc.). For a specified opening of these orifices, the flow out of them will be dependent on the pressure maintained in the system (which in turn is dependent on the setting of the demand orifices) ..."

Thus, unlike demand-driven analysis, the head-driven approach treats demands as unknown variables and recognizes the primacy of pressures by defining a relationship between the outflow and the pressure at each node. This relationship is given, in its general form, by (Tanyimboh et al., 1999)

$$H_j = H_j^{min} + K_j Q_j^{n_j} \qquad (13.3)$$

where H_j = head ay node j corresponding to demand Q_j
K_j = flow-resistance coefficient
n_j = exponent

H_i^{min} is the minimum required nodal head below which the outflow at node j is unsatisfactory or zero. Since demands are unknown, Eq. (13.3) can be rearranged as

$$Q_j = \left(\frac{H_j - H_j^{min}}{K_j} \right)^{1/n_j} \qquad (13.4)$$

The head-driven analysis requires substitution of Q_j into Eq. (13.2) in the place of D_i. The solution algorithms for solving the resulting problem have been described by Gupta and Bhave (1996) and Tabesh (1998).

As explained in the previous section, demand-driven analysis may produce unrealistic results (i.e., pressures being negative or less than satisfactory), especially under partially failed conditions of a system. In such cases, head-driven analysis is superior to demand-driven approach in that the former is able to determine the nodes with insufficient supply and the respective magnitudes of the shortfalls. On the other hand, performing a head-driven analysis requires extensive field data collection and calibration to determine K and n values, possibly for each node (Gupta and Bhave, 1997). In addition to these practical difficulties, lack of robust methods for computational solution of constitutive equations in head-driven analysis makes the demand-driven network models the primary choice for hydraulic analysis or reliability assessments.

Node Flow Analysis. In order to overcome the weaknesses of demand-driven analysis, Bhave (1991) developed a method called node flow analysis (NFA) to predict performance when a distribution network is deficient. The method was based on the assumption that when a network is deficient and thus unable to satisfy the nodal demands, the network will try to meet the demands, as far as possible, under the given conditions. In other words, the flow in the network will adjust such that the total supply is maximized under deficient conditions. Thus, the network flow analysis problem is considered as an optimization problem in the NFA subject to certain constraints that relate hydraulic gradeline at each junction to available flow at that junction.

The complete optimization problem is as follows:

$$\text{Maximize:} \quad \text{Total Outflow} = \sum_j q_j \quad (13.5a)$$

where q_j is the flow available at node j, subject to

$$H_m = H_{om} \quad \text{for all } m \quad (13.5b)$$

where m is the number of source nodes and H_{om} is the specified hydraulic grade line (HGL) value at source node m.

$$q_j = q_j^{req} \quad \text{if } H_j \geq H_j^{min} \quad (13.5c)$$

or

$$0 < q_j < q_j^{req} \quad \text{if } H_j = H_j^{min} \quad (13.5d)$$

or

$$q_j = 0 \quad \text{if } H_j \leq H_j^{min} \quad (13.5e)$$

$$\sum_i Q_{ij} - q_j = 0 \quad \text{for all } j \quad (13.5f)$$

where i is the set of all nodes connected to node j and assuming flow is positive if it is into node j.

$$\sum_{loop} h_L = 0 \quad \text{for all loops} \quad (13.5g)$$

The basic differences from the usual demand-driven formulation are the additional constraints defined by Eqs. (13.5c) to (13.5f). Unlike Eqs. (13.3) and (13.4), a discrete flow-head relationship is defined. The method still requires defining a minimum pressure threshold at each node.

The network-solvability rule known as the unknown-number rule requires that in order for a network problem to be feasible, the total number of unknowns must be equal to the total number of nodes in the network. Let the number of source nodes and junctions in a network be S and J, respectively. For source nodes, the nodal flows are the only unknowns. On the other hand, neither the flows nor the HGL values at junctions are known in the NFA described above. Depending on the number of demand junctions, the number of these unknowns (junction heads + junction flows) will be between $J + 1$ (the network must have at least one demand node) and $2J$ (all junctions are assigned nonzero demands). Thus, the total number of unknowns for the entire network lies between $M + N + 1$ and $M + 2N$, which is greater than the total number of nodes in the network, $M + N$.

In order to bring down the total number of unknowns, NFA analysis treats either the demand or the HGL as unknown at each demand node. In other words, if constraint (13.5c) is chosen, then nodal flow is treated as known, being equal to the nodal demand, and the nodal head as unknown. On the other hand, choosing constraint (13.5d) implies a known nodal head but an unknown nodal flow, which is also constrained to being less than the nodal demand. Finally, choosing constraint (13.5e) indicates a strictly zero nodal flow and an unknown nodal head subject to being less than the pressure threshold.

The solution procedure to the NFA described above begins by imposing constraint (13.5c) on all demand junctions. Note that this is equivalent to the usual demand-driven analysis. Once the available HGL values are determined at all demand nodes, it is determined which one of the constraints (13.5c) to (13.5e) actually applies to each demand node and a new solution is obtained. The solution should normally be stopped here if all the nodal constraints are indeed applicable, which can be verified from the results of the new solution. However, as Tanyimboh and Tabesh (1997) have also found, when a network with locally insufficient heads is simulated using the demand-driven approach, the deficiency appears to be far more serious and widespread than it is in reality. The result is that there are almost always some demand junctions in the network that are pressure deficient from an initial demand-driven analysis but then violate the constraint given by Eq. (13.5d), because of $q_j > q_j^{req}$, at the end of the next solution. Thus, the solution proceeds in an iterative manner until a set of demand nodes are found in the deficient network that are truly pressure deficient. In other words, when no discrepancies are found between each nodal head and nodal flow and the constraint assigned to that node, the final solution is achieved.

13.3 SEMI-PRESSURE-DRIVEN ANALYSIS

This section describes the formulation and implementation of a semi-pressure-driven approach developed by Ozger and Mays (2003). The method formulates the network flow problem similar to the NFA of Bhave (1981) and can be used to predict deficient network performance under the following partially failed conditions:

- Simulation of fire-fighting demands at various points of a network
- Simulation of future demand loadings for better targeting and prioritization of network-upgrading schemes
- Simulation of failure of various network mechanical components such as pipes, valves, and pumps

The semi-pressure framework described below is developed around the EPANET2 software using EPANET Toolkit Functions and C++ programming language. The terms "minimum required pressure," "critical pressure," and "threshold pressure" are used interchangeably in the following discussion. By definition, critical pressure at a junction is a pressure value below which the system fails to supply the full demands attributed to the given junction.

The semi-pressure-driven method starts with a hydraulic analysis of the deficient network using EPANET2. In other words, the network is first simulated from a demand-driven point of view to identify junctions of the network with pressure values below the critical pressure. Threshold values within a network may vary from junction to junction depending on the characteristics of the service connection and the type of development served by that node. Thus, ideally, critical pressure value is a unique number for each junction in the network and must be determined empirically by field testing. However, in the absence of such field data, approximate values of critical pressures can be determined using one or more of the guidelines discussed in Section 13.2.1.

Once the pressure-deficient nonzero demand junctions are identified from an initial demand-driven analysis, the next step is to determine the available flows at those nodes, presumably knowing that the remaining demand nodes are fully satisfactory in terms of both pressure and demand, and the zero demand nodes have pressures above the cavitation

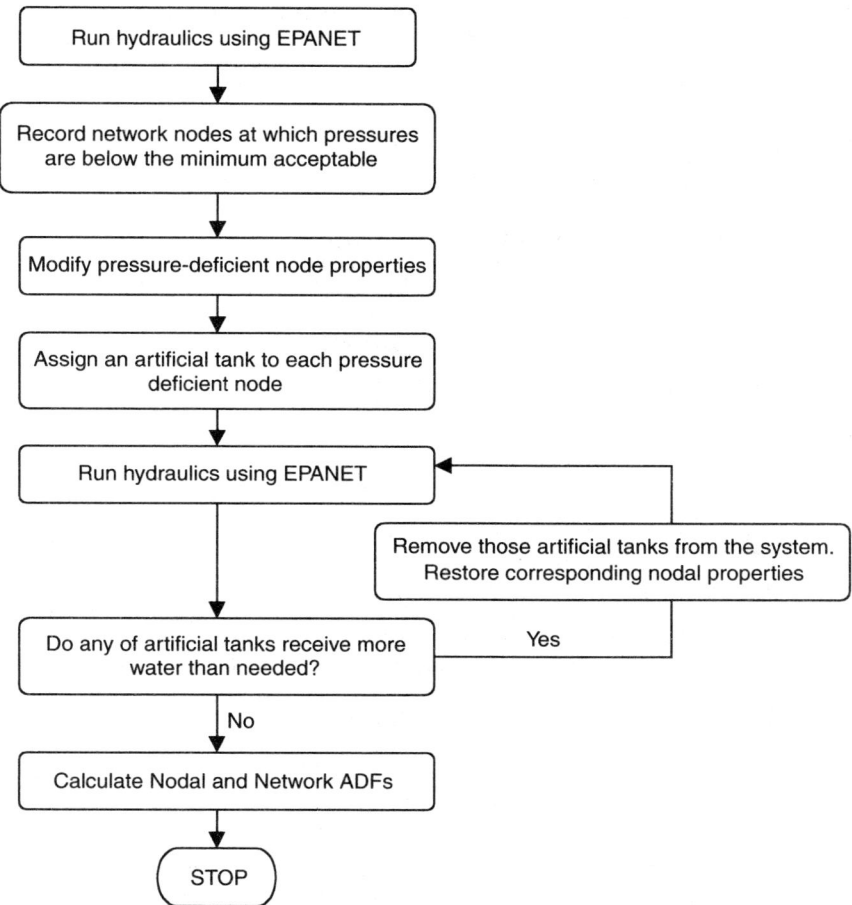

FIGURE 13.1 Flowchart of semi-pressure-driven algorithm.

limit. For this purpose, the following modifications are made to each pressure-deficient nonzero-demand node:

1. {New node elevation} = {Original node elevation} + {Threshold pressure head}
2. Set demand to zero.
3. Connect an artificial reservoir to the node by an infinitesimally short CV pipe that allows flow only from the node to the reservoir.
4. {Artificial tank elevation} = {New node elevation}.

With these modifications, demand at each pressure-deficient junction in the algorithm is treated as an unknown while a pressure threshold is imposed. Figure 13.1 shows a flowchart of the semi-pressure-driven algorithm. Note that the algorithm proceeds in an iterative manner. That is, if one or more artificial reservoirs receive more water than their nodes demand, those artificial reservoirs are removed from the network and the original elevations and demands at the corresponding nodes are restored. A performance index, later to be used for a network reliability/availability assessment, can be defined at each node as

$$\text{ADF}_j = \frac{Q_j^{avl}}{D_j} \tag{13.6}$$

for a steady-state analysis where ADF_j is the available demand fraction at node j, Q_j^{avl} is the available flow to node j, and D_j is the total consumer demand allocated to node j. A network-wide-available-demand fraction is defined as

$$\text{ADF}_{net} = \frac{\sum_{\text{all nodes}} Q_j^{avl}}{\sum_{\text{all nodes}} D_j} \tag{13.7}$$

Similarly, for an extended period simulation (EPS) analysis, nodal ADFs are given by

$$\text{ADF}_j = \frac{\sum_{t=1}^{nt} Q_{jt}^{avl}}{\sum_{t=1}^{nt} D_{jt}} \tag{13.8}$$

where t is the time step and nt is the total number of time steps from the time of component failure to when the network starts its normal operation following restoration of the malfunctioned equipment. Random pipe breaks typically take from several hours to a day to repair. It is difficult, on the other hand, to estimate the repair/replacement time for network components damaged as a result of a deliberate attack.

Systemwide ADF for an EPS analysis is then given by

$$\text{ADF}_{net} = \frac{\sum_{t=1}^{nt} \sum_{\text{all nodes}} Q_j^{avl}}{\sum_{t=1}^{nt} \sum_{\text{all nodes}} D_j} \tag{13.9}$$

It can be shown, mathematically, that the systemwide ADF is equivalent to the demand weighted average of nodal ADFs. Thus, unlike conventional reliability models based on pressure deficiencies, there is a deterministic relationship between nodal reliabilities and the system reliability in this approach.

13.3.1 Applications of Semi-Pressure-Driven Analysis

The semi-pressure-driven methodology described in the previous section can be useful for the following purposes. First, it can be used as a descriptive tool to predict the deficiency of network performance in terms of supply deficiency for what-if scenarios. Some of the most critical what-if scenarios are as follows:

1. One or more sources in the system are shutdown.
2. A pump station is out of service.
3. A control valve that has logistic importance in terms of flow delivery is broken.
4. A transmission line is damaged and taken out of service for repair.
5. A given portion of the network has to be isolated because of contamination.

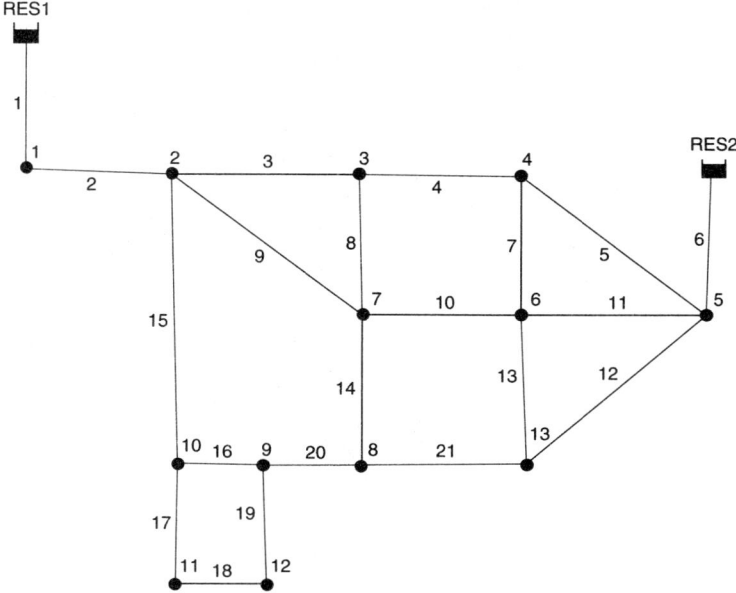

FIGURE 13.2 Application network for semi-pressure-driven analysis.

The input requirements for any what-if scenario are an accurate description of the equipment/network component that will be out of service, the network isolation valve scheme, and the duration of repair/replacement process only if an EPS analysis is needed. Once a semi-pressure-driven analysis is performed for any what-if scenario, the following questions can be answered using the results:

1. Are there enough isolation valves so that the isolation can be reduced to as small an area as possible?
2. How much of the demand can be supplied at each junction?
3. How much of the networkwide demand can be satisfied?
4. What is the temporal variation of nodal and networkwide ADFs?
5. What portions of the network are the most vulnerable to the given what-if scenario?
6. What operational changes could lessen the deficiency in network performance?
7. How long does it take for the network to fully restore its normal operational ranges once the repair/replacement work is finished?

13.3.2 Steady-State Application

To provide a more effective description of the proposed semi-pressure-driven approach, an analysis was conducted on the distribution network presented in Fig. 13.2, with the pipe and node characteristics of the given network listed in Tables 13.1 and 13.2, respectively. First, a fully functional network (i.e., no pipe failures) is analyzed. Then, assuming no simultaneous pipe failures can occur, all single pipe failure scenarios are simulated. The threshold pressure head, below which there begins a shortfall in fulfillment of nodal demand, is taken as 15 m for all junctions of the application network.

TABLE 13.1 Pipe Characteristics

Pipe ID	Length (m)	D (mm)	C (H-W)
1	609.60	762	130
2	243.80	762	128
3	1524.00	609	126
4	1127.76	609	124
5	1188.72	406	122
6	640.08	406	120
7	762.00	254	118
8	944.88	254	116
9	1676.40	381	114
10	883.92	305	112
11	883.92	305	110
12	1371.60	381	108
13	762.00	254	106
14	822.96	254	104
15	944.88	305	102
16	579.00	305	100
17	487.68	203	98
18	457.20	152	96
19	502.92	203	94
20	883.92	203	92
21	944.88	305	90

Table 13.3 is a comparison between the usual demand-driven analysis and the semi-pressure-driven results when pipe 3 in the application network fails. Note that there are nine demand nodes at which the minimum pressure threshold of 15 m cannot be satisfied to fully supply their demands. Table 13.4 summarizes the results from all single pipe failure scenarios that are simulated using the proposed semi-pressure-driven approach. Note that the total network demand is 3146.4 m^3/h and Eq. (13.7) is used to calculate network ADFs.

TABLE 13.2 Node Characteristics

Node ID	Elevation (m)	Demand (CMH)*
1	27.43	0.0
2	33.53	212.4
3	28.96	212.4
4	32.00	640.8
5	30.48	212.4
6	31.39	684.0
7	29.56	640.8
8	31.39	327.6
9	32.61	0.0
10	34.14	0.0
11	35.05	108.0
12	36.58	108.0
13	33.53	0.0
RES1	60.96	N/A
RES2	60.96	N/A

*CMH = cubic meter/hour

TABLE 13.3 Comparison Between Demand-Driven and Semi-Pressure-Driven Analyses

	Demand-driven		Semi-pressure-driven		
Node ID	Demand (CMH)	Pressure (m)	Available demand (CMH)	Pressure (m)	ADF
Junc 1	0.00	32.96	0.00	33.16	—
Junc 2	212.40	26.62	212.40	26.91	1.000
Junc 3	212.40	5.77	212.40	17.91	1.000
Junc 4	640.80	2.76	165.77	15.00	0.259
Junc 5	212.40	11.83	212.40	19.97	1.000
Junc 6	684.00	3.40	497.97	15.00	0.728
Junc 7	640.80	6.67	640.80	17.01	1.000
Junc 8	327.60	4.77	274.74	15.00	0.839
Junc 9	0.00	16.39	0.00	20.94	—
Junc 10	0.00	17.34	0.00	20.87	—
Junc 11	108.00	11.40	108.00	16.48	1.000
Junc 12	108.00	9.35	66.25	15.00	0.613
Junc 13	0.00	5.32	0.00	14.83	—
RES 1	−1480.75	0.00	−1168.45	0.00	—
RES 2	−1665.65	0.00	−1222.28	0.00	—

When pipe 3 failure is simulated using usual demand-driven analysis, there are nine nodes that appear to be pressure deficient (i.e., pressure <15 m). This indicates that demands at those nodes cannot be fully satisfied. The shortfalls in nodal supplies are later found using the semi-pressure-driven algorithm. Note that although nodes 3, 5, 7, and 11 are found to be pressure-deficient from the demand-driven analysis, they are not pressure- or supply-deficient using the semi-pressure-driven analysis. Tanyimboh and Tabesh (1997) have also found that when a network with locally insufficient heads is simulated using the demand driven approach, the deficiency appears to be far more serious and widespread than it is in reality. In fact, the semi-pressure-driven algorithm proceeds in an iterative manner for the same reason until no violations of the constraint given by Eq. (13.5d) are found.

Network available demand fractions for single pipe breaks are shown in increasing order in Table 13.4. Note that the network cannot fully satisfy the given demand loadings even when there are no pipe breaks. An even more interesting result is that there are four pipe break scenarios in which the system performance is better than that of the fully functional network.

13.3.3 What-If Scenarios

The purpose of this section is to illustrate the application of the semi-pressure-driven method for extended period simulation (EPS) of what-if scenarios while comparing the results to that of a usual demand-driven analysis.

What-if scenarios that may cause deficient performance in water distribution networks can be divided into several phases:

Phase I: Detection and isolation/removal of equipment/component

Phase II: Physical and Hydraulic Deficiencies

Phase III: Recovery phase

TABLE 13.4 Network ADFs Resulting from Single Pipe Breaks

Broken pipe ID	Number of iterations	Total available supply (CMH)	Network ADF
2	3	1233.81	0.3921
1	3	1233.81	0.3921
6	3	1309.00	0.4160
3	3	2390.72	0.7598
4	3	2825.85	0.8981
9	3	2846.44	0.9047
15	3	2930.40	0.9314
21	3	3015.84	0.9585
17	2	3027.60	0.9622
16	3	3038.40	0.9657
12	3	3039.12	0.9659
19	3	3046.67	0.9683
11	3	3062.98	0.9735
14	3	3092.60	0.9829
8	3	3095.90	0.9839
7	3	3097.30	0.9844
5	3	3102.57	0.9861
No breaks	3	3102.64	0.9861
13	3	3103.00	0.9862
10	3	3103.72	0.9864
18	3	3108.48	0.9879
20	2	3145.15	0.9996

Detection of an abnormality in network operation can be in several ways. In general, it can be said that the better the network is monitored, the faster the detection is. Thus, remote monitoring and control systems such as SCADA systems play an important role in finding about an abnormality in network operation. Depending on the type of threat, detection can be water quality based if a chemical is introduced into the system, and/or hydraulics based such as very high or more likely very low pressures, abnormal changes in tank levels, etc. Calls from the consumers can also be a factor here. Sometimes, a network performance can also become deficient even though it may not be the real target of the threat. For example, an extensive power outage may deprive the system of pumping capacity.

Isolation and/or removal of equipment also depends on the type of threat. For example, a chemical threat may require part(s) of the network to be isolated via isolation valves. Similarly, a water main break requires the nearest valves to be closed for proper isolation. Thus the number and spatial distribution of isolation valves is the main factor that would determine how quick and effective the isolation can be done. The more the valves situated in a network, the less the impact of a failure since the isolation can be restricted to a smaller portion of the system. A typical problem faced by utilities is that isolation valves malfunction or break when repair crews try to close them for purpose of isolation. Therefore, regular valve exercising must be an essential part of maintenance programs to keep the isolation valves in good working condition.

A system is physically deficient until the cause of abnormal network operation is permanently mitigated. Most physical deficiencies are mechanical such as component breaks, failures, malfunctions, and power outages. On the other hand, nonmechanical physical deficiencies such as a chemical intrusion are also possible. It is also during this interval that a hydraulic deficiency may first appear. From a supply point of view, a system is said to be hydraulically deficient if anywhere in the system pressures are insufficient to fully supply consumer demands. The repair/replacement process to mitigate a mechanical deficiency may take from several hours to weeks depending on the type and severity of threat, the available repair/replacement technologies, and the availability of appurtenances to be replaced if there has been a physical damage to network components such as pipes, valves, reservoirs, tanks, and pump stations. Most natural pipe breaks take from several hours to a day for repair. Chemical treats may require flushing of the system until the concentrations are well below the limits set by agencies.

Recovery phase begins when physical deficiency of the system is fully removed and ends when the system regains its normal operation. Tank and reservoir levels, pump flows, pressures throughout the system are a few of the indicators that can be used to mark when the system returns to its normal operation.

13.3.4 Example EPS Application

The second application of the semi-pressure driven method to predict a deficient network performance uses the hypothetical network Net3.net that comes with the EPANET software installation. This network, shown in Fig. 13.3, is composed of two sources (lake and river), 117 pipes, 2 pumps, 3 tanks, and 92 junctions.

The layout (see Fig. 13.3) for this example network is not highly looped and would be an interesting case for a reliability assessment considering possible failure scenarios. The original simulation duration of 24 h is increased to 240 h to show the network

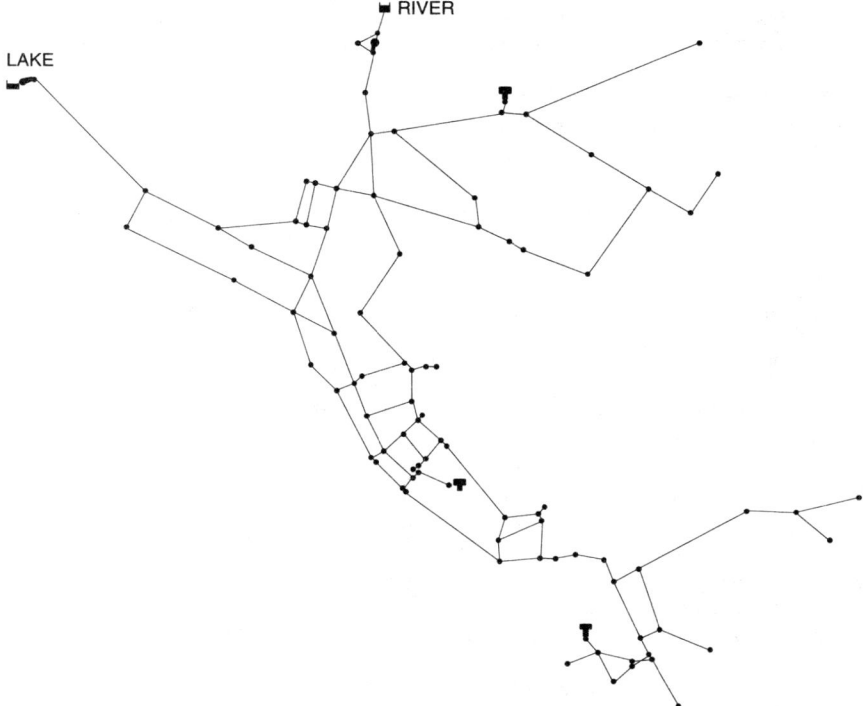

FIGURE 13.3 Schematic of example network Net3.net.

performance during prefailure, failure, recovery, and postrecovery phases. Five different demand patterns have been used throughout the network. Hazen-Williams is used as the headloss formula. Flows are in gpm, elevations and HGLs are in feet, and the pressures calculated are in pounds per square inch (psi). Other details of the network characteristics such as pipe diameters, lengths, demand patterns, pump curves, etc. can be found with an EPANET program installation and opening the Net3.net example file. The following simple controls are specified in the original input file for simulating normal network operation:

Lake source operates only part of the day
PUMP 10 OPEN AT CLOCKTIME 1 AM
PUMP 10 CLOSED AT CLOCKTIME 3 PM

Pump 335 controlled by level in Tank 1
When pump is closed, bypass pipe is opened
PUMP 335 OPEN IF TANK 1 BELOW 17.1 ft
PUMP 335 CLOSED IF TANK 1 ABOVE 19.1 ft
PIPE 330 CLOSED IF TANK 1 BELOW 17.1 ft
PIPE 330 OPEN IF TANK 1 ABOVE 19.1 ft

With those controls, percentages of daily supplies from the lake source, river source (pumping only), and river source (gravity only) are approximately 17, 28, and 55 percent, respectively. Next, the following failure scenarios will be simulated to assess the network redundancy from a supply point of view using the semi-pressure-driven algorithm:

Scenario 1: River source is shutdown from 72 to 168 h
Scenario 2: Lake source is shutdown from 72 to 168 h
Scenario 3: Network cannot be pumped due to a power outage from 72 to 168 h

In addition, a comparison between the usual demand-driven analysis and the semi-pressure-driven analysis will be made for Scenario 1. Graphical results showing tank level variations, supply rates from all the sources, and the resulting networkwide available demand fraction for all the phases will be presented for each scenario.

The following assumptions are made for each failure scenario described above:

1. Although a minimum pressure threshold value can be defined for each non-zero demand junction, a global value of 30 psi is assumed networkwide for the sake of simplicity. For most real networks, a global value will not be representative of the entire network. For example, a downtown area with high-rise buildings will more likely have higher pressure thresholds than the rest of the network.
2. Reasonable changes to the network operation will be assumed to minimize the effect of outage for each scenario. Any changes to compensate for reduced supply capacity will be explained under the relevant scenario section.
3. It is assumed that the demand variation of the network remains the same during the outage. In other words, the same demand patterns are maintained during all the phases of given failure scenario, namely pre-failure, failure, recovery, and post-recovery.

Scenario 1: River Source Shutdown. The first scenario that will be simulated using the usual demand-driven and the proposed semi-pressure-driven analyses assumes that the river source that supplies 55 percent and 28 percent of the daily demand via gravity flow and pumping, respectively, is shutdown from 72 to 168 h. Because this source is completely shutdown, it is assumed that the Lake source is continuously utilized during the outage. Original controls are maintained at other times. After these changes, the rule-based controls in EPANET look like

RULE 1
IF SYSTEM TIME >= 72
AND SYSTEM TIME < = 168
THEN PUMP 335 STATUS IS CLOSED
AND PIPE 330 STATUS IS CLOSED
AND PUMP 10 STATUS IS OPEN

RULE 2
IF TANK 1 LEVEL BELOW 17.1
THEN PUMP 335 STATUS IS OPEN
AND PIPE 330 STATUS IS CLOSED

RULE 3
IF TANK 1 LEVEL ABOVE 19.1
THEN PUMP 335 STATUS IS CLOSED
AND PIPE 330 STATUS IS OPEN

RULE 4
IF SYSTEM CLOCKTIME = 1 AM
THEN PUMP 10 STATUS IS OPEN

RULE 5
IF SYSTEM CLOCKTIME = 3 PM
THEN PUMP 10 STATUS IS CLOSED

Figures 13.4 and 13.5 show the results from a usual demand-driven analysis of the given scenario. It should be noted that demands are fully satisfied even during the interval when the river source is shutdown. The first tank to completely drain is Tank 1 at 81 h, while Tank 2 is emptied before 82 h. Finally, Tank 3 is out of water at 84 h. Note that because of the demand-driven approach, Lake pump starts to feed the system without storage in an instantaneous manner following 84 h. In other words, the pump simply produces whatever the entire network demands at the given hour regardless of its flow limits. According to the specified pump curve, Pump 10 can produce at most 6850 gpm before the head drops to zero. This problem is addressed by the following warning messages immediately following program execution.

WARNING: Negative pressures at t hours.

WARNING: Pump 10 open but exceeds maximum flow at t hours, where t is the simulation time the warning message refers to and happens to be after 84 h in this case.

All this and the fact that the analysis results in negative pressures indicates that usual-demand-driven approach cannot cope well with such abnormal conditions.

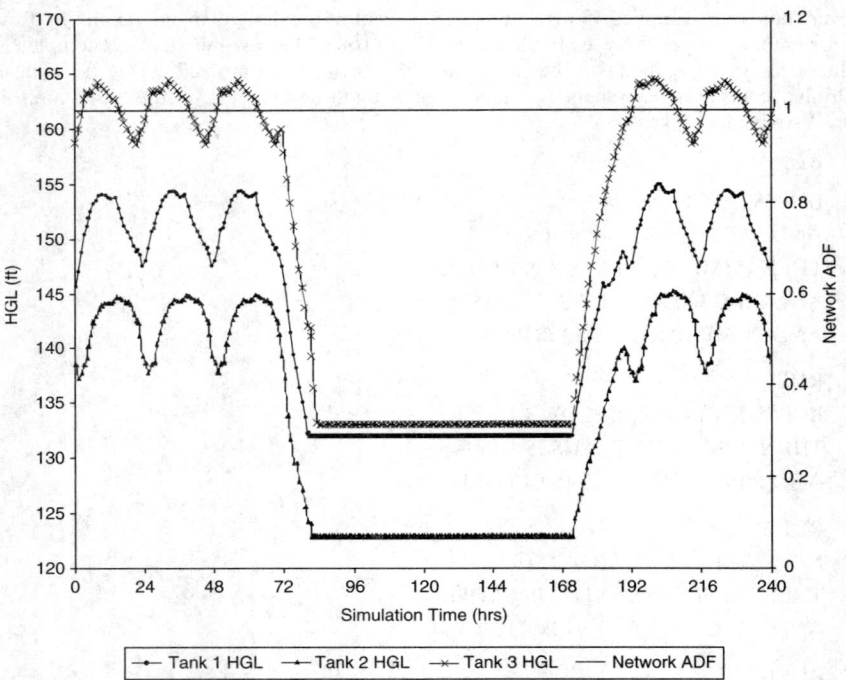

FIGURE 13.4 Demand-driven analysis of Scenario 1: tank levels and network ADF.

The semi-pressure-driven approach, on the other hand, produces the more realistic results shown in Figs 13.6 and 13.7. The network tanks drain at about the same hours as before while network ADF starts decreasing slowly at 81 h. Note that this is also the beginning of the hydraulic deficiency phase. Finally, when the last tank is emptied at 84 h, network ADF drops abruptly from about 0.984 to 0.365. From then on, the network relies on the lake supply via Pump 10, whereas the pump can only provide 33 to 47 percent of the total demand depending on the time of the day. The recovery phase of this scenario starts at 168 h when the river source returns to normal operation. Network ADF recovers almost instantaneously and hydraulic deficiency phase ends at 168.1 h. Overall, only 42.33 percent of the total network demand can be satisfied during this phase. The tanks, on the other hand, return to their normal operational range at about 191 h.

Scenario 2: Lake Source Shutdown. The second failure scenario to be simulated is the Lake source shutdown between 72 and 168 h. It should be remembered that the Lake source provides about 17 percent of the daily consumption during normal operation of the network. The following set of simple and rule-based controls are used to simulate the current scenario.

Simple controls are as follows:

Pump 335 controlled by level in Tank 1

When pump is closed, bypass pipe is opened

Pump 335 OPEN IF Tank 1 BELOW 17.1

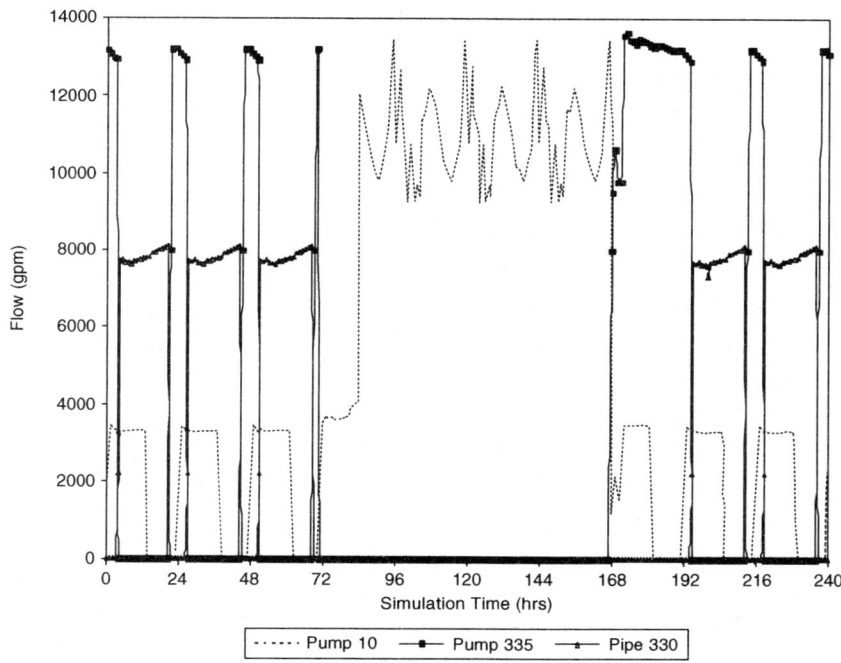

FIGURE 13.5 Demand-driven analysis of Scenario 1: pump and gravity flowrates.

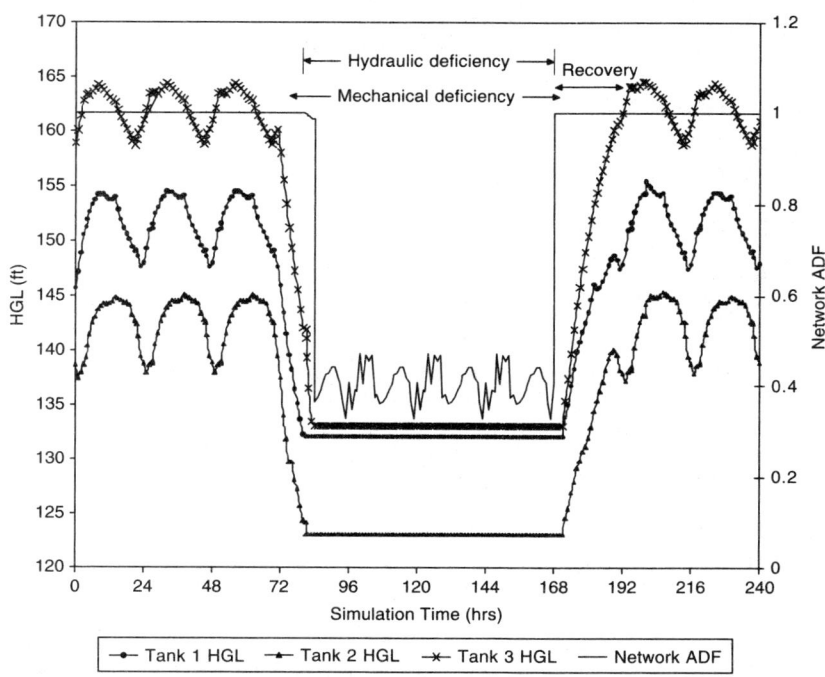

FIGURE 13.6 Semi-pressure-driven analysis of Scenario 1: tank levels and network ADF.

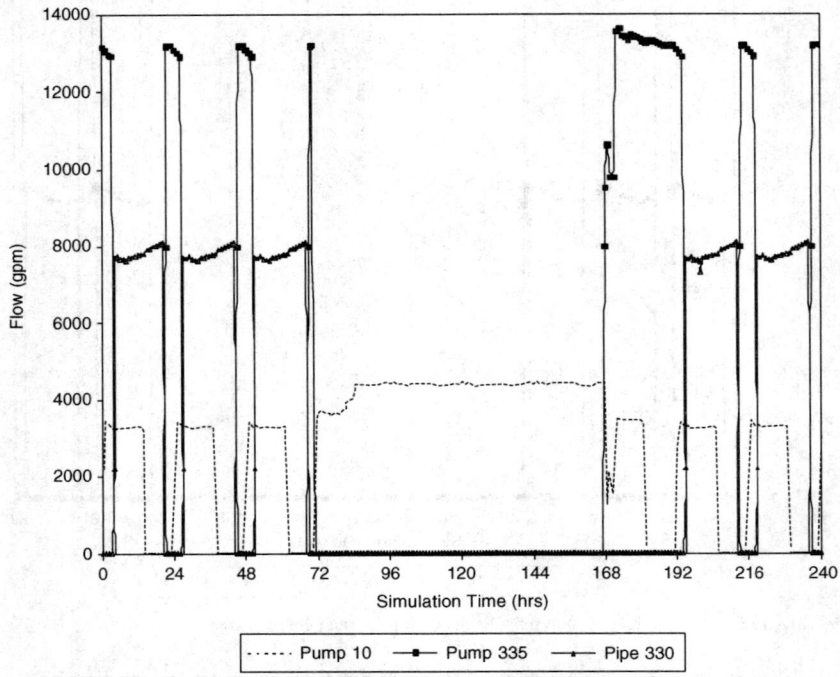

FIGURE 13.7 Semi-pressure-driven analysis of Scenario 1: pump and gravity flowrates.

Pump 335 CLOSED IF Tank 1 ABOVE 19.1
Pipe 330 CLOSED IF Tank 1 BELOW 17.1
Pipe 330 OPEN IF Tank 1 ABOVE 19.1

and the rule-based controls are as follows:

RULE 1
IF SYSTEM TIME >= 72
AND SYSTEM TIME <= 168
THEN PUMP 10 STATUS IS CLOSED

RULE 2
IF SYSTEM CLOCKTIME = 1 AM
THEN PUMP 10 STATUS IS OPEN

RULE 3
IF SYSTEM CLOCKTIME = 3 PM
THEN PUMP 10 STATUS IS CLOSED

The results from the semi-pressure-driven analysis, Figs. 13.8 and 13.9, indicate that the demands from network junctions subject to minimum pressure thresholds would still be fully

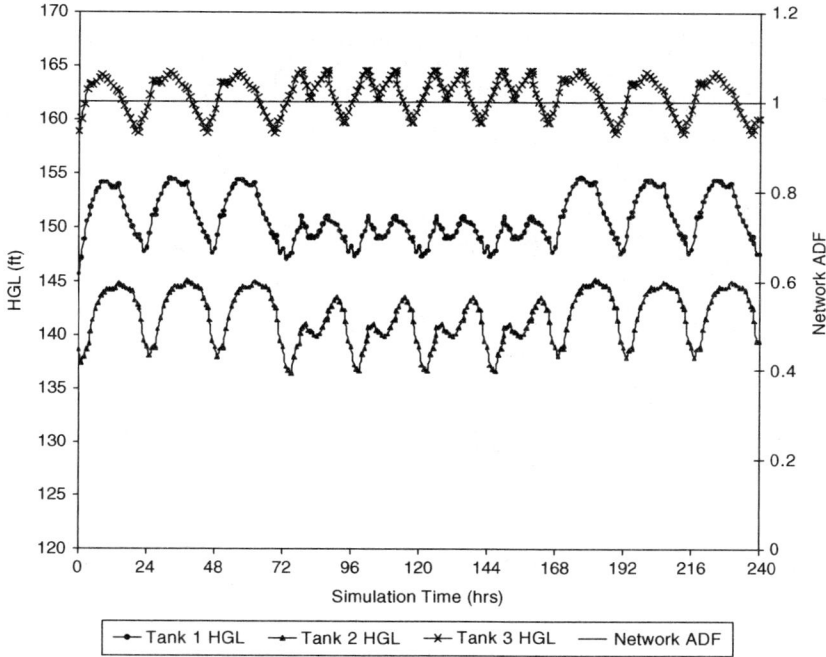

FIGURE 13.8 Semi-pressure-driven analysis of Scenario 2: tank levels and network ADF.

satisfied in such a scenario although tank levels follow different patterns during the outage. In other words, unlike Scenario 1, no hydraulic deficiency is expected in this scenario. In fact, a usual demand-driven analysis yields the same results as the semi-pressure-driven approach. Network recovery is almost instantaneous following 168 h for this scenario.

Scenario 3: No Pumping Due to Power Outage. The last failure scenario for our example network will analyze a power outage from 72 to 168 h. In other words, we will try to address the following question—what happens when the network is deprived of pumping capacity and has to rely on gravity flow only. Rule-based controls for this scenario are as follows:

RULE 1
IF SYSTEM TIME > = 72
AND SYSTEM TIME < = 168
THEN PUMP 10 STATUS IS CLOSED
AND PUMP 335 STATUS IS CLOSED
AND PIPE 330 STATUS IS OPEN

RULE 2
IF SYSTEM CLOCKTIME = 1 AM
THEN PUMP 10 STATUS IS OPEN

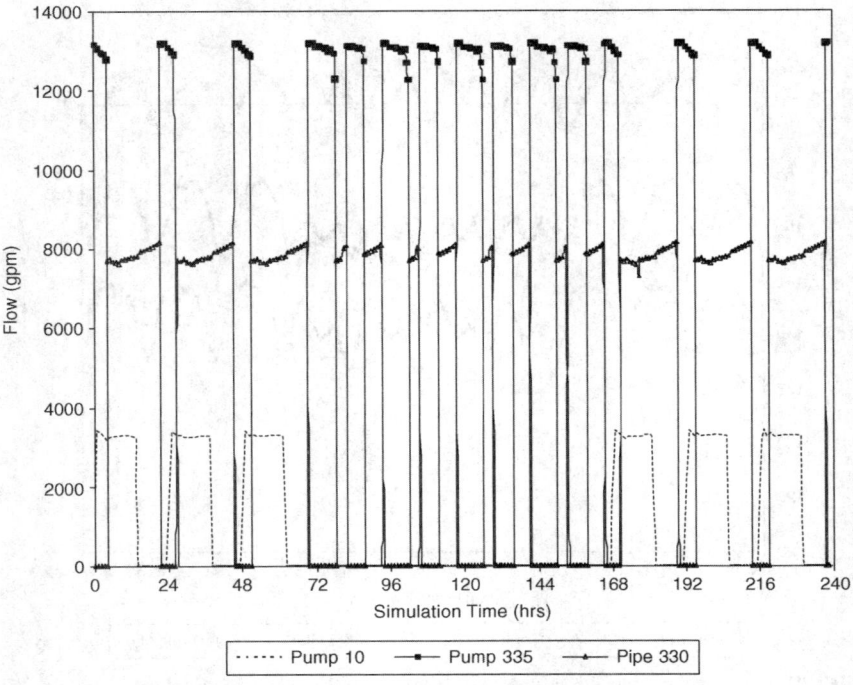

FIGURE 13.9 Semi-pressure-driven analysis of Scenario 2: pump and gravity flowrates.

RULE 3
IF SYSTEM CLOCKTIME = 3 PM
THEN PUMP 10 STATUS IS CLOSED

RULE 4
IF TANK 1 LEVEL ABOVE 19.1
THEN PUMP 335 STATUS IS CLOSED
AND PIPE 330 STATUS IS OPEN

RULE 5
IF TANK 1 LEVEL BELOW 17.1
THEN PUMP 335 STATUS IS OPEN
AND PIPE 330 STATUS IS CLOSED

Figures 13.10 to 13.13 summarize the results for this failure scenario. Similar to Scenario 1, demand-driven analysis forces the gravity line to supply the instantaneous network demand at the expense of negative/insufficient pressures throughout the system. The semi-pressure-driven analysis, on the other hand, considers demands and pressures simultaneously and estimates that Tank 1 becomes empty at 96 h. Tank 2 runs out of water one hour later. It is

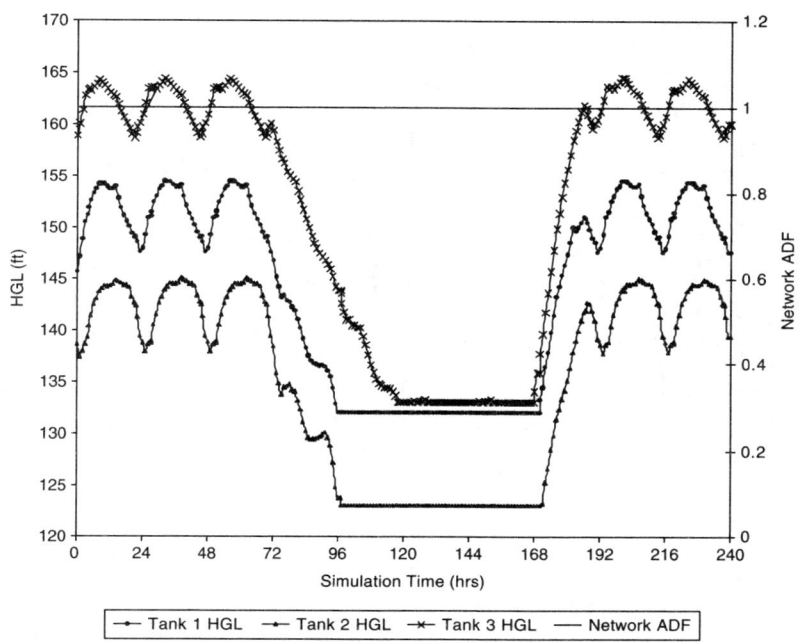

FIGURE 13.10 Demand-driven analysis of Scenario 3: tank levels and network ADF.

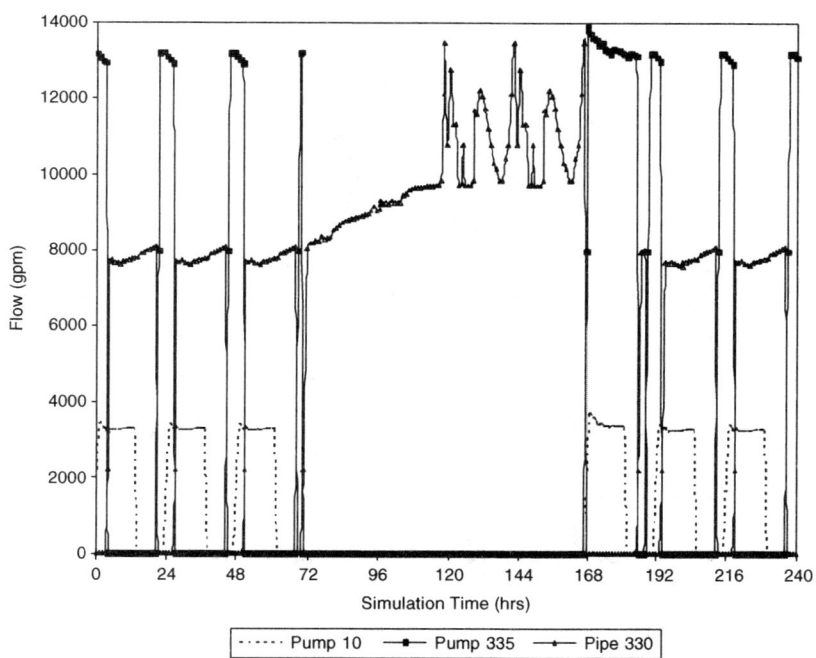

FIGURE 13.11 Demand-driven analysis of Scenario 3: pump and gravity flowrates.

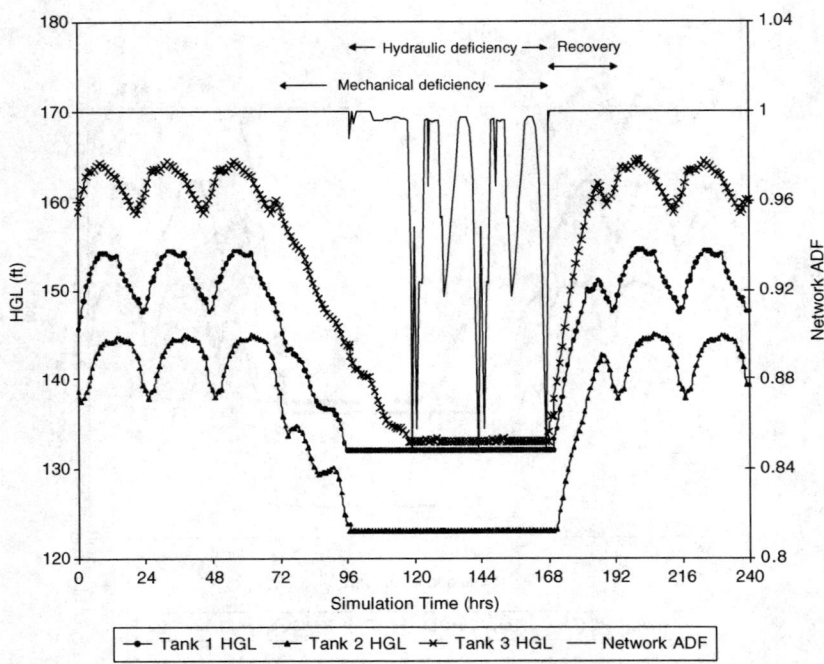

FIGURE 13.12 Semi-pressure-driven analysis of Scenario 3: tank levels and network ADF.

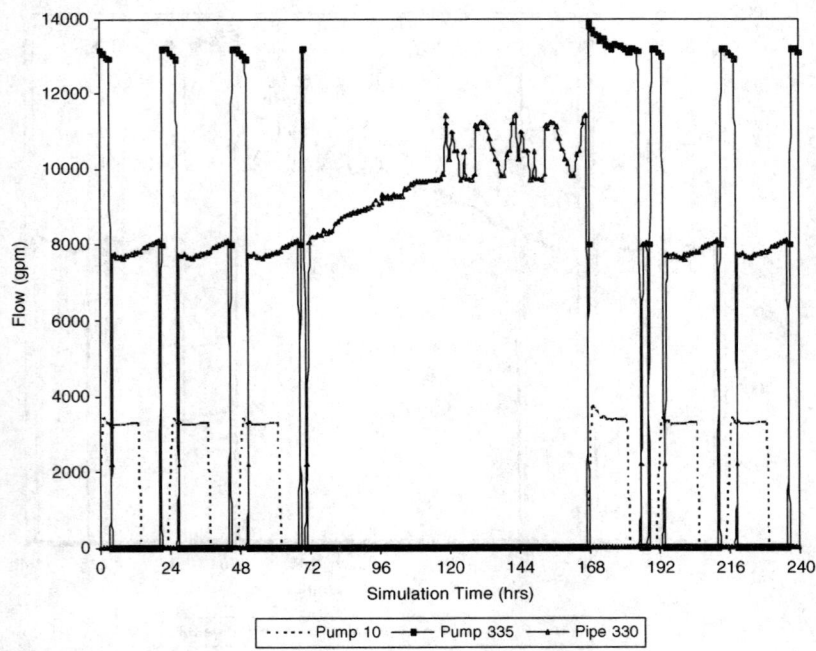

FIGURE 13.13 Semi-pressure-driven analysis of Scenario 3: pump and gravity flowrates.

also at this hour that the hydraulic deficiency starts and the network ADF drops below 1 for the first time. At 118 h, Tank 3 also becomes empty. During the hydraulic deficiency phase, network ADF varies between 0.847 and 1.000 depending on the time of the day. In fact, later during this phase, after Tank 3 empties, a pattern in ADF variation can be seen until the recovery phase starts. Overall, 96.56 percent of the total system demand can be supplied during the hydraulic deficiency phase. Network is fully recovered in terms of available supply at 169 h while full recovery of tank levels is achieved at about 193 h.

It should be noted that ADF variation is also available for each node in the system from the results of the semi-pressure-driven analysis. This would reveal which part(s) of the system are the most vulnerable for each scenario.

Three failure scenarios have been investigated in this section using both demand-driven and the proposed semi-pressure-driven analyses. The results clearly indicate that demand-driven analysis cannot cope well with abnormal operating conditions of a water distribution network.

13.3.5 Reliability Models

It has been shown in the previous sections that the proposed semi-pressure-driven method can be used as a descriptive tool to predict deficient network performances under possible failure scenarios. Network components vulnerable to terrorist attacks include, but are not limited to, pipes, pump stations, reservoirs, tanks, valves, etc. Possible demand variations can also be considered when part(s) of a network are damaged and shut down following a terrorist attack. Considering that a scenario may include one or more of network components perhaps an infinite number of different scenarios can be thought of for a given network. On the other hand, if a subset of those scenarios can be defined that are critical to the network operation, then a network reliability measure can be defined, following the work of Fujiwara and De Silva (1990), as

$$R_{net} = 1 - \sum_{i=1}^{N} \left(1 - ADF_{net}^i\right) v_i \quad (13.10)$$

where ADF_{net}^i = network available demand fraction resulting from scenario i
N = number of failure scenarios in the critical subset
v_i = probability of scenario i occurring

The above formulation can be used both for steady-state and extended-period analyses. However, extended-period applications require estimating the time and the duration of failure to which the results may be sensitive. A long enough duration should be chosen so that parameters such as ADF, tank levels, etc., attain cyclic patterns and the aforementioned sensitivity is minimized. Although, the probability, v_i is another fuzzy parameter in the above formulation, it should perhaps be decided by the utility managers and/or operators depending on the level of risk they think different components of their system possess. A few of the guidelines for defining a critical subset of failure scenarios are listed in Section 13.3.1.

13.4 EFFECT OF VALVING ON SYSTEM RELIABILITY

A valve is defined as a mechanical device by which the flow of fluid may be started, stopped, or regulated by a movable part that opens or obstructs passage. Valves are used in water distribution systems for a variety of purposes. Accordingly, different types of valves can be grouped into four types (Ysusi, 2000) (1) isolation valves, (2) control valves, (3) blow-offs, and (4) air-release and vacuum-release valves.

Control valves are used to regulate flow or pressure at different points of the system by creating headloss or pressure differential between upstream and downstream sections. The mission of isolation valves is to isolate a portion of the system whenever system repair, inspection or maintenance is required at that segment. The two most common types of isolation valves are gate valves and butterfly valves. Unlike the control valves, little attention has been paid to the layout and operation of isolation valves. Yet, from a reliability point of view, isolation valves are of great interest since they determine the extent of isolation should a portion of the system need repair, inspection, or maintenance. As an extreme example, a system with no valves would have to be shut down completely at the source for any maintenance. On the other hand, if there is sufficient valving in the system the outage or isolation can be limited to a small portion of the system.

Little is known about the reliability of isolation valves. The reason for that is such valves are fully opened under normal operating conditions and fully closed when isolation of a portion of the system is needed. Valves that remain in either position for extended periods of time become difficult (or even impossible) to operate (Ysusi, 2000). Therefore, they should be exercised at least once a year or more often if the water is corrosive or dirty. According to Walski (2000), valve exercising is the single most important form of preventive maintenance for improving reliability.

Valves are typically located around junctions in a distribution system. Figure 13.14 shows a typical valve scheme for a typical city block. Their locations and numbers are usually determined using the rules of thumbs that have evolved over the years. The most commonly used rule of thumb is to install a minimum of $n - 1$ valves around a junction to which n links (pipes) are connected. In other words, a minimum of three valves should be placed at every cross-intersection and a minimum of two valves is needed around every T-section.

Although regular exercising programs may increase their reliability, valves are not 100 percent reliable whenever they need to be closed for isolation of a distribution segment. Therefore, another consideration for valve placement is that no more than four valves should need to be operated when an isolation is needed. Consider, for example, pipe 1 in Fig. 13.14. Should this pipe fail, six valves (valves a, b, c, d, e, and f) need to be closed in order to isolate the break from the remainder of the network. If the isolation is achieved through closing of all six valves, only the consumers whose service connection is linked to pipe segment 1 will be without water during the repair work. However, if there is a percent probability, p, that each valve can be operated (closed) successfully, the mission can only be accomplished with probability p^n. For example, for $p = 0.90$, the chance of operating all six is 53 percent ($0.90^6 = 0.53$). For example, if valve c fails to operate among the six,* three additional valves (k, l, and m) need to be closed for complete isolation. That means customers receiving water from pipe segment 4 will also be without water during the repair work on pipe segment 1.

Walski (2000) presented a few more rules of thumbs for locating isolation valves in distribution systems. The AWWA's Introduction to Water Distribution (1986) states that: "Isolating valves in the distribution system should be located less than 500 ft apart (150 m) in business districts and less than 800 ft (240 m) apart in other parts of the system. It is a good practice to have valves located at the end of each block so that only one block will be without water during repair work." Similarly, The Ten States Standards states "Where systems serve widely scattered customers and where future development is not expected, the valve spacing should not exceed one mile." Finally, Walski (2000) points out that good practice requires placement of an isolation valve for every fire hydrant lateral.

---

*The probability of that scenario occurring is $(1 - p)(p^5) = (1 - 0.90)(0.90^5) = 5.9$ percent.

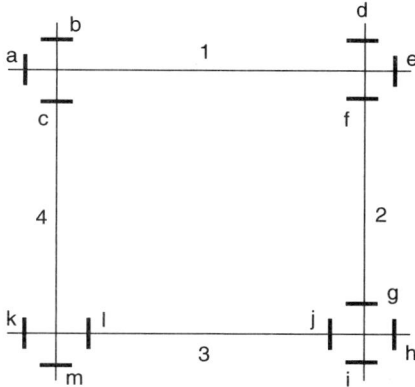

FIGURE 13.14 Example valve layout around a city block.

The effect of valving on distribution system reliability has two aspects. First, some consumers will be out of service when a pipe break occurs in the system. The spatial extent of isolation is mostly determined by the locations of valves in the system. It is theoretically possible to determine the impact of an isolation, induced by a link failure, by calculating the volume of water which consumers located in the isolated portion of the system are without during the repair work. Let us call this the shut-off volume. Bouchart and Goulter (1991) studied this aspect of valving and expressed the reliability in terms of the expected volume of deficit along the isolated segment(s).

It should be noted, however, that network hydraulic models rarely possess the level of detail to determine the shut-off volume accurately. Typically, demands from city blocks are assigned to nearby junctions while in reality demands are withdrawn via the service connections located along the supply lines. The concept of lumping the continuous demands along the supply lines into single "point" loads at the network junctions using various demand allocation techniques is required in many cases (e.g., hydraulic simulation) for mathematical tractability (Bouchart and Goulter, 1991).

The second implication of a valve scheme, besides determining the extent of direct shut-off (or volume of deficit), is the effect of isolation on system connectivity. The connectivity of a network is significantly affected by the valving scheme. For example, when pipe 1 fails in the valve scheme in Fig. 13.15, connectivity among pipe segments 2, 3, and 4 would be preserved; whereas connectivity would be lost in the valve scheme in Fig. 13.16. The same goes for the links 5, 6, and 7. Unlike the expected volume of deficit, the effect of a valve scheme on a network connectivity can be assessed only with hydraulic simulation of the deficient network.

13.4.1 Optimal Valve Scheme: Problem Formulation and Solution Space

Perhaps the single most important factor affecting the extent of isolation resulting from a pipe failure in a water distribution network is the valve layout in the vicinity of the break. A network with no isolation valves would have to be shut down at the source for any maintenance work. The more valves situated in a network, the less the impact of a failure since the isolation can be restricted to a smaller portion of the system. Thus, only by knowing the valve layout in the vicinity of a pipe break is it possible to determine which links are to be closed and the deficient performance of the remaining network.

A fully valved network from a modeling approach is a network where each pipe in the network has isolation valves at each end. Since demands are assumed lumped into junctions

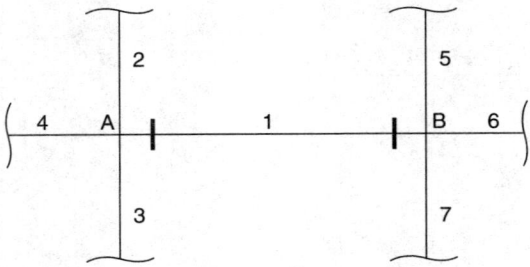

FIGURE 13.15 Valve scheme with pipe 1 fully valved.

for mathematical tractability in conventional network modeling approaches, it would not make any difference if there were intermediate valves or not, from a modeling point of view. In real life, on the other hand, demands are almost continuously distributed along a pipeline (withdrawn via service connections). Thus, the more the number of intermediate valves located on a pipe, the smaller the number of out-of-service consumers during the repair work following a pipe break.

The optimum valve location problem can be formulated for any given network as follows. A decision variable can be defined for each junction and the pipe that connects that junction to the rest of the system as $x_{l,j}$: decision on whether to place a valve at the end of pipe l connected to junction j, $x_{l,j}$ is 1 if the valve should be placed and 0 otherwise.

Thus, the total number of decision variables is equal to the number of junctions for any network. Note that a solution to the problem is a binary string resulting in a combinatorial optimization problem. Due to complex nature of the problem it is impossible to write the objective function explicitly in terms of the decision variables. All this suggests that heuristic based optimization techniques are the most appropriate to find an optimum solution to the valve layout problem.

Since decision variables for this problem are binary (two outcomes are possible for each variable), the solution space for a network with n pipes contains 2^{2n} solutions. The multiplier 2 of n comes from the fact that each pipe carries the decision, whether or not to place a valve at each end of a pipe. Let us consider a very small size water distribution network with 50 pipes. There would be $2^{100} = 1.26765 \times 10^{30}$ solutions in the solution space driven simulation takes 0.001 s. It would take as the number of pipes reaches the thousands (typical for even mid-size networks), the problem solution space and the time required for complete enumeration becomes extremely large. However, note that if the objective is to

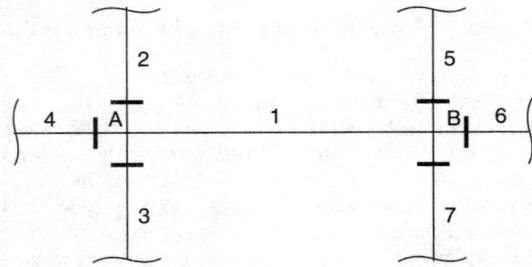

FIGURE 13.16 Valve scheme with no valves on pipe 1.

maximize only network reliability, then the solution becomes trivial as the optimal solution would be a fully valved network. Thus, the objective function must also consider the cost of the valve scheme for the network.

Let us now try to add the following rule of thumb as a constraint to our problem—install $n - 1$ valves around a junction to which n links (pipes) are connected. Then, the total number of solutions, NS, in the solution space is reduced to

$$\text{NS} = \prod_{j=1}^{J} n_j \qquad \text{for } n_j > 1 \qquad (13.11)$$

where J is the total number of junctions and n_j is the number of pipes connected at junction j.

Thus adding the given constraint to our problem formulation would result in a dramatic shrinkage of the problem solution space. Depending on network size and geometry (e.g., the level of looping), it may even be feasible to try total enumeration to find the optimal valve layout in some problems. If total enumeration is impossible, a local search heuristic such as simulated annealing or genetic algorithms can be used to find a good solution (possibly the global optima) to the optimum valve location problem. Ozger (2003) developed algorithms to combine semi-pressure-driven based reliability/availability analyses and genetic algorithms to find optimal valve schemes in water distribution networks.

REFERENCES

AWWA, *Introduction to Water Distribution*, American Water Works Association, Denver, CO, 1986.

Bhave, P. R., *Analysis of Flow in Water Distribution Networks*, Technomic Publishing, Lancaster, PA, 1991.

Bouchart, F., and I. Goulter, "Reliability Analysis for Design," in L. W. Mays (ed.), *Water Distribution Systems Handbook*, McGraw-Hill, New York, Chap. 18, 2000.

Bouchart, F., and I. Goulter, "Reliability Improvements in Design of Water Distribution Networks Recognizing Valve Location," *Water Resour. Res. American Geophysical Union* 27(12): 3029–3040, 1991.

Chin, D. A., *Water Resources Engineering*, Prentice Hall, Upper Saddle River, New Jersey, 2000.

Fujiwara, O., and A. U. De Silva, "Algorithm for Reliability-Based Optimal Design of Water Networks," *J. Environ. Eng.* 116(3): 575–587, 1990.

Gupta, R., and P. R. Bhave, "Closure of 'Comparison of Methods for Predicting Deficient Network Performance,'" *J. Water Resour. Planning Management, ASCE* 123(6): 370, 1997.

Gupta, R., and P. R. Bhave, "Comparison of Methods for Predicting Deficient Network Performance," *J. Water Resour. Planning Management, ASCE* 122(3): 214–217, 1996.

Ozger, S. S., "A Semi-Pressure-Driven Approach to Reliability Assessment of Water Distribution Networks," Ph.D. Dissertation, Department of Civil and Environmental Engineering, Arizona State University, Tempe, Arizona, 2003.

Rossman, L. A. (2001). "Thread: Modeling demands." MIKE NET support forum. http://www.boss-intl.com/ forums/

Rossman, L. A., "Computer Models/EPANET," in L. W. Mays (ed.), *Water Distribution Systems Handbook*, McGraw-Hill, New York, Chap. 12, 2000a.

Rossman, L. A., *EPANET Users Guide*, Drinking Water Research Division, Risk Reduction Engineering Laboratory, Office of Research and Development, U.S. Environmental Protection Agency, Cincinnati, OH, 2000b.

Rossman, L. A., *EPANET Users Manual*, Drinking Water Research Division, Risk Reduction Engineering Laboratory, Office of Research and Development, U.S. Environmental Protection Agency, Cincinnati, OH, 1994.

Tabesh, M., "Implications of the Pressure Dependency of Outflows on Data Management, Mathematical Modeling and Reliability Assessment of Water Distribution Systems," Ph.D. thesis, University of Liverpool, Liverpool, England, 1998.

Tanyimboh, T. T., and M. Tabesh, "Discussion of 'Comparison of Methods for Predicting Deficient Network Performance,'" *J. Water Resour. Planning Management, ASCE* 123(6): 369–370, 1997.

Tanyimboh, T. T., M. Tabesh, and R. Burrows, "Appraisal of Source Head Methods for Calculating Reliability of Water Distribution Networks," *J. Water Resour. Planning Management, ASCE* 127(4): 206–213, 2001.

Tanyimboh, T. T., R. Burd, R. Burrows, and M. Tabesh, "Modeling and Reliability Analysis of Water Distribution Systems," *Water Science Tech.* 39(4): 249–255, 1999.

Walski, T. M., "Maintenance and Rehabilitation/Replacement," in L. W. Mays (ed.), *Water Distribution Systems Handbook*, McGraw-Hill, New York, Chap. 17, 2000.

Wood, D. J., *User's Manual—Computer Analysis of Flow in Pipe Networks Including Extended Period Simulation*, Department of Civil Engineering, University of Kentucky, Lexington, KY, 1980.

Ysusi, M. A., "System Design: An Overview," in L. W. Mays (ed.), *Water Distribution Systems Handbook*, McGraw-Hill, New York, Chap. 3, 2000.

CHAPTER 14
REMOTE MONITORING AND NETWORK MODELS: THEIR POTENTIAL FOR PROTECTING U.S. WATER SUPPLIES

Robert M. Clark
Environmental Engineering and Public Health Consultant, Cincinnati, Ohio

Srinivas Panguluri
Shaw Environmental Inc., Cincinnati, Ohio

Roy C. Haught
National Risk Management Research Laboratory, USEPA
U.S. Environmental Protection Agency
Cincinnati, Ohio

14.1 INTRODUCTION

The events of September 11, 2001 have raised concerns over the safety and security of U.S. critical infrastructure including water and waste water systems. Shortly after September 11, the U.S. EPA's Administrator appointed a Water Protection Task Force (WPTF). The WPTF has established a research agenda emphasizing the need to understand the factors that affect the security of drinking water distribution systems. Conducting research and field studies that would lead to the development of early warning systems (EWS) is high on the WPTF's agenda.

Although there is little available in the literature, the Water Supply and Water Resources Division (WSWRD) of the U.S. EPA has, for many years, been conducting research that could be utilized to support the development of an EWS. This chapter will provide an overview of the U.S. EPA's early research to advance the use of remote monitoring and control technology (RMCT) coupled with the use of conventional water quality sensors to monitor and control the water quality at various locations (source water, treatment, and distribution) within the drinking water infrastructure. The bulk of the technical information presented in this chapter is based on the authors' experience working with or for the U.S. EPA in various capacities. Information presented in this chapter should not be presumed to represent the U.S. EPA policy or official endorsement.

Three case studies are presented to demonstrate the selection and implementation methodologies of RMCT systems. For purposes of this chapter, RMCT implies real-time monitoring and control, and includes the sampling/monitoring instrumentation, Supervisory Control and Data Acquisition (SCADA) system, and the associated communication media. Advanced remote monitoring devices that can potentially be applied to water systems security are briefly discussed, including sensor evaluation studies and network modeling research. The authors believe that monitoring systems can be integrated with water quality network modeling as part of an EWS.

14.2 REMOTE MONITORING AND CONTROL TECHNOLOGY

The U.S. EPA's first research project, which incorporated real time monitoring at a remote location, was conducted at its Testing & Evaluation (T&E) facility in Cincinnati, Ohio and focused on evaluating real-time RMCT systems for small drinking water package plant systems. It was assumed that implementation of RMCT systems could reduce manpower requirements for operating package drinking water plants by providing the ability to monitor and control several systems remotely from a centralized location (Haught and Panguluri, 1998).

The SCADA component of RMCT systems is widely used in industrial environments and by larger water utilities to control and monitor their individual facility operations. However, water utilities typically do not utilize available SCADA systems for conventional water quality monitoring. They monitor water quality parameters by employing "sampling crews" on a scheduled or random basis that provides a periodic snap-shot of the overall system quality. Current drinking water regulations require all conventional treatment system operators to provide water quality monitoring to ensure that good quality water is provided to the consumers (U.S. EPA, 1996). However, since the regulations do not clearly specify that real-time monitoring of water quality is required, utilities have been reluctant to install and operate such devices.. However, after the events of 9/11, utilities may become more interested in the potential for implementing SCADA for monitoring water quality. SCADA can fulfill this function by constantly monitoring water quality within treatment and distribution systems. SCADA systems can potentially reduce the risk of security or even non-security related threats and detect undesirable water quality changes within a system (Meckes et al., 1998).

14.2.1 RMCT System Components

An RMCT system can be established by networking a variety of commercially available apparatus, instruments, hardware and software. The three major components of an RMCT system are online sampling and control devices, SCADA, and communications media. These components are described below.

Online Sampling/Control Devices. Depending upon the actual field instrumentation utilized in the system, this can be the most expensive component of an RMCT system. The types of instruments (sensors, switches, monitors, and controllers) used in RMCT systems vary widely depending upon the parameters that need to be controlled and/or observed. The cost for online sampling devices can range form $300 to $85,000. Control units such as pumps or shut-off valves are less expensive (Panguluri et al., 1999). Periodic sensor/instrument calibration and system checks are essential to ensure the quality of data collected. Therefore, the costs associated with maintenance and calibration of the online sensors should also be considered when planning the acquisition and implementation of an RMCT system. If electric power is not readily available at the desired location where an RMCT system is to be installed then the costs for installing a suitable power apparatus (such as a solar panel and battery pack) should be considered.

SCADA Systems. Each site should be evaluated individually for appropriate SCADA system selection. Recent advances in electronic hardware and software technologies have resulted in several cost-effective SCADA alternatives. Larger utilities typically use some type of SCADA system for water distribution system control which can easily be integrated to include online sampling instrumentation in a cost-effective manner. These systems are typically centralized with a "master station" that periodically, electronically "polls" the "slave" units for data, or issues controls based on a sequence of events. A microprocessor-based "smart" SCADA system could be used in remote locations and by small systems where online communication is expensive. Smart systems have higher initial costs, but overall costs are reduced since the communication costs (e.g. long-distance phone costs, etc.) are negligible because most of the operating functionality is transferred from the main computer to the individual SCADA unit at the remote site (Panguluri et al., 1999). Newer SCADA units are fairly inexpensive with capital costs ranging between $500 and $5,000. The basic features and options for selecting a SCADA system are discussed in more detail in the Section entitled, "Selection of a SCADA System."

Communication Media. The primary options available for communication include direct wire, phone line, cellular phones, radio and/or satellite systems. Direct wire and phone line communication media are the most inexpensive. Factors to be considered for selecting communication media include the cost, geographic location, and ease of access to the site. The primary limitations associated with selecting the communication media include installation and operating costs. Installation can vary between $200 (for a simple telephone line) to greater than $75,000 for a satellite-based system (Battelle Memorial Institute, 2000) and the ongoing monthly operating costs can range from $25 for a phone line to over $1,000 per month for satellite service.

14.2.2 Engineering and Designing an RMCT System

When engineering and designing an RMCT system, it is important to understand the specific needs of the treatment and distribution systems being considered. The following points should be considered carefully before purchase and setup of an RMCT system (Panguluri et al., 1999).

- What are the near- and long-term design objectives?
- What is the type of treatment system being controlled?
- What are the complexities of the distribution system (size, location)?
- What locations are best suited for sampling and/or control system installation?
- Is sufficient flow and water pressure available for online instruments?
- Is there an existing SCADA system available?
- What types of communication media are available at the selected locations?
- Are there plans for an automated monitoring and/or control system at the selected location?

- How many parameters are going to be monitored and/or controlled?
- Are there any specific security concerns?
- Other site-specific information?

Foresight into engineering and design approaches will enhance the capabilities of the RMCT. It should include all the desired options up-front (such as monitoring, control, security, recordkeeping, and reporting options); a design change is usually cost prohibitive especially if SCADA hardware has to be replaced. Mistakes in the initial selection of SCADA could also cause the loss of the initial capital investment. The sampling system design is not so "critical" as long as there is an allowance for sufficient space for addition of instruments and sensors. The type of communication media used varies widely depending upon the purpose and geographic location of the monitored site.

14.2.3 Selection of a SCADA System

Prior to the purchase of a SCADA system, it is essential to understand the variability that exists within such systems. Many SCADA systems may not provide the results needed to justify their purchase. Therefore, once the need for a SCADA system is justified and the suitable monitoring and control technologies are identified, the following factors should be carefully evaluated (Haught and Panguluri, 1998):

- Cost (initial, training, service agreements, and operation and maintenance)
- Ease of operation (user-friendly to the operator)
- Ease of customization (programmability)
- Networking ability (connecting to several remote systems)
- Remote operability (ease of remote technical diagnosis)
- Scalability (ease of adding monitoring and control devices to the system)
- Vendor support (hardware and software upgrades and remote diagnosis)

Currently, there are several commercially available small-scale SCADA systems which should be further evaluated for the use of open standards. A SCADA system must be capable of allowing future growth with respect to the number of input and output channels. These input and output channels are used to communicate with various monitoring and control devices. The SCADA hardware must also contain sufficient memory to store the monitored data for extended periods of time. In case of brownouts or blackouts, the system should normally self-boot upon resumption of the power supply. Along with these basic features, the selected system should have some advanced features, which include the following:

- Call-out feature—This feature allows the system's software to notify appropriate personnel if problems develop with a treatment system or water quality. This feature can greatly enhance operator response in emergency situations and prevent costly shutdowns and loss of water and/or water quality.
- Security feature—This feature allows the system to be able to implement security levels of access to the treatment plant. Security levels with passwords can deny or allow monitoring and/or control access.
- Open Data Base Connectivity (ODBC)—This feature allows for open communication with other databases and tools that can be integrated to provide additional features. The data then can also be used for network modeling.

It is important that each site is evaluated individually for appropriate SCADA system selection. Whenever possible, a microprocessor-based unit or a "smart" system should be chosen. Smart systems are slower than PLC-based systems but, can greatly reduce the cost of online communication. Smart systems are also typically more expensive, but the payoff is in savings associated with communication, operation and maintenance, and travel/repair costs. Using smart systems, a majority of the operating functionality can be transferred to the remote site. Smart systems can also minimize the need for a dedicated online central computer. In such an implementation, the main computer is used only for periodic monitoring, transfer of monitoring and reporting data, troubleshooting, and modifications of control parameters.

Capital costs can be further reduced by utilizing existing hardware, e.g., most utilities have a computer that can be used as the main computer for remote monitoring and control. The cost of the SCADA units can range anywhere between $500 to $30,000 depending upon the size and features desired. There are a number of companies developing PC-based SCADA units for small systems that can be purchased for less than $500. However, these systems only include three or four input/output (I/O) channels for monitoring and control. Some of the SCADA system hardware includes free software. However, the software provided may not be user-friendly.

The cost of SCADA software has plummeted over the past few years. For example, the U.S. EPA purchased a commercially available graphical (Windows-based) SCADA system development software package in 1999 at a cost of approximately $30,000. The newest software version of this system costs just under $6,000. These packages are designed to work with a variety of SCADA hardware and provide tremendous flexibility. Generally, the cost of software increases with flexibility desired. The price of commercially available software usually depends on the number of I/O channels licensed for use along with the number of computers from which the system is operated. Based on a recent review, the cheapest single user 75-channel key would cost around $1000 from a major SCADA software vendor. It is also important to note that commercial software is typically designed to work with high-end or high-use hardware. "Software drivers" may not be available for cheap hardware and custom software drivers may need to be developed in certain cases.

14.2.4 Implementation of RMCT System—Case Studies

The U.S. EPA's National Risk Management Research Laboratory's Water Supply and Water Resources Division has developed, tested and/or evaluated off-the-shelf RMCT systems within the United States—the Battelle Project, three sites in Washington D.C. (Meckes et al., 1998; Panguluri et al., 1999), one at the U.S. EPA's T&E facility and at three locations in rural West Virginia (Haught and Panguluri, 1998, Panguluri et al., 1999). The DC sites were implemented for monitoring only and they represent an example of a large-scale remote monitoring network. The T&E site represents a medium-scale RMCT network whereas the rural WV sites represent a small-scale RMCT network. The following is a brief discussion of the case studies along with some of the lessons learned from the implementation of RMCT systems at each of these sites.

Battelle Project. The U.S. EPA began a research program for evaluating RMCTs in the 1990s in collaboration with the Battelle Columbus Laboratories (Battelle Memorial Institute, 2000). The project was to evaluate and recommend various types of off-the-shelf sensors that could be monitored remotely and then used in conjunction with an off-the-shelf SCADA unit to monitor and control a small package treatment plant. The emphasis was to automate as many functions as possible (including the treatment unit operations and the calibration of sensors) using commercially available equipment.

Using sensors for turbidity, residual chlorine, nitrate, and fluoride, a commercially available SCADA unit was adapted and configured to monitor an ultrafiltration package plant (UFPP) located at the T&E facility. A prototype water (sample) delivery system was designed using a distribution-hub, solenoids, and a stepper pump that allowed selective sampling from several locations. The four locations on the UFPP sampled by the RMCT system included (1) finished water, (2) post bag filter water, (3) raw water, and (4) reject water from the filtration membrane. These four locations were selected for measurement because data from these sampling points not only provided the mandated water quality measurements, but also performance information on both the UFPP and the RMCT system.

Washington, D.C. (DC) Remote Monitoring Network. Following a number of coliform violations, EPA's Region 3 office directed the DC Water and Sewer Authority (DCWASA) to implement a number of corrective actions to its water distribution system (Clark et al., 1999). Remote monitoring of water quality parameters within the distribution system was identified as being one possible method for identifying water quality problems. Consequently in 1997, the U.S. EPA initiated a research study to install a remote network at various locations in DC to monitor water quality within the distribution system (Meckes et al., 1998). The DCWASA staff teamed with the U.S. EPA to select appropriate locations within the distribution system for installation of online sampling stations. Several objectives were established as follows:

- Development of methods to monitor real-time water quality at various locations within the distribution system.
- Field evaluation of sensors and remote monitoring technologies for inclusion in the network.
- Development of effective methods to publish the real-time data to enhance consumer confidence.
- Evaluation of costs associated with implementing such systems.
- Identification of problems and possible corrective actions for implementing remote monitoring networks.

Free chlorine, pH, temperature, and turbidity were selected as the monitored parameters based on the availability of online sensor technologies (and associated analyzer instruments). It was believed that these water quality parameters could be reliably monitored continuously and require only limited maintenance.

After suitable location(s) were identified, customized sampling and monitoring systems were built. DCWASA used the existing SCADA system to track various operating parameters within the distribution system. Using the existing system minimized long-term on-site support costs. Figure 14.1 is a schematic representation of the raw water collection and treatment systems which are owned and operated by the Army Corps of Engineers, Washington Aqueduct (WAD).

Figure 14.2 is a schematic diagram that shows the relationship between the WAD and the DCWASA distribution system. The remote-monitoring system in DC was implemented in three phases. In the first phase, a remote-monitoring system was installed at the Fort Reno No. 2 tank (Fig. 14-2) which provided security and easy access to the distribution system. Subsequently, based on initial success at this location, two other sites (Bryant Street and Blue Plains) were selected and added to the remote-monitoring network in the second phase. The third phase involved the development of a Web-based application to publish the real-time data to enhance consumer confidence. During the evaluation, it was clear that use of the existing SCADA system to manage the monitored data provided clear advantages over other available systems. Some of these advantages are listed below:

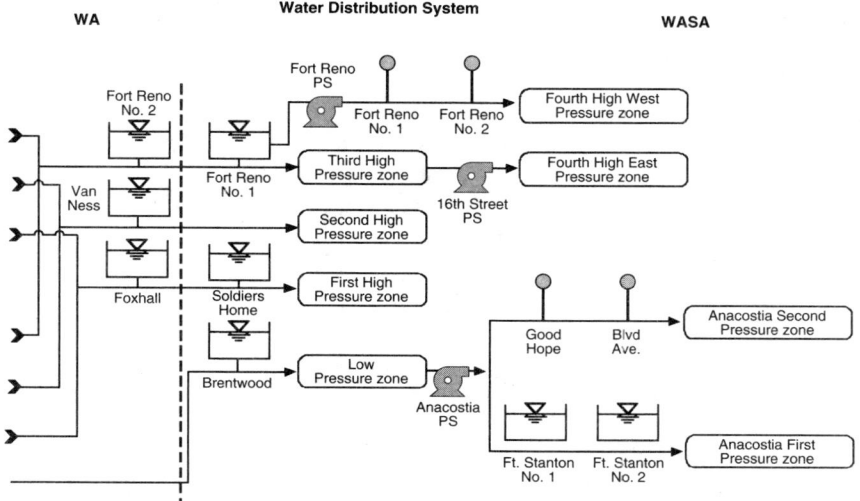

FIGURE 14.1 District of Columbia water system—major distribution zones.

- Easy expansion and setup. Since the SCADA system was operational, it was only necessary to add smaller SCADA "slave" units at the monitoring point.
- Maintenance support. The DCWASA personnel were well trained in using and maintaining the existing system.
- PC support. Although the DCWASA's SCADA used a proprietary operating system, it provided a PC-link support which could be used to *dump* data into a regular PC for further distribution. The availability of this feature was critical so that data could be made available to the U.S. EPA users in a PC format (MSDOS or Windows). This feature was utilized in the overall SCADA implementation. Figure 14.3 shows the relationship between the SCADA system and the transmission of the data through the Internet. The added benefit of this feature was tight security; an authorized end user can only copy the relevant data *published* on the PC and cannot directly access the SCADA system. This feature also eliminated any potential interference between the sampling system data and the distribution system data.

Based on these advantages, the existing SCADA system was therefore selected to be used as a pipeline to channel the online sampling data obtained from the various instruments. The data downloaded from DCWASA's SCADA system were published on a website operated in Cincinnati. Since the project has ended the systems and the website are currently not operational. The overall project, however, did demonstrate that such systems could be developed. The lessons learned from this research are documented in Sec. 14.2.6 of this chapter. Figure 14.4 shows some of the output data for the Fort Reno Tank.

West Virginia RMCT Network. In May 1991, the U.S. EPA provided funding to support a research project titled "Alternative Low Maintenance Technologies for Small Water Systems in Rural Communities" (Goodrich et al., 1993). This project involved the installation of a

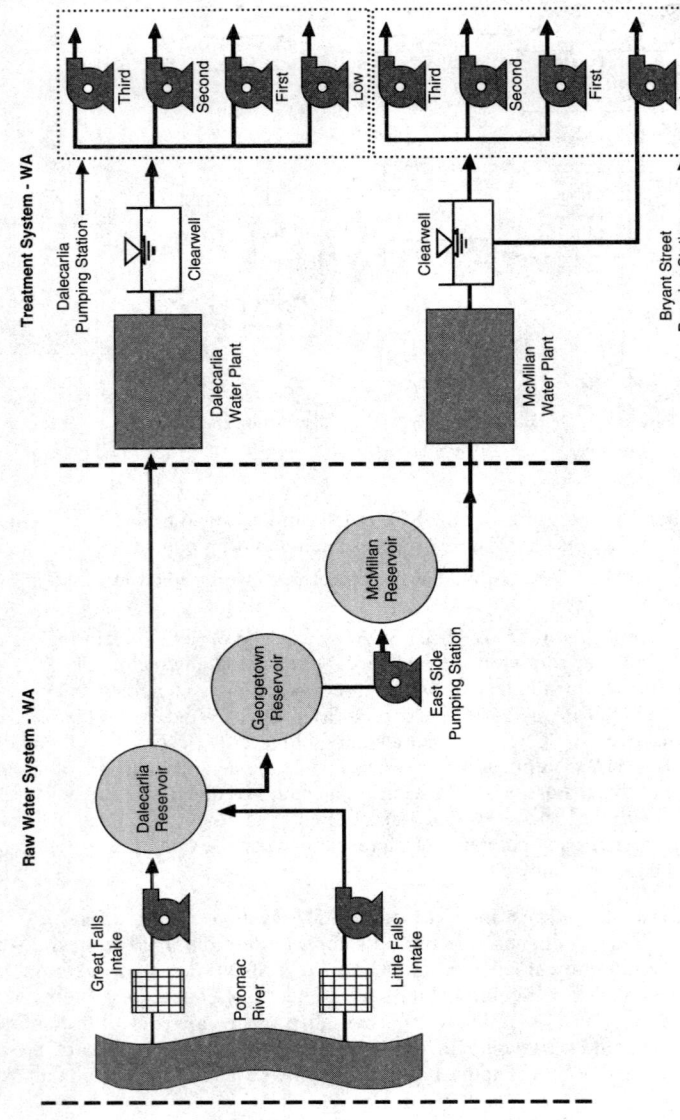

FIGURE 14.2 District of Columbia water system—major facilities.

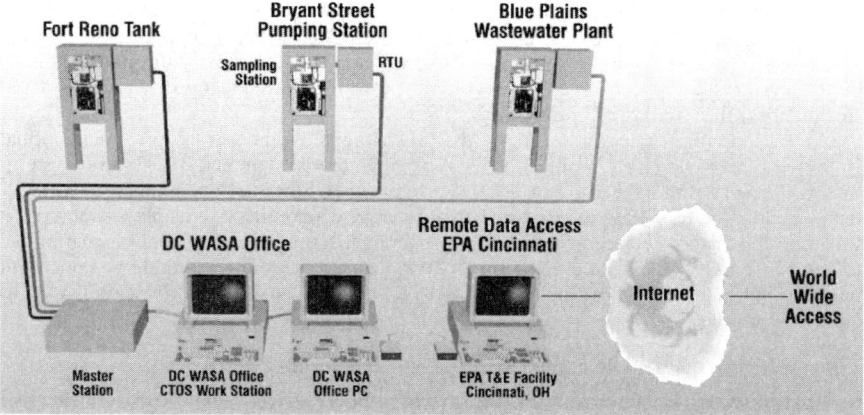

FIGURE 14.3 Relationship between the SCADA system and the transmission of the data through the Internet.

small drinking water treatment package plant in a rural location in West Virginia in order to evaluate the cost-effectiveness of package plant technology in removing microbiological contaminants. Secondary objectives of this project included automation of the system to minimize the operation and maintenance (O&M) costs, assessment of the community's acceptance of such a system, "ability-to-pay," and the effect of the distribution system on water quality at the tap. The following is a brief history of the overall project.

The treatment system is located in rural Coalwood (McDowell County), WV approximately 12 mi from the McDowell County Public Services Division (MCPSD) office in the Appalachian Mountains. Prior to 1994, an aerator combined with a slow sand filter was being used for water treatment at this site. This combined unit had been operational for over

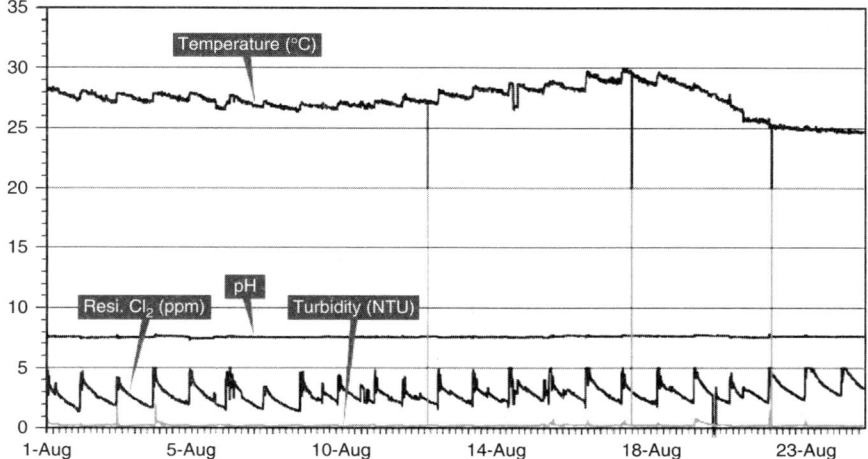

FIGURE 14.4 Output data for the Fort Reno Tank.

30 years and needed substantial repairs. The water flowed by gravity from an abandoned coal mine to an aerator built over a 6-ft diameter slow sand filter. A hypochlorinator provided disinfection to the treated water, and the water flowed by gravity through the distribution system to the consumer. The volume of water from the mine was considered sufficient for the small rural community.

An engineering study conducted by MCPSD estimated the cost of a new conventional water treatment system (to replace the existing treatment system and distribution system) to be $328,000, resulting in a cost of $10,933 per customer. Since this did not represent a practical solution, the U.S. EPA investigated an alternative economically feasible technology. It was essential that the replacement technology operate in a rugged environment with minimal maintenance. Also, the treated water quality characteristics were required to be consistent with the Surface Water Treatment Rule (SWTR) and the Total Coliform Rule (TCR) as described below:

- No more than one sample per month may be total coliform positive.
- Turbidity at all times must be less than 5NTU and normally cannot be greater than 0.5 NTU
- Heterotrophic Plate Count (HPC) must be less than 500 mL if the chlorine residual value is less than 0.2 mg/L

Based on a review of existing technology, the U.S. EPA determined that a packaged UF system would be ideally suited for this location and in 1992 such a unit was purchased and installed at this site. The UF system has been in operation since 1992, and has provided water of good quality to the community.[*]

The packaged UF system, as initially installed, used a programmable logic controller (PLC), along with PLC controllable hardware for automation. The UF system included an online pH sensor, online chlorine sensor, pressure gauges, etc. Initially, the UF system operating and water quality parameters were logged manually from the instrument's analog/digital displays. However, after the completion of the initial project it became apparent that the UF system would require additional capabilities to comply with the SDWAA, SWTR, and the WV Department of Health monitoring and reporting guidelines. These regulations require the treatment unit operator(s) (in this case the MCPSD) to maintain process and water quality records on a daily basis, resulting in significant costs for associated time and travel. Furthermore, during inclement weather conditions, performing these routine tasks became extremely difficult. Similar monitoring requirements at other remotely located sites required the operators to dedicate considerable amounts of manpower to routine tasks.

Because of these routine requirements the operators sometimes had to delay troubleshooting activities at other problematic locations causing service disruptions. Even though these regulatory requirements are designed to improve the quality of the water and protect the consumer, they inadvertently caused troubleshooting delays, loss of water, and customer dissatisfaction. To address these issues, in 1996, the U.S. EPA developed, installed, and tested a remote-monitoring system at the site. The system used commercially available hardware along with proprietary U.S. EPA developed software. The software was a MSDOS[†] based system that was hardware specific, not user-friendly, and the overall cost of ownership was very high. Therefore, in 1998, the U.S. EPA updated the SCADA system with a scalable commercially available off-the-shelf, user-friendly, Windows[‡]-based SCADA system. The capital cost for the hardware and instrumentation cost approximately

---

[*]In 2002, this unit was shut down as the distribution network was tied to a larger centrally operated treatment plant by MCPSD.
[†]MSDOS is a registered trademark of Microsoft Corporation.
[‡]Windows is a registered trademark of Microsoft Corporation.

$12,000 (of which, the SCADA hardware was only about $1500). The total cost including technical support, training, and setup was about $33,000. After the success of this project in 2000, the U.S. EPA installed similar SCADA systems at Bartley and Berwind sites in McDowell County for remote monitoring of the water quality. The lessons learned from this research are documented in Sec. 14.2.6 of this chapter.

T&E RMCT Network. U.S. EPA currently operates a SCADA system at the T&E Facility to remotely monitor and control the Distribution System Simulator (DSS). The DSS is designed to simulate conditions encountered in a typical drinking water distribution system and consists of two sets of three individual pipe loops (a total of six loops). Some of the monitoring and control devices installed in the DSS system were designed to be operated remotely. These include the main recirculation pumps, chemical addition pumps, flow meters, and several online sensors that are used for drinking water quality monitoring. The number and types of components managed in the DSS are similar to a small or mid-size utility.

The RMCT devices installed at this location include various on-line sensors and automated controllers such as pH sensors, flow meters, temperature sensors, turbidimeters, and variable frequency pump controllers. The SCADA system hardware is relatively old and unsophisticated although the software system components are relatively new (about 6 years) old.

The industry standard analog (4 to 20 mA) and digital inputs/outputs (IOs) to/from the RMCT devices are transmitted and/or received at the central station PC via slave SCADA units. The slave SCADA units convert the electrical signals to data signals that are communicated to a PC program and vice-versa. Each pipe loop has a dedicated slave SCADA unit that is capable of handling 16 analog IOs and additional 16 - digital IO channels. These IO channels require a separate plug-in SCADA module that performs the required IO function. Each module/channel can serve only one function and must be configured appropriately to process its specific IO function prior to its use.

The SCADA units are preconfigured to communicate using RS-485 protocol. The signals from each SCADA unit are sent to the central PC through the RS-485 to RS-232C protocol converter. This protocol converter device is necessary to facilitate the communication between the PC and multiple SCADA units; the PC serial communication ports use only the RS-232C protocol to communicate with other serial port devices. It acts as a gateway for processing the signals transmitted between the multiple SCADA slave units and the PC serial port.

A data server program is necessary to process data received to and from the SCADA software program (Human Machine Interface—HMI) because the newer software program did not have a direct "driver" (a software program) that can directly communicate with the protocol converter. The HMI program is a Windows-based program used by the DSS operator to communicate with various electromechanical devices. There are additional software-based alarming and remote access programs that are used to provide the additional functionality. The pros and cons of this type of implementation are discussed in Sec. 14.2.6 of this chapter.

14.2.5 Advanced Remote Monitoring Devices

The case studies presented in the previous subsection of this chapter focus on SCADA technology and conventional water quality monitoring devices (such as, pH, ORP, turbidity, temperature, residual chlorine) that can be adapted for remote monitoring. These parameters are well established as surrogate indicators of water quality. However, the events of September 11, 2001 have shifted the U.S. EPA's focus on identifying monitoring devices that can be used to monitor for the presence of harmful substances in the water at various locations starting from source water to the location of the last consumer. Advanced monitoring technologies for the purposes of water security are being studied. These studies are discussed in the following sections.

Biosentinels. The advancement in computer technologies in the 1990s has led to the development of a variety of real-time biosentinels that can be used to detect the presence of toxic compounds in water much like the "canary in the coalmine."Unlike conventional sampling/detection methods which are typically integrated on a random or scheduled basis, these instruments can be used to monitor the real-time changes in water quality. To be effective, conventional sampling/detection would require a wide-array of analyses on a sample in order to identify the presence of various potentially toxic substances or the type and/or level of contaminants must be known prior to the analysis.

Biosentinel instruments are typically designed to use a specific organism to detect changes in water quality. These instruments monitor and analyze behavioral changes in the specific organism under the influence of the sampled water in real-time. If any "toxic" substance is present in the sampled water, it is expected that they would induce a behavior change in the organism that can be analyzed to produce instrument "alarms." The organisms (e.g. Daphnia, fish, clams) used in these instruments are typically sensitive to many harmful substances. It is believed that these instruments have great potential for source water monitoring. Information obtained from the instruments can be used to undertake measures such as water intake diversion and/or further treatment of source water, thereby enhancing source water protection. Biosentinels currently being evaluated at the T&E facility include the BBE-Daphnia Toximeter, BBE-Algae Toximeter, Clam Monitor, and Fish Monitor.

Multiparameter Microchip Sensors. U.S. EPA is also evaluating the effectiveness of microchip sensors. These sensors are essentially a "lab-on-a-chip," and can perform a variety of chemical analyses in real-time based on well-known electrochemical principles. Typically, an array of electrochemical sensors is mounted on a chip that performs the analysis. Several vendors are currently developing multiparameter chips. The U.S. EPA is currently evaluating a six-parameter measuring chip (Six-CENSE).

Advanced Spectrophotometers. Spectrophotometers have been designed to measure absorbance, reflection, transmission and/or fluorescence of various substances under specific conditions. The measured photometric response is then translated to quantify parameter values. The U.S. EPA is currently evaluating the multiparameter Zero Angle Photon Spectrometer (ZAPS) unit at the T&E facility.

14.2.6 Points to Consider for Effective Implementation of RMCT and Advanced Devices

This section presents a summary of lessons learned from the various research activities discussed earlier in this chapter.

Small-to-Medium RMCT Implementations. Managers considering implementation of RMCT in small-to-medium-sized utilities should consider these recommendations:

- Evaluate the components of the treatment and distribution system that are candidates for RMCT implementation. The cost of upgrading the treatment system for remote operations can be significant.
- For new SCADA implementation, follow the generic selection methodology presented earlier in this chapter (or other appropriate methodology).
- Carefully and methodically integrate various components of an RMCT system into the utility service's standard operating procedures (SOP), business management plan, quality assurance quality control program, and regulatory reporting.

- Provide appropriate training to management, office, and technical personnel for operating the associated SCADA system.
- Identify and address maintenance and calibration of the switches, sensors, and controllers in standard operating practices.
- Assure that the quality of the data retrieved from the SCADA system is high so that operators, managers, and regulators develop confidence in the results.

Medium-to-Large RMCT Implementations. Managers of medium-to-large utilities should consider the following guidance:

- Implement RMCTs in a phased manner in a few locations first before expanding the program to the entire utility.
- Study the potential integration of the monitoring data with network hydraulic/water quality models to enhance the application of the RMCT.
- Develop procedures for responding to alarm "events" at various locations in a network.

Sensor Device Selection and Implementation. For managers considering the installation of new sensors the following guidelines apply:

- Consider testing and evaluating a variety of sensors prior to any large-scale implementation.
- Evaluate the suitability of each instrument at the installed location considering both security and maintenance.
- Recognize that biosensors are maintenance intensive and can trigger "false positive" alarms even with normal and routine changes in source water quality.
- Consider outsourcing the maintenance and calibration of advanced sensors to qualified companies to minimize maintenance issues.

In addition to enhancing water security long-term, real-time monitoring can provide data that can be used to significantly improve treatment and distribution system operation and reduce system downtime. This should improve customer satisfaction, improve consumer relations, and result in health benefits. Data from remote monitoring systems can potentially be used to satisfy regulatory recordkeeping and reporting requirements. Used effectively, RMCTs can substantially reduce labor costs (associated with time and travel) for small system operators. Remote monitoring can instantly alert qualified operators to undesirable water quality and/or other changes in treatment and distribution system(s). Using SCADA systems, an operator at a central location can monitor, respond, and adjust water treatment performance in a matter of seconds and troubleshooting can be performed remotely, reducing downtime and increasing repair efficiency. The higher initial cost typically associated with the implementation of RMCTs and advanced sensors can result in long-term benefits if the system is properly planned, designed, and integrated into the standard operating procedures and the business system of the utility.

14.3 EARLY WARNING SYSTEMS

As discussed previously, there is a heightened level of concern to protect water and wastewater systems. In response the U.S. EPA's WPTF has recommended that the U.S. EPA's Office of Research and Development initiate research into the development of Early Warning systems (EWSs) for both source and finished waters.

In June 2002, the Center for Information Management, Integration and Connectivity (CIMIC) at Rutgers University and EPA's Region II Office, in response to national and regional concerns over infrastructure security, convened a workshop entitled "Monitoring and Modeling Drinking Water Systems for Security and Safety." Attendees from Industry, Local, State and Federal Agencies, and members from academia discussed the state-of-the-art in the area of water system security. A strong recommendation was made by the workshop to develop an early warning system. In this context, an EWS is an integrated system of monitoring stations located at strategic points in a water utility's source waters or in its distribution system, designed to warn against contaminants that might threaten the health and welfare of drinking water consumers. An EWS should be integrated or packaged with appropriate sensors and predictive modeling capability. A follow-up workshop held by the same parties in December, 2002 reinforced this need. Much of the past and ongoing research being conducted by the U.S. EPA's WSWRD (a division of the National Risk Management Research Laboratory), presented in the previous sections of this chapter will contribute to the objective of EWS development.

EWS are intended to reliably identify low-probability/high-impact contamination events in source or distributed water. The following objectives were identified by an International Life Sciences Institute (ILSI) working group report on EWSs for hazardous events in water (ILSI, 1999):

- Provide warning in sufficient time to respond to a contamination event and prevent exposure of the public to the contaminant.
- Has the capability to detect all potential contamination threats.
- Can be operated remotely.
- Can identify the point at which the contaminant was introduced.
- Has a low rate of false positive and false negative results.
- Provides continuous, year-round surveillance.
- Produces results with acceptable accuracy and precision.
- Requires low skill and training.
- Is affordable to the majority of public water systems.

A key aspect of an effective EWS will be the capability to operate in a remote monitoring mode. The characteristics of an EWS are discussed in the following sections. It is clear that many of the projects conducted by the U.S. EPA provide support for development of an EWS.

14.3.1 Characteristics of Early Warning System

A contamination event in source or distributed water must be identified in time to allow for an appropriate response that mitigates or eliminates its adverse impact (ILSI, 1999). The features of an ideal EWS include the following:

Rapid Response Time. The response time for an EWS is the time period from the point at which the contaminant contacts the sensor to the point when a result is reported and a response is initiated. Depending on the nature of the threat, in order to prevent damage, the response time must be such that it allows measures to be taken to prevent or minimize exposure of the public to the contaminant. The required response time will depend on a number of factors, such as the point at which the contaminant is introduced into the system; plant characteristics and detention times; and the nature of the contaminant itself. In some cases, a response time of several hours is sufficient, while in other cases, the response time would have to be on the order of minutes. In most cases, EWSs will likely utilize field deployable units or biomonitoring systems.

Fully Automated. Ideally, an EWS would require little or no operator intervention and would allow for 24-h operation. The system should also allow for remote operation such that a response at a remote sensor would be immediately relayed to an RMCT system. In the event that a contaminant is detected, the EWS should immediately trigger an alarm and contact a plant operator via page, phone, and/or fax.

Screens for a Range of Contaminants. As discussed above, there are a number of agents that could pose a potential threat if introduced into a water supply. It is, however, impossible to know in advance which agent would be used. Therefore, EWSs should have the capability to screen for a range of potential agents. Frequently, the problem is that methods that are effective for screening a large number of agents are not specific enough to distinguish between harmful and benign substances. Thus, the ability to screen for a range of contaminants must be balanced against the need for specificity.

Specific for the Contaminants of Concern. Specificity refers to the ability of an EWS to identify specific biological, chemical or radiological agents. Ideally, an EWS would positively identify specific agents that pose a threat to public health, and would be capable of differentiating between these substances and closely related, yet benign substances.

Sufficient Sensitivity. Sensitivity refers to the lowest level of detection and quantification that an EWS can achieve. An EWS should be sufficiently sensitive to provide quantification of a specific agent at the lowest level that poses a threat to public health.

Low Occurrence of False Positives and False Negatives. False positives occur when an EWS indicates the presence of a contaminant that is actually absent from the sample. False negatives occur when a contaminant is present at levels of concern, but is not detected by the EWS. Both types of errors are problematic, and ideally, the rates of false positives and false negatives would be zero. However, given the fact that there is always a probability of either type of error occurring, it is necessary to characterize the rate at which they occur and include this information in the decision making process.

High Rate of Sampling. Assuming that a monitoring device is online, it is desirable to have the device monitor for contaminants as frequently as is possible. A high rate of sampling would be crucial in the event that the agent is introduced in one defined event and is entrained in the water system in plug flow. Detection of "contaminant plugs" would offer an excellent way to distinguish sabotage from natural changes in water quality.

Reliable and Rugged. The systems must be able to withstand field conditions and still perform reliably. Challenges that a remote monitoring system may face include power outages, extreme environmental conditions, potential vandalism or theft, and fouling of sensor components (biological, chemical, or particulate fouling).

Requires Minimal Skill and Training. The equipment used in the EWS should not require excessive skill or training to operate and maintain the equipment or to interpret the results.

Affordable Cost. The more affordable the monitoring system the more available it will be to plants with limited resources, and the more widely it can be deployed within a system.

14.4 REMOTE MONITORING AND NETWORK MODELING

During the second workshop sponsored jointly by U.S. EPA Region II and CIMIC, a number of papers were presented that emphasized the need to integrate the results from monitoring and sensing platforms into a stable and robust decision support system. A critical part of this decision support system will be incorporation of distribution system models with RMCT systems. This chapter has addressed ongoing research that might provide the basis for this type of integrated system development. In addition to emergency response, such systems may also play a role in achieving compliance with existing and future regulations under the Safe Drinking Water Act (SDWA). This section will discuss the status of the SDWA and the evolution of water quality models and how they might assist utilities in achieving compliance with the Act as well as provide enhanced security.

14.4.1 The Safe Drinking Water Act

The SDWA of 1974 and its Amendments of 1986 (SDWAA) require that the U.S. EPA establish maximum contaminant level goals (MCLGs) for each contaminant which may have an adverse effect on the health of persons. Each goal must be set at a level at which no known or anticipated adverse effects on health occur, allowing for an adequate margin of safety (Clark, Adams, and Miltner, 1987). Maximum Contaminant Levels (MCLs), which are the endorsable standards, must be set as close to MCLGs as feasible.

The SDWA and its amendments have posed a major challenge to the U.S. drinking water industry due to the large number of regulations implemented over a short time frame. Most of the regulations promulgated under the SDWAA have focused on treated water although there is substantial evidence that water quality can deteriorate between the treatment plant and the point of consumption. Factors that can influence the quality of water in distribution systems include chemical and biological quality of source water; effectiveness and efficiency of treatment processes; adequacy of treatment facilities, storage facilities, and distribution system; age, type, design, and maintenance of the distribution network; and quality of treated water (Clark and Coyle, 1990). Initially these regulations were promulgated with little understanding of the effect that the distribution system can have on water quality.

However, the SDWAA has also been interpreted as meaning that some MCLs shall be met at the consumer's tap, which in turn, has forced the inclusion of the entire distribution system when considering compliance with a number of the MCLs, rules and regulations. Consequently, there is growing awareness of the possibility that drinking water quality can deteriorate between the treatment plant and the consumer. For example, SDWAA regulations emphasizing system monitoring include the SWTR, the total coliform rule (TCR), the Lead and Copper Rule (LCR), and the Trihalomethane Regulation. Both the SWTR and the TCR specify treatment and monitoring requirements that must be met by all public water suppliers. The SWTR requires that a detectable disinfectant residual be maintained at representative locations in the distribution system to provide protection from microbial contamination. The TCR regulates coliform bacteria which are used as "surrogate" organisms to indicate whether or not system contamination is occurring. Monitoring for compliance with the LCR is based entirely on samples taken at the consumer's tap. The recent Stage 1 Disinfectants and Disinfectants By-Products Rule (DDBPR) has lowered the standard for trihalomethanes from 0.1 mg/L to 0.08 mg/L. This standard applies to all community water supplies in the United States. Monitoring and compliance is required at selected points in the distribution system. Some of these regulations may, however, provide contradictory guidance. For example, the SWTR and TCR recommend the use of chlorine to minimize risk from microbiological contamination. However, chlorine or other disinfectants interact with natural organic matter in treated water to form disinfection by-products. Raising the pH of treated water will assist in controlling corrosion but may increase the formation of trihalomethanes.

Distribution systems are extremely complex and difficult to study in the field. Therefore, interest has been growing in the use of hydraulic and water quality models as a mechanism for evaluating the various factors that influence the deterioration of water quality in drinking water distribution systems. Recent new and emerging regulations under the SDWA have created an even higher awareness of distribution system related issues. The recent promulgation of the Stage II DDBPR and the establishment of the Initial Distribution Systems Evaluation requirement will only enhance this interest.

14.4.2 Distribution System Design in the United States

Distribution systems in the United States are generally designed to ensure hydraulic reliability, which includes adequate water quantity and pressure for fire flow as well as domestic and industrial demand. In order to meet these goals large amounts of storage are usually incorporated into system design, resulting in long residence times, which in turn may contribute to water quality deterioration. Many water distribution systems in the United States are approaching 100 years of existence and an estimated 26 percent of the distribution system pipe in this country is unlined cast iron and steel and is in poor condition. At current replacement rates for distribution system components, a utility will replace a pipe every 200 years (Kirmeyer, Richards, and Smith, 1994).

Conservative design philosophies, aging water supply infrastructure, and increasingly stringent drinking water standards are resulting in concerns over the viability of drinking water systems in the United States. Questions have been raised over the structural integrity of these systems as well as their ability to maintain water quality from the treatment plant to the consumer.

Many distribution systems serve communities with multiple sources. A factor infrequently considered that may influence water quality in a distribution system is the effect of mixing of water from these different sources. Water distribution systems frequently draw water from multiple sources, such as a combination of wells, and/or surface sources. The mixing of waters from different sources that takes place within a distribution system is a function of complex system hydraulics (Clark, Grayman, and Males, 1988; Clark, Grayman, and Goodrich, 1991; Clark et al., 1991).

It is difficult to study the problems of system design and the effects of long residence times in full-scale systems. Constructing specially designed pipe loops is one approach to simulate full scale systems. However, properly configured and calibrated mathematical hydraulic models can be used to effectively study water quality in situ. Such models may also be used to assess various operational and design decisions, determine the impacts resulting from the inadvertent introduction of a contaminant into the distribution system, and assist in the design of systems to improve water quality.

Water Quality in Networks. Significant transformations may take place in the bulk water phase and at the pipe wall as water moves through a network. Cross connections, failures at the treatment barrier, and transformations in the bulk phase can all degrade water quality. Corrosion, leaching of pipe material, and biofilm formation and scour can occur at the pipe wall to degrade water quality. Contaminants may be treated as either conservative or may experience decay or growth as they move through the system. Many investigators have attempted to understand the possible deterioration of water quality once it enters the distribution systems. Bacteriological quality changes may cause aesthetic problems involving taste and odor development, discolored water, slime growths, and economic problems including corrosion of pipes and biodeterioration of materials (Water Research Centre, 1976). Bacterial numbers tend to increase during distribution and are influenced by a number of factors including bacterial quality of the finished water entering the system, temperature, residence time, presence or absence of a disinfectant residual, construction materials, and availability of nutrients.

The relationship of bacteriological quality to turbidity and particle counts in distribution water was studied by McCoy and Olson (1986). An upstream and a downstream sampling site in each of three distribution systems (two surface water supplies and a ground water supply) were sampled twice per month over a 1-year period. Turbidity was found to be related in a linear manner to total particle concentration, but not to the number of bacterial cells. Degradation of bacterial water quality was shown to be the result of unpredictable intermittent events that occurred within the system.

Hydraulic and Water Quality Models. Cross (1936) proposed the use of mathematical methods for analyzing the flow in networks over half a century ago. Computer based models for performing this type of analysis were first developed in the 1950s and 1960s and greatly expanded and made more available in the 1970s and 1980s. Currently, dozens of such models are readily available on computers ranging from micro computers (Wood, 1980a) to supercomputers (Sarikelle, Chang, and Loesch, 1989).

Hydraulic models developed to simulate flow and pressures in a distribution system either under steady state conditions or under time varying demand and operational conditions are generally referred to as extended period simulation (EPS) models. Hydraulic models may also incorporate optimization components which aid the user in selecting system parameters which result in the best match between observed system performance and model results (Gessler and Walski, 1985). The theory and application of such hydraulic models is thoroughly explained in many widely available references (Gessler and Walski, 1985; American Water Works Association, 1989).

Hydraulic models are also required to provide flow information used in distribution system water quality models. Hydraulic models and water quality models can be tightly bundled into a single entity or a stand alone hydraulic model can be used to generate a file containing hydraulic flow conditions, which is then utilized by a stand alone water quality model. The usefulness and reliability of a water quality model is totally dependent on the proper hydraulic characterization of the network.

Steady State Water Quality Models. The use of models to determine the spatial pattern of water quality in a distribution system resulting from sources of differing quality was suggested by Wood (1980b) in a study of slurry flow in a pipe network. He presented an extension to a steady state hydraulic model in which a series of simultaneous equations are solved for each node.

A similar formulation was later used by Chun and Selznick (1985) in a 166-link representation of the Alameda County, California Water District with three sources of water of differing hardness. Metzger (1985) proposed a similar approach.

In a generalization of this formulation, Males et al. (1985) used simultaneous equations to calculate the spatial distribution of variables that could be associated with links and nodes such as concentration, travel times, costs, and other variables. This model, called SOLVER, was a component of the Water Supply Simulation Model (WSSM), an integrated data base management, modeling, and display system that was used to model water quality in networks (Clark and Males, 1986; Clark and Goodrich, 1993, Clark, 1993a).

An alternative steady-state "marching out" solution was introduced by Clark et al. (1988) for calculating spatial patterns of concentrations, travel times, and the percentage of flow from sources. In this approach, links are hydraulically ordered starting with source nodes and progressing through the network until all nodes and links are addressed (Grayman, Clark, and Males, 1988a).

Wood and Ormsbee (1989) investigated alternative methodologies for predicting water quality and determining the source of delivered flow under steady-state conditions. They found an iterative cyclic procedure to be both effective and efficient. This procedure was similar to the "marching out" solutions described previously for networks that they identify as *source-dependent* (networks where the nodes can be hydraulically sequenced starting from sources). However, for non source-dependent networks, which are rare, their algorithm iterates until a unique solution is found.

Dynamic Water Quality Models. Although steady state water quality models proved to be useful tools, the need for models that would represent the dynamics of contaminant movement was recognized. In the mid 1980s, several models that simulated the movement and transformation of contaminants in a distribution system under temporally varying conditions were developed and applied. Three such models were initially introduced at the American Water Works Association Distribution Systems Symposium in 1986 (Clark et al., 1986; Liou and Kroon, 1986; Hart, Meader, and Chiang, 1986). Grayman, Clark and Males (1988b) developed and applied a water quality simulation model that used flows previously generated by a hydraulic model and a numerical scheme to route conservative and nonconservative contaminants through a network. In this model, each pipe link was represented as a series of "sub-links" and "sub-nodes" with the length of each sub-link selected to approximate the distance that a contaminant will travel during each time step. The number of sub-links varied with the velocity of flow in a link (Grayman, Clark, and Males, 1988a). Kroon and Hunt (1989) developed a similar model originally implemented on a mini-computer (Liou and Kroon, 1986). This model is directly tied to a hydraulic model and generates both tabular and graphical output displaying the spread of contaminants through a network. Hart et al. (1991) developed a model using the GASP IV simulation language.

14.4.3 Early Applications of Water Quality Modeling

One of the first projects to investigate the feasibility of modeling water quality in drinking water distribution systems was conducted under a cooperative agreement initiated between the North Penn Water Authority (NPWA) in Lansdale, PA, and the U.S. EPA. It focused on the mixing of water from multiple sources. The project investigated the feasibility for development and application of a steady state water quality model. As the study progressed, it became obvious that the dynamic nature of both demand patterns and water quality variations required the development of a dynamic water quality model. In addition, techniques for semicontinuous monitoring of volatile organic contaminants were explored (Clark et al., 1988). The concept of water quality modeling was extended to the South Central Connecticut Regional Water Authority in a study which also documented the possible negative impact of storage tanks on water quality in distribution systems. A water quality model called the Dynamic Water Quality Model (DWQM) resulted from these studies. The DWQM was later applied to a waterborne outbreak in Cabool, Missouri and was extended to become EPANET which is a state-of-the-art water quality/hydraulic model developed by the U.S. EPA

Interest has been growing in the use of hydraulic and water quality models as a mechanism for evaluating the various factors that influence the deterioration of water quality in drinking water distribution systems. Recent new and emerging regulations under the Safe Drinking Water Act have created an even higher awareness of distribution systems related issues. Sophisticated hydraulic and water quality models that integrate Geographic Information Systems (GIS) and modeling are currently being developed and marketed.

14.5 SUMMARY AND CONCLUSIONS

U.S. EPA's WPTF, established after the events of September 11, 2001, has established a research agenda with a strong emphasis on the need to understand the factors that affect the security of drinking water distribution systems. Two important aspects of distribution system security are remote monitoring and sensing to detect the presence of contaminants and the application of hydraulic/water quality models to track contaminants once they are introduced into the system. The WSWRD of the U.S. EPA has, for many years, conducted a variety of research in both of these areas. This chapter has provided an overview of the U.S. EPA's early research to advance the use of RMCT coupled with the use of conventional water quality sensors to monitor and control the water quality at various locations (source water, treatment, and distribution) within the drinking water infrastructure. Also discussed are the use of water quality models and the potential for combining water quality modeling and remote monitoring and sensing as the basis for an EWS.

It is clear that deploying RMCT systems alone will not provide genuine protection against water security threats, although the outputs from these systems may indicate the presence of contaminants of concern. To have a truly effective EWS, a utility will require the following supporting elements—a hydraulic/water quality model; information architecture that will allow for the reliable, secure, and error-free transfer of data from the monitor to a centralized database; and some type of decision support system. These needs should provide the basis for an active research program in the area of water security for the future.

ACKNOWLEDGMENTS

For the case studies presented in this chapter, the authors express their thanks to the DCWASA and MCPSD representatives for their assistance in installation and operation of these RMCT systems. Specifically, we wish to thank Dr. Jim Goodrich (U.S. EPA), Mark Meckes (U.S. EPA), Radha Krishnan (Shaw), Majid Dosani (Shaw), Gary Lubbers (Shaw), Dave Elstun (Shaw), George Papadopolous (DCWASA), John Mattingly (DCWASA), Dave Morton (DCWASA), Wally Haider (DCWASA), Bill Barrett (MCPSD), and the vendors of the SCADA systems for their assistance in design and operation of these systems. We also wish to thank other U.S. EPA, Shaw, DCWASA and MCPSD personnel for their assistance in implementing this project.

REFERENCES

American Water Works Association, "Distribution Network Analysis for Water Utilities," *AWWA Manual M-32*, Denver, CO, 1989.

Chun, D. G., and H. L. Selznick, "Computer Modeling of Distribution System Water Quality," *ASCE Spec. Conf. on Comp. Applic. in Water Res.*, Buffalo, NY, June, 1985.

Clark, R. M., and J. A. Coyle, "Measuring and Modeling Variations in Distribution System Water Quality," *J. Am. Water Works Assoc.* 82(8): 46–53, 1990.

Clark, R. M., G. S. Rizzo, J. A. Belknap, and C. Cochrane, "A Water Quality and the Replacement and Repair of Drinking Water Infrastructure: The Washington, DC Case Study," *J Water SRT-Aqua* 48(3): 106–114, 1999.

Clark, R. M., J. A. Coyle, W. M. Grayman, and R. M. Males, "Development, Application, and Calibration of Models for Predicting Water Quality in Distribution Systems," in *Proceedings of the AWWA Water Quality and Technology Conference*, St. Louis, MO, 1988.

Clark, R. M., J. Q. Adams, and R. M. Miltner, "Cost and Performance Modeling for Regulatory Decision Making," *Water* 28(3): 20–27, 1987.

Clark, R. M., W. M. Grayman, and R. M. Males, "Contaminant Propagation in Distribution Systems," *J. Environ. Eng.*, ASCE 114(2): 1311–1316, 1988.

Clark, R. M., W. M. Grayman, R. M. Males, and J. Coyle, "Modeling Contaminant Propagation in Drinking Water Distribution Systems," *J. Water SRT- Aqua* 37(3): 137–151, 1988.

Cross, H., Analysis of Flow in Networks of Conduits or Conductors," *Univ. of Ill. Eng. Experiment Station Bulletin 286*, Urbana, IL, 1936.

Gessler, J., and T. M. Walski, *Water Distribution System Optimization, TREL 85 11, WES*, Corps of Engineers, Vicksburg, MS, 1985.

Goodrich, J., J. Adams, and B. Lykins, Jr., "Ultrafiltration Membrane Application for Small System," U.S. EPA National Risk Management Research Laboratory, 1993.

Grayman, W. M., R. M. Clark, and R. M. Males, "A Set of Models to Predict Water Quality in Distribution Systems," in *Proceedings of the International Symposium on Computer Modeling of Water Distribution Systems*, Lexington, KY, May 1988b.

Grayman, W. M., R. M. Clark, and R. M. Males, "Modeling Distribution-System Water Quality: Dynamic Approach," *J. Water Resour. Planning Management, ASCE*, 114(3): 295–312, May 1988a.

Hart, F. L., "Applications of the NET Software Package," *Proceedings of the AWWARF/EPA Conference on Water Quality Modeling in Distribution Systems*, Cincinnati, OH, February 1991.

Hart, F. L., J. L. Meader, and S. N. Chiang, "CLNET—A Simulation Model for Tracing Chlorine Residuals in a Potable Water Distribution Network," in *AWWA Distribution System Symposium Proceedings*, Minneapolis, MN, 1986.

Haught, R. C., and S. Panguluri, "Selection and Management of Remote Telemetry Systems for Monitoring and Operation of Small Drinking Water Treatment Plants," in *Proceedings of the First International Symposium on Safe Drinking Water in Small Systems*, Washington DC, USA, May 10–13, 1998.

International Life Sciences Institute, Risk Science Institute, *Early Warning Monitoring to Detect Hazardous Events in Water Supplies*. ILSI PRESS, Washington DC, 1999.

Kirmeyer, G. J., W. Richards, and C. D. Smith, "An Assessment of the Condition of North American Water Distribution Systems and Associated Research Needs," *AWWA Research Foundation*, Denver, CO, 1994.

Kroon, J. R., and W. A. Hunt, "Modeling Water Quality in the Distribution Network," in *AWWA Water Quality Technical Conference*, Philadelphia, PA, November 1989.

Liou, C. P., and J. R. Kroon, "Propagation and Distribution of Waterborne Substances in Networks," *AWWA Distribution System Symposium Proceedings*, Minn., MN, September 1986.

Males, R. M., R. M. Clark, P. J. Wehrman, and W. E. Gates, "Algorithm for Mixing Problems in Water Systems," *J. Hyd. Eng.*, ASCE 111(2): 201–211, February 1985.

McCoy, W. F., and B. H. Olson, "Relationship Among Turbidity Particle Counts and Bacteriological Quality Within Water Distribution Lines," *Water Res.* 20: 1023–1029, 1986.

Meckes, M. C., J. S. Mattingly, G. J. Papadopoulos M. Dosani, and S. Panguluri, "Real Time Water Quality Monitoring of a Water Distribution System Using Remote Telemetry," in *Proceedings of the American Water Works Association (AWWA) Distribution System Symposium*, Austin, Texas, September 20–22, 1998.

Metzger, I., "Water Quality Modeling of Distribution Systems," in *ASCE Specialty Conference on Comp. Application in Water Research*, Buffalo, NY, June 1985.

Panguluri, S., R. C. Haught, M. C., Meckes, and M. Dosani, "Remote Water Quality Monitoring of Drinking Water Treatment Systems," in *Proceedings of AWWA Water Quality Technology Conference*, Denver, CO, November 1999.

Sarikelle, S., Y. Chuang, and G. A. Loesch, "Analysis of Water Distribution Systems on a Supercomputer," in *AWWA Comp. Specialty Conf.*, Denver, CO, April 1989.

U.S. EPA, Office of Water, "Drinking Water Regulations and Health Advisories," U.S. EPA 822-B-B-96-002, October 1996.

Water Research Centre, "Deterioration of Bacteriological Quality of Water During Distribution," Notes on Water Research, No. 6, 1976.

Wood, D. J., "Slurry Flow in Pipe Networks," *J. Hydraul., ASCE* 106(1): pp. 12–18, May 1980b.

Wood, D. J., and L. E. Ormsbee, "Supply Identification for Water Distribution Systems," *J. AWWA* 81(7): pp. 56–81, July 1989.

Wood, D. J., *Computer Analysis of Flow in Pipe Networks*, Dept. of Civil Eng., Univ. of Kentucky, Lexington, KY, 1980a.

CHAPTER 15
HYDRAULIC AND WATER QUALITY MODELING FOR CONTAMINATION RESPONSE

Francois J.-C. Bouchart
James W. Davidson
University of Calgary
Calgary, Alberta

15.1 OPERATIONAL NEEDS FOR RESPONDING TO CONTAMINATION EVENTS

The contamination of a drinking water supply, whether accidental or deliberate, can pose serious risks to public health and industries which require a supply of high quality water. An appropriate and effective response is therefore required to minimize these risks while also keeping disruptions to water supplies at a minimum. When a contaminant, chemical or biological, enters the distribution system the operations staff is presented with the challenging tasks of delineating the extent of the contamination spread, isolating the contaminated sections of the network, flushing the contaminant from the affected zones, and bringing the system back to full operations at the end of the event. These tasks are made more difficult by the high level of interconnectivity inherent in typical pipe network layouts and the low level of observability of the distribution system.

Water distribution systems typically are designed with a high level of redundancy, thereby assuring high levels of consumer service. The multiple flow paths created by a highly interconnected network allow a continued supply of water to consumers even when a segment of pipe is isolated. Maintenance and upgrades to the distribution system can then be conducted with minimal impact on consumers. For example, a residential subdivision supplied by a single main connected to the rest of the distribution system is vulnerable to disruptions in supply. The isolation of this main for maintenance would cut off supply to the whole subdivision—a completely unacceptable situation in most locales. Hydraulic redundancy can be achieved by connecting the subdivision to the rest of the system with a second main. The looped pipe network created by this second main ensures that the isolation of one main will cause limited disruptions to supply. Only in the event that both mains are isolated will the subdivision be completely cut off from its water supply. For large pressure zones, hydraulic redundancy is achieved through multiple connections to the network's backbone of transmission mains, thereby keeping disruptions to supply to a minimum.

Unfortunately, the provision of network redundancy may prove to be a liability during a contamination incident. When a single main connects a subdivision to the rest of the water distribution system, this main is the only pathway for the contaminant. A contaminant emanating from the subdivision can be quickly contained by either isolating the main from the rest of the distribution system, or maintaining a positive flow into the subdivision (for example, by opening a hydrant). The contaminant cannot leave the sub-division and affect other parts of the network. Similarly, a contaminant can be easily prevented from entering this subdivision by isolating the single main. The benefits of the non-redundant design go beyond the operational simplicity of having to close a single valve to isolate and contain the contaminant. The detection and delineation of contamination is also simplified by the use of a single main. The presence of the contaminant in a water sample taken from the main confirms that the contaminant has entered the subdivision and the subdivision is being supplied by contaminated water.

In a looped network, all potential pathways into a subdivision would have to be monitored to establish whether contamination has spread to the subdivision. Pressure zones supplied from multiple connecting mains then become vulnerable to contamination due to the large number of potential pathways into the zone, and the increased likelihood that intrusions into the zone will remain undetected. Flow through one zone to another (or one subdivision to another) compounds the problem of clearly delineating the spread of a contaminant, and isolating this contaminant. For example, two adjacent subdivisions are connected to a transmission main, each with a dedicated main. To increase the hydraulic reliability of the system, a third connecting main is installed between the two subdivisions. During a contamination incident, changes in water consumption patterns of the subdivisions might cause a flow reversal in the connecting main. This flow reversal might either help contain the contaminant, or allow further spread.

The last illustration, where a flow reversal in the connecting pipe can either help contain the contaminant or allow its further spread, highlights a second feature of water distribution systems, the lack of observability. Observability refers to the ability to observe or measure the state of the system. Water distribution systems typically have a low level of observability due to the sparsity of the supervisory control and data acquisition (SCADA) system. Water flows are measured only at a limited number of locations, normally at pumping stations. Similarly, pressure information is restricted to discrete locations in the distribution system, selected more on the basis of access than maximizing the information derived from these measurements. The result is a sparse set of measurements that provide a very limited picture of the state of the distribution system. This lack of observability is even greater where water quality is concerned. Water quality monitoring is concentrated at water treatment plants. Once water leaves the treatment facility, knowledge of how the quality changes within the distribution system is restricted to grab samples. Some utilities have invested in online water quality samplers (usually limited to chlorine measurements), but these monitoring networks lag far behind the use of flow meters and pressure probes.

When the potential for the introduction of a contaminant is superimposed on a water distribution for which little real-time information is available, it quickly becomes apparent that the response to such an event is very challenging. Operations staff must first be able to detect the presence of the contaminant. This detection requires an effective monitoring network across the distribution system capable of quickly detecting the presence of critical contaminants (Berry et. al., 2003; Ostfeld, 2003). Once this contaminant has been detected, a plan or strategy must be initiated to determine the extent of the contamination spread, the concentrations of the contaminant, and the risk posed to the consumers in the affected sections of the network. To prevent any further spread of the contaminant, the operations staff must be able to isolate the affected area while keeping disruptions to a minimum. Health concerns typically override the potential inconvenience of water supply interruptions.

With very limited knowledge of the condition of the distribution system, duty of care suggests the delineation of the maximum potential area contaminated, and the establishment of an isolation perimeter. This area is revised and reduced as additional water samples are taken, and knowledge of the contamination spread is developed. The challenge presented by this task is that institutional knowledge of the distribution system may not be sufficient to delineate the maximum potential spread of the contaminant. Tools are therefore needed to supplement this knowledge to ensure that the contaminant does not spread beyond the isolation perimeter. Once properly isolated, the contaminant must be flushed from the affected area before bringing the distribution system back to full and normal operations.

15.2 ROLE OF HYDRAULIC AND WATER QUALITY MODELS

Hydraulic and water quality models have come to the forefront of tools to assist in the operations of water supply utilities, providing a complete estimate of the state of a water distribution system. Operators can then use the information to modify operational settings, or in the case of a contaminant incident, formulate a response.

Hydraulic network models are normally formulated such that for a given demand loading condition, known pipe characteristics (length, diameter, roughness), and known water levels in tanks and reservoirs, the solver computes the flow in each link and the pressure at each node (Larock et al., 2003). This solution provides an estimate of the state of the distribution system. Of particular importance during a contamination event is the complete determination of flow directions and flow rates, whereby it is possible to track how a contaminant might arrive at a location and where this contaminant might propagate from this location.

Information on water movements generated by hydraulic network solvers can be used in several ways to formulate appropriate response strategies. Prior to a contamination incident, hydraulic models can be used to investigate and evaluate alternative response strategies for hypothetical events. Using risk management techniques, critical contamination scenarios can be developed and simulated by the hydraulic solver. Alternative response actions, including closing valves, changing pump settings and opening hydrants, can be introduced into the simulation sequence, thereby affording insights into the best way of responding to the hypothetical contamination event. The full sequence of actions necessary to bring the system back to full and normal operations can then be formulated *a priori* and enacted at a moment's notice. The shortcoming of this off-line use of hydraulic models is the potential for the incident to deviate too much from the limited set of scenarios included in the emergency response planning exercise.

The alternative to emergency response planning is to use hydraulic models in real (or near real) time. Rather than modeling hypothetical events, a hydraulic model of the water distribution system is maintained and updated continuously. Upon detection of a contaminant in the distribution system, the hydraulic model is used to formulate an appropriate response to the event. This ensures that appropriate responses can be developed for incidents not included in emergency planning exercises. The main problem with this approach is the dependence on real-time data. SCADA data are available at limited locations in the water supply system. However, this information does not map well with the input needs of most hydraulic solvers. Rather than pressures and flow rates at discrete locations, hydraulic network solvers typically require data on water consumption throughout the system. Therefore, intermediate models are required that estimate the water consumption at individual locations in the distribution system. A mass balance of the distribution system, whereby the volume of water entering the system minus any increase in storage in the system is equal to the water consumed, can be combined with historical consumption data to arrive at an estimate of water consumption at a given location (or node). Once this information is available, an estimate of the current state of the system can be computed. With all flow directions (and flow rates) defined, points downstream of a known point of contamination can be subsequently identified and the contamination spread delineated. Overlays of the SCADA data can be used to confirm the accuracy of the model estimates, and confirm the actual response of the network to operational actions. As with the scenarios developed in an emergency planning exercise, alternative actions can be evaluated through simulation and a strategy developed in real time to bring the distribution system back to full and normal operations. However, in this case the simulations are done at the same time as operations staff are dealing with the incident, and hence, the number of alternatives is greatly reduced.

A second level of modeling involves the determination of water quality throughout the network. Using the solution to the hydraulic network solver to define the convective term, water quality models can compute the concentration of a constituent in the water. Growth and decay kinematics for the constituent are included in the formulation. In the case of a contamination event, the contaminant is the constituent modeled. The results are concentration estimates in each link/pipe and at each node/junction of the network. This concentration information can then be used to delineate the segments of the pipe network with concentrations above a critical threshold, beyond which health and/or industries are placed at risk. A response strategy can then be developed on the basis of bringing concentrations below the critical threshold as quickly as possible, followed by a more thorough flushing of the contaminant out of the distribution network.

15.3 CONVENTIONAL MODELING

The extent to which hydraulic and water quality models can assist operators during a contamination incident is defined by the specifics of their model formulations, the availability of real-time data, and the limitations on the time available to perform repeated simulations. The first of these elements, the model formulation, is of critical importance since it defines the capabilities of the models to yield suitable information, and identifies the data required to solve the model.

15.3.1 Definition of Hydraulic Network Solver Formulation

A hydraulic model of a water distribution network formulates the fundamental laws of physics which define how water under pressure travels within this network of pipes. The first of these laws is the conservation of mass, which states that mass can neither be created nor destroyed. The result is a system of equations that ensures that the flow into a node (or location) of the network must equal the flow of water out of that node, including any addition or abstraction of water at that location. The second law relates to energy in the water system, expressed in terms of head (which combines the elevation of the water and its pressure). Water travels from locations of high energy to locations of low energy. Recirculation at an instantaneous point in time, where flow from one location returns to this same location, is not possible since the energy levels are the same at the start and end of this flow path. This second law yields a second set of equations that defines the interrelationships between the energy levels at the nodes of the distribution network. The two resulting systems of equations are linked together by a constitutive law that defines the loss in energy as a flow travels along a pipe; for example, the Hazen-Williams equation defines energy losses as a function of flow in the pipe. While the two systems of equations are linked by a nonlinear constitutive law, it is possible to converge to a single solution.

The solution to the hydraulic network formulation consists of the pressure at each node, and the flow rate in each link of the network of pipes, pumps, and valves which make up a water distribution system. The input parameters of the model include the network topology, the hydraulic characteristics of the components that make up the network (in the case of a pipe, these characteristics include length, diameter, and roughness), and the demand load across the network. The need to know the water demand at each node in the distribution system leads to what is termed a demand-driven model, whereby the hydraulic solution is computed to meet this water demand load. This requirement becomes important in real-time operations, and specifically during contamination events, due to the lack of knowledge regarding actual water consumption at individual nodes. Estimates of these nodal demands are therefore necessary if conventional hydraulic models are to be used in an operational setting.

15.3.2 Estimation of Nodal Demands

Conventional hydraulic modeling of water distribution systems must begin with good estimates of water consumption at individual nodes. Historically, hydraulic network solvers are used to design water distribution systems to meet specific hydraulic performance criteria. A worst-case scenario can be devised based on population projections, historical per capita consumption rates, and observed temporal variations in water demand. No real-time consumption data is needed since the design demand load is expected to occur only very rarely, if ever. Real-time modeling of contamination events changes this condition, requiring the replication of a real event rather than a hypothetical event. The validity of the resulting simulation will then be only as good as the estimates of water consumption used within the conventional demand-driven hydraulic solver.

The measurement of water consumed at individual connections in a water distribution in real (or even near real) time is currently not possible. While the technology needed to measure these flows exists, the cost to implement such a network-wide SCADA system is prohibitive. Extensive data quality controls and procedures would also be needed to ensure that measurement errors and probe failure could be detected. The alternative is to employ estimates of water consumption based on mass balance across the distribution system. The net flow into the distribution system then corresponds to the actual consumption in the network. Flow meters located at strategic locations in the system can be used to refine these estimates by allowing mass balances to be computed for individual portions of the distribution network. However, once again this approach cannot be scaled down to individual nodes since this would require the metering of all pipes connected to the given node. Estimates of the population served at each node, plus historical consumption patterns, are then used to redistribute the actual water consumption derived from mass balance to individual nodes in the model.

Water consumption estimates based on mass balance and historical consumption patterns may be adequate for modeling normal system operations. However, water consumption patterns may significantly deviate from these norms during a water contamination incident. Consider a contamination incident that remains undetected in the first few hours. During this time, it is possible that water consumption patterns will not deviate substantially from those patterns observed historically. However, once detected, and depending on whether it is detected by consumers and/or the water supply utility, a number of scenarios are possible. If the contaminant is easily detected by consumers, their consumption may dramatically drop. Such a drop in consumption would be detected from the available SCADA flow data and a correction made to the mass balance. However, the operations staff would lack the necessary information to deduce whether the reduction is system-wide, or concentrated in a specific portion of the network. Contrast this scenario with one where the contaminant is not easily detected by the consumers, and a reduction in consumption is triggered by a public health notice issued to the affected areas. Under this second scenario, the water utility should have a better idea of where the reductions in consumption are likely to occur in the distribution system. However, increases and/or decreases in consumption in other parts of the distribution system, as consumers react to the event, may mask the rate at which these reductions occur as well as affect the actual magnitude of these changes in consumption. The dynamics of the system provides significant uncertainties in the nodal demand estimates, which are critical to the accurate simulation of the distribution system. The diesel spill that occurred in Glasgow (U.K.) in 1997 illustrates the danger of not recognizing the potential for water consumption patterns to change during a contamination incident, and that operational knowledge of a system may be insufficient for dealing with such an incident (Fraser, 1998).

15.3.3 Connectivity and Spread Potential

Conventional hydraulic solvers provide a solution that includes both the direction and magnitude of flow in every pipe. As explained previously the solution to the hydraulic model can deviate substantially from the actual behavior of the distribution system due to inaccuracies in the demand estimates, which are at best informed guesses. One way of reducing the complexity of the analysis is to ignore the magnitude of the flows and concentrate on flow directions. This reduction in the model precludes the ability to compute concentrations of the contaminant since travel times and mixing at junctions play an important role in the fate of the contaminant. However, if a precautionary principle is adopted, whereby the mere presence is sufficient to deem a point contaminated, then the actual concentrations found at that location become secondary to delineating the spread of contamination. Under this principle, deviations in flow magnitude have no impact on the contamination-spread pattern in terms of which nodes are at risk, whereas a few deviations in flow direction can alter the contamination-spread pattern completely.

The connectivity approach to contamination spread is concerned solely with flow direction and seeks to establish the worst-case aggregate effect of all flow direction patterns that are feasible subject to available SCADA measurement data. Since the magnitude of the flow is not to be computed, the actual demand loads at individual nodes are no longer needed. Going one step further, since shifting consumption patterns could cause significant flow reversals in the network, the approach foregoes the use of any demand estimates. Instead, the connectivity approach relies solely on available SCADA data, and more specifically measured flow direction (not magnitude) and relative head differences. The connectivity approach does not make use of hydraulics equations defined in the conventional hydraulic solver. Instead, inference rules are used to determine flow directions. Flow magnitudes, regarded as unimportant when modeling contaminant spreads, are not provided by the technique.

15.4 DEFINITIONS

The connectivity technique to delineating contamination spread necessitates the creation of a new notation, working with flow directions rather than flow magnitudes, and operating modes rather than the single hydraulic solution generated by conventional solvers (Davidson and Bouchart, 2003). An operating mode is defined as a pattern in which *all* pipes in the network are assigned a flow direction. In a network with m pipes there are 2^m operating modes. However, for any network only a very small portion of the 2^m operating modes will be feasible and a smaller subset of these will be consistent with the available SCADA information. Figure 15.1 illustrates the concept of feasible operating modes.

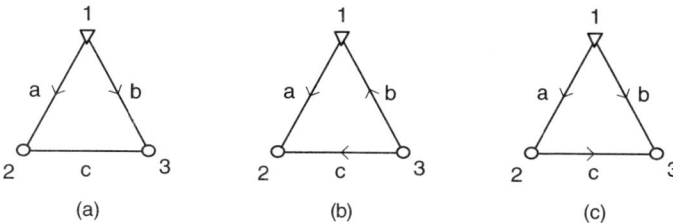

FIGURE 15.1 Flow patterns, infeasible, and feasible operating modes.

Figure 1a does not represent an operating mode because pipe c has no flow direction assigned to it. Figures 15.1b and 15.1c are both operating modes. However, the operating mode in Fig. 15.1b is infeasible because flows in all the pipes connected to junction node 3 are directed away from that node. [In Fig. 15.1 the source node is represented as a triangle, and demand or junction nodes are represented as circles.] The conservation of mass law precludes having a demand node which is not supplied by at least one pipe. The network in Fig. 15.1c represents a feasible operating mode; one of only two feasible modes for that network.

At any given time the network will be operating in one of its feasible operating modes. The solution provided by a conventional hydraulic solver represents only one of many feasible operating modes, and not necessarily the current operating mode of the system due to the potential for errors in the demand estimates mentioned previously. Furthermore, the solution provided by the hydraulic solver will often contradict the available SCADA data due to the inability of conventional hydraulic solvers to incorporate SCADA measurements in the problem formulation. The connectivity approach recognizes these shortcomings of conventional solvers, and yields a spread potential than encapsulates all feasible operating modes. The delineation of the contamination spread is then based on ensuring that all potential spread is captured, and that no spread is missed due to uncertainties in the demand estimates.

15.5   THE CONNECTIVITY MATRIX

The contamination-spread pattern is a function of the upstream/downstream hierarchy formed between nodes. In a single operating mode, such as the solution produced by conventional hydraulic solvers, this hierarchy is simple and immediately obvious. The relationship between any two nodes fits in one of three mutually exclusive categories: (1) upstream, (2) downstream, or (3) neither. If a node is upstream of a contaminated node the node can be considered a potential source of the contamination. If a node is downstream of a contaminated node the node will become contaminated as the contaminant spreads. When the effects of several feasible operating modes are considered simultaneously, a fourth possible relationship emerges, a node can be both upstream and downstream of another node by being upstream in one mode and downstream in another. Neither relationship can be ruled out on the basis of the available SCADA data.

A connectivity matrix is a mechanism that can encapsulate the combined effects of all operating modes that are both feasible and consistent with the current SCADA data. The connectivity matrix is a square matrix of dimension $n \times n$, where n is the number of nodes in the network. Elements of the connectivity matrix are equal to either 0 or 1. Element c_{ij} is equal to 1 if a flow path exists from node i to node j in at least one operating mode that is both feasible and consistent with SCADA data. Therefore node i is potentially upstream of node j if c_{ij} is equal to 1. Node i is potentially downstream of node j if c_{ji} is equal to 1. If both c_{ij} and c_{ji} are equal to 1, feasible modes exist in which nodes i and j are alternatively upstream and downstream of each other and the worst-case contamination spread scenario must consider both relationships.

15.6 ENUMERATION OF OPERATING MODES

Unlike the single operating mode produced by hydraulic solvers, no assumptions about nodal demands are required to find the full set of operating modes required for the connectivity matrix. Using a small set of inference rules it is possible to distinguish modes that are feasible hydraulically and consistent with available SCADA data from those modes that are not feasible. However, enumeration of the full set of 2^m operating modes becomes intractable for anything but trivially small networks. The solution is to derive the connectivity matrix directly from vastly more efficient path enumeration algorithms that do not exhibit the exponential increase in computational effort with problem size.

There are two tests for the feasibility of any operating mode. (1) All junctions or demand nodes must be examined to determine if at least one incident link is directed toward the node. (2) All loops must be examined to ensure that no loop exists in which all links are directed clockwise or counterclockwise about the loop, resulting in flow that circulates around the loop. If the operating mode passes both tests, it is feasible.

The two feasibility tests lead to two inference rules that can be used to deduce flow directions, a procedure used in the efficient construction of connectivity matrices (Davidson et al., 2003). (1) If all incident links at a junction node are directed away from the junction node except one link with unknown direction, that link must be directed to supply the junction node. (2) If all links in a loop are directed clockwise or counter-clockwise about a loop except one link with unknown direction that link must be directed in the opposite direction to avoid circulation.

Experience has shown that two additional inference rules are required for efficient construction of connectivity matrices. (3) For any link with unknown flow direction, test the effect of flow assumed in either direction. If only one of the two directions is feasible, assume flow in the feasible direction. The fourth rule is a more complicated variation of the third rule that states a flow direction is feasible only if it can be shown that there exists at least one feasible mode that supports that flow direction.

15.7 DELINEATION OF MAXIMUM SPREAD POTENTIAL

The examination of individual operating modes is unwieldy for large networks. It is also true that the connectivity matrix becomes difficult to view and interpret for larger networks. It is essential that any computer software package that implements the algorithms described in this section incorporates extensive capabilities to display potential contamination spread based on the connectivity matrix. Software of this type can be easily integrated with conventional hydraulic analysis packages that yield single operating mode scenarios based on best-guess demand estimates. The spread scenarios from best-guess estimates can thereby be compared with the more conservative worst-case scenario based on the connectivity matrix. This approach minimizes the danger associated with overconfidence in demand-driven scenarios.

The resulting approach, combining the strengths of conventional hydraulic modeling where concentrations and travel times can be readily computed with the ability to define the spread potential that is fully consistent with the available real-time SCADA data, provides a powerful tool to operations staff responsible for an appropriate response to a contamination incident. Rather than relying on institutional knowledge of the hydraulics of the network, it is then possible to delineate to the full potential extent of a contamination incident. The result is an approach that is consistent with the precautionary principle of containing the spread, and then reducing the size of the isolated area as additional information becomes available.

REFERENCES

Berry, J., L. Fleischer, W. Hart, and C. Phillips, "Sensor Placement in Municipal Water Networks," in *Proceedings of the World Water & Environmental Resources Congress 2003*, Philadelphia, Pennsylvania, 10 p., on CD-ROM, 2003.

Davidson, J. W., and F. J.-C. Bouchart, "Operating Modes and Connectivity Matrices for Water Distribution Systems," in *Proceedings of the World Water & Environmental Resources Congress 2003*, Philadelphia, Pennsylvania, 10 p., on CD-ROM, 2003.

Davidson, J. W., F. J.-C. Bouchart, and S. Cavill, "Rule-Based Technique for Inference of Flow Directions in Water Distribution Networks," in *Proceedings of the First World Congress on Information Technology in Environmental Engineering*, ITEE 2003, Gdansk, Poland, 2003.

Fraser, R., *The Burncrooks Inquiry: Report on the Disruption to Public Water Supplies in the Area Served by Burncrooks Waterworks, December 1997*. The Scottish Office, Edinburgh, Scotland, 71 p., 1998.

Larock, B. E., R. W. Jeppson, and G. Z. Watters, *Hydraulics of Pipeline Systems*, CRC Press, Boca Raton, FL, 2003.

Ostfeld, A., "An Early Warning Detection System (EWDS) for Drinking Water Distribution Systems Security," in *Proceedings of the World Water & Environmental Resources Congress 2003*, Philadelphia, Pennsylvania, 7 p., on CD-ROM, 2003.

CHAPTER 16
OPTIMAL MONITORING STATIONS ALLOCATIONS FOR WATER DISTRIBUTION SYSTEMS SECURITY

Avi Ostfeld
Technion—Israel Institute of Technology, Haifa, Israel

16.1 INTRODUCTION

Following the events of September 11, 2001, the Administrator of the U.S. Environmental Protection Agency (U.S. EPA) created the Water Protection Task Force (WPTF), which identified water distribution systems as a major area of vulnerability. The need to predict the spread of contaminants in distribution systems and to monitor their concentrations at various system locations so as to keep the water supplied to the public safe, is of a major concern.

Tracking pollutant movement and concentrations within a water distribution system is a complex task which requires (1) a mathematical quantity-quality model for conducting an accurate extended period hydraulic water quantity and quality simulation and (2) the ability to monitor real-time pollutants' concentrations. The optimal allocation problem of water quality monitoring stations that provide a real-time early warning detection system for keeping the potable water supplied safe, subject to a given budget, available monitoring locations, level of service, and technology constraints, ties (1) and (2), and is the subject of this chapter.

Looking at the above three mutually connected topics (1) a mathematical model for extended period hydraulic water quantity and quality simulation, (2) real-time water quality monitoring, and (3) optimal allocation of monitoring stations, the following is stated:

1. In 1990 the U.S. EPA promulgated rules requiring that water quality standards must be satisfied at the consumer taps rather than at the sources treatment plants. This initiated the need for water quality modeling in distribution systems, and the EPA to develop EPANET (U.S. EPA, 2003). EPANET is a public sector model that performs extended period simulation of hydraulic and water quality behavior within pressurized pipe networks consisting of wells, tanks, pipes, consumer nodes, pumping units, and valves. EPANET tracks the flow of water in each pipe, the pressure at each node, the level of water in each tank, and the concentration of chemical constituents throughout the distribution network, during a simulation period. Following the EPANET development, commercial spin-offs of EPANET that even extended the EPANET capabilities also became available.

2. Commercial activity, EPA, and independent research groups are conducting intensive research into the development of sensors and monitoring systems for measuring, or providing "finger prints," in real time of contaminants intrusions into water distribution systems—so far with partial success. Monitoring stations/sensors exist for free chlorine, chloramines, dissolved oxygen, pH, conductivity, Redox/ORP, temperature, and others. For threats of bioweapons agent's intrusions (e.g., Bacteria types like Anthrax, Salmonella, or Biotoxin types like Botulinum toxins, Saxitoxin) commercial sensors do not exist. However, this situation is changing rapidly because of the need for online water quality hazards monitoring equipment from one side, and the advances of research technology prototype products, from the other.

3. Water distribution systems are spatially diverse. As such, they are inherently vulnerable to physical, chemical, or biological threats. In general, physical disruptions can result in significant economic cost inconvenience, but the direct threat to human health is limited. Contrary to that is contamination intrusion—chemical and/or biological, which is one of the most serious potentional threats to water distribution systems.

A contaminant can be dumped directly into tanks, wells, water intakes, or treatment plants; or can be injected at any connection to a water distribution system using a pump or a mobile pressurized tank, capable of overcoming the system pressure. Backflow preventers provide an obstacle. However, they do not exist at all connections; and some may not be functional. Thus, identification of an intrusion must be made with information acquired within the distribution system itself. The ability to monitor water quality within the system is a significant concern and no formal procedure or guidelines exist at present on where and how to locate water quality monitoring stations for drinking water distribution systems security, subject to extended hydraulic and water quality unsteady conditions.

A methodology to enhance water distribution system security that links extended period water quantity and quality simulation with real-time monitoring for providing a real-time early warning detection system (EWDS) against deliberate terrorist hazard intrusions is the subject of this chapter. The methodology, implemented in a noncommercial program entitled optiMQ, was developed by Ostfeld and Salomons (2003). The description of optiMQ followed by an example application is described below.

16.2 SCIENTIFIC BACKGROUND

Water quality may vary within a water distribution system due to internal degradation/growth or external intrusion. Efficient water quality monitoring is thus one of the most important tools to provide adequate and reliable water supply. A decline in pressure at one or more of the system nodes can cause a reduction of the quantities supplied, while an accidental entry of a contaminant, or self deterioration of water quality within the network itself is a severe damage to public health (Geldreich, 1991).

For providing maximum protection to public health all system nodes would need to be monitored, which obviously is an impossible task. On the other hand, if it is assumed that water quality does not change as it is distributed through the network, only the sources would need to be monitored. Since phenomena in the network itself (e.g., a deliberate terrorist intrusion of hazards, corrosion, or THM formations) cause water quality to change, there is a need to monitor in the network, and not just at the sources.

An early warning detection system should reliably identify contamination events (accidental or deliberate) in source water or distribution systems, to allow an effective response. Sensor and information technologies that can provide the tools needed to continuously monitor water quality variables, transmit monitoring data in real-time, validate, display, and interpret the data, are rapidly developing.

In 1990 the U.S. EPA announced rules requiring that water quality standards must be satisfied at the consumer taps rather than at the source treatment plants. This initiated the need for water quality modeling and raised other corresponding problems and issues.

The control of pressure is achieved by hydraulic devices such as booster pumps and pressure reducing valves, while water quality by booster chlorine injections (Pool and Lansey, 1997; Boccelli et al., 1998, 2001) and by monitoring. The problem of booster chlorine injections is to add minimum disinfectant amounts at different system locations at different times, such that the total disinfectant mass injected is minimized, while retaining a minimum threshold disinfectant concentration level at each of the consumers' nodes. Since the injection process does not affect water movement, it can be shown that if chlorine decay follows a first (or zero) order reaction then the response at a node is linear with respect to an injection at the booster (Boccelli et al., 1998). This linearity relationship has been used as the basis of a linear programming problem to determine the booster injection rates that minimize the total mass of the chlorine injected, while maintaining desired chlorine levels at all nodes for all times (Pool and Lansey 1997; Boccelli et al., 1998). The decision variables in this problem were the amounts of chlorine to be injected at each of the in-situ boosters.

Optimal booster locations are an extension of the booster operation problem. Here, the decisions are the location of boosters and their injection rates for a typical demand cycle (e.g., a day). The objective of the location/operation problem is to minimize the mass of chlorine injected for a desired number of in-situ boosters or to minimize the cost of new booster stations while limiting the injected chlorine mass. This problem can be considered as an inverse type of the monitoring location problem in that the sources of injected material are to be identified to satisfy desired levels at all locations. The monitoring problem is to identify the source of the contaminant from measurements throughout the system.

Pool (2002) formulated and solved the location problem as a mixed integer linear programming problem using linear response functions. Tryby and Uber (2001) solved a mixed integer linear programming model that uses water age as the basis for determining sample locations and scheduling.

Water quality monitoring varies through a distribution system and verifying that acceptable water is delivered to customers is not straightforward. The U.S. EPA regulations require that samples should be taken at locations that are representative of the water quality in the system. "Representative" has not been explicitly defined, and several approaches have been developed to quantify representativeness.

Lee and Deininger (1992) developed a procedure based on steady-state flow under one or more demand patterns. The authors used the notion of coverage to denote an assumption that water quality at particular upstream nodes can be inferred by water sampling at some downstream nodes. Using pathway analysis coupled with integer programming, the layout of the monitoring stations was found, maximizing the system coverage. Nodal contributions were determined using the steady state source algorithm of Boulos and Altman (1993).

Following Lee and Deininger (1992), Kumar et al. (1997) applied a greedy heuristic-based algorithm to the same problem to overcome the inherent dimensional difficulty of integer programming. The algorithm provided the same results for the small example solved by Lee and Deininger but no proof of global optimality was shown. Al-Zahrani and Moied (2001) applied a genetic algorithm to solve the same problem.

The demand coverage method first introduced by Lee and Deininger (1992) has several limitations. First, it only considers steady-state water quality conditions that are rarely achieved. Second, the method does not consider the time water age of the system and water quality variations. For example, in case of a long pipe with several consumption outlets the methodology will result in a single monitoring station at the downstream pipe end, regardless of the deterioration/contamination intrusion locations along the pipe. Harmant et al. (2001) modified the objective function to introduce time dependence and water quality into the demand coverage model. The new form weighs the sampling towards bigger flows and "older" water. To emphasize nodes with lower water quality, Woo et al. (2001) further modified the objective by applying weights at each term by normalizing the concentrations by the source values. Thus, nodes with lower water quality received higher weights in the objective function.

All of the above models used the notion that if downstream water quality is acceptable then water supplied before reaching that node must be acceptable. The tendency in the optimal solution then was to install meters at downstream locations where a mixture of flows exists. This basic idea holds for an early warning detection system except that installing monitors at distant points will miss or delay detection of intrusions until many customers have been affected.

Trying to cope with this deficiency, Kessler et al. (1998), and Ostfeld and Kessler (1997, 1999, 2001) presented and applied a design methodology for detecting random accidental contamination intrusions into municipal water distribution systems. The methodology developed was capable of identifying an optimal set of monitoring stations for a given level of service which allows capturing an accidental contamination intrusion into the system. The level of service was defined as the maximum volume of consumed contaminated water prior to detection. The proposed algorithm involved the establishment of an auxiliary network that represents all possible flow directions for a typical demand cycle (e.g., a day), an "all shortest paths" algorithm to identify domains of pollution, and a "set covering" algorithm to optimally allocate the monitoring stations. The algorithm outcome was a minimal set of monitoring stations that satisfies a given level of service. The main shortcoming of Kessler et al. (1998), and Ostfeld and Kessler (1997, 1999, 2001) is in not taking into account the water dilution and water quality variations as they are distributed in the network, and in not addressing explicitly the unsteady state of the contamination flow.

This limitation, together with those of Lee and Deininger (1992) and Kumar et al. (1997) is addressed below through optiMQ, developed by Ostfeld and Salomons (2003).

16.3 optiMQ

Below is the description of optiMQ—definitions, methodology, main program components, and an example application.

16.3.1 Definitions

The following definitions are used herein:

- *Level of service (LOS).* A predefined maximum volume of polluted water exposure to public at a concentration higher than a minimum hazard level (MHL).
- *Pollution event (PE).* A deliberate terrorist injection(s) of a pollutant at one or several nodes of the system, including the consumer nodes, tanks, wells, and treatment plant intakes. Up to three possible pollutant injections at different times are considered to assemble a pollution event (PE).
- *Domain of pollution event (DOPE).* The set of all contaminated nodes up to the LOS due to a pollution event.

- *Domain of detection of a node (DODN).* The set of all detectable pollution events by a single monitoring station.
- *Redundancy (R).* The relative number of pollution events detectable by at least two monitoring stations.

16.3.2 Methodology

The methodology is composed of two main stages.

Randomized Pollution Matrix (RPM) Construction. The pollution matrix (PM) concept was first introduced by Kessler et al. (1998). The RPM constructed within optiMQ is an extension of the PM. It is an nn by m 0 – 1 matrix, where nn refers to the water distribution system nodes (i.e., wells, tanks, reservoirs, consumer nodes), m to a set of randomized pollution events, and "1" and "0" to contaminated/noncontaminated nodes, respectively. The jth column lists all contaminated nodes at an accumulated volume equal to the level of service, due to a randomized pollution event; the ith row: all randomized contamination events detectable by a monitoring station located at node i, assuming the monitoring equipment provides real-time data.

The pollution events considered within optiMQ are a maximum of three short hazard injections (5 min) at three different randomly selected system nodes. The RPM thus provides a stochastic representation of the pollutants intrusion consequences, resulting from a randomized set of pollution events. Figure 16.1 is a schematic representation of the RPM.

Maximum Row Matching Using a Genetic Algorithm (GA). On the RPM, given a number of monitoring stations that comprise a candidate detection system, a genetic algorithm is applied to find the set of rows with maximum column matching. Genetic algorithms are adaptive search techniques introduced by Holland (1975), and further implemented by Goldberg (1989). A genetic algorithm is a domain heuristic independent global search technique that imitates the mechanics of natural selection and natural genetics of Darwin's evolution principle. The fundamental idea is to simulate the natural evolution mechanisms of

Node	Randomized pollution events		
	1	*	m
1	1	0	0
DODE → *	0	1	1
nn	0	1	1

↑
DOPE

Legend:
nn = total number of water distribution system nodes; m = total number of randomized pollution events considered; DOPE = domain of pollution event; DODE = domain of detection event; * = symbol for the ith DOPE, DODE, respectively; 0, 1 = noncontaminant, contaminant nodes, respectively.

FIGURE 16.1 The randomized pollution matrix (RPM).

TABLE 16.1 The GA Parameters Used in optiMQ

GA parameter	Description
String	Integer string equal to the number of additional monitoring stations considered; each bit of the string can receive any number between 1 to the total number of the water distribution system nodes (i.e., the monitoring stations candidate locations)
Selector	Roulette
Crossover type	Blending: a linear combination of the selected two string parents
Elitism	The best chromosome in each generation is delivered unchanged to the next generation
Crossover probability	0.95
Mutation probability	0.02
Number of generations	100
Population size	50

chromosomes (represented by string structures) involving selection, crossover, and mutation. During the last decade, GAs became one of the more robust optimization techniques used in water distribution systems management (e.g., Simpson et al., 1994; Savic and Walters, 1997; Savic et al., 1999).

Within optiMQ each GA chromosome (string) is an integer string of size NOMS, where NOMS represents the number of suggested additional monitoring stations, that together with the set of existing monitoring stations (e.g., at tanks, wells) comprise a detection system. Each integer number within the string can receive any integer value corresponding to a node label of the water distribution system (i.e., a candidate location of a monitoring station), excluding the nodes at which monitoring stations exist.

The fitness of each GA string corresponds to the columns coverage, termed the *detection likelihood* (DL), with a maximum value of 1 if all pollution events are detectable by a set of monitoring stations.

The outcome of this stage is an optimal detection system for a given number of monitoring stations and a level of service. Table 16.1 summarizes the GA parameters used in optiMQ.

16.3.3 The optiMQ Program

The methodology is implemented in optiMQ, tailoring optiGA, a commercial genetic algorithm engine, with EPANET. Figure 16.2 shows the front screen of optiMQ, built of seven stages:

1. First the appropriate EPANET [name.inp] input file is selected.
2. Existing monitoring stations, if any, are defined (e.g., at wells, tanks).
3. Uneven nodes probabilities to be polluted, if any, are specified (e.g., if a node in the system is highly secured, then its probability to be injected can be set a small number).
4. The pollution event parameters are defined: the number of contamination injection events (up to a maximum of three); a time interval between consecutive contamination events; the LOS (i.e., the maximum contaminated volume exposure to public prior to detection); the MHL (i.e., the concentration above which the water is considered contaminated); and the pollutant injection rate (i.e., mass per time of the pure injected pollutant rate).

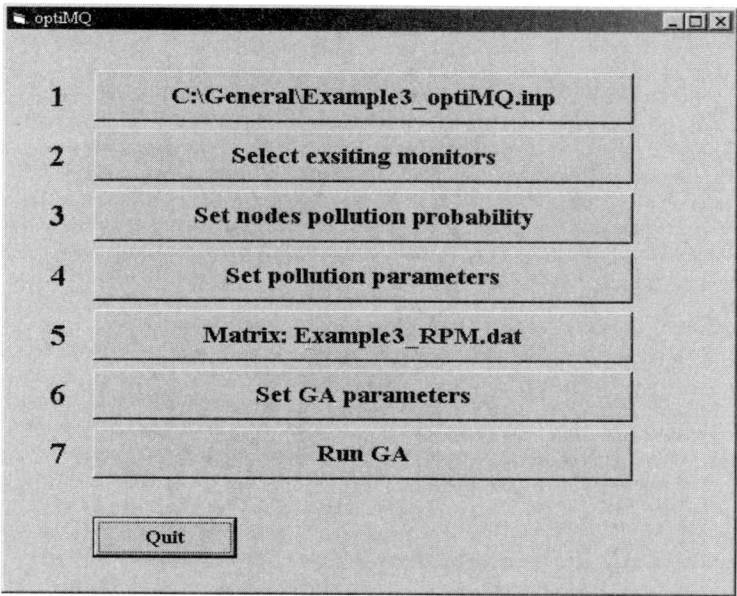

FIGURE 16.2 The optiMQ main screen.

5. The RPM is constructed, with the user defining the size of the RPM for pollution events of two or three random injections (for a one injection pollution event, each node in the system is injected at a resolution of 5 min up to a total of 24 h—the demand cycle considered).
6. GA parameters are specified—number of generations, population size, mutation probability.
7. The number of the candidate monitoring stations are defined and optiGA is activated.

16.3.4 Example Application

The example application is EPANET Example 3 (Fig. 16.3). The system consists of two constant head sources—a lake and a river; three elevated storage tanks, 117 pipes, 92 nodes (consumers and internal nodes), and two pumping units. The system is subject to a 24-h demand pattern. The total system flow production and consumption over time is shown in Fig. 16.4. The remaining data (i.e., detailed consumptions, demand patterns, pipes and pumping units characteristics, tank volumes, operational rules, etc.) are exactly as those used in Example 3 of EPANET (U.S. EPA, 2003). Example 3 is freely available from the U.S. EPA website when downloading EPANET, and thus the remaining data is not repeated herein.

The analysis was performed on a 2.4 GHz/512 K RAM personal computer, with the following computational durations: *Stage A—RPM construction* lasted 9.8 and 16.5 h for one (i.e., 27936 pollution events—a resolution of one pollutant injection at each system node each 5 min, up to 24 h—the demand cycle considered); and for two or three injections

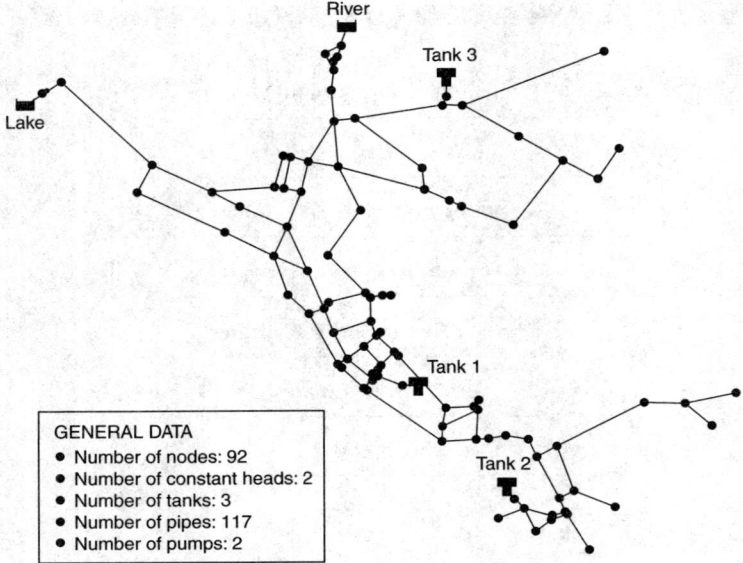

FIGURE 16.3 General layout of EPANET Example 3 (U.S. EPA, 2003).

(i.e., 47045 randomized pollution events equal to $5nn^2$, where nn = 97 is the total number of system nodes), respectively. The $5nn^2$ figure, is a heuristic quantity found for the number of the random realizations needed (i.e., Monte Carlo simulations) to be performed for constructing the RPM for two or three pollution events injections, so as to receive a robust (i.e., not fluctuating) solution. In case of a pollution event of one injection, each of the nodes

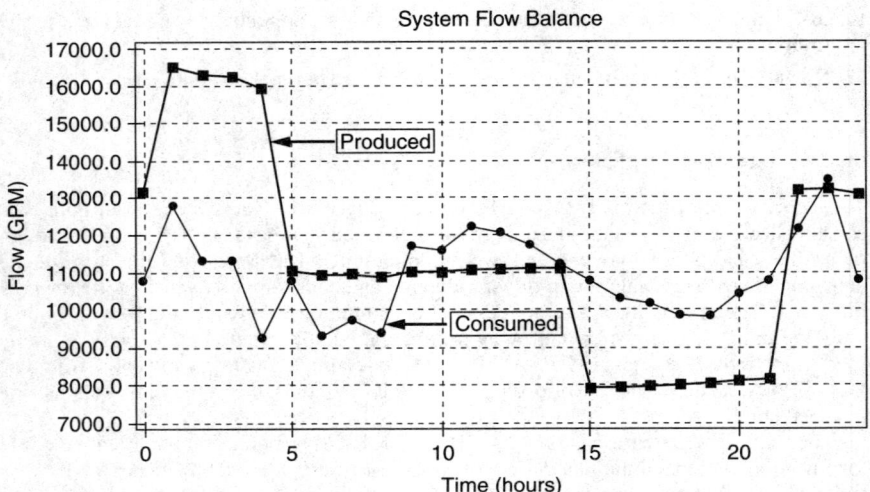

FIGURE 16.4 System total flow production and consumption over time for EPANET Example 3.

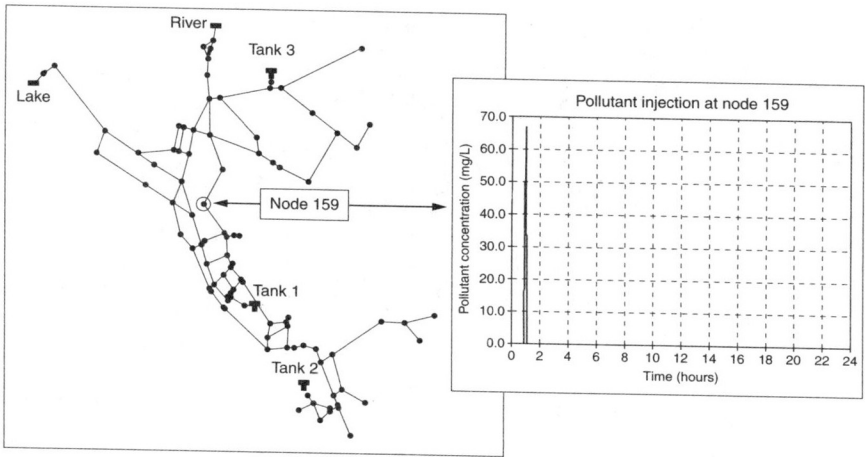

FIGURE 16.5 Pollutant injection example event at node 159 at 1 a.m.—2 kg/min of a pure pollutant (%100 concentration) injection for 5 min (i.e., a total of 10 kg).

was injected at a resolution of 5 min, up to a total of 24 h. Thus, the RPM size was deterministic. *Stage B—GA search* lasted 20 to 30 min for each GA run, with 50 strings at each population, and 100 generations. The computational durations at this stage were dependent on the RPM size (i.e., 27936 or 47045), and on the number of the candidate monitoring stations considered at a specific GA run.

Figures 16.5 and 16.6 illustrate the nature of a pollution event which comprises one contamination injection occurring at node 159 at 1:00 a.m. The pollutant is injected at a rate of

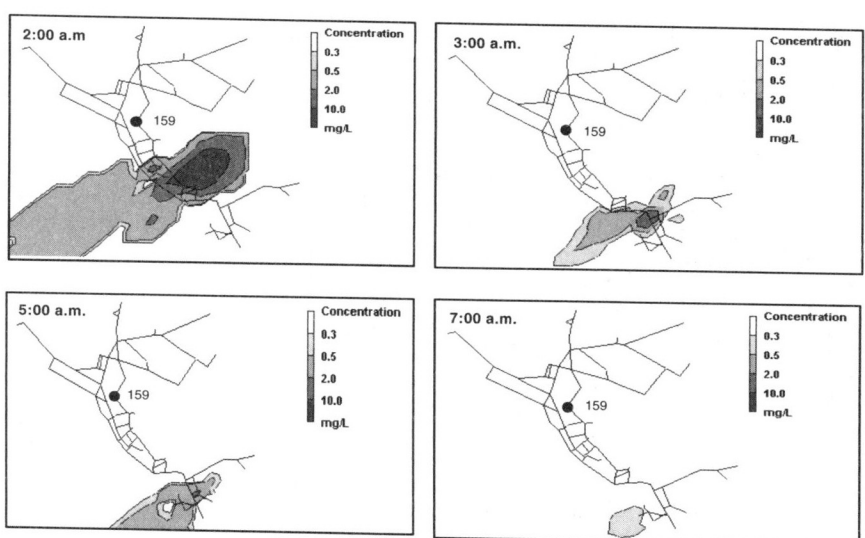

FIGURE 16.6 The resulted contamination spreading of the pollutant as of the pollution event at node 159 at 1 a.m.

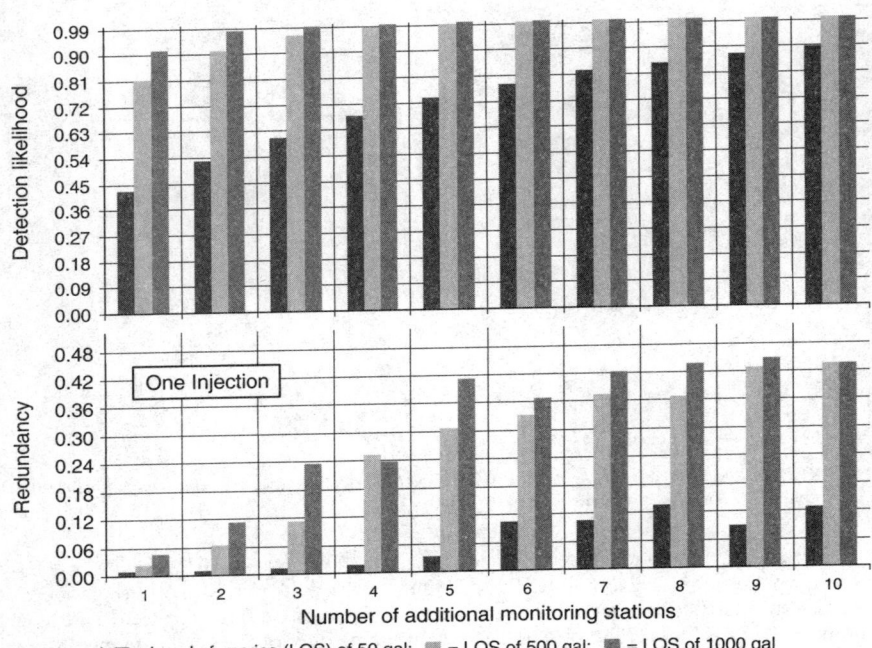

FIGURE 16.7 Tradeoff between the level of service (LOS) and the number of additional monitoring stations for a one injection pollution event.

2 kg/min for 5 min. Thus a total of 10 kg of a pure pollutant is injected. The result of that is a sharp increase of the pollutant concentration at node 159 (Fig. 16.5), and a subsequent pollutant spread downstream, as shown in Fig. 16.6. The event is completed at around 7:00 a.m.

Figures 16.7 to 16.9 describe the main results of this model. The figures show the tradeoffs between the detection likelihood (DL) and the redundancy (R), versus the number of monitoring stations (NOMS), for pollution events (PE) of one, two, and three injections, and for a level of service (LOS) of 50, 500, and 1000 gal, respectively. The durations between consecutive contamination injections were assumed to be 15 min (i.e., in cases of two or three injections comprising a pollution event); the minimum hazard level (MHL) was 0.3 mg/L; the pollution injection rate was 2 kg/min lasting 5 min; monitors were assumed to exist at the sources and at the tanks (i.e., a total of five); and the nodes probabilities to be injected—even.

It can be seen from Fig. 16.7 that as the LOS improves the DL decreases; as the NOMS increases so does the DL and the R; and that for six additional monitoring stations (i.e., a total of 11 monitoring stations including the existing ones at the sources and at the tanks) the DL is approximately one, for a LOS of 500 or 1000 gal. Increasing the NOMS above six yields mainly an increase of the R, less the DL. The net computational duration for constructing the data for Fig. 16.7 was about 40 h.

Figures 16.8 and 16.9 show similar behaviors for two and three injections, respectively, compared to those shown in Fig. 16.7. The net computational duration for constructing the data for each figure was about 60 h.

Figure 16.10 shows the monitoring stations layout for a LOS of 500 gal (see also Fig. 16.8); six additional monitoring stations; two injections; and even probabilities of each node in the

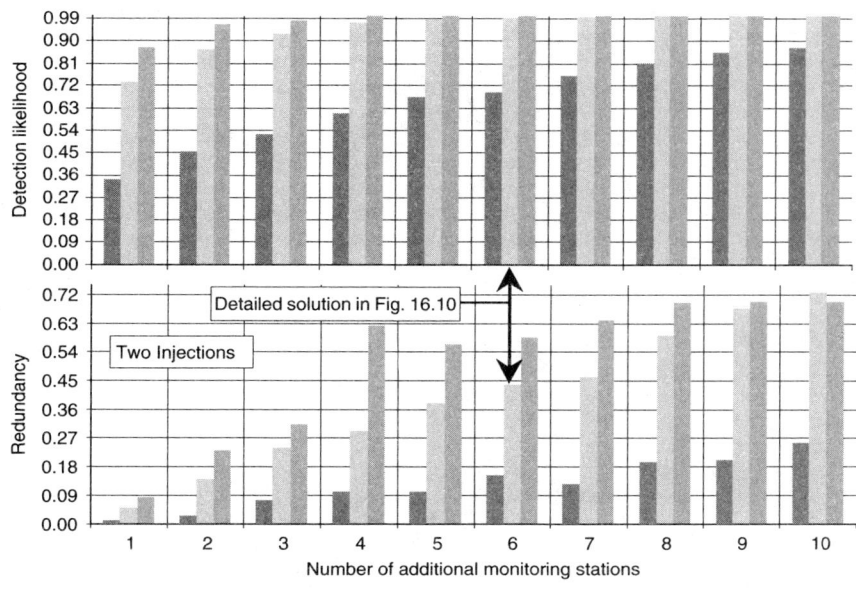

FIGURE 16.8 Tradeoff between the level of service (LOS) and the number of additional monitoring stations for a two injection pollution event.

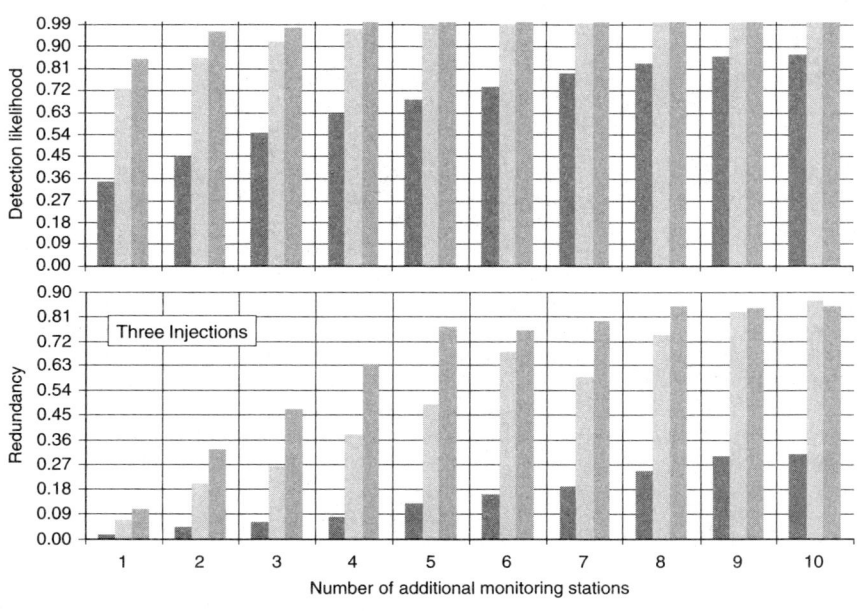

FIGURE 16.9 Tradeoff between the level of service (LOS) and the number of additional monitoring stations for a three injection pollution event.

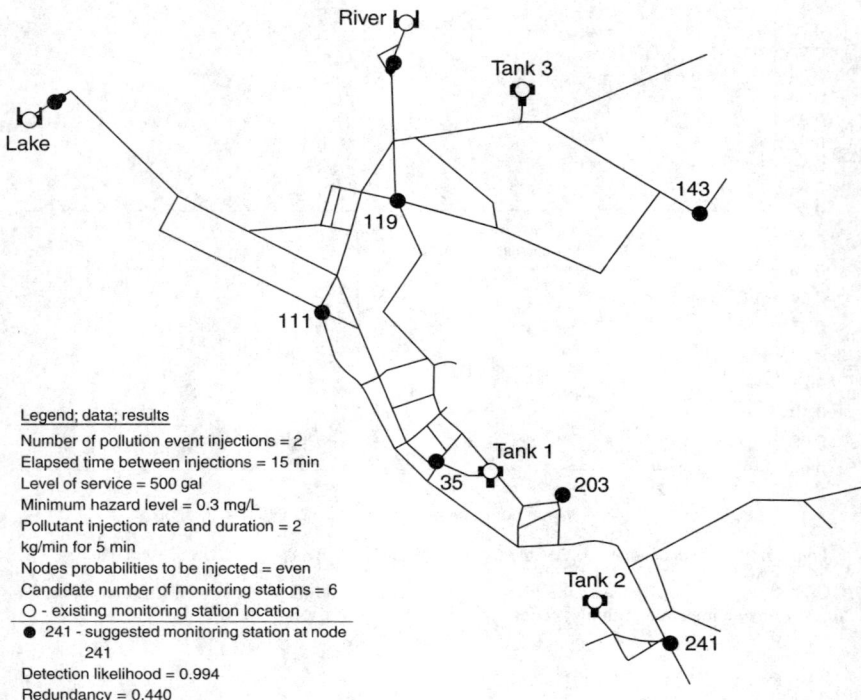

FIGURE 16.10 Example of a monitoring station allocation solution.

system to be injected. It can be seen from Fig. 16.10 that the six monitoring stations are stretched along the entire system, with three (143, 203, and 241) being close to the system edge nodes.

Figure 16.11 describes a sensitivity analysis to the results presented in Fig. 16.10. The water distribution system is divided into four security zones with each having a different relative probability to be injected, ranging from one (the lowest probability zone) to ten (the highest). The results show a slight change of the detection likelihood (0.996 compared to 0.994), but a high increase of the redundancy: 0.54 compared to 0.44. In terms of the monitoring stations location, the monitoring stations at nodes 111, 143, 203, and 241 remain in place; monitoring station 35 is switched with monitoring station 181; and monitoring station 119 is replaced with monitoring station 184, reinforcing the higher vulnerable zone of the system.

16.4 CONCLUDING REMARKS

A deliberate contamination intrusion is generally viewed as the most serious potential terrorist threat to water distribution systems. The spread of chemical or biological agents throughout a distribution system can result in severe consequences of sickness or death among the people consuming the water.

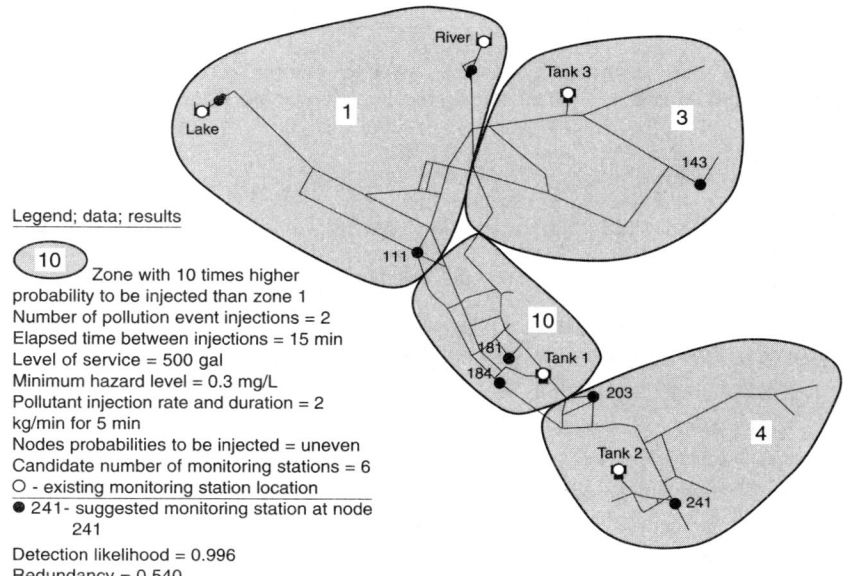

FIGURE 16.11 Sensitivity analysis—Example 3 with different security level zones.

Many locations within the overall water distribution system are vulnerable to a deliberate intrusion of a chemical or a biological agent. These can be surface water sources, treatment intakes, tanks, or direct injections through connections to distribution system mains or consumer nodes.

Real-time pollutants monitoring when combined with network modeling can play an important role in tracking and containing the spread of contaminants throughout the distribution system and thus providing an effective early warning system for enhancing drinking water distribution systems security.

optiMQ addresses this topic by linking EPANET and a genetic algorithm in an overall model for optimally allocating a set of monitoring stations aimed at capturing deliberate external terrorist hazard intrusions through water distribution system nodes like sources, tanks, treatment plant intakes, and consumers—subject to extended period unsteady hydraulics and water quality conditions, and for a given defending level of service to public— a maximum volume of polluted water exposure at a concentration higher than a minimum hazard level. The methodology, made up of two main stages—construction of an RPM and maximum column coverage of the RPM using a GA, was demonstrated using EPANET Example 3.

The example application showed tradeoffs between the number of monitoring stations allocated versus the monitoring stations detection likelihood and redundancy, for different pollution event scenarios and service levels provided to the consumers.

The main limitation, to date, of the methodology proposed is the real-time assumption of the monitoring equipment. Online monitors exist for turbidity, residual chlorine, pH, etc., serving as limited indirect indicators to pollutant levels. However, this situation is changing rapidly because of the need for online water quality hazard monitoring equipment on one hand, and the advances of research technology prototypes products, on the other.

ACKNOWLEDGMENTS

This research was supported by Technion V.P.R. Fund—Technion Res. and Dev. Foundation and Landau/Ben-David, and by the Technion Grand Water Research Institute (GWRI).

NOTATION

The following abbreviations are used in this chapter:

   DL = detection likelihood
   DODN = domain of detection of a node
   DOPE = domain of pollution event
   GA = genetic algorithm
   LOS = level of service
   MHL = minimum hazard level
   nn = number of water distribution system nodes
   NOMS = number of monitoring stations
   PE = pollution event
   PM = pollution matrix
   RPM = randomized pollution matrix

REFERENCES

Al-Zahrani, M. A., and K. Moied, "Locating Optimum Water Quality Monitoring Stations in a Water Distribution System," in Bridging the Gap: Meeting the World's Water and Environmental Resources Challenges, *Proceedings of the ASCE Annual Conference on Water Resources Planning and Management*, Sec. 1, Chap. 266, 2001.

Boccelli, D. L., and J. G. Uber, "Evaluation of Multi-Component Model of Chlorine Residual in Water Distribution Systems," in U. Bogumil, B Coulbeck, and J. Rance (eds.), *Proceedings of CCWI 2001*, Water Software Systems, vol. 1, pp. 195–210, 2001.

Boccelli, D. L., M. E. Tryby, J. Uber, L. A. Rossman, M. L. Zierolf, and M. M. Polycarpou, "Optimal Scheduling of Booster Disinfection in Water Distribution Systems," *J. Water Resour. Planning Management Division, ASCE* 124(2): 99–111, 1998.

Boulos, P., and T. Altman, "Explicit Calculation of Water Quality Parameters in Pipe Distribution Systems," *Civil Eng. Syst.* 10: 187–206, 1993.

Geldreich, E. E., "Investigating the Outbreak in Cabool, Missouri for a Water Supply Connection," in *Proceedings of the AWWARF/EPA Conference on Water Quality Modeling in Distribution Systems*, Cincinnati, Ohio, pp. 55–56, 1991.

Goldberg, D. E., "*Genetic Algorithms in Search, Optimization, and Machine Learning*," Addison-Wesley, New York, 1989.

Harmant, P., A. Nace, L. Kiene, and H. Fotoohi, "Optimal Supervision of a Drinking Water Distribution Network," in Bridging the Gap: Meeting the World's Water and Environmental Resources Challenges, *Proceedings of the ASCE Annual Conference on Water Resources Planning and Management*, Sec. 1, Chap. 324, 2001.

Holland, J. H., *Adaptation in Natural and Artificial Systems*, The University of Michigan Press, Ann Arbor, 1975.

Kessler, A., A. Ostfeld, and G. Sinai, "Detecting Accidental Contaminations in Municipal Water Networks," *J. Water Resour. Planning Management Division, ASCE* 124(4): 192–198, 1998.

Kumar, A., M. L. Kansal, and G. Arora, "Identification of Monitoring Stations in Water Distribution Systems," *J. Environ. Eng., ASCE* 123(8): 746–752, 1997.

Lee, B. H., and R. A. Deininger, "Optimal Location of Monitoring Stations in Water Distribution Systems," *J. Environ. Eng., ASCE* 118(1): 4–16, 1992.

optiGA, available at: http://www.optiwater.com

Ostfeld, A., and A. Kessler, "Closure on Detecting Accidental Contaminations in Municipal Water Networks," *J. Water Resour. Planning Management Division, ASCE*, September/October, 125(5): 308–310, 1999.

Ostfeld, A., and A. Kessler, "Detecting Accidental Contaminations in Municipal Water Networks: Application," *Proceedings of the WRPLM Annual Conference, ASCE*, Chicago, Illinois, USA, pp. 272–278, 1997.

Ostfeld, A., and A. Kessler, "Protecting Urban Water Distribution Systems against Accidental Hazards Intrusions," in *Proceedings of the IWA Second Conference*, Berlin, 2001.

Ostfeld, A., and E. Salomons, "An Early Warning Detection System for Drinking Water Distribution Systems Security," in *Proceedings of the Annual WRPLM ASCE Conference*, Philadelphia, Pennsylvania, June 2003, Published on CD, 2003.

Pool, S., "Optimal Operation and Location of Chlorine Boosters," MS thesis, the University of Arizona, Tucson, AZ, USA.

Pool, S., and K. Lansey, "Locating and Operating Disinfection Boosters in Water Networks," in *Proceedings of the WRPLM Annual Conference, ASCE*, Chicago, Illinois, USA, pp. 738–743, 1997.

Savic, D. A., and G. A. Walters, "Genetic Algorithms for the Least-Cost Design of Water Distribution Networks," *J. Water Resour. Planning Management Division, ASCE* 123(2): 67–77, 1997.

Savic, D. A., G. A. Walters, R. M. Atkinson, and S. M. Randall, "Genetic Algorithm Optimization of Large Water Distribution System Expansion," *J. Measurement Control* 32(4): 104–109, 1999.

Simpson, A. R., G. C. Dandy, and L. J. Murphy, "Genetic Algorithms Compared to Other Techniques for Pipe Optimization," *J. Water Resour. Planning Management Division, ASCE* 120(4): 423–443, 1994.

Tryby, M., and J. Uber, "Representative Water Quality Sampling in Water Distribution Systems," in Bridging the Gap: Meeting the World's Water and Environmental Resources Challenges, *Proceedings of the ASCE Annual Conference on Water Resources Planning and Management*, Sec. 1, Chap. 369, 2001.

U.S. EPA, "EPANET," 2003, available at: http://www.epa.gov/ORD/NRMRL/wswrd/epanet.html

Woo, H.-M., J.-H. Yoon, and D.-Y. Choi, "Optimal Monitoring Sites Based on Water Quality and Quantity in Water Distribution Systems," in Bridging the Gap: Meeting the World's Water and Environmental Resources Challenges, *Proceedings of the ASCE Annual Conference on Water Resources Planning and Management*, Sec. 1, Chap. 318, 2001.

CHAPTER 17
CONTINGENCY PLANNING FOR EMERGENCY WATER SUPPLY IN NON-CONVENTIONAL TIMES

Hendrik J. Bruins
Jacob Blaustein Institute for Desert Research
Ben-Gurion University of the Negev
Sede Boker Campus, Israel

17.1 INTRODUCTION

Modern water supply systems in urban and most rural areas conveniently deliver potable water through complex piping networks to each and every home. Such a situation is common daily routine in many countries and regions. The advantages of centralized water supply systems are manifold in normal circumstances. Water intake, treatment, quality control, and distribution are taken care of by water utility companies and/or the authorities.

However, centralized water supply systems can be vulnerable to natural disasters (Alexander, 1993) and human-made catastrophes such as terrorism and war. The powerful San Francisco earthquake of 1906, which had a magnitude of 8.3 on the Richter scale, caused much damage to buildings and also wrought havoc to the water distribution network (Harris, 1997; Putnam, 2000; Bruins, 2000). Many water pipes in the city were ruptured. Small local fires that started in the wake of the earthquake could spread unimpeded, as the fire brigade had hardly any water at its disposal. A terrible firestorm developed that destroyed large parts of San Francisco.

A contingency may be defined in very general terms as an adverse development or an alternative future (Dror, 1986; Rosenthal and Kouzmin, 1993). Before April 18, 1906, the municipality and inhabitants of San Francisco had undoubtedly certain expectations of the future and ideas of city development in the years to come. A vague, general awareness of the risk of earthquakes was probably present. Yet, the strength of the quake that struck the city and the subsequent firestorm fits the general pattern of many disasters: "unexpected, unscheduled, unprecedented, and almost unmanageable" (Rosenthal et al., 1989).

The coordinated multiple terrorist attacks on the Twin Towers in New York and the Pentagon in Washington DC, the 11th of September 2001, also fit the above pattern.

The incredible nature of these events and the shock of the planned catastrophe—a few thousand people were suddenly killed, some jumping in agony from their offices in the Twin Towers, driven out by smoke and intense heat—rumbled around the world. The unexpectedness and the character of these attacks led to comparisons with the sudden Japanese strike at Pearl Harbor on December 7, 1941, which drew the United States into the Second World War.

The still unfolding ramifications of the September 11 attacks cast a dark shadow on the future. The risk of nuclear and biological terrorism is considered to be present (Foxell, 1997). Governments struggle to come to grips with the new threats that are both potent and elusive. The international political situation is far from stable, while the spread of atomic weapons continues. Pakistan became the first Islamic nation to join the nuclear club in 1998, following renewed atomic bomb-tests by India. Declarations from North Korea boldly stated in April 2003 that the country has built nuclear weapons (Niksch, 2003). Iran is suspected to be in the process of developing weapons of mass-destruction.

Future terrorist attacks, nuclear accidents, and war situations may cause unprecedented pollution of the human environment, including water resources. Clean potable water is a necessary requirement to sustain human life, whether in normal circumstances or in disaster situations. Water supply systems constitute a critical infrastructure in any nation, but may be vulnerable to sabotage or bombing (Clark and Deininger, 2000; Bruins, 2000).

The crucial role of contingency planning is its proactive task to provide the coping mechanisms and potential solutions to make the ramifications of natural disasters and human-made catastrophes somewhat more manageable (Bruins, 2000). This article explores the looming threats of nonconventional risks, such as nuclear accidents, terrorism, or war, which may lead to contamination of water resources and/or destruction of water supply systems. The approach is generic in the context of multiple future scenarios, though specific water monitoring and water purification elements are mentioned that may be used in proactive contingency planning, either by government authorities, water utility companies, or individual citizens.

17.2 THE BIRTH OF THE ATOMIC AGE AND WORLD WAR II—HIROSHIMA AND NAGASAKI

The discovery by Henri Becquerel in 1896 in Paris that penetrating rays were produced by certain uranium salts was an event of great historical significance, which led to developments that placed humanity in the Atomic Age. The peculiar property was termed *radioactivity* by Marie Curie. She and her daughter Irène eventually died from overexposure to radiation, accumulated over the years in the course of their scientific research (Schubert and Lapp, 1957).

The discovery of the neutron by James Chadwick in 1932 became the key to unlock the enormous energy in atomic matter. Two German physicists, Otto Hahn and Friedrich Strassmann, discovered in 1938 that a neutron could split a nucleus of uranium, whereby atomic energy is released. Great concern that Nazi Germany would be the first to develop atomic bombs led to cooperation by the United States and Britain in the Manhattan Project, started in August 1942. The first time a self-sustaining chain-reaction was initiated, using natural uranium and graphite, was on a squash court at the University of Chicago on December 2, 1942 (Barnaby, 1971). The unlocking of atomic energy enabled the development of weapons so powerful that their destructiveness dwarfs anything else in historical perspective. The first ever atom-bomb test, code-named Trinity, took place in the United States on July 16, 1945 in the Alamogordo desert in New Mexico. Three weeks later the new weapon was used against Japan, three years and eight months after Pearl Harbor.

Early in the morning of August 6, 1945, the inhabitants of Hiroshima obviously had no idea that above their city a completely new type of weapon was being positioned, which would suddenly strike with enormous power. At 8:15 a.m. a scorching flash of light and heat, about 600 times stronger than the sun, hit the center of Hiroshima, as "Little Boy"—equivalent to 15,000 tons TNT—exploded at a height of 580 m above the city. Initial radiation at the hypocenter was about 100,000 rad.* The subsequent air blast destroyed all buildings within 2 km from ground zero, as pressures within this radius ranged from 7 to 3 t/m^2. The immense heat ignited multiple fires, accumulating into a large firestorm. The resulting convective motions of the warm humid August air produced a cumulonimbus cloud with thunder and rainfall—at times black and oily, and somewhat radioactive. Approximately 120,000 people were killed outright and by the end of 1945 about 140,000 people had died, i.e., 40 percent of the population of Hiroshima (Barnaby and Rotblat, 1982; Pittock et al., 1986).

Three days after the destruction of Hiroshima by an uranium bomb followed the attack on Nagasaki by a plutonium bomb. On August 9, 1945, at 11.02 a.m., "Fat Man" exploded above Nagasaki at an altitude of 500 m. The explosive yield was equivalent to about 21,000 tous TNT. The hills around Nagasaki gave some protection for the regions beyond, reducing the impact of thermal irradiation and blast. However, the larger yield of the explosion, as compared to Hiroshima, caused destructive damage to buildings at a radius of 3 km from ground zero. Spreading fires also caused a firestorm in Nagasaki followed by *black rain*. An estimated 74,000 people had died by the end of 1945, out of a population of 280,000, i.e. 26 percent (Barnaby and Rotblat, 1982; Pittock et al., 1986).

Looking from the opposite angle, it can be said that 60 percent of the population of Hiroshima and 74 percent of the population of Nagasaki survived the first ever nuclear attacks on cities. How bad were the environmental conditions for the survivors in terms of radioactive pollution? Both atomic bombs were detonated as air bursts. Although direct nuclear radiation of gamma rays and fast neutrons caused death and biological damage, relatively little radioactive fallout was produced, because the fireballs—*ca* 400 m in diameter—did not reach the earth surface. The black rain that fell in both cities washed out some airborne radioactive aerosols. It has been estimated by Shimazu (1985) that total whole-body gamma ray doses accumulated by survivors were about 13 rad in Hiroshima and 42 to 129 rad in Nagasaki. Physiological effects resulting from such doses are not readily identifiable (Pittock et al., 1986), as can be seen in Table 17.1.

17.3 THREE MILE ISLAND AND CHERNOBYL

The danger of radioactive contamination is not limited to war conditions. An accident in a nuclear power station may lead to massive contamination of the environment, including water resources. On March 28, 1979 something went wrong in the Three Mile Island water-cooled nuclear reactor, situated in Pennsylvania along the Susquehanna River. A chain of mishaps caused the loss of cooling water and the reactor came unpleasantly close to a meltdown. However, there was apparently no escape of radioactive particles and the environment and the surrounding residents did not suffer harm. Nevertheless, the reactor was a total-loss and cleaning the site took more than 10 years at a cost of about one billion dollars. Since the accident at Three Mile Island, no new reactors have been ordered in the United States (Rees, 1994; Hopkins, 2001).

The worst nuclear accident so far in the world occurred at the Chernobyl nuclear plant in the former Soviet Union. On April 26, 1986, a power surge in the reactor led to an

---

*Rad = Radiation absorbed dose. The unit, equivalent to the absorption of 100 ergs per gram, signifies the absorbed energy from ionizing radiation.

TABLE 17.1 The Observable Diagnostic Effects—Within a Period of 30 Days—in Relation to Various Amounts of Accumulated Whole-Body Radiation

Whole-body radiation (rad)	Observable effects within a period of 30 days
0–25	No observable effects.
25–50	A dose of 25 rad is the minimum known to produce a significant blood change as manifest by a blood count. No other body changes noted in this range.
50–100	About 10 percent of people exposed would be expected to exhibit symptoms of radiation sickness.* Temporary blood changes. Full recovery of body functions within a few days.
100–200	A dose of 150 rad produces radiation sickness in 25 percent of people. Possible disability due to radiation effects.
200	50 percent of people exposed will suffer radiation sickness. About 2 percent of cases may result in death.
300	Over 90 percent radiation sickness and 25 percent radiation death. Radiation injury and disability very probable.
400–450	Lethal dose to 50 percent of people exposed.
600 or more	Nearly all exposed people expected to die within 30 days.

*Radiation sickness is characterized by a sensation of nausea followed by vomiting and sometimes diarrhea. These symptoms may disappear within a few days. Those who recover may show in between the second and fourth week, loss of hair, small hemorrhages in the skin and mouth, ulceration in the bowels and diarrhea, loss of appetite, loss of weight, sustained high fever, decline in red and white blood cells (Schubert and Lapp, 1957).

Source: Schubert and Lapp (1957); original unit given in Roentgen (R) which is similar to rad (1 R = 1 rad) in water and in soft biological tissues (Hutchinson et al., 1985).

explosion that blew the lid off Pandora's Box (Rich, 1991). The graphite core of the reactor burned and radioactive material went up into the atmosphere for a period of 10 days.

Timely and reliable information to the public in the region of Chernobyl and to the surrounding countries would have been of great importance and a mark of responsible crisis management. Quite the opposite happened, as the disaster was probably first evaluated as bad publicity for the communist system, marked by secrecy. It took almost 3 days until an announcement was made on Soviet television on the evening of April 28. The town of Prypiat, situated at a distance of only 3 km from the nuclear reactor, was evacuated after a long time interval of 60 h following the explosion. It took another week before the people living within a 30 km radius from Chernobyl were evacuated. People living outside the 30 km exclusion zone hardly received attention from the authorities for almost 3 years after the explosion (Rich, 1991).

Rainwater heavily contaminated with radioactive material came down in parts of Scandinavia and Britain, depending on the winds and meteorological conditions. Also, here the government failed in appropriate crisis management. The reaction of the British authorities was initially characterized by inaction and a "don't worry" approach. Decisive measures were only taken after 6 weeks. A temporary ban was imposed on grazing in certain areas, while sheep could not be slaughtered, which affected about a million animals. Cow milk in certain areas of Britain reached levels of 150-200 Becquerel* of radioactive iodine-131 (^{131}I) per liter in the period 5 to 8 May, 1986, which is slightly above the level permitted for infants.

The Prypiat River is the most important natural drainage channel in the Chernobyl region, flowing into the Dnieper River. The latter river provides drinking water to about 35 million people including those living in the capital Kiev. Concerning radioactive pollution

*The *Becquerel* (*Bq*) is a special unit of radioactivity in the International System of Units (SI). One Becquerel is equal to one disintegration per second (U.S. EPA, 2000).

of rivers in Ukraine, it was found that runoff from large watersheds had greater impact on secondary river pollution—within the first few years (3 to 4 years) after the Chernobyl accident—than surface river runoff from highly contaminated territories with relatively small watersheds. Measurements at the end of the Braginka and Senna Rivers show that since 1992 to 1993 the concentration of strontium-90 has begun to exceed that of caesium-137.

Worldwide, there were 438 reactors in use at the end of the year 2000, generating electricity in 33 countries. About 16 percent of all electricity in the world is produced by nuclear power. The share of nuclear electricity in the European Union countries is 35 percent. France is the largest single user of nuclear power, producing about 77 percent of all electricity in nuclear power plants. Nuclear power in the Unites States only accounts for 22 percent of its total energy consumption. The United States is in absolute terms the largest producer of nuclear energy, generating 31 percent of the world total. There are 110 commercial reactors in 32 U.S. states. Six states rely on nuclear power for more than 50 percent of their energy requirements.

17.4 RECENT TRENDS AND FUTURE RISKS

The end of communism in the Soviet Union in the 1990s and the end of the Cold War between NATO and the Warsaw Pact led to a distinct relaxation in terms of nuclear risk. However, in May 1998, India decided to renew its nuclear bomb tests for the first time since 1974. Tension immediately increased in the South Asia region and Pakistan promptly reacted with its own atomic bomb tests, becoming the first Islamic State to cross the nuclear threshold.

The year 1999 witnessed the first military action ever by NATO. The attack against Milosevic' Yugoslavia, paradoxically, brought about a rude awakening of the nuclear might of post-communist Russia. Following the attack, President Yeltsin of Russia dramatically warned in April 1999 about the danger of a new European war and possibly a third world war, as a result of the unilateral actions taken by NATO. Russian missiles with nuclear warheads were reportedly re-targeted at cities of countries participating in the NATO action against Yugoslavia. The warning was not taken seriously by the west. Former U.S. general Alexander Haig, in a television interview, jokingly attributed Yeltsin's remark to Vodka (Bruins, 2000).

However, it is not fully realized that both Russia and China may reserve the same logic and right to act unilaterally, as NATO did, in a situation they perceive as critical. Both countries are superpowers in terms of nuclear weapons and missile systems. The economic difficulty for Russia to maintain a conventional army equal in quality to that of the United States may have contributed to a fundamental shift in Russia's defense policy that occurred in the years 1999–2000. Russia decided to make it easier to press the nuclear button in an international crisis, as outlined in a 21-page national security doctrine published on January 14, 2000 (BBC, 2000; Bruins, 2000).

Iran is continuing its nuclear program with materials supplied by Russia and is suspected to be in the process of developing nuclear weapons. Iran also received assistance in the past from Pakistan and China. The International Atomic Energy Agency (IAEA) put strong pressure on Iran in 2003 to allow detailed monitoring of its reactors. The issue of nuclear developments in Iran and the role of Russia were apparently high on the agenda in a September 2003 meeting at Camp David between the U.S. President Bush and the Russian President Putin. Following the September 11 attacks on the United States, the possible scenario of non-conventional weapons getting into the hands of terrorists has obviously become a major source of concern for the United States, as well as for other countries.

North Korea defied the political pressures of the United States in the period 2002–2003, restarted nuclear installations at Yongbyon and withdrew itself from the Nuclear Non-Proliferation Treaty. North Korea announced in April 2003 that it had developed nuclear weapons (Niksch, 2003). It also tested medium to long-range missiles that can carry such weapons. North Korea has sold its weapon systems in the past to various clients and there is a future risk that other nations or terrorist organizations may get hold of North Korean, nonconventional weaponry. As the spread of nuclear weapons continues, the risk increases that one day, they may be used again in terrorism, conflict or war, for the first time since the 9th of August 1945.

17.5 POSSIBLE CONTAMINATION OF WATER RESOURCES IN NUCLEAR CONTINGENCIES

The lessons from Chernobyl and past atmospheric nuclear bomb tests show that rainwater—the basis of the hydrological cycle—can become highly contaminated with airborne radioactive material. Radioactive dust and aerosol particles remain in suspension in the atmosphere for a certain time, in relation to particle size, weight, air flow patterns and gravity. A division is made between early fallout deposited within 24 h of a nuclear explosion and global or delayed fallout occurring afterwards (Barnaby and Rotblat, 1982). A falling raindrop tends to catch any tiny particle in the air along its path of movement. Therefore, rain will wash radioactive particles out of the air on to the ground, thereby concentrating and increasing radioactive levels in certain places, as compared to dry fallout.

A very uneven pattern of radioactive fallout was deposited in Britain after the Chernobyl disaster. Meteorological conditions at the time and rainfall caused heavy deposition of radioactive cesium in a specific geographic band (Rich, 1991) across Scotland, northern England, northern Ireland, and Wales.

The differentiation between surface water and groundwater constitutes the principal division in the assessment of water resources. According to a theoretical scenario of a nuclear missile attack (Ambio, 1982) and the possible effects on water resources (Wetzel, 1982), it was concluded that the worst radioactive contamination would be in rainwater, followed by rivers, lakes, and groundwater. The principal reason for this differentiation is the fact that rainwater falls usually first on the ground before entering streams and lakes. Many types of soil tend to absorb the most dangerous nuclear fission products. Following nuclear weapons tests in 1963, it was found in central Europe that rainwater had a radioactivity of 500 pCi/L* of residual beta activity, whereas stream water in rivers in the same region measured a much lower amount of about 30 pCi/L.

However, in certain climatic and geological regions the soil filtering of rainwater or the dilution effect is reduced for various reasons, including lack of drainage. Thus surface water may become highly contaminated with radioactive fallout in karstic, permafrost, desert, and marshland regions.

17.6 CONTINGENCY PLANNING FOR WATER SUPPLY SYSTEMS IN NONCONVENTIONAL TIMES

Learning the lessons from the 1906 San Francisco disaster, water engineering systems have been developed in California that are designed in several ways to cope better with earthquakes. Key elements in this example of water-supply contingency planning include

---

*The *Curie* is a unit of radioactivity equal to a nuclear transformation rate of 37 billion (3.7×10^{10}) disintegrations or decays per second. One *picoCurie* is equal to 10^{-12} curies, which is approximately two nuclear disintegrations per minute. One gram of radium is said to have 1 Curie (1 Ci) of activity, as defined historically. The concentration units for radio-nuclides in water is reported as the activity per liter (usually pCi/L) (U.S. EPA, 2000).

redundancy, spare, and substitute structures (Putnam, 2000; Bruins, 2000). Flexible polyurethane hoses have been designed that can be fixed quickly across ruptured and displaced water pipes in order to restore the flow of water through the water distribution network. Moreover, a Standardized Emergency Management System (SEMS) has been developed and imposed by the California state government so that terminology and interagency coordination during emergencies will follow a common template (Bradford-Benini, 1998; Putnam, 2000).

Contingency planning and SEMS emergency management proved valuable during the 1989 Loma Prieta earthquake and the 1994 Northridge earthquake (Comfort, 1994). Management deals with action in real time, overseeing the proper implementation of plans, which is as important as the planning process itself (Bruins and Lithwick, 1998). There has to be an intimate relationship between contingency planning and crisis management (Rosenthal and Kouzmin, 1993).

The great problem of contingency planning is, of course, the uncertainty of the future. Rational planning, in general, cannot be perfectly rational as no human being can ever know all of the planning alternatives open to him at any moment or all the consequences that would follow from any planning action (Faludi, 1987, p. 315). Indeed, scholars such as Wildavsky (1979) argue that the rational planning model is inherently flawed. In classic planning there was a role for the *creative leap* (Faludi, 1987) from specialists or people, who came up with an idea or proposal that proved valuable or even crucial. Given the general problem of uncertainty, Rosenthal and Kouzmin (1993) argue that contingency analysis and planning ought to consider *multiple* future scenarios, in view of the lack of clear knowledge about *the* future.

Facing a range of possible future disaster scenarios, two principal situations of contaminated water environments need to be considered in a generic approach that should lead, proactively, to specific actions in the framework of contingency planning. One principal scenario involves nonconventional contamination of water resources—nuclear accident, terrorist attack, war—while water supply systems in a certain geographic area are not damaged physically and continue to have energy for relatively normal functioning. Another more complex principal scenario involves destroyed water supply systems, nonconventional contamination of water resources, and mass movements of people—refugees or displaced persons—that require emergency water provision, besides food and shelter.

The first "stationary" scenario is less complex, but still requires crucial planning elements and emergency mechanisms to be installed *before* the nonconventional die is cast. Ongoing real-time monitoring of selected biological and chemical substances, as well as radioactivity in general, are obviously key elements. The issue of water quality monitoring network design in general terms is treated in detail by Harmancioglu et al. (1999). Only real-time monitoring can set off alarm bells at the onset of the contamination to enable emergency management measures to prevent people from drinking polluted water and the provision of alternative water.

Clark and Deininger (2000) discuss the potential dangers of bio-terrorism against water supply systems. They argue that maintenance of residual chlorine in the water at a concentration of 0.5 mg/L is an important line of defense against many poisons and biological substances. A sudden decline in the level of chlorine in the water can be used as an alarm bell, as it may indicate a bio-chemical attack.

> On July 10, 1986, the water supply to the presidential rooms of the White House was cut off after a monitor indicated a lack of chlorine. President Reagan got his morning coffee anyhow, using bottled water. The supply to the West Wing was not shut off, but the staff was warned not to drink the water. (Clark and Deininger, 2000:77; New York Times, 1986).

The continuous monitoring of chlorine residuals in a water supply system is important and can be done at minimal cost. However, a drop in chlorine levels does not give information about the substance involved that causes the decline. The direct and quick detection of dangerous toxins in water is also crucial in contingency planning. A recent scientific breakthrough in the identification of biochemical warfare contaminants was developed by biologist Robert Marks of the Ben-Gurion University of the Negev using state-of-the-art analytical devices known as biosensors (Marks et al., 1997; Polyak et al., 2000). Marks developed a device having the size of a shoebox that can be installed in water supply systems or taken anywhere as a mobile system to spot toxins in water. The key biosensor is an *E. Coli* bacteria to which the light-producing enzyme luciferase has been attached through genetic engineering by Shimshon Belkin of the Hebrew University of Jerusalem. When the bacteria become exposed to poisonous chemicals, they emit light.

The innovation by Marks is that he succeeded in gluing these genetically engineered bacteria to the end of a fiber-optic probe, using polysaccharides produced by algae. The patented probe can be placed in water. If the water contains chemicals that damage their genetic material (genotoxins), the bacteria light up. The optical fiber transmits the light to a sensitive photo detector in the box, which enhances and analyzes the signal (Schechter, 2001; Siegel, 2003). The device was successfully demonstrated, on September 2001, to the New Jersey Water Board.

The installment of any testing and monitoring equipment to detect pathogenic contaminants in public water supplies has not yet been implemented in most systems (Michel-Kerjan, 2003). Water supply systems are often owned and operated by local water companies. State or federal legislation may be required to bring about the incorporation of critical monitoring equipment, as well as emergency management regulations, in public water supply systems. The SEMS developed in California could serve as a general example of a state-wide approach.

In addition to the monitoring of water quality, safe water resources should be identified and selected in every province, state or nation as critical water reserves. Groundwater is usually better protected against contamination by terrorism and warfare, as compared to surface water. Therefore, groundwater is to be regarded as a key resource in contingency planning for nonconventional times.

Hydrogeological studies ought to be conducted in each region to select groundwater resources that are most safely protected from possible contamination (Bruins, 2000). Limestone regions with karst may enable highly polluted rainwater to flow underground through natural fissures and dissolution channels without sufficient soil filtering. Underground aquifers in such limestone areas may, therefore, become severely contaminated. Nuclear atmospheric tests in the atmosphere, conducted in the twentieth century, greatly increased the tritium content of rain and surface water (Davies and DeWiest, 1970). Records and present values of the relative amount of tritium in various groundwater aquifers may be an indicator of the residence-time dynamics of each aquifer and the related safety of the water resource.

The "mobile" scenario, involving destroyed or incapacitated water supply systems, nonconventional contamination of water resources, and mass movements of people requiring emergency water provision, is obviously complex for contingency planning. How can water be provided in such emergency circumstances? Will there be electricity or fuel for generators to operate water pumps? Various aid organizations and NGOs have been involved for many years in the supply of relief to large numbers of refugees or displaced persons. The supply of water and sanitation in emergencies (Chalinder, 1994) is not a new subject. However, the requirements in a nonconventional disaster may differ substantially from the wealth of good practice experience accumulated by various organizations.

Drinking-water supply based on rainwater, collected in cisterns or water tanks, during or shortly after airborne radioactive, biological, or chemical contamination, is obviously problematic. Also river water may be seriously contaminated. The direct deposition of fallout on the surface of freshwater reservoirs would be relatively minor, according to Wetzel (1982).

If the radioactive fallout particles sink to the bottom, the water quality may not be too bad, but purification methods are advisable, as discussed below. However, dams with artificial lakes could be attacked, whereby the surface water would disappear. Surface water in karstic, permafrost, desert, and marshland regions may also be unreliable, as noted above.

Can contaminated surface water or groundwater be treated in emergencies? New methods of detection by portable systems, described above, may enable identification of the pollutants involved. A purification technique, developed by Marks (Ben-Gurion University of the Negev) and Sackler, is capable of cleaning water contaminated by organic compounds from a chemical attack. The polluted water is pumped into a reactor where ultraviolet rays purify the water. After 10 h of exposure to the UV light, less than one percent of the contaminants remain in the water (Siegel, 2003).

Mobile water desalination and purification systems may be transported to areas where people require emergency water supply. Clean Water Technology, Inc. (Issaquah, Washington; http://www.halcyon.com/rclark/eci/) has developed a mobile water treatment system that can be moved to disaster regions. The system fits into a 20-ft cargo container for transport and can treat up to 72,000 gal (275,000 L) of water a day to US federal safe drinking water standards (U.S. Water News Online, 1995). Another company, Lifestream Water Purification Equipment (Huntington Beach, CA; http://www.lifestreamwater.com/products.html) has developed various mobile systems that can supply water in the range of 200 gal/day (757 L) to 1,000,000 gal/day (3,785,000 L). A Swiss company (Sunwater; info@sunwater.ch) has produced both a fixed water supply system to purify or desalinate water that can be installed in homes or buildings (aqua station 1800) and a mobile system (aqua mobile 1800). The output of both systems is 1800 L/day. The system allows autarkic water purification at almost every place. It can be used for desalination of seawater and brackish water, as well as sterilization of polluted water.

Some water purification and desalination systems operate on solar energy and thus have a self-contained independent power supply (Mathew et al., 2001), which can be very important in emergencies. The output of these solar energy systems is usually lower as compared to diesel powered mobile systems. Solar powered systems for water treatment are even available on a personal or family scale. Two main systems can be distinguished—solar distillation (solar stills) and the use of reverse osmosis powered by photovoltaic electricity generated by solar panels.

Mass production of solar stills took place during the second World War for the U.S. Navy to provide 200,000 inflatable plastic stills for use in life-crafts. Thus ocean water could be distilled to provide freshwater to survivors of naval battles or shipping accidents. Solar distillation (Malik et al., 1982) produces distilled water from seawater, brackish water, or any other water source. Hence the system is functional for general water purification. The solar still system is most economic for individual households. A commercial solar distillation unit has, for example, been developed by Rosendahl System (RSD; http://www.rsdsolar.de/kompakt1_e.htm). The system (F6-250C) has a collector surface of 2.5 m^2 and produces 15 to 20 L/day. The transport weight of the entire system is about 90 kg, including collector, crude water tank, drinking water tank, solar cell, rechargeable battery, injection pump, drinking water pump, etc.

For water supply outputs above 1000 L/day by single units, solar distillation is not advisable, but reverse osmosis is preferable. The latter system may be powered, on a small scale, by photovoltaic electricity from solar panels. Reverse osmosis (RO) filters water by squeezing it through a semi-permeable membrane with openings of about 0.0001 μ. RO can be considered the most convenient and effective method of water filtration. RO purifies water as it filters heavy metals, chlorine, pesticides, pathogens, bacteria, virus, and even radioactive materials (Table 17.2).

Unlike conventional filtering devices, the substances left behind in front of the RO membrane are automatically diverted to a waste drain so they do not accumulate in the system.

TABLE 17.2 Approximate Removal Rate of Common Impurities by Reverse Osmosis Systems, Using a Typical Thin Film Composite (TFC) Membrane.

Aluminum	97–98%	Nickel	97–99%
Ammonium	85–95%	Nitrate	93–96%
Arsenic	94–96%	Phosphate	99+%
Bacteria	99+%	Polyphosphate	98–99%
Bicarbonate	95–96%	Potassium	92%
Bromide	93–96%	Pyrogen	99+%
Cadmium	96–98%	Radioactivity	95–98%
Calcium	96–98%	Radium	97%
Chloride	94–95%	Selenium	97%
Chromate	90–98%	Silica	85–90%
Chromium	96–98%	Silicate	95–97%
Copper	97–99%	Silver	95–97%
Cyanide	90–95%	Sodium	92–98%
Ferrocyanide	98–99%	Sulphate	99+%
Fluoride	94–96%	Sulphite	96–98%
Iron	98–99%	Zinc	98–99%
Lead	96–98%	*Virus	99+%
Magnesium	96–98%	*Insecticides	97%
Manganese	96–98%	*Herbicides	97%
Mercury	96–98%	*Detergents	97%
*Total Dissolved Solids	95–99%		

The percentage of removal shown is a conservative estimate.
Source: PURE PAK Water Purification Systems (http://www.purepak.com/understandro.htm)

Part of the unprocessed water (feed water) is used for this purpose to carry away the rejected substances to the drain, thereby keeping the membrane clean. The membrane needs to be replaced once in a period ranging from about 1 to 4 years, depending on the water source.

RO systems seem capable of purifying water that has become polluted by nonconventional contaminants. Therefore, reverse osmosis may form a crucial component in water-supply contingency planning. Reverse osmosis systems may range from very large units, supplying water to cities, to very small units for family or individual use. For example the Pure Pak company produces stationary RO units suited for installation at home.

Small mobile systems powered by solar energy are also available. For example, Lushan, Inc. (http://lushaninc.com/solar.htm) produces a unit that contains a seven-stage water purification system, including a polypropylene sediment filter (10 μ), carbon block filter (5 μ), TFC reverse osmosis membrane, an ultraviolet lamp for disinfection, and finally a filter that polishes the taste of the water. The self-contained system, powered by a solar panel, is available in two sizes, delivering 50 or 100 gal/day (189 or 379 L). These mobile systems can be very useful in emergency situations for small groups or individuals, even in remote locations.

Finally, an ultraportable water filtration system, weighing only 1.6 lb. (0.7 kg) can be of crucial value in "mobile" disaster scenarios with people on the move. The improved Katadyn pocket filter (info@MajorSurplusNSurvival.com) is the only portable water purifier approved by the International Red Cross for field work. The filter has a ceramic element that is combined with a silver ion charged inner core. The Katadyn pocket filter is said to remove 97 percent of all radioactivity. The device, looking a bit like a small bicycle pump, is 10 in (25 cm) long and 2 in (5 cm) in diameter. Hence it is ideal for backpacking and use in emergencies. The flow rate is 1 quarter of a gallon (0.95 L) per min, while the service life of the filter element is up to 13,000 gal (almost 50,000 L).

17.7 CONCLUSIONS

- The Chernobyl nuclear disaster in 1986 demonstrated that different government systems, both in the former Soviet Union and in the United Kingdom, can be very slow in crisis management and disaster response in a nonconventional emergency situation.
- The spread of nonconventional weapons continues. Future terrorist attacks, nuclear accidents and war situations may cause unprecedented pollution of the human environment, including water resources.
- Water-supply systems constitute a critical infrastructure in any nation. Proactive contingency planning actions ought to be taken by the authorities and water utility companies in order to be able to detect nonconventional contamination of water resources in real-time.
- Suitable water purification systems should be installed that are capable of cleaning contaminated water in emergencies.
- Individual households may install compact reverse osmosis water purification systems in their homes. Reverse osmosis systems are capable of removing pollutants that slip through the treatment or lack of treatment by water utility companies. Nonconventional pollutants are also removed by reverse osmosis.
- Multiple contingency scenarios ought to be considered for water supply in nonconventional times, in order to develop redundancy, spare, and substitute structures for alternative water supply, in case existing systems are destroyed or incapacitated.
- Mobile water purification systems that have an independent power supply for their operation should form an important component in contingency planning.
- Safe groundwater reserves should be identified by hydrogeological studies and kept in reserve as alternative water-supply resources in emergencies.
- A disaster situation that affects water supply may be totally unexpected. It can take considerable time before the authorities or aid organizations are able to get their act together to provide assistance. Therefore, people may have to take care of themselves in the short or medium term. Mobile, portable water purification units that have an independent power supply (solar energy, muscle power) are crucial in such contingencies.

REFERENCES

Alexander, D., *Natural Disasters*, University College London Press, London, 1993.

Ambio, "Reference Scenario: How a Nuclear War Might to be Fought, *Ambio* 11(2-3): 94–99, 1982.

Barnaby, F., *Man and the Atom*. Thames and Hudson, London & Funk and Wagnalls, New York, 1971.

Barnaby, F., and J. Rotblat, "The Effects of Nuclear Weapons," *Ambio* 11(2-3): 84–93, 1982.

BBC, "Russia Lowers Nuclear Threshold," BBC News, Europe, January 14, 2000.

Bradford-Benini, J., "Getting Organized Pays Off for Disaster Response," *J. Contin. Crisis Management* 6(1): 61–63, 1998.

Bruins, H. J., "Proactive Contingency Planning vis-à-vis Declining Water Security in the 21st Century," *J. Contin. Crisis Management* 8(2): 63–72, 2000.

Bruins, H. J., and H. Lithwick, "Proactive Planning and Interactive Management in Arid Frontier Development," in H. J. Bruins and H. Lithwick (eds.), *The Arid Frontier—Interactive Management of Environment and Development*, Kluwer Academic Publishers, Dordrecht/Boston/London, Chap. 1, pp. 3–29, 1998.

Chalinder, A., *Water and Sanitation in Emergencies. Good Practice Review 1*. Relief and Rehabilitation Network, Overseas Development Institute (ODI), London, 1994.

Clark, R. M., and R. A. Deininger, "Protecting the Nation's Critical Infrastructure: the Vulnerability of U.S. Water Supply Systems," *J. Contin. Crisis Management* 8(2): 73–80, 2000.

Comfort, L. K., "Risk and Resilience: Inter-Organisational Learning Following the Northridge Earthquake of 17 January 1994," *J. Contin. Crisis Management* 2(3): 157–170, 1994.

Davies, S. N., and R. J. M. DeWiest, *Hydrogeology*, Wiley, Chichester, 1970.

Dror, Y., *Policymaking Under Adversity*, Transaction Books, New Brunswick, 1986.

Faludi, A., *A Decision-centred View of Environmental Planning*, Pergamon Press, Oxford, 1987.

Foxell, J. W., "The Prospect of Nuclear and Biological Terrorism," *J. Contin. Crisis Management* 5(2): 98–108, 1997.

Harmancioglu, N. B., O. Fistikoglu, S. D. Ozkul, V. P. Singh, and M. N. Alpaslan, *Water Quality Monitoring Network Design*, Water Science and Technology Library, vol. 33, Kluwer Academic Publishers, Dordrecht/Boston/London, 1999.

Harris, S. L., "Unstable Lands: the Terror of Temblors and Volcanoes," in *Restless Earth*, The Book Division, National Geographic Society, Washington, DC, pp. 146–215, 1997.

Hopkins, A., "Was Three Mile Island a 'normal accident,'" *J. Contin. Crisis Management* 9(2): 65–72, 2001.

Hutchinson, T. C., M. A. Harwell, W. P. Cropper, and H. D. Grover, "Additional Potential Effects of Nuclear War on Ecological Systems," in M. A. Harwell and T. C. Hutchinson (eds.), *Environmental Consequences of Nuclear War. Volume II. Ecological and Agricultural Effects*, SCOPE 28, Wiley, Chichester, pp. 173–267, 1985.

Malik, A. S., et al., *Solar Distillation*, Pergamon Press, Oxford, 1982.

Marks, R. S., E. Bassis, A. Bychenko, and M. M. Levine, "Chemiluminescent optical fiber immunosensor for detecting cholera antitoxin," *Opt. Eng.* 36(12): 3258–3264, 1997.

Mathew, K., S. Dallas, G. Ho, and M. Anda, "Innovative Solar-Powered Village Potable Water Supply," in *Women Leaders on the Uptake of Renewable Energy Seminar*, Perth, June 2001, pp. 97–105, 2001.

Michel-Kerjan, E., "New Challenges in Critical Infrastructures: A U.S. Perspective," *J. Contin. Crisis Management* 11(3): 132–141, 2003.

New York Times, White House Water Cut Off Temporarily, July 10, p. 16, 1986.

Niksch, L. A., "North Korea's Nuclear Weapons Program," *Congressional Research Service*, The Library of Congress, IB91141, pp. 1–18, updated August 27, 2003.

Pittock, A. B., T. P. Ackerman, P. J. Crutzen, M. C. MacCracken, C. S. Shapiro, and R. P. Turco, *Environmental Consequences of Nuclear War. Volume I. Physical and Atmospheric Effects*. SCOPE 28, Wiley, Chichester, 1986.

Polyak, B., E. Bassis, A. Novodvorets, S. Belkin, and R. S. Marks, "Optical Fiber Bioluminescent Whole-Cell Microbial Biosensors to Genotoxicants," *Water Sci. Technol.* 42(1-2): 305–311, 2000.

Putnam, D. R., "Earthquakes and Water Security: Contingency Planning in California," *J. Contin. Crisis Management* 8(2): 103–108, 2000.

Rees, J., *Hostages of Each Other: The Transformation of Nuclear Safety since Three Mile Island*, University of Chicago Press, Chicago, 1994.

Rich, V., "An Ill Wind from Chernobyl," *New Scientist* 20 April, vol. 130, No. 1765, pp. 26–29, 1991.

Rosenthal, U., and A. Kouzmin, "Globalizing an Agenda for Contingencies and Crisis Management: An Editorial Statement," *J. Contin. Crisis Management* 1(1): 1–12, 1993.

Rosenthal, U., M. T. Charles, and P. 't Hart, "Introduction: The World of Crises and Crisis Management," in U. Rosenthal, M. T. Charles, and P. 't Hart (eds.), *Coping with Crises: The Management of Disasters, Riots and Terrorism*. Charles C. Thomas, Springfield, pp. 3–33, 1989.

Schechter, E., "Israeli Scientist Perfects Rapid Test to Detect Water Poisoning," *The Jerusalem Report*, December 31; vol 12, No. 18, p. 6, 2001.

Schubert, J., and R. E. Lapp, *Radiation: What It is and How It Affects You*, Viking Press, New York, 1957.

Shimazu, Y. (ed.), "Lessons from Hiroshima and Nagasaki," SCOPE-ENUWAR Report HI.01.85, 1985.

Siegel, J., "BGU Scientist Develops Method to Detect, Clean Contaminated Water," *The Jerusalem Post*, March 3, vol 71, No. 21438, p. 4, 2003.

U.S. EPA, "Radionuclides Notice of Data Availability–Technical Support Document," Targeting and Analysis Branch, Standards and Risk Management Division, Office of Ground Water and Drinking Water, United States Environmental Protection Agency. Prepared by U.S. EPA Office of Ground Water and Drinking Water, in collaboration with U.S. EPA Office of Indoor Air and Radiation, and United States Geological Survey, 2000.

U.S. Water News Online, "Chemical-Free Mobile System Treats Water in Emergencies," 1995, available at: http://www.uswaternews.com/archives/arcquality/5chefree.html.

Wetzel, K. G., "Effects on Global Supplies of Freshwater," in *Nuclear War: The Aftermath*, Royal Swedish Academy of Sciences, Stockholm, *AMBIO*, vol. 11(2-3): pp. 126–131, 1982.

Wildavsky, A., *Speaking Truth to Power*, Little-Brown, Boston, 1979.

CHAPTER 18
DEVELOPMENT OF THE NEXT GENERATION MICROBIOLOGICAL AND CHEMICAL DETECTION CAPABILITIES FOR WATER SUPPLIES

Richard Skaggs
Tim Straub
Bob Wright
Cynthia Bruckner-Lea
Scott Harvey
Pacific Northwest National Laboratory
Richland, Washington

18.1 INTRODUCTION

In discussing the role of science and technology, the National Research Council (NRC) emphasizes that "... well-reasoned science and technology program will be a vital component of strategies for countering terrorism"(NRC, 2003). Their review of water security research needs highlights improved analytical methodologies and monitoring systems for drinking water. Looking beyond current capabilities and given the potentially large number of terrorist targets, the next generation detection and early warning systems must involve wide-spread deployment of low-cost, robust, yet highly sensitive and selective detection and measurement systems. This challenge is depicted in Fig. 18.1. Traditional sensors that can meet low cost and durability requirements generally lack the sensitivity and selectivity needed for the early detection for defense against widespread exposure. The required sensitivity and selectivity can be obtained by using laboratory equipment, but that approach falls far short of providing the speed, cost, durability, and deployability needed.

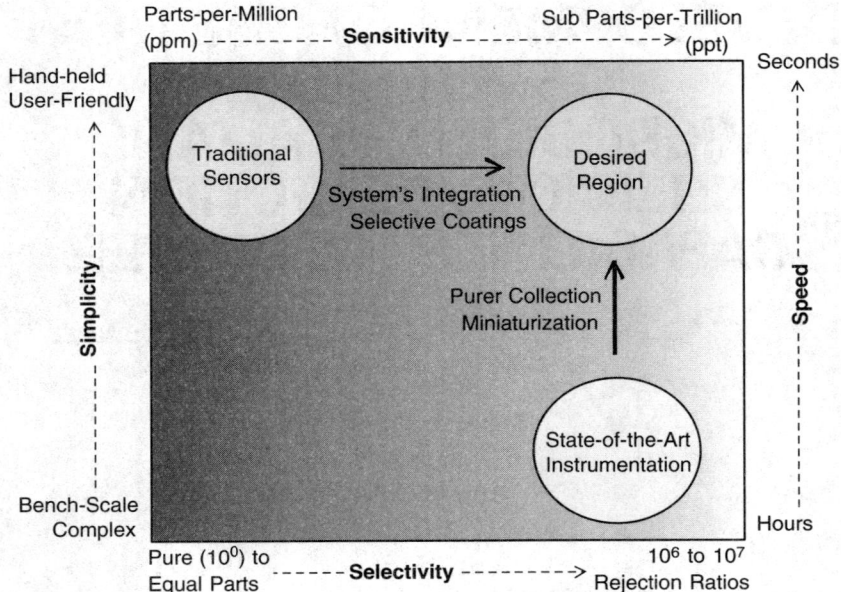

FIGURE 18.1 Illustration of the next generation detection and early warning system challenge.

This chapter describes representative research activities that illustrate the scientific tools and approaches that will lead to development of the next generation biological and chemical detection capabilities for water systems. The information provided is not intended as a comprehensive review of these activities. As noted by the NRC (2003), water security research is being conducted broadly across many government agencies and organizations (e.g., EPA, DoD, and universities). Nonetheless, the issues addressed by the research areas presented do reflect the promise for significant improvements in future detection and monitoring systems.

18.2 MICROBIOLOGICAL DETECTION CAPABILITIES

18.2.1 Current State of Detection of Microbiological Pathogens in Water Supplies

Regulatory Drivers. In the United States, current methods for the detection of microorganisms in water supplies are driven primarily by existing and future United States Environmental Protection Agency (U.S. EPA) regulations. One of the major changes to the 1996 Safe Drinking Water Act mandates that the EPA develop a list of additional waterborne contaminants every 5 years. This candidate contaminant list (CCL) not only identifies new threats, but also identifies (a) whether the contaminant can be controlled by existing water treatment, (b) the known and/or suspected health risks associated with the contaminant, and (c) if there are adequate methods to detect the contaminant (U.S. EPA, 1998). For microbiological contaminants, listing on the CCL may be due, in part, to known waterborne disease outbreaks caused by newly recognized organisms. The two most notable outbreaks in North America in recent years were the *Cryptosporidium parvum* outbreak in Milwaukee, Wisconsin in 1993, and the 2000 *E. coli* O157:H7 outbreak in Walkerton, Ontario, Canada. The development of U.S. EPA Method 1623 (U.S. EPA, 2001) is an example of a method that was developed in response to known waterborne disease outbreaks caused by this organism and its subsequent listing on the CCL. This method will now be used to monitor for *Giardia* and *Cryptosporidium* in water supplies under the Long Term Enhanced Surface Water Treatment Rule (LT2).

Shortcomings of the Current Methods for the Detection of Microorganisms in the Context of a Bioterrorism Attack. With both the September 11, 2001 terrorist attacks, and the Anthrax attacks on the U.S. Postal System shortly thereafter, water utilities have since been on a heightened state of awareness that their supplies are vulnerable to infrastructure, radiological, chemical, and biological attack. Methods currently used for the detection of microorganisms in water have several serious shortcomings that make them relatively unsuitable to respond to a natural event, let alone a bioterrorism attack (Allen, Clancy, and Rice, 2000). Some of the more serious shortcomings include, but are not limited to the following:

Too Many Methods. There is no single process train that allows for the co-concentration of viruses, bacteria, and protozoan parasites from large volumes of water (>10 to 10,000 L of water), further process the concentrate to remove inhibitors to yield a sample suitable for reliable biodetection, and multiplexed pathogen detection (Straub and Chandler, 2003). Most methods developed to date have viewed waterborne pathogen detection as a "disease of the day" approach. In the context of a bioattack on the water supply, it is unlikely that there will be a priori knowledge of the agent released, where it was released, and when *or if*, the agent was released.

Extremely Large Volumes of Water are Needed. Large volumes (>10 to 10,000 L of water) are necessary because even in the most polluted environmental waters, pathogen concentrations are relatively low (Table 18.1) (Gerba, 1996; U.S. EPA, 1988). Unless the drinking water supply is intentionally contaminated by raw sewage, it is highly unlikely that the terrorists have sufficient quantities of an agent to achieve levels seen in these waters. Nonetheless, many of the agents have very low infectious doses; 1 to 10 agents needed as opposed to 10^6 needed for *Salmonella* food poisoning.

Concentration of Inhibitors. Concentrating large volumes of water to volumes appropriate for assay tends to co-concentrate inhibitors that cause problems for end point detection (Schwab, DeLeon and Sobsey,1996; Tebbe and Vahjen, 1993). Manifestations of these problems can include cell-culture toxicity (viral assays U.S. EPA, 1993) and failure of nucleic acid detection methods like polymerase chain reaction (PCR) and hybridization (Small et al., 2001; Tsai et al., 1994; Tebber and Vahjen, 1993). Sample processing for the analytical separation of pathogens from inhibitors has not been adequately addressed and continues to be problematic in the context of rapid detection of biological agents in water

TABLE 18.1 Expected Concentrations of Indicator Bacteria and Pathogens in Polluted Water

	Concentration per liter	
Organism	Raw sewage	Polluted stream water
Coliform bacteria	10^{10}	10^6
Enteric viruses	10^3	10–100
Giardia	100	1–10
Cryptosporidium	100–10,000	1–1000

Sourec: Gerba (1996)

supplies. Promising new methods for automated processing of environmental samples will be discussed in later sections of this chapter.

Cumbersome, Time-Consuming Methods. With the exception of some of the presence/absence tests for coliform bacteria (APHA, AWHA, and WEF, 1998), methods for viruses and protozoan pathogens in water require a high degree of operator skill and time. For example, detection of enteric viruses using in-vitro cell culture techniques requires at least 1 day to process the sample and several weeks to several months to detect infected cells (U.S. EPA, 1993). Two further complications for viral threats in drinking water are (1) the processing and detection methods are not conducive to rapid high-throughput screening that would be needed in the event of a terrorism incident, and (2) some viruses do not replicate in cell culture.

18.2.2 Desired Capabilities of a Microbiological Detector for the Detection of Pathogens in Water

Whether the pathogens that have entered a drinking water distribution system are the result of a natural event (e.g., floods, fires, etc.), inevitable failure of the multibarrier approach to remove pathogens, or an intentional contamination event by a terrorist, a number of issues need to be considered to build appropriate detectors. Some of the issues to be considered are as follows:

How Soon After an Event Do We Need to be Alerted? On the surface, this seems to be a relatively simple question to answer—e.g., immediately. The reality is that

1. "Detect to warn" systems do not exist for drinking water. In the military concept, detect to warn means that the utility knows that agent X was released at point Y in the distribution system, and that the agent can be detected within 2 to 5 min. In this time frame, the utility can isolate that node in the distribution system to quickly prevent further contamination. This is a huge, and possibly an unrealistic assumption in a terrorist attack.
2. In a "detect-to-protect/treat" mode, the timescale between release and detection is much longer, usually a couple of hours. In this scenario, the water utility may have autonomous systems at various points in the distribution system. If contamination is noted, the public served by the distribution system is notified and given prophylactic treatment (if it exists). Next generation systems for water borne pathogens are likely to fall into the category of "detect-to-protect/treat."

Performance Characteristics. "Box" systems designed for use for the detection of bioterrorism agents in water supplies should have performance characteristics similar to or better than existing methods for pathogen detection in water.

Sensitivity. Because of the concentration issue, the system should be capable of detecting very few pathogens dispersed in very large volumes of water. Many of the current rapid detectors that are in use today for aerosol monitoring have method detection limits of approximately 10^3 of a particular agent. These levels would not meet existing EPA regulatory requirements for drinking water where the maximum contaminant level targets are often ≤ 1 per 100 mL (*E. coli*) to ≤ 1 per 10,000 L (viruses, protozoa, etc.).

Specificity. In the biological context this means that the method used to detect a pathogen of interest only detects the pathogen of interest and not its close relatives. In the context of automated biodetection it means low false positive/low false negative rates. Designing systems to meet these requirements is exceedingly difficult because minimizing false positives will increase the false negative rate and vice versa.

Operator Training. The ultimate users of the technology should be able to run the equipment with little to no training or "hands-on" involvement. As discussed previously, today's advanced microbiological methods for the examination of drinking water are slow and require much training and practice to become proficient in the method. Detection of potential bioterrorism agents is even more complex and may overtax a utility that is still becoming proficient in new monitoring methods required by the new EPA regulations.

What is the Ultimate Goal of the Detector? Very little discussion has been given to the ultimate detection needs of the end user. From the EPA and water utility context there are several needs for detection:

Presence/Absence. Bioterrorism agents should not be present in a water supply. presence/absence (PA) methods are the easiest to deploy in automated systems, but depending on the method used to determine presence or absence, it reveals little to any information beyond its "signature."

Enumeration. This is useful for determining the extent of the contamination in the distribution system, or if consequence management efforts have been successful in restoring the distribution system water quality to meet EPA compliance. Methods such as real-time PCR and plate counts can indicate numbers of organisms. The problem is that reliable automated methods for multi agent presence/absence detection have not been developed, let alone enumeration of each agent present. Consider the case where the terrorists contaminate the distribution system with raw sewage. The number of different kinds of pathogens present in sewage is tremendous. In this scenario, enumeration would be of little value.

Viability/Infectivity. One of the criticisms of nucleic acid and even antibody-based detection of an infectious agent is that it provides little to no information on its viability or infectivity. This is particularly important if disinfection efforts are evaluated and signatures of these organisms are still found by these methods. Does this mean that the disinfection procedures did not work?

Speciation. Bioterrorism agents often have very close relatives. For example, the vaccine strain of anthrax is relatively harmless. It is different from its virulent relative by the absence of a plasmid (extra chromosomal DNA found in some bacteria) that contains the pathogen genes. There is a need for detectors to be able to differentiate between these close relatives to determine the relative risks of a released agent.

18.2.3 Methods Currently Available for the Detection of Waterborne Pathogens

This section of the chapter discusses current, off-the-shelf technology that can be used to build first generation early warning systems for the detection of waterborne pathogens. A system is more than the end-point detector. It involves the entire process from large volume sample collection, processing to separate microbiological agents of interest, and finally end-point detection. Figure 18.2 shows a conceptual diagram of the components of a system. The integration and automation of these systems remains one of the biggest challenges for building these early warning systems.

Large Volume Concentration Methods. Filtration remains one of the most viable options for large volume concentration of water samples. Different types of filters work for one group of pathogens (e.g., viruses, bacteria, and protozoans), or multiple groups of waterborne pathogens (all three groups). Depth filters and absolute porosity filters are suitable for larger organisms such as bacteria and protozoans. Depth filters trap organisms by the tortuous path that water must take through the filter. Yarn-wound filters, originally used for the concentration of *Giardia* and *Cryptosporidium* (Gerba, 1996) were essentially depth filters. The fibers were then manually cut and the organisms were eluted using detergents such as SDS and Tween. Absolute porosity filters are usually composed of polycarbonate. Lasers

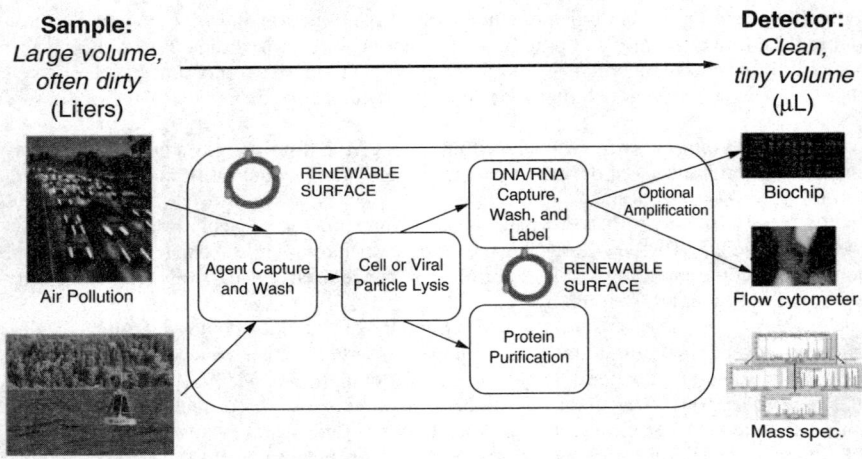

FIGURE 18.2 The biodetection process. Samples from the environment need to be large because pathogens are often too dilute. But, concentration methods tend to co-concentrate inhibitors. Analytical separations are needed to render tiny, clean volumes suitable for today's detection platforms.

are used to create pores of uniform size. One of the filtration options for Method 1623 employs these kinds of filters. Absolute filters can also be used to trap bacteria for subsequent analysis using the membrane filtration technique (APHA, AWWA, and WEF, 1998). Waterborne viruses are concentrated from water samples using a positively charged filter (U.S. EPA, 1993). Viruses are then eluted from the filter using complex protein solutions or defined amino acids at basic pH.

Each of the above methods works by trapping organisms in the filter material that requires the user to elute the organisms of interest in a smaller volume of a buffered solution. Both depth and absolute pore size filters concentrate particulate matter that is too large to pass through the filter. Clogging can become an issue with highly turbid water samples. Recently, more attention has been given to hollow fiber filtration (Evans-Strickfaden et al., 1996; Juliano and Sobsey, 1997; Kuhn and Oshima, 2001; Morales-Morales et al., 2003; Olszewski, Winona, and Oshima, 2001; Oshima et al., 1995). These filters are probably the most viable option for simultaneously concentrating viruses, bacteria and protozoan parasites from large volumes of water. Water is circulated through the filter system, and molecules whose molecular weight is greater than the molecular weight cutoff (MWCO) are retained (retentate), while molecules smaller than the MWCO pass through the filter and discarded (permeate). The tangential flow through these filters combined with buffered solutions prevents clogging of the filters (Morales-Morales et al., 2003). It also keeps organisms from irreversibly absorbing to the filter material. Recoveries of viruses, bacteria, and protozoans often exceed 60 percent for most types of matrix water. The primary disadvantage of these filters is the cost and size of the apparatus. For very large volume sampling (in excess of 100 L) the units are quite large (approximately 3 to 6 ft long) such that their principle use would be relegated to water leaving the treatment plant or a major pump station in the distribution system. Costs are contained by reusing the filters, and Morales—Morales et al. (2003) gave specific protocols for cleaning and conditioning the filters to eliminate sample carry over from positive samples.

Sample Processing Techniques. Large volume filtration can concentrate samples from thousands of liters to several mL, but these processes tend to concentrate inhibitors and nontarget organisms as well as the organisms of interest. Even treated drinking water has residual chemicals that can inhibit nucleic acid methods like PCR (Abbaszadegan et al., 1997). A number of methods have been developed for dealing with these secondary concentration and processing issues, and several strategies are discussed below. While some methods have been incorporated into today's methods for the examination of water (U.S. EPA, 2001), other methods, most notably the next generation and automated methods that are being developed for biological defense applications (Bavykin et al., 2001; Bruckner-Lea et al., 2000, 2002; Burns et al., 1998; Chandler et al., 1999, 2000a,b) have not been specifically tested for waterborne pathogens.

Antibody Separation Techniques. Magnetic beads derivatized with antibodies that are specific to a pathogen are used for the secondary concentration and processing of water samples for *Giardia* and *Cryptosporidium* using EPA Method 1623 (U.S. EPA, 2001; Rochelle et al., 1997; Bukhari et al., 2000). These methods reduce a majority of matrix that can interfere with the immunofluorescent staining (IMS) and microscopic examination steps of these samples. IMS techniques have also been widely used for a number of other applications including separation of pathogens from liquid and solid foods (Blake and Weimer, 1997; Chandler et al., 2000a; Grant, Ball, and Rowe, 1998; Heckotter, Bulte, and Lucker, 1998; Rijpens et al., 1999).

Automation of these IMS methods for high throughput screening has also been reported. Scientists at Pacific Northwest National Laboratory have developed a system called BEADS (Biodetection Enabling Analyte Delivery System). The key feature of this system is the renewable surface flow through columns. These columns are automatically packed with material of the user's choice. Analytical separations are then performed by perfusing samples over the column. Trapped analytes are either eluted for downstream examination or can be examined on column. The BEADS platform for analytical separation was combined with other components to produce an automated and integrated flow-through system that combines the processes of sample concentration, IMS as the renewable surface column, and PCR for the specific and sensitive detection of *E. coli* O157:H7 from poultry carcass rinsate and water samples (Bruckner-Lea et al., 2004; Chandler et al., 2000a). Recently completed experiments for river water samples (Fig. 18.3) that were spiked with low numbers of *E. coli* O157:H7 demonstrated that

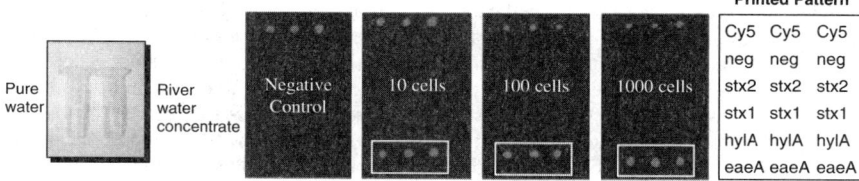

FIGURE 18.3 BEADS biobox results for one replicate concentration run. The fluorescently labeled PCR product from the system (20 μL) was manually transferred to a microarray for hybridization and detection. The three hybridized spots on the bottom of the microarray are triplicate probes targeting the *eaeA* gene of *E. coli* O157:H7. The three spots on the upper side of each image are positive control markers for positional reference to the array. Three other genes are represented on this array, but are not hybridized because the PCR reaction only contained the primer pair specific for the *eaeA* gene.

1. Sampling, capture, and PCR were performed in approximately 2h.
2. There was *no operator involvement* once the sample was started.
3. *Reproducible 10 cell* detection from turbid matrix water (*ca* 10 NTU).
4. *No sample carryover*. PCR was performed using dUTP and UDG to prevent carryover. In addition, rigorous cleaning protocols provided a system that was ready to process the next sample.
5. The *entire process* from sample to read-out on the array could be completed in 4 h.

The problem with (micro) fluidic automated sample processing devices is that very small volumes are assayed and may only focus on one pathogen target at a time. For the BEADS system, researchers are investigating both the ability to sample and process much larger volumes of water (>10 L) and simultaneously purify multiple pathogenic agents that may be present.

Other immuno separation methods are available. Lawrence Livermore National Laboratory's APDS (Automated Pathogen Detection System) uses antibodies conjugated to fluorescent carboxylated beads, e.g., Luminex 100 system. In this system, there are 100 different color bead sets. Each bead set is composed of a specific ratio of red and near infrared dyes. A specific antibody pool is then conjugated to a specific bead set. Analytes of interest are trapped on their specific conjugated bead set and stained with a fluorescent (phycoerythrin dye) secondary antibody. When the samples are analyzed, the red laser classifies each bead set in the system whereas the green laser reads the fluorescent intensity of analytes bound to the beads. Their system is primarily designed for aerosol monitoring. Thousands of liters of air are sampled and fluidized. The fluidized particulates are then perfused over a bead trap column that has been packed with antibody-coated beads. The beads are then flushed into the Luminex system for readout. This system may be amenable to pathogen detection in water supplies if the aerosol monitor is replaced with a large volume water concentration system. Miniaturization of this system for field deployment will be needed as the APDS instrument is about the size of an automated teller machine and the water filtration systems needed for very large volume sampling will require additional space.

Alternative Whole Cell Capture Methods. Immuno-separation techniques are perhaps the most widely recognized and proven method for separating cells of interest from their surrounding matrix. Systems built around this method require a priori knowledge of the agents that may be present in a water supply. Emerging or unusual pathogens could potentially be missed if the sample processing methods did not take these risks into account. Two methods have been proposed to provide broader spectrum pathogen capture—hydroxyapatite and lectin affinity.

Hydroxyapatite. For general cell capture (both pathogenic and nonpathogenic organisms), the presence of anionic polymers on cell surfaces can be used to capture both gram-positive and gram-negative eubacteria. In the case of the gram-positive organisms, teichoic and teichuronic acids are present with repeating units of glycerol or ribitol and phosphate. In the case of the gram-negative organisms, the presence of keto-octadecanoic acids and other acidic carbohydrates in the core and lipid A region of the major outer membrane polymer lipopolysaccharide (LPS) likewise gives these cells an anionic nature. This polymer is thought to make up ~40 percent of the gram negative cell outer membrane (Takayama, Qurashi, and Mascagni, 1983). As a demonstration of this, whole bacterial cells separated by isoelectric focusing formed peaks at low pI values (Armstrong et al., 1999). The use of hydroxyapatite (a form of calcium phosphate) to bind bacterial cells with high affinity has been demonstrated in complex food and fecal matrices (Berry and Siragusa, 1997). This is consistent with the belief that one of the roles of teichoic acids is binding of cations from the environment (Lang, Glassey, and Archibald, 1982).

For general cell capture methods based on this principle, Berry and Siragusa (1997) showed that positively charged hydroxyapatite (HA) particles could be used to concentrate and purify bacteria from suspensions of samples such as ground beef and bovine feces prior to PCR analysis for cell identification (Berry and Siragusa, 1997). They screened a total of 11 different common food pathogens and 14 different types of *Salmonella* and found that HA was suitable for cell purification prior to PCR analysis (Berry and Siragusa, 1997, 1999). Since the cell capture is due to van der Waals and electrostatic interactions between the cells and the particles, the adherence of bacteria to hydroxyapatite particles depended upon the specific cell type. The efficiency of capture varied from 46 percent for *E. coli* 0157:H7 to 99 percent for *Yersinia enterocolitica*, with two trials run for each organism. Regardless of the capture efficiency, this approach resulted in sample purification method from complex matrices. The purified sample was immediately ready for PCR amplification.

Lectin and Carbohydrate Affinity. Another approach that can be used for microbial cell capture is to use lectins that target the carbohydrate rich cell envelope polymers of microbes (Bundy and Fenselau, 1999). The cell envelope (including cell walls) of all eubacterial cells as well as those of plants, fungi, and protozoans is known to be carbohydrate rich. In addition to the anionic external carbohydrate polymers for both gram negative and gram positive eubacteria (LPS, teichoic acids), the chitinous cell wall (N-acetyl glucosamine polymer) of *Giardia* cysts (Ward et al., 1988) and the oligosaccharides of the rotavirus VP4 and VP7 outer glycoproteins all provide attractive targets for affinity purification. The particularly attractive aspect of using lectin affinity to recognize carbohydrate moieties is that these are often fundamental structural elements of cell walls or proteins that may be less likely to vary than protein sequences that can mutate or be expressed at different levels depending on environmental conditions. Lectin based capture of several eubacteria including pathogenic *Salmonella spp.* and *E. coli* has been demonstrated in combination with different analytical technologies such as electrochemical biosensors and mass spectrometry (Bundy and Fenselau, 1999; Ertl and Mikkelsen, 2001).

An analogous approach to capturing microorganisms is to take advantage of the carbohydrate binding properties of the microorganisms themselves. Adherence to the gut epithelial surface is normally critical for colonization and eventually pathogenicity of bacteria in the gut (Neeser et al., 2000; Schembri et al., 2002). In some cases, such as for pathogenic *E. coli*, colonization of the gut epithelial surface alone, without toxin production, causes illness (Wanke and Guerrant, 1987). In fact, inhibition of pathogen colonization in the gut by normal flora competing for carbohydrate binding sites is thought to be a mechanism of host defense (Neeser et al., 2000). The best-studied examples of carbohydrate binding in enteropathogenic bacteria are the adhesins used by *S. flexneri* and *E. coli*, which bind to Man-Man, Neu-Gal, or Gal-Gal disaccharides (Lis and Sharon, 1998; Wizemann, Adamou, and Langermann, 1999). In another example, for *Campylobacter spp.*, the carbohydrate rich mucins of the gut have been shown to serve as structures for adhesion (Kelty, 1997).

The necessity for host cell recognition through carbohydrate binding has also been described for rotavirus (Rolsma et al., 1998). The ability of the virus to adhere to cells by recognition of N-glycolylnueraminic acid ganglioside could potentially be used as a means to separate rotavirus from other microorganisms within fecal samples. The trophozooites of *G. lamblia* also use lectin-like affinity for glucose-mannose and mannose-6 phosphate moieties, among other interactions, to adhere to cells (Sousa et al., 2001). This affinity interaction will not be useful in capturing these microbes in water samples as only the cyst form of the organism is shed (Ward et al., 1998). However, lectins can be used to bind surface carbohydrates on *Giardia* cysts. Therefore, a collection of lectin and/or carbohydrate moieties could be rationally chosen to bind organisms in a semiselective nature in order to concentrate and purify the organisms of interest for detection.

Both hydroxyapatite and lectin/carbohydrate affinity have particular advantages over traditional immuno-separation methods. Immobilization of HA and lectins onto magnetic or polystyrene beads for automated capture may be challenging. In addition, tests to characterize the capture efficiency of the agents of interest will need to be conducted. Also experiments will need to be performed to determine if these less specific methods will also co-concentrate inhibitors rendering these methods of little use for downstream detection.

Alternative Approaches: Nucleic Acid Isolation. Sample processing methods for nucleic acid isolation are also available. Kits for the extraction of nucleic acids from whole cells or virus particles include trade names such as Qiagen's DNEasy and RNEasy kits (for DNA and RNA isolation, respectively), MoBio, Bio101, and PrepMan. All of these methods employ benchtop manipulation by the operator and require some training to properly use them.

Semi- and full-automated methods for DNA isolation have been developed at Argonne National Laboratory (ANL) and Pacific Northwest National Laboratory (PNNL). The ANL system is entirely field portable. Small volume water samples are passed through a syringe column containing the necessary chemicals for cell lysis, RNA capture, fractionation of the RNA, and subsequent fluorescent labeling of the RNA. The eluted samples are then hybridized to a microarray and read with a field portable microarray reader. The PNNL approach uses the BEADS renewable surface column. In this case, cells are lysed in appropriate buffer and perfused over a separation matrix. This separation matrix is either composed of silica for total DNA capture, or affinity capture of ribosomal RNA using polystyrene beads that have been derivatized with universal 16S rRNA capture sequences (Bruckner-Lea et al., 2000, 2002; Chandler et al., 1999, 2000b). In both the ANL and PNNL systems, the nucleic acid is ready for hybridization to a microarray and/or PCR amplification.

Multiplexed Detection. After secondary concentration and sample processing, multiplexed detection is needed for detection of pathogens in water. Whether the sample processing methods yield purified viral particles and cells or nucleic acids, multiplexing for sensitive and specific detection remains a huge challenge. For example, the Livermore APDS system does an excellent job of specific multiagent pathogen and biological toxin detection. However, sensitivity data suggests that this system requires pathogen concentrations to be several orders of magnitude greater than levels needed to achieve EPA regulatory compliance. Thus, there is a high potential for false negatives that could result in many people becoming infected. BAX PCR systems reviewed by States et al. (2003) have both the sensitivity and specificity needed to detect pathogens in water samples, but (a) are limited by the number of different pathogens that can be tested and (b) the system is not easily multiplexed.

New methods are being developed to meet the needs of highly multiplexed detection, but these methods are still in the very early stages of testing and will require significant optimization to be useful for detecting waterborne pathogens. Among the most promising of these methods are DNA microarrays and DNA suspension arrays. DNA microarrays are generated using biorobotic instruments that can deliver specific DNA sequences for each pathogen on a standard glass microscope slide. Figure 18.4 illustrates the concept of what microarrays can do for waterborne pathogen detection.

DNA suspension arrays work in a similar manner as the Luminex flow cytometer for the APDS system. The bead sets are now derivatized with oligonucleotide sequences. For instance, one bead set may be derivatized with the *eaeA* gene for detection of *E. coli* O157:H7 and another bead set derivatized with the *stx2* gene.

Community DNA or RNA can be labeled with specialized kits from commercial vendors. The nucleic acids are hybridized to the array and can be visualized with a scanner such as the one described by Bavykin et al. (2001) or a Luminex system in the case of the suspension arrays. If community nucleic acids are labeled and hybridized to the array, multiplexed detection is a relatively simple task. Unfortunately, the limit of detection for these direct hybridization approaches is approximately 10,000 copies of a specific gene of interest. Depending on the gene of interest, this may represent 10 to 10,000 cells.

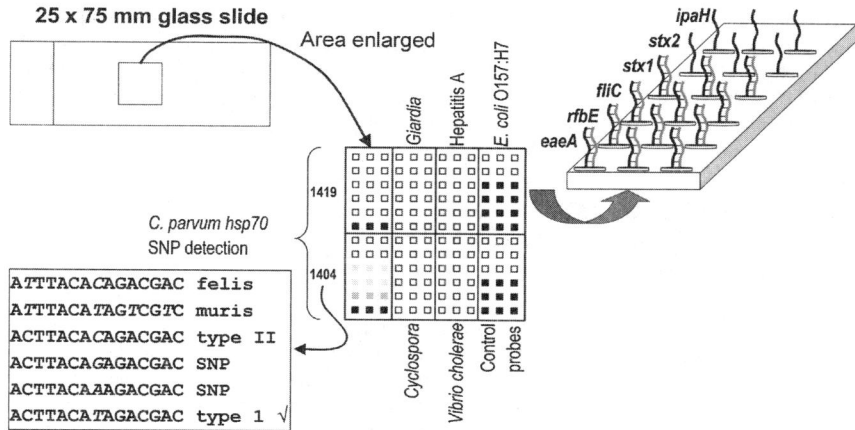

FIGURE 18.4 Conceptual illustration of a DNA microarray for the detection of microbial pathogens in a water supply. From very slight differences in DNA sequences between closely related pathogens (*Cryptosporidium parvum*) to gross differences between organisms (*E. coli* vs. hepatitis A), microarrays can provide a wealth of information about pathogens in a sample.

The alternate approach to labeling community nucleic acids is to generate labeled PCR products. This may be accomplished by purchasing end-labeled PCR primers (Call, Brockman, and Chandler, 2001a; Straub et al., 2002) or incorporating fluorescent nucleotides into the PCR reaction (Chizhikov et al., 2001, 2002). The PCR products are then hybridized to the array. The advantages of this approach are that the sensitivity and specificity are high. The primary disadvantage is the optimization of highly multiplexed PCR. Simply adding more PCR primers to the reaction is not as easy as it appears. Two possible outcomes of adding more primer pairs to detect more pathogens are (1) failure of the entire reaction (no amplification of any gene) or (2) loss of PCR specificity. The upper limit of multiplexed PCR is approximately 6 to 8 primer pairs. Recently, scientists at PNNL demonstrated a successful 6-plex reaction for the specific detection of *E. coli* O157:H7, *Salmonella*, and *Shigella*. River water samples (1 L) were spiked with 1000 cells of each organism and processed using a DNEasy kit from Qiagen. The 6-plex PCR reaction was performed and the products were hybridized to the array. As shown in Fig. 18.5, amplification and subsequent

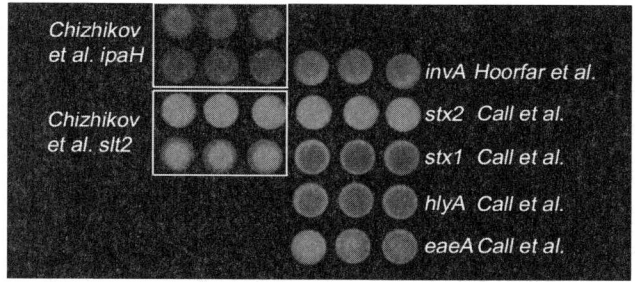

FIGURE 18.5 Multiplexed gene detection of *E. coli* O157:H7, *Salmonella*, and *Shigella*. The *invA* gene is specific for the detection of *Salmonella*. The *ipaH* gene is specific for the detection of *Shigella*. The remainder of the genes on the array are for the specific detection of *E. coli* O157:H7.

detection using microarrays was successful. Repeated attempts to increase the number of gene sequences for even greater multiplexed detection failed. Attempts to increase the method limit of detection also failed. Thus, even though multiplexed gene detection was possible, the sensitivity of the method would be worse than methods currently available for the detection of these organisms without multiplexing.

18.3 CONCLUDING REMARKS FOR MICROBIOLOGICAL SYSTEMS OF THE FUTURE

Building a system for continuous monitoring of natural and intentionally introduced microorganisms in the distribution system will require at least several years of research to build a first generation prototype system. Of the off-the-shelf technology reviewed in this chapter, there have been no public reports of engineering these components together into an integrated system suitable for water quality monitoring. Systems that are built for responding to potential terrorism incidents should also be capable of meeting the daily monitoring requirements a water utility must perform to meet EPA compliance. Lessons learned from the *Cryptosporidium* incident in Milwaukee and the *E. coli* O157:H7 incident in Walkerton, Ontario show that waterborne outbreaks due to natural causes and failures in the multiple barrier approach have occurred. This means that the systems will need to have validated performance characteristics similar to or better than current EPA approved methods. This is a much more difficult challenge than responding to a huge release scenario. In the huge release scenario, detector sensitivity may not be an issue or is it? There are very few, if any modeling studies that have been released to the public that show dilution figures or dispersion patterns once a pathogen is introduced into the system. Risk assessment studies have been performed for certain waterborne pathogens that extrapolated data from infectious dose and actual number of cases in a waterborne disease incident. Some of these data suggest that detector sensitivity does need to be on the order of 1 organism per 10 to 100 L of water. Then, there is the problem of "putting all the eggs in one basket" approach to building a system. More evidence is indicating that orthogonal detection or indicator systems are needed to corroborate information obtained from the primary system. For example, water utilities already monitor their distribution system for turbidity. Sudden spikes in turbidity may alert officials that there is a potential problem where monitoring may be needed.

While several systems for airborne contamination are being tested at various facilities around the country, systems for detecting microbiological pathogens in public water supplies lag far behind. The military and national laboratories are currently conducting much needed research in this area, but most of the results of this research may not be available to water utilities for some time due to the sensitive and classified nature of the work. Partnerships between organizations like the American Water Works Association Research Foundation, the U.S. EPA, and the national laboratories will greatly assist in getting this new technology transferred to utilities as it becomes available.

18.4 CHEMICAL DETECTION SYSTEMS OF THE FUTURE

18.4.1 General Approach

Trace analysis of signature compounds in air or water is usually accomplished by concentrating target species from a large volume environmental sample on an appropriate sorbent.

Unfortunately, the organic matrix components are also concentrated along with the signature compound. This necessitates separating the signature compound from this complex mixture in sufficient purity to allow identification and quantification. Due to the complexity of the matrix mixture, sophisticated laboratory-based multidimensional instrumentation is often required to provide adequate separation power (Harvey and Wright, 1993). Next generation instrumentation will likely be based on highly selective sorbents that allow collection of relatively pure analyte fractions during the initial sampling step. Due to the high purity of the initial fraction, subsequent analytical steps leading to high-integrity identifications at trace concentrations can be greatly simplified. The required analytical instrumentation will reflect this simplification by being compact, lightweight, and field portable. Simplified systems may consist of a single separation step to remove minor impurities followed by detection with a sensor-type detector. Key to this approach is the high degree of matrix discrimination accomplished during the initial sampling step.

Several approaches toward providing selective sampling involve use of selective sorbents for compound class isolation or methods to target specific compounds based on molecular recognition. Two molecular recognition mechanisms that might be useful are immunorecognition and molecularly imprinted polymers. Depending on the intended use, it is desirable for each of these to operate in either the aqueous or nonaqueous modes.

Immunosorbents. Immunochromatographic enrichment is ideally suited for analysis of aqueous samples. This approach is currently limited by the scarcity of antibodies toward relevant target compounds. Often custom antibodies need to be prepared to address specific detection objectives. Traditional antibody production results in monoclonal IgG from hybridoma cell lines or polyclonal IgG from serum. Antibodies can be isolated and covalently bound in either a random or oriented geometry to various stationary supports (Matson and Little, 1988). The site-oriented approach places the antibody recognition sites out towards solution, well removed from the sluggish diffusion layer adjacent to the support material. This results in a higher capacity sorbent with more desirable diffusion characteristics. More recently, genetic engineering has allowed production of relevant antibody fragments (Dooley, 1998; Feldhaus et al., 2003; Harris, 1999; Molloy et al., 1995). For example, a yeast display library can be used in combination with a technique called molecular evolution to produce relevant single-chain antibodies with high affinity towards target species under specific analysis conditions (Feldhaus et at., 2003). Other techniques exist to engineer antibody fragment variants that have exceptional stability toward extreme conditions (including exposure to organic solvents) (Dooley, 1998; Harris, 1999; Molloy et al., 1995). Owing to the small size of these fragments, an extremely stable high capacity immunosorbent can be prepared. Although genetic engineering allows production of relevant antibodies in less time, the amount of antibody produced is only adequate for limited quantities of immunosorbent.

Molecularly Imprinted Polymers (MIP). These polymers were first synthesized by using a functional monomer that could form a reversible covalent bond with a template molecule (Wulff, 1995; Steinke, Sherrington and Dunkin, 1995). The polymer displayed affinity toward the template due to the reversible formation of the reaction product. The field of molecular imprinting was revolutionized in 1981 when Arsahdy and Mosbach described noncovalent imprinting techniques (Arshady and Mosbach, 1981). In this approach, the template is associated by noncovalent interactions with the monomer. The polymer is formed in the presence of the template. The final step is the removal of the template resulting in a complementary cavity in the polymer that displays high affinity and selectivity for the template (Kriz, Ramstrom, and Mosbach, 1997; Lanza and Sellergren, 2001; Remcho and Tan, 1999). One of the advantages of the noncovalent method is that associations are kinetically fast and fully compatible with chromatography (Remcho and Tan, 1999).

MIP polymerization is carried out in a nonpolar solvent that accentuates association between the functional monomer and the template. Recognition of the template is, therefore, optimal in the organic solvent used for synthesis. MIPs have several advantages including the relative ease of synthesizing bulk quantities, stability towards extreme conditions of pH, temperature, and ionic strength, as well as tolerance of exposure to a wide range of both aqueous and organic solvents.

Other sorbents have been specifically designed to recognize basic species. One of these polymers, BSP3, was designed to recognize CW nerve agents for sensor array applications. BSP3 consists of hydrogen-bond acidic hexafluorobisphenol groups alternating with oligo(dimethylsiloxane) segments. Nerve agents interact with this polymer through the basic phosphoral group (Grate, Kaganove, and Nelson, 2001). Although the selectivity of these phases may not be as high as molecular recognition phases, there are numerous examples where the added selectivity of this phase can significantly enhance existing instrumental formats (Grate, Patrash, and Kaganove, 1999; Harvey et al., 2002). Another series of selective sorbents has been described that is based on metal complexes of β-diketonate ligands. These polymers have proven useful for selective retention of gas-phase oxygenated and nitrogen-containing species from extremely complex samples such as cigarette smoke (Picker and Sievers, 1981; Wenzel, Bonasia and Brewitt, 1989; Wenzel et al., 1987).

Aqueous Analysis. Immunochromatographic trace enrichment is ideally suited for analysis of aqueous solutions. Numerous investigators have described impressive results with this approach. For example, Hennion et al. describes the analysis of Seine River water for a series of chlorotriazines on a mixed immunochromatographic column that contained antibodies toward both atrazine and simazine (Pichon et al., 1998, 1999). Typically, detection limits on the order of 100 ng/L were obtained from analyzing 50 mL of river water. This same research group has performed similar work with phenylurea herbicides (Pichon et al., 1998, 1999).

Although MIPs are usually used for analysis in the nonaqueous mode, analysis of aqueous samples can be approached with a typical MIP support using some experimental modifications. Organics from large volume water samples can be concentrated on octadecylsilica or XAD-2 sorbent prior to online elution with an organic solvent followed by transfer to the MIP column. The target compound will be selectively retained on the MIP. Alternatively, the large volume water sample can be processed on a restricted-access material (RAM) column that contains hydrophilic coating on the surface with a lipophilic coating in the pores where only smaller molecules can gain access (Fleischer and Boos, 2001). These phases are particularly good for concentrating hydrophobic analytes in the presence of large polar molecules (such as fulvic and humic acids) that might otherwise foul the MIP sorbent. Once the RAM sorbent has collected the signature compound, this column can be eluted with an organic solvent and the analyte transported to the MIP selective sorbent.

Although some MIP polymers are active in the aqueous mode, these usually are not used in aqueous solvents since hydrogen bonding with water competes with the typical noncovalent template recognition mechanism. Therefore, imprinting in aqueous media is effective only if the associations are particularly strong. There is an expanding interest in aqueous imprinting since successful methodologies would make MIPs a more suitable replacement for antibodies. Several improvements in aqueous imprinting have been described, including the use of cooperative recognition using dual functional monomers within the same polymer (Koniyama et al., 2003). Another approach involves template recognition at an aqueous-air interface where hydrogen bonding with the template can effectively occur (Koniyama et al., 2003). Yet another approach has been to use vinyl substituted cyclodextrins as functional monomers (Asanuma et al., 2001). Here, hydrophobic regions of the template are associated with the hydrophobic toroidal center of the cyclodextrin in the aqueous environment before being incorporated at precise geometries into the polymer. Refinement of approaches like these can be expected to expand the application of MIPs to general aqueous applications.

Future selective analysis from aqueous media may very well exploit new materials. For example, one area that looks particularly promising is the incorporation of lipocalins into chromatographic stationary phases. Lipocalins are proteins that contain a lipophilic pocket that can be genetically engineered to have antibody-like specificities (Schlehuber, Beste and Skerra, 2000; Skerra, 2000, 2001). This approach has several advantages over the use of antibodies including the fact that once engineered to recognize a particular target, large quantities of lipocalin can be readily produced.

Analysis of Air. The primary focus of this section is chemical detection in aqueous systems. However, gas-phase sampling is important since many techniques for determining organics in aqueous samples depend on trapping from a purge gas (purge-and-trap) or statically sampling the headspace gas above an aqueous sample. Therefore, the analytical problem and technical approach are essentially the same in either medium, although the analytical interfaces may differ somewhat. Again, a theme that is likely to emerge in future instrumentation is the performance enhancement made possible by incorporating highly selective sorbents that display high affinity for the target analyte. As with water, analysis can proceed using a selective stage as a primary stage for air sampling. Alternatively, a more traditional nonselective sorbent could be used for collecting the organics out of air followed by transferring the complex organic matrix along with the signature compound to a secondary selective sorbent stage using supercritical fluids or an organic solvent. Nonaqueous immunochromatography remains an intriguing possibility for this application. However, the ability of antibodies to serve in this capacity is controversial. Several studies describe gas-phase immunointeractions used in combination with piezoelectric sensors (Guilbault and Luong, 1988; Ngeh-Ngwainbi et al., 1986). Another study, however, claims immunointeractions do not occur in the gas phase at levels above nonspecific interaction with the IgG protein (Rajakovic, Ghaemmaghami and Thompson, 1989). Interest in immunoactivity in organic solvents parallels successful nonaqueous enzymology studies. Two literature reports describe immunointeractions that are less robust but more selective in organic solvents than those observed in aqueous systems (Russell et al., 1989; Penalva, Puchades, and Maquieira, 1999).

Relevant research at Pacific Northwest National Laboratory (PNNL) (Harvey, submitted) has focused on demonstrating the ability of MIP stationary phases for providing discrimination against complex matrix components while providing selective retention of a CW agent surrogate target compound. This proof-of-principle experiment was designed to mimic the use of an MIP stage as a secondary stage for air sampling. The top of Fig. 18.6 is a capillary gas chromatogram of a gasoline sample that has been spiked with diisopropyl methylphosphonate (DIMP), a Sarin surrogate. The general region of DIMP elution is marked above the chromatogram. The profile is so highly complex that one cannot locate the DIMP peak with any certainty and quantification is not possible due to the abundance of overlapping peaks. The bottom chromatogram is the result of applying the mixture to the DIMP-specific MIP, eluting with pentane, and collecting the retained fraction. Enormous matrix discrimination has been achieved while near quantitative recovery of the analyte is maintained (Harvey, submitted). This selective retention is useful for incorporating into simplified next generation instrumentation for air analysis of CW agents.

Another method that would be useful for analysis of air would be solid-phase microextraction (SPME) sampling followed by gas-chromatographic analysis (Harvey et al., 2002). In this approach, a fiber that contains a polymer coating is exposed to the atmosphere (or aqueous head space) for a defined period of time. This technique can also be used for direct aqueous sampling by emerging the fiber directly in the solution. Usually a nonselective nonpolar polymer, such as polydimethylsiloxane (PDMS), is used for the fiber coating. After sampling, the fiber is thermally desorbed into the heated injection port of a gas chromatograph and sampled compounds analyzed. This arrangement can provide near-real-time data when automated for continuous unattended analysis. SPME analysis would be particularly

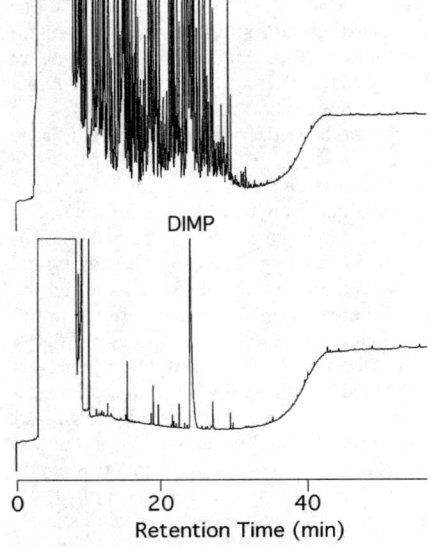

FIGURE 18.6 Capillary gas chromatograms illustrating the total DIMP-spiked gasoline sample (top) compared to the retained fraction from the DIMP-specific MIP (bottom).

useful for monitoring battlefields or CW storage bunkers. Significant improvement in this technique could be obtained by substituting selective polymers for the PDMS coating. Recently, we have demonstrated SPME sampling of Sarin using the selective BSP3 polymer (Harvey et al., 2002). This offers at least a 20-fold higher affinity towards Sarin and at least a twofold lower affinity toward interfering hydrocarbons. Presently this technique is limited by the temperature stability of the BSP3. Future advancement can be expected to enhance the thermal characteristics of this stationary phase to the point where the entire G-nerve agent series can be analyzed. Next generation instrumentation will utilize these improved polymers in combination with fast GC to provide near-real-time monitoring for nerve agents.

Recently PNNL has demonstrated deposition of a silane-based monolayer on a SAMMS (self-assembled monolayers on mesoporous supports) ceramic material from a supercritical fluid (Zemanian, et al., 2001). Deposition from supercritical fluids rather than from solution offers the advantage of higher surface coverage with fewer defects produced in a reduced time. Functional monolayers that lead to high capacity sorbents that selectively bond metals through coordination with the functionalized ligands can be incorporated. Various ligands can selectively target metals such as mercury (through thiols) or actinides [through 1,2-hydroxypyridinone (HOPO)]. Additionally, it is possible to form two-dimensional imprints on the surface of SAMMS that enhance analyte recognition. Future developments may target more advanced SAMMS material that incorporates three-dimensional imprints. This material may offer advantageous flow and sample capacity properties for selective air sampling.

Sensor-like Detectors. In general, sensors lack the specificity to directly analyze highly complex samples. Next generation analytical systems are, however, likely to use the highly effective combination of selective sampling along with sensor-like detection. If the initial fraction is not pure enough, it is possible to insert a short separation column between the selective sorbent and the sensor-like detector to remove minor impurities. Numerous sensors can be used including those based on MIPs and immunointeractions. One area undergoing rapid development is the construction of MIPs using fluorescent monomers (Wandelt, et al., 2002; Zhang et al., 2001; Rathbone and Gee, 2001; Turkewitsch et al., 1998). Transduction is achieved by polymer fluorescence quenching upon binding of the template molecule. Selective sorbent/detector combinations should adhere to the principle of orthogonality to optimize instrument selectivity. Microfabricating MIP and immunoreagents into miniaturized detectors using microfluidics is a rapidly developing area of research and can be expected to result in highly sensitive detection devices that operate with minimal reagents (Bernard, Michel, and Delmarche, 2001; Yan and Kapua, 2001). Possible sensor-like enzyme detectors include amperometric, channel thermistors, and fluorescent evanescent wave fiber optics (i.e., pH optrodes) (Hundeck et al., 1990; Majors, 1995). Further development of gold nanoparticles containing specific surface chemistries combined with surface plasmon spectroscopy can be expected to provide significant advancements (Haes and Van Duyne, 2002). Additionally, future development of microcantilever sensing that incorporates molecular recognition surfaces can be expected to produce highly sensitive and selective sensors (Pinnaduwage, et al., 2003). When combined with selective sampling, these sensors will form the basis for the next generation analytical instrumentation for trace detection of high priority biological/chemical/nuclear signatures and threats.

REFERENCES

Abbaszadegan, M., M. Huber, C. Gerba, and I. Pepper, "Detection of Viable Giardia Cysts by Amplification of Heat Shock-Induced mRNA," *Appl. Environ. Microbiol.* 63: 324–328, 1997.

Allen, M. J., J. L. Clancy, and E. W. Rice, "The Plain Hard Truth About Pathogen Monitoring," *J. Am. Water Works Assoc.* 92: 64–76, 2000.

APHA, AWWA, and WEF (ed.), "Standard Methods for the Examination of Water and Wastewater," 20th ed., APHA, AWWA, WEF, Washington DC, 1998.

Armstrong, D. W., G. Schulte, J. M. Schniederheinze, and D. J. Westenberg, "Separating Microbes in the Manner of Molecules. I. Capillary Electrokinetic Approaches," *Anal. Chem.* 71: 5465–5469, 1999.

Arshady, R., and K. Mosbach, "Synthesis of Substrate-Selective Polymers by Host-Guest Polymerization," *Makromol. Chem.* 182: 687–692, 1981.

Asanuma, H., T. Akiyama, K. Kajiya, T. Hishiya, and M. Komiyama, "Molecular Imprinting of Cyclodextrin in Water for the Recognition of Nanometer-Scaled Guests," *Anal. Chim. Acta* 435: 25–33, 2001.

Bavykin, S. G., J. P. Akowski, V. M. Zakhariev, V. E. Barsky, A. N. Perov, and A. D. Mirzabekov, "Portable System for Microbial Sample Preparation and Oligonucleotide Microarray Analysis," *Appl. Environ. Microbiol.* 67: 922–928, 2001.

Bernard, A., B. Michel, and E. Delmarche, "Micromosaic Immunoassays," *Anal. Chem.* 73: 8–12, 2001.

Berry, E. D., and G. R. Siragusa, "Hydroxyapatite Adherence as a Means to Concentrate Bacteria," *Appl. Environ. Microbiol.* 63: 4069–4077, 1997.

Berry, E. D., and G. R. Siragusa, Integration of hydroxyapatite concentration of bacteria and seminested PCR to enhance detection of *Salmonella typhimurium* from ground beef and bovine carcass sponge samples. *J. Rapid Methods Automation Microbiol.* 7: 7–23, 1999.

Blake, M., and B. Weimer, "Immunomagnetic Separation of *Bacillus stearothermophilus* Spores in Food and Environmental Samples," *Appl. Environ. Microbiol.* 63: 1643–1646, 1997.

Bruckner-Lea, C. J., B. P. Dockendorff, M. D. Quinonez-Diaz, J. W. Grate, and T. M. Straub, "Integrated Systems for Pathogen Detection," in press, 2004.

Bruckner-Lea, C. J., N. C. J. Anheirer, D. Holman, T. Tsukuda, M. T. Kingsley, F. J. Brockman, J. M. Price, J. W. Grate, and D. P. Chandler, "Integrated systems for DNA sample preparation and detection in environmental samples," *SPIE Proc.* 4200: 74–81, 2000.

Bruckner-Lea, C. J., T. Tsukuda, B. Dockendorff, J. C. Follensbee, M. T. Kingsley, C. Ocampo, J. R. Stults, and D. P. Chandler, "Renewable Mcrocolumns for Automated DNA Purification and Flow Through Amplification: From Sediment Samples Through Polymerase Chain Reaction," *Anal. Chim. Acta.* 469: 129–140, 2002.

Bruno, J., and H. Yu, "Immunomagnetic-Electrochemiluminescent Detection of *Bacillus anthracis* Spores in Soil Matricies," *Appl. Environ. Microbiol.* 62: 3474–3476, 1996.

Bukhari, Z., M. M. Marshall, D. G. Korich, C. R. Fricker, H. V. Smith, J. Rosen, and J. L. Clancy, "Comparison of Cryptosporidium parvum Viability and Infectivity Assays Following Ozone Treatment of Oocysts," *Appl. Environ. Microbiol.* 66: 2972–2980, 2000.

Bundy, J., and C. Fenselau, "Lectin-Based Affinity Capture for MALDI-MS Analysis of Bacteria," *Anal. Chem.* 71: 1460–1463, 1999.

Burns, M. A., B. N. Johnson, S. N. Brahmasandra, K. Handique, J. R. Webster, M. Krishnan, and T. S. Sammarco, "An Integrated Nanoliter DNA Analysis Device," *Science* 282: 484–487, 1998.

Call, D. R., F. J. Brockman, and D. P. Chandler, "Detecting and Genotyping *E. coli* O157:H7 Using Multiplexed PCR and Nucleic Acid Microarrays," *Int. J. Food Microbiol.* 67: 71–80, 2001.

Chandler, D. P., B. L. Schuck, F. J. Brockman, and C. J. Bruckner-Lea, "Automated Nucleic Isolation and Purification from Soil Extracts Using Renewable Affinity Microcolumns in a Sequential Injection System," *Talanta* 49: 969–983, 1999.

Chandler, D. P., J. Brown, D. R. Call, J. W. Grate, D. A. Holman, L. Olson, and M. S. Stottlemyer, "Continuous, Automated Immunomagnetic Separation and Microarray Detection of *E. coli* O157:H7 from Poultry Carcass Rinse," *Int. J. Food Microbiol.* 70: 143–154, 2000a.

Chandler, D. P., D. A. Holman, F. J. Brockman, J. W. Grate, and C. J. Bruckner-Lea, "Renewable Microcolumns for Solid-Phase Nucleic Acid Separations and Analysis from Environmental Samples," *Trends Anal. Chem.* 19: 314–321, 2000b.

Chizhikov, V., A. Rasooly, K. Chumakov, and D. D. Levy, "Microarray Analysis of Microbial Virulence Factors," *Appl. Environ. Microbiol.* 67: 3258–3263, 2001.

Chizhikov, V., M. Wagner, A. Ivshina, Y. Hoshino, A. Z. Kapikian, and K. Chumakov, "Detection and Genotyping of Human Group A Rotaviruses by Oligonucleotide Microarray Hybridization," *J. Clin. Microbiol.* 40: 2398–2407, 2002.

Dooley, H., S. D. Grant, W. J. Harris, and A. J. Porter, "Stabilization of Antibody Fragments in Adverse Environments," *Biotechnol. Appl. Biochem.* 28: 77–83, 1998.

Ertl, P., and S. R. Mikkelsen, "Electrochemical Biosensor Array for the Identification of Microorganisms Based on Lectin-Lipopolysaccharide Recognition," *Anal. Chem.* 73: 4241–4248, 2001.

Evans-Strickfaden, T. T., K. H. Oshima, A. K. Highsmith, and E. W. Ades, "Endotoxin Removal Using 6000 Molecular Weight Cut-Off Polyacrylonitrile (PAN) and Polysulfone (PS) Hollow Fiber Ultrafilters," *PDA J. Pharm. Sci. Technol.* 50: 154–157, 1996.

Feldhaus, M. J., R. W. Siegel, L. K. Opresko, J. R. Coleman, J. M. Feldhaus, Y. A. Yeung, J. R. Cochran, P. Heizelman, D. Colby, J. Swers, C. Graff, H. S. Wiley, and K. D. Wittrup, "Flow-Cytometric Isolation of Human Antibodies from a Nonimmune Saccharomyces Cerevisiae Surface Display Library," *Nat. Biotech.* 21: 163–170, 2003.

Fleischer, C. T., and K.-S. Boos, "Highly Selective Solid-Phase Extraction of Biofluids Using Restricted-Access Materials in Combination with Molecular Imprinted Polymers," *Am. Lab.* 33(10): 20–25, 2001.

Gerba, C. P., "Pathogens in the Environment," in I. L. Pepper, C. P. Gerba, and M. L. Brusseau (eds.), *Pollution Science*, Academic Press, New York, pp. 279–299, 1996.

Grant, I. R., H. J. Ball, and M. T. Rowe, "Isolation of *Mycobacterium paratuberculosis* from Milk by Immunomagnetic Separation," *Appl. Environ. Microbiol.* 64: 3153–3158, 1998.

Grate, J. W., S. J. Patrash, and S. N. Kaganove, "Hydrogen Bond Acidic Polymers for Surface Acoustic Wave Vapor Sensors and Arrays," *Anal. Chem.* 71: 1033–1040, 1999.

Grate, J. W., S. N. Kaganove, and D. A. Nelson, "Carbosiloxane Polymers for Sensors," *Chem. Innov.* 30: 29–37, 2001.

Guilbault, G. G., and J. H. Luong, "Gas Phase Biosensors," *J. Biotechnol.* 9: 1–10, 1988.

Haes, A. J., and R. P. Van Duyne, "A Highly Sensitive and Selective Surface-Enhanced Nanobiosensor," in K. J. Shea, M. Yan, and M. J. Roberts (eds.), *Molecularly Imprinted Materials—Sensors and Other Devices,* Materials Research Society Symposium Proceedings, Warrendale, PA, vol. 723, pp.133–138, 2002.

Harris, B., "Exploiting Antibody-Based Technologies to Manage Environmental Pollution," *Trends Biotechnol.* 17: 290–296, 1999.

Harvey, S. D., and B. W. Wright, "Multidimensional Chromatography of 2,2'-Thiodiethanol, 2-Chloroethyl Ethyl Sulfide, and Methamidophos Chemical Warfare Surrogates in Environmental Matrices," PNL-8433, Prepared for the Office of Arms Control and Nonproliferation, March 1993.

Harvey, S. D., "Molecularly Imprinted Polymers for Selective Analysis of Chemical Warfare Surrogate and Nuclear Signature Compounds in Complex Matrices," *J. Sep. Sci. A* (submitted).

Harvey, S. D., D. A. Nelson, B. W. Wright, and J. W. Grate, "Selective Stationary Phase for Solid-Phase Microextraction Analysis of Sarin (GB)," *J. Chromatogr. A* 954: 217–225, 2002.

Heckotter, S., M. Bulte, and E. Lucker, "Detection of *Escherichia coli* Serogroup O157 in Foods by Immunomagnetic Separation," *Archiv fur Lebensmittelhybiene* 49: 25–48, 1998.

Hundeck, H.-G., A. Sauerbrei, U. Hubner, T. Scheper, K. Schugerl, R. Koch, and G. Antranikian, "Four-Channel Enzyme Thermistor System for Process Monitoring and Control in Biotechnology," *Anal. Chim. Acta* 238: 211–221, 1990.

Juliano, J., and M. D. Sobsey, "Simultaneous Concentration of Cryptosporidium, Bacteria, and Viruses by Hollow Fiber Ultrafiltration," *Presented at the Water Quality Technology Conference*, Denver, CO, 1997.

Kelty, J. M, "Pathogenesis of Enteric Infection by *Campylobacter*," *Microbiology* 143: 5–12, 1997.

Koniyama, M., T. Takeuchi, T. Mukawa, and H. Asanuma, "Molecular Imprinting: From Fundamentals to Applications," Wiley-VCH, Weinheim, pp. 119–128, 2003.

Kriz, D., O. Ramstrom, and K. Mosbach, "Molecular Imprinting. New Possibilities for Sensor Technology," *Anal. Chem.* 69: 345A–349A, 1997.

Kuhn, R. C., and K. H. Oshima, "Hollow Fiber Filtration of Cryptosporidium Parvum Oocysts from 10 Liters of Surface Water," *Presented at the Water Quality Technology Conference*, Nashville, TN, 2001.

Lang, W. K., K. Glassey, and A. R. Archibald, "Influence of Phosphate Supply on Teichoic Acid and Teichuronic Acid Content of *Bacillus subtilis* Cell Walls," *J. Bacteriol.* 151: 367–375, 1982.

Lanza, F., and B. Sellergren, "Molecularly Imprinted Extraction Materials for Highly Selective Sample Cleanup and Analyte Enrichment," *Adv. Chromatogr.* 41: 137–173, 2001.

Lis, H., and N. Sharon, "Lectins: Carbohydrate-Specific Proteins that Mediate Cellular Recognition," *Chem. Rev.* 98: 637–674, 1998.

Majors, R. E., "Sample Preparation and Handling for Environmental and Biological Analysis," *LC-GC* 13: 542–552, 1995.

Matson, R. S., and M. C. Little, "Strategy for the Immobilization of Monoclonal Antibodies on Solid-Phase Supports," *J. Chromatogr.* 458: 67–77, 1988.

Molloy, P., L. Brydon, A. J. Porter, and W. J. Harris, "Separation and Concentration of Bacteria with Immobilized Antibody Fragments," *J. Appl. Bacteriol.* 78: 359–365, 1995.

Morales-Morales, H. A., G. Vidal, J. Olszewski, C. M. Rock, D. Dasgupta, K. Oshima, and G. B. Smith, "Optimization of a Reusable Hollow-Fiber Ultrafilter for Simultaneous Concentration of Enteric Bacteria, Protozoa, and Viruses from Water," *Appl. Environ. Microbiol.* 69: 4098–4102, 2003.

National Research Council, "Review of the EPA Water Security Research and Technical Support Action Plan," National Academy Press, Washington, D.C., 2003.

Neeser, J. R., D. Granato, M. Rouvet, A. Servin, S. Teneberg, and K. A. Karlsson, "*Lactobacillus johnsonii* Ls1 Shares Carbohydrate-Binding Specificities with Several Enteropathogenic Bacteria," *Glycobiology* 10: 1193–1199, 2000.

Ngeh-Ngwainbi, J., P. H. Foley, S. S. Kuan, and G. G. Guilbault, "Parathion Antibodies on Piezoelectric Crystals," *J. Am. Chem. Soc.* 108: 5444–5447, 1986.

Olszewski, J., L. Winona, and K. Oshima, "Hollow Fiber Filtration to Concentrate Viruses from Water," Presented at the Water Quality Technology Conference, Nashville, TN, 2001.

Oshima, K. H., T. T. Evans-Strickfaden, A. K. Highsmith, and E. W. Ades, "The Removal of T1 and PP7 and Poliovirus from Fluids with Hollow-Fiber Ultrafilters with Molecular Weight Cut-Offs of 50,000, 13,000, and 6000," *Can. J. Microbiol.* 41: 316–322, 1995.

Penalva, J., R. Puchades, and A. Maquieira, "Analytical Properties of Immunosensors Working in Organic Media," *Anal. Chem.* 71: 3862–3872, 1999.

Pichon, V., M. Bouzige, and M.-C. Hennion, "New Trends in Environmental Trace-Analysis of Organic Pollutants: Class-Selective Immunoextraction and Cleanup in One Step Using Immunosorbents," *Anal. Chim. Acta* 376: 21–35, 1998.

Pichon, V., M. Bouzige, C. Miege, and M.-C. Hennion, "Immunosorbents: Natural Molecular Recognition Materials for Sample Preparation of Complex Environmental Matrices," *Trends Anal. Chem.* 18: 219–235, 1999.

Picker, J. E., and R. E. Sievers, "Lanthanide Metal Chelates as Selective Complexing Sorbents for Gas Chromatography," *J. Chromatogr.* 203: 29–40, 1981.

Pinnaduwage, L. A., J. E. Hawk, V. Boiadjiev, D. Yi, and T. Thundat, "Use of Micocantilevers for the Monitoring of Molecular Binding to Self-Assembled Monolayers," *Langmuir* 19: 7841–7844, 2003.

Rajakovic, L., V. Ghaemmaghami, and M. Thompson, "Adsorption on Film-Free and Antibody-Coated Piezoelectric Sensors," *Anal. Chim. Acta* 217: 111–121, 1989.

Rathbone, D. L., and Y. Ge, "Selectivity of Response in Fluorescent Polymers Imprinted with N^1-Benzylidene Pyridine-2-Carboxamidrazones," *Anal. Chim. Acta* 435: 129–136, 2001.

Remcho, V. T., and Z. J. Tan, "MIPs as Chromatographic Stationary Phases for Molecular Recognition," *Anal. Chem.* 71: 248A–225A, 1999.

Rijpens, N., L. Herman, F. Vereecken, G. Jannes, J. De Smedt, and L. De Zutter, "Rapid Detection of Stressed *Salmonella spp.* in Dairy and Egg Products Using Immunomagnetic Separation and PCR," *Int. J. Food Microbiol.* 46: 37–44, 1999.

Rochelle, P., D. Ferguson, T. Handojo, R. De Leon, M. Stewart, and R. Wolfe, "An Assay Combining Cell Culture with Reverse Transcriptase PCR to Detect and Determine the Infectivity of Waterborne Cryptosporidium Parvum," *Appl. Environ. Microbiol.* 63: 2029–2037, 1997.

Rolsma, M. D., T. D. Kulenschmidt, H. B. Gelberg, and M. S. Kulenschmidt, "Structure and Function of a Ganglioside for Porcine Rotavirus," *J. Virol.* 72: 9079–9091, 1998.

Russell, A. J., L. J. Trudel, P. L. Skipper, J. D. Groopman, S. R. Tennenbaum, and A. M. Klibanov, "Antibody-Antigen Binding in Organic Solvents," *Biochem. Biophys. Res. Comm.* 158: 80–85, 1989.

Schembri, M. A., D. W. Ussery, C. Workman, H. Hasman, and P. Kelm, "DNA Microarray Analysis of *fim* Mutations in *Escherichia coli*," *Mol. Genet. Genom.* 267: pp. 721–729, 2002.

Schlehuber, S., G. Beste, and A. Skerra, "A Novel Type of Receptor Protein, Based on the Lipcalin Scaffold, with Specificty for Digoxigenin," *J. Mol. Biol.* 297: 1105–1120, 2000.

Schwab, K., R. De Leon, and M. Sobsey, "Immunoaffinity Concentration and Purification of Waterborne Enteric Viruses for Detection by Reverse Transcriptase PCR," *Appl. Environ. Microbiol.* 62: 2086–2094, 1996.

Skerra, A., "Anticalins: A New Class of Engineered Ligand-Binding Proteins with Antibody-Like Properties," *Rev. Mol. Biotech.* 74: 257–275, 2001.

Skerra, A., "Lipocalins as a Scaffold," *Biochim. Biophys. Acta* 1482: 337–350, 2000.

Small, J., D. R. Call, F. J. Brockman, T. M. Straub, and D. P. Chandler, "Direct Detection of 16S rRNA in Soil Extracts by Using Oligonucleotide Microarrays," *Appl. Environ. Microbiol.* 67: 4708–4716, 2001.

Sousa, M. C., C. A. Goncalves, V. A. Vairos, and J. Poires-de-silva, "Adherence of *Giardia lamblia* Trophozoites to int-407 Human Intestinal Cells," *Clin. Diag. Lab. Immunol.* 8: 256–265, 2001.

States, S., M. Scheuring, J. Kuchta, J. Newberry, and L. Casson, "Utility-Based Analytical Methods to Ensure Public Water Supply Security," *J. Am. Wat. Works Assn.* 95: pp. 103–115, 2003.

Steinke, J., D. C. Sherrington, and I. R. Dunkin, "Imprinting of Synthetic Polymers Using Molecular Templates," *Adv. Polym. Sci.* 123: 81–125, 1995.

Straub, T. M., D. S. Daly, S. Wunshel, P. A. Rochelle, R. DeLeon, and D. P. Chandler, "Genotyping Cryptosporidium Parvum with an hsp70 Single-Nucleotide Polymorphism Microarray," *Appl. Environ. Microbiol.* 68: 1817–1826, 2002.

Straub, T. M., and D. P. Chandler, "Towards a Unified System for Detecting Waterborne Pathogens," *J. Microbiol. Meth.* 53: 185–197, 2003.

Takayama, K., N. Qureshi, and P. Mascagni, "Fatty Acyl Derivatives of Glucosamine 1-Phosphate in *Escherichia coli* and Their Relation to Lipid A. Complete Structure of a Diacyl GlcN-1-P Found in a Phosphatidylglycerol-Deficient Mutant," *J. Biol. Chem.* 258: 7379, 1983.

Tebber, C., and W. Vahjen, "Interference of Humic Acids and DNA Extracted Directly from Soil in Detection and Transformation of Recombinant DNA from Bacteria and a Yeast," *Appl. Environ. Microbiol.* 59: 2657–2665, 1993.

Tsai, Y., B. Tran, L. Sangermano, and C. Palmer, "Detection of Poliovirus, Hepatitis A Virus, and Rotavirus from Sewage and Ocean Water by Triplex Reverse Transcriptase PCR," *Appl. Environ. Microbiol.* 60: 2400–2407, 1994.

Turkewitsch, P., B. Wandelt, G. D. Darling, and W. S. Powell, "Fluorescent Functional Recognition Sites Through Molecular Imprinting. A Polymer-Based Fluorescent Chemosensor for Aqueous cAMP," *Anal. Chem.* 70: 2025–2030, 1998.

U.S. EPA, "Comparative Health Effects Assessment of Drinking Water," United States Environmental Protection Agency, 1988.

U.S. EPA, "U.S. EPA Manual of Methods for Virology EPA/600/4-84/0-13," United States Environmental Protection Agency, 1993.

U.S. EPA, "Drinking Water Candidate Contaminant List EPA-815-F-98-002," United States Environmental Protection Agency, 1998.

U.S. EPA, "Method 1623 *Cryptosporidium* and *Giardia* in Water by Filtration/IMS/FA EPA-821-R-01-025," United States Environmental Protection Agency, 2001.

Wandelt B., P. Turkewitsch, S. Wysocki, and G. D. Darling, "Fluorescent Molecularly Imprinted Polymer Studied by Time-Resolved Fluorescence Spectroscopy," *Polymer* 43: 2777–2785, 2002.

Wanke, C. A., and R. L. Guerrant, "Small-Bowel Colonization Alone is a Cause of Diarrhea," *Infect. Immun.* 55: 1924–1926, 1987.

Ward, H. D., J. Alroy, B. I. Lev, G. T. Keusch, and M. E. A. Pereira, "Biology of *Giardia lamblia*," *J. Exp. Med.* 167: 73–88, 1988.

Wenzel, T. J., P. J. Bonasia, and T. Brewitt, "Application of Metal β-Diketonate Polymers as Selective Sorbents in Complex Mixture Analysis and for Sulfur-Containing Compounds," *J. Chromatogr.* 463: 171–176, 1989.

Wenzel, T. J., L. W. Yarmaloff, L. Y. St. Cry, L. J. O'Meara, M. Donatelli, and R. W. Bauer, "Metal Chelate Polymers as Selective Sorbents for Gas Chromatography," *J. Chromatogr.* 396: 51–64, 1987.

Wizemann, T. M., J. E. Adamou, and S. Langermann, "Adhesins as Targets for Vaccine Development," *Emerg. Infect. Dis.* 5: 395–403, 1999.

Wulff, G., "Molecular Imprinting in Cross-Linked Materials with the Aid of Molecular Templates–A Way Towards Artificial Antibodies," *Angew. Chem. Int. Ed. Engl.* 34: 1812–1832, 1995.

Yan, M., and A. Kapua, "Fabrication of Molecularly Imprinted Polymer Microstructures," *Anal. Chim. Acta* 435: 163–167, 2001.

Zemanian, T. S., G. E. Fryxell, J. Liu, S. Mattigod, J. A. Franz, and Z. Nie, "Deposition of Self-Assembled Monolayers in Mesoporous Silica from Supercritical Fluids," *Langmuir* 17: 8172–8177, 2001.

Zhang, H., W. Verboom, and D. N. Reinhoudt, "9-(Guanidinomethyl)-10-Vinylanthracene: A Suitable Fluorescent Monomer for MIPs," *Tetrahedron Lett.* 42: 4413–4416, 2001.

INDEX

Page numbers followed by italic *f* or *t* denote figures or tables, respectively

Acanthamoeba, 2.10
Access control, 12.16–12.26
  access control attributes, 12.17
    combination, 12.17
    knowledge basis, 12.17
    physical attribute, 12.17
    possession basis, 12.17
  building access control concepts, 12.25
  card access technologies, 12.21
    magnetic strip, 12.21
    proximity, 12.21
    smart card, 12.21
    Weigard card, 12.21
  card reader systems, 12.19–12.21
    installation detail for single door card reader, 12.20*f*
  clear-zone areas, 12.23
  delivery access control, 12.24
  effective site access control, 12.23
  guard tour system, 12.21
  interior access control concepts, 12.25
  layered access control example, 12.26*t*
  layered security systems, 12.16–12.17
  locking systems, 12.17–12.18
    electrical keypads, 12.18
    electrified locking systems, 12.18
    fail-safe locking systems, 12.19
    fail-secure locking systems, 12.19
    key locks, 12.17
    mechanical keypads, 12.18
  site perimeter access control, 12.21–12.23
    clear-zone areas, 12.23
    delivery access points, 12.24
    standoff distance, 12.23
    vehicle access control barriers, 12.24
    vehicle checkpoints, 12.24
    standoff distance, 12.23
    vehicle access control barriers, 12.24
    vehicle checkpoints, 12.24
Access control applications for surveillance, 12.29–12.31
  CCTV integration, 12.30
  email alarm notification, 12.30
  graphical map tool, 12.30
  threat level adjustments, 12.30
  video badging, 12.20
Access control attributes, 12.17
  combination, 12.17
  knowledge basis, 12.17
  physical attribute, 12.17
  possession basis, 12.17
Active infrared sensors, 12.6
Adenovirus, 2.7–2.8
Advanced early warning systems, 11.23–11.25
  definition of, 11.23
  summary of, 11.24*t*, 11.25*t*
Advanced remote monitoring devices, 14.11–14.12
  advanced spectrophotometers, 14.12
  biosentinels, 14.12
  multiparameter microchip sensors, 14.12
Advanced spectrophotometers, 14.12
Advection-dispersion equation, 6.11–6.12
American Gas Association (AGA), 5.7
American Water Works Association (AWWA), 3.3, 3.9, 4.3
  manual on emergency planning (M19), 4.3
AMSA asset-based vulnerability analysis, 3.5*f*
AMSA response planning approach, 3.5*f*

Association of Metropolitan Sewage
    Agencies (AMSA), 3.5, 3.9
Association of State Drinking Water Administ-
    rators (ASDWA), 3.2, 3.6, 3.9, 3.15–3.27
  NRWA/ASDWA Guide for Security
    Decisions, 3.15–3.27
Astroviruses, 2.8
Atomic Age, 17.2
Atomic bomb test (Trinity), 17.2
Attack tree, 5.10
Automated Pathogen Detection System
    (APDS), 18.8, 18.10

Bacillary dysentery, 2.4
Bacilli, 2.3
Backflow prevention devices, 9.3
Bacterial pathogens, 2.2–2.7
  campylobacter, 2.6–2.7
  escherichia coli O157:H7, 2.5
  legionella, 2.7
  salmonella, 2.3
  shigella, 2.3–2.4
  vibrio, 2.6
  vibrio cholerae, 2.6
  yersinia, 2.5
Baseline demands, 4.5
Bioalams, 11.13
Biodetection Enabling Analyte Delivery
    System (BEADS), 18.7, 18.7*f*
Biological agents:
  biological warfare agents, 6.2, 6.3*t*
  nonwarfare agents, 6.4, 6.5*t*
  threat potential, 6.3*t*, 6.4*t*
Biological threats, 1.6, 4.4–4.5, 8.4*t*
BIOS crackers, 5.13
Biosensors, 17.8
Biosentinels, 14.12
Biotection process, 18.6*f*
Bioterrorism Act of 2002, 1.9, 3.1, 3.6, 3.7,
    3.12, 8.2, 9.1
  amendments to, 1.9*t*
  compliance dates, 3.2*t*
  requirements, 1.9
Black rain, 17.3
Building access control, 12.25
Buried line sensors, 12.5
Butterfly valves, 13.26

Campylobacter, 2.6–2.7
Campylobacter jejuni, 2.1, 6.4
Campylobacter spp., 18.9

Candidate contaminate list (CCL), 18.2
Carbohydrate affinity, 18.9
Card access technologies, 12.21
  magnetic strip, 12.21
  proximity, 12.21
  smart card, 12.21
  Weigard card, 12.21
Card reader systems, 12.19–12.21
  installation detail for single door card
    reader, 12.20*f*
CCTV integration, 12.30
Chemical agents, 6.6–6.7
  chemical warfare agents, 6.7
  chemical nonwarfare agents, 6.7
  threat potential, 6.8*t*, 6.9*t*, 6.10*t*
Chemical analysis for early warning moni-
    toring 11.8–11.12
  chemical and radioactive constituents,
    methods for detecting, 11.9*t*, 11.10*t*
  dissolved oxygen, 11.8
  general organic chemical parameters,
    11.11
  ion specific electrodes for monitoring raw
    water, 11.11*t*
  metals, 11.8
  nitrate and ammonia, 11.8
  oil and petroleum, 11.11
  on-line analytical probes, 11.8
  organic chemicals, 11.12
  oxidant demand and oxidant residual,
    11.11
  pesticides, 11.12
  radioactivity, 11.12
Chemical detection, 18.12–18.17
  immunosorbents, 18.13
  molecularly imprinted polymers (MIP),
    18.13–18.17
    analysis of air, 18.15–18.16
    aqueous analysis, 18.14–18.15
    sensor-like detectors, 18.16–18.17
Chemical detection in air, 18.15–18.16
  gas-chromatographic analysis, 18.15
  gas-phase sampling, 18.15
  MIP stationary phase, 18.15
  nonaqueous immunochromatography,
    18.15
  solid-phase microextraction (SPME) sam-
    pling, 18.15
Chemical threats, 1.6, 1.7*t*, 4.4–4.5
Chemical warfare agents, 1.7*t*
Chernobyl, 17.3–17.5

Cholera, 2.6
Chlorine inactivation of microbes, 2.4*t*
Closed circuit television (CCTV),
    12.11–12.16
  system components, 12.12–12.16
  special considerations, 12.16
Coagulation, 11.6
Coaxial strain sensors, 12.6
Cocci, 2.2
Cold War, 17.5
Coliform bacteria, 18.3
Communication protocols, 5.6
  control networks, 5.6
  device networks, 5.6
  enterprise networks, 5.6
  sensor networks, 5.6
Communism, 17.5
Computer outlaw, 5.2
Computer security incident handling guide, 5.15
Computer system infrastructure, 5.3–5.6
Conditional logistic regression, 10.12
Conditional probability approach, 7.15–7.17
  conditional probability rule, 7.15
  example of approach, 7.16
Connectivity technique, delineating contamination spread, 15.6–15.8
  connectivity matrix, 15.7–15.8
  operating modes, 15.8
  maximum spread potential, 15.8
Conservation of mass:
  flow, 4.6
  storage tanks, 4.12, 9.6, 9.8
  substance concentrations, 4.12, 9.5–9.6
Consortium water systems, 8.7–8.10
Contamination by radiation, 17.2–17.5
  Chernobyl, 17.3–17.5
  dangers of, 17.2–17.5
  diagnostic effects, 17.4*t*
  Hiroshima, 17.2–17.3
  Nagasaki, 17.2–17.3
  Three Mile Island, 17.3–17.4
Contamination events:
  confirmation and characterization of, 11.3–11.4
  historical reconstruction of, 10.1–10.52
  types of, 10.3–10.4
    long-term chronic, 10.4
    short-term acute, 10.3
Contamination events, historical reconstruction of, 10.1–10.52

Dover Township (Toms River), New Jersey, 10.22–10.43
  assessment of contaminated public water supply, 10.26–10.29
  findings and conclusions, 10.42–10.43
  Monte Carlo simulation, 10.49
  sensitivity analysis, 10.40–10.42
  simulation methods, 10.29–10.37
Gideon, Missouri, 10.4–10.9
  Gideon outbreak, 10.5–10.7
  systems analysis, 10.7–10.8
Redlands, California, 10.43–10.52
  background information, 10.44–10.45
  data development, 10.46–10.48
  general methodology, 10.45–10.46
  modeling procedure, 10.48–10.49
  types of, 10.3–10.4
    long-term chronic, 10.4
    short-term acute, 10.3
Walkerton, Ontario, 10.10–10.22
  conclusions, 10.22
  conditional logistic regression, 10.12
  exposure scenarios, 10.11*t*
  hazard model, 10.13
  methodology, 10.10–10.13
  results, 10.13–10.22
Contamination threats, 4.4–4.5, 8.4, 9.3–9.4
Contaminant transport, 4.5–4.20
  modeling of, 4.5–4.20
Contingency, 17.1
Contingency planning, 17.1–17.13
  crisis management, 17.7
  critical role of, 17.2
  disaster scenarios, 17.7–17.10
    mobile scenarios, 17.8
    stationary scenarios, 17.7
  nonconventional times, 17.6–17.10
  problem of, 17.7
Control network protocols, 5.6
Control valves, 13.26
Cryptosporidium, 2.1, 6.3, 8.4, 18.3, 18.5, 18.7, 18.12
Cryptosporidium parvum, 2.1, 2.10, 6.4, 6.5, 18.3
Currie, definition of, 17.6
Cyber attack, 5.1, 5.10–5.17
  attack tree, 5.10
  business continuity, 5.16–5.17
  definition, 5.1
  incident response, 5.15–5.17
  preparation, 5.15

Cyber attack (*Conti.*):
   response process phases, 5.16
      SCADA specific incident response, 5.16
   information links, 5.17
   mitigation methods, 5.14–5.15
   scenarios, 5.10–5.11
   tools for, 5.12–5.14
      BIOS crackers, 5.13
      Hackers' Swiss knife, 5.13
      ICH hacking tools, 5.13
      key loggers, 5.13
      mail bombs, 5.13
      nukers, 5.13
      password crackers, 5.13
      scanners, 5.12
      Trojan horse, 5.12
      viruses, 5.12
      Windows NT hacking tools, 5.13
      worms, 5.12
Cyber security:
   SCADA networks, 5.9–5.10
   twenty-one step guide, 5.9–5.10
Cyber threats, 1.6

Daphnia, 11.13
Darcy-Weisbach equation, 4.7
Decision support system for early warning system, 8.10–8.16
Demand multipliers, 4.5
Demand (water) simulation, 4.20–4.35
Detection:
   of chemical compounds, 18.12–18.17
   of microorganisms, 18.2–18.12
Device network protocols, 5.6
Digital encryption standard (DES), 5.7
Digital video management, 12.27–12.28
   alarm review utility, 12.28
   camera control, 12.27
   live viewing module, 12.27
   recording setup, 12.27
   searching recorded video, 12.27
   web browser application, 12.28
Digital video recorders, 12.14
Disaster scenarios, contingency planning, 17.7–17.10
   mobile scenarios, 17.8
   stationary scenarios, 17.7
Disinfection, 11.6
DNA isolation, 18.10
DNA microarrays, 18.10

DNA suspension arrays, 18.10
Dual technology sensors, 17.8
Dynamic Water Quality Model (DWQM), 14.19

Early warning decision support system (EWDSS), 8.10–8.16
   information management research, 8.14–8.16
      alert management system, 8.15
      data validation, 8.15
      real-time data acquisition information network systems, 8.16
      security enforcement, 8.15
      sensor data management system, 8.15
   modeling research, 8.12–8.14
   project management, 8.17
   public health surveillance, 8.16–8.17
   specific tasks, 8.12
Early warning detection systems (EWDS), 16.2
Early warning monitoring methods, 11.7–11.14
   biological monitoring (bioalarms), 11.13–11.14
   chemical analysis, 11.8–11.12
      chemical and radioactive constituents, methods for detecting, 11.9t, 11.10t
      dissolved oxygen, 11.8
      general organic chemical parameters, 11.11
      ion specific electrodes for monitoring raw water, 11.11t
      metals, 11.8
      nitrate and ammonia, 11.8
      oil and petroleum, 11.11
      on-line analytical probes, 11.8
      organic chemicals, 11.12
      oxidant demand and oxidant residual, 11.11
      pesticides, 11.12
      radioactivity, 11.12
   emerging monitoring methods, 11.14
   microbiological analysis, 11.12–11.13
   physical analysis, 11.7–11.8
Early warning system (EWS), 8.6–8.7, 9.4, 11.1–11.30, 14.1, 14.13–14.15
   characteristics of, 8.6–8.7, 14.14–14.15
   design and operation, 11.29
   elements of, 11.2f

framework for assessing, 11.2–11.7
  communication linkages, 11.5
  contamination events, 11.3–11.4
  detection mechanisms, 11.3
  institutional issues, 11.4
  response to an early warning, 11.5–11.7
monitoring methods, 11.7–11.14
  biological monitoring (bioalarms), 11.13–11.14
  chemical analysis, 11.8–11.12
  emerging monitoring methods, 11.14
  microbiological analysis, 11.12–11.13
  physical analysis, 11.7–11.8
overview of, 11.1–11.2
requirements for, 8.6
Early warning system, case studies, 11.23–11.27
  advanced early warning systems, definition of, 11.23
  advanced early warning systems, summary of, 11.24t, 11.25t
  ORANCO Ohio River EWS, 11.23
  Paris EWS, 11.26
  Rhine River, 11.23
  River Dee EWS, 11.23
  River Trent (UK), 11.26–11.27
  St. Clair River (Canada), 11.27
  Yodo River EWS, Japan, 11.26
Early warning system, risk-based analysis, 11.27–11.29
  risk-based decision making process for EWS, 11.30f
  spill risk model, 11.28–11.29
    application to Ohio River, 11.29
    model formulation, 11.28
    Monte Carlo simulation, 11.28
Early warning system, surface water models, 11.14–11.22
  incorporating a model, 11.15–11.16
  river oil spill models, summary of, 11.22t
    NRDAM, 11.22t
    RiverSpill, 11.22t
    ROSS, 11.22t
    ROSS2, 11.22t
    ROSS3, 11.22t
    WPMB, 11.22t
  selected models, 11.16–11.22
    oil spill models, 11.21
    REMM model, 11.20
    Rhine Alarm-model, 11.18
    Riverine spill modeling (RSMS), 11.16
    River oil spill models, summary of, 11.22t
    RiverSpill, 11.21
    R-TOT model, 11.10
  spill models, 11.15
    components of, 11.15
    fate module, 11.15
    flow module, 11.15
    water quality transport module, 11.15
E. coli, see *Escherichia coli*
Efficient hydrologic tracer-test design (EHTD), 6.10–6.20
  basic design of, 6.10–6.11
  experimental example, 6.13–6.19
  range of capabilities, 6.11
  toxic release outcome prediction, 6.11–6.13
  methodology, 6.12–6.13
Electrified locking, 12.18
Electro mechanical vibration sensing, 12.5
Electronic noses, 11.8
Email alarm notification, 12.30
Emergency action procedures (EAPs), 3.39–3.41
  incident-specific, 3.40–3.41
Emergency response planning, 3.6–3.7, 3.36–3.41, 15.3
  ASDWA plan outline, 3.8t
  emergency action procedures (EAPs), 3.39–3.41
  incident-specific EAPs, 3.40–3.41
  large systems 3.7
  needs, 3.6–3.7
  outline for, 3.36–3.41
  planning process, 3.28
  policies, 3.37–3.39
  references and links for, 3.41
  small water systems, 3.7
Emergency response plan outline, 3.36–3.41
Entamoeba histolytica, 2.11
Entrance detection, pump stations, 12.10
  Raw water intake station, 12.11
  Water treatment station, 12.10
Enteric fever, 2.3
Enteric viruses, 18.3
Enterprise network protocols, 5.6
EPANET, 4.12–4.17, 8.13–8.14, 9.7–9.10, 9.11–9.23, 10.7, 10.32–10.37, 13.8, 14.19, 16.1–16.2

EPANET (*Conti.*):
  applications, 10.7, 10.32–10.37
  case studies using EPANET, 9.11–9.23
    Gideon waterborne outbreak, 9.13–9.23
    North Marin Water Authority, 9.11–9.13
  convective transport model, 4.17
  hydraulic simulation equations, 4.13–4.14
  other features, 4.16
    travel time to a node, 4.16
    source tracing, 4.16
    water age, 4.16
  reaction rate model, 4.16
    bulk phase reaction, 4.16
    wall reaction, 4.16
  toolkit functions, 13.8
  water quality simulation equations, 4.15
Escherichia coli, 2.1, 2.2t, 2.5, 6.4, 10.10–10.22, 18.4, 18.7, 18.9
  Walkerton, Ontario outbreak of, 10.10–10.22
Escherichia coli O157:H7, 2.5
Etiological groups, 2.2–2.12
Event driven method, 4.3
Extended period simulation (EPS), 4.8, 13.15–13.25, 14.18

Fail-safe locking systems, 17.19
Failure probability:
  for complex system, 7.12
  for parallel system, 7.5
  series system, 7.4
Fault trees, 5.9, 7.18–7.22
  definition, 7.18
  evaluation of, 7.20–7.22
  example of, 7.18f
  fault tree symbols, 7.19f
  SCADA, 5.9
Fault tree analysis, 7.17–7.22
  advantages of, 7.17
  disadvantages, 7.17
  example fault tree, 7.18f
  for pumping system, 7.21f
  objective of, 7.17
  system availability, 7.22
  system reliability, 7.22
  tree symbols, 7.19t
    event symbols, 7.19
    gate symbols, 7.19
Fence mounted cable sensors, 12.3
Fence mounted electric field sensors, 12.6
Fibre-optic buried cable system, 12.5

Fiber-optic strain sensors, 12.6
Fire demands, 4.5
Friction, 4.7

Gate valves, 13, 26
Genetic algorithms (GA), 10.29, 13.29, 16.5
Genetic engineering (threat), 6.6
Genotoxins, 17.8
Giardia, 2.1, 18.3, 18.5, 18.7, 18.9
G. Lambia, 2.1, 18.9
Glass break sensors, 12.8
Gradient algorithms, 4.14
Graphical map tool, 12.30

Hackers' Swiss knife, 5.13
Hazard model, 10.13
Hazen-Williams equation, 4.7
Hazen-Williams roughness coefficient, 4.7
Hepatitis A, 2.8
Hepatitis E, 2.8–2.9
Hiroshima, Japan, 17.2
Historical reconstruction of contamination events, 10.1–10.52
Homeland Security Act, 1.10
Human machine interface (HMI), 5.3
Hydrants, 3.22
Hydraulic modeling, role of, 14.18, 15.3–15.4
Hydraulic network modeling (*see* Network hydraulic modeling)
Hydraulic redundancy, 15.1–15.2
Hydroxyapatite, 18.8–18.9

ICH hacking tools, 5.13
Immunofluorescent staining (IMS), 18.7
Immuno separation methods, 18.8
  automated pathogen detection system (APDS), 18.8
Inactivation of microbes using chlorine, 2.4t
Industrial chemical poisons, 1.7t
Information analysis and infrastructure protection (IAIP), 1.10
Information management research for EWDSS, 8.14–8.16
  alert management system, 8.15
  data validation, 8.15
  real-time data acquisition information network systems, 8.16
  security enforcement, 8.15
  sensor data management system, 8.15
Information Storm Center, SANS Institute, 5.2

Information technology (IT) infrastructure, 5.2, 5.4*t*
  computerized maintenance management system, 5.2
  human resources, 5.2
  laboratory information management system, 5.2
  supervisory control and data acquisition system, 5.2
Instrumentation, Systems, and Automation (ISA) Society, 5.7
Intelligent electron devices (IED), 5.7
Intelligent video applications, 12.28–12.29
  access control authentication, 12.28
  behavior tracking, 12.29
  digital video motion detection (DVMD), 12.29
  facial character recognition, 12.28
  object/target tracking, 12.28
Interior access control, 12.25
Interior boundary penetration sensors, 12.8
Interior volumetric sensors, 12.7
International Life Sciences Institute (ILSI), 14.14
Intrusion detection, 12.1–12.11
  alarm activation time, 12.2
  definition, 12.1
  design concepts and goals, 12.3–12.4
  general recommendations, 12.9–12.11
    pump stations, 12.10
    raw water intake stations, 12.11
    water reservoirs and elevated tanks, 12.9
    water treatment stations, 12.10
Intrusion detection sensor categories, 12.4, 12.4*t*
  exterior sensor types, 12.5
  fence-mounted cabling sensors, 12.5
  freestanding exterior sensors, 12.6
  interior boundary penetration sensors, 12.8
  interior sensor types, 12.7
  intrusion detection technologies, 12.4*t*
  issues to consider, 12.1–12.2
  performance characteristics, 12.2–12.3
    false alarm rate, 12.3
    probability of detection, 12.3
    vulnerability to defeat, 12.3
  video motion detection, 12.9
  wireless sensors, 12.9
Intrusion detection sensor categories, 12.4, 12.4*t*
  exterior sensor types, 12.5
  fence-mounted cabling sensors, 12.5
  freestanding exterior sensors, 12.6
  interior boundary penetration sensors, 12.8
  interior sensor types, 12.7
Islands of automation, 5.3
Isolation valves, 13.26
  butterfly valves, 13.26
  gate valves, 13.26

Key locks, 12.17
Key loggers, 5.13

Layered security systems, 12.16–12.17
Lead and copper rule, 14.16
Leakage, 4.22*t*
Lectin, 18.9
Legionella, 2.7
Linear beam sensors, 12.8
Lipopolysaccharide (LPS), 18.8
Locking systems, 12.17–12.18
  electrical keypads, 12.18
  electrified locking systems, 12.18
  fail-safe locking systems, 12.19
  fail-secure locking systems, 12.19
  key locks, 12.17
  mechanical keypads, 12.18
Logic trees, 7.1
Long Term Enhanced Surface Water Treatment rule (LT2), 18.3

Magnetic field sensors, 12.5
Mail bombs, 5.13
Manhattan Project, 17.2
Manning's equation, 4.8
Manning's roughness factor, 4.8
Maximum contaminant level, 11.29
Maximum spread potential, 15.8
Microbiological pathogen detection, 18.2–18.12
  biotection process, 18.6*f*
  current methods, 18.5–18.12
    alternative whole cell capture methods, 18.8–18.10
    antibody separation techniques, 18.7–18.8
    biodetection enabling analyte delivery system (BEADS), 18.7, 18.7*f*
    filtration, 18.5–18.7
    immunofluorescent staining (IMS), 18.7
    immuno separation methods, 18.8

Microbiological pathogen detection (*Conti.*):
    large volume concentration methods, 18.5–18.7
    multiplexed detection, 18.10–18.12
    nucleic isolation, 18.10
    sample processing techniques, 18.7
  current state of the art of detection, 18.2–18.4
    shortcomings of methods, 18.3–18.4
  desired capabilities of detection, 18.4–18.5
    ultimate goal, 18.5
    ultimate utility, 18.4–18.5
Micro fluidic automated sample, 18.8
Microsporidia, 2.11
Microwave sensors, 12.7
MIP polymerization, 18.14
Multiplexed detection, 18.10–18.12
  DNA microarrays, 18.10
    conceptual illustration of, 18.11*f*
  DNA suspension arrays, 18.10
  of *E. Coli*, 18.11*f*
Minor losses, 4.9
Mobile water desalination and purification systems, 17.9–17.10
Model calibration, 4.11
Molecular weight cutoff (MWCO), 18.6
Monitor/control systems for surveillance, 12.27–12.31
  access control applications, 12.29–12.31
    CCTV integration, 12.30
    email alarm notification, 12.30
    graphical map tool, 12.30
    threat level adjustments, 12.30
    video badging, 12.20
  digital video management, 12.27–12.28
    alarm review utility, 12.28
    camera control, 12.27
    live viewing module, 12.27
    recording setup, 12.27
    searching recorded video, 12.27
    web browser application, 12.28
  intelligent video applications, 12.28–12.29
    access control authentication, 12.28
    behavior tracking, 12.29
    digital video motion detection (DVMD), 12.29
    facial character recognition, 12.28
    object/target tracking, 12.28
Monitoring, 4.20, 9.4–9.10, 17.7
  contingency planning, 17.7

Monitoring methods, 11.7–11.14
  biological monitoring (bioalarms), 11.13–11.14
  chemical analysis, 11.8–11.12
    chemical and radioactive constituents, methods for detecting, 11.9*t*, 11.10*t*
    dissolved oxygen, 11.8
    general organic chemical parameters, 11.11
    ion specific electrodes for monitoring raw water, 11.11*t*
    metals, 11.8
    nitrate and ammonia, 11.8
    oil and petroleum, 11.11
    on-line analytical probes, 11.8
    organic chemicals, 11.12
    oxidant demand and oxidant residual, 11.11
    pesticides, 11.12
    radioactivity, 11.12
  emerging monitoring methods, 11.14
  microbiological analysis, 11.12–11.13
  physical analysis, 11.7–11.8
Monitoring systems, 9.4–9.10
Monitor clearwells, 12.10
Monitor hatches, 12.10
Monitor ladder, 12.10
Monitor vaults, 12.10
Monte Carlo simulation, 10.49, 11.28
Multiparameter microchip sensors, 14.12

Naegleria, 2.11–2.12
Nagasaki, Japan, 17.2
National Center for Homeland Security research (NCHSR), 8.1
National Rural Water Association (NRWA), 3.2, 3.3, 3.6, 3.9, 3.15–3.27
  NRWA/ASDWA Guide for Security decisions, 3.15–3.27
Network characterization, 4.9
Network hydraulic modeling, 4.6–4.11, 13.3–13.25, 14.18, 15.3–15.6
  connectivity, 15.6
  conservation of mass, 4.6
  Darcy-Weisbach equation, 4.7
  demand-driven analysis, 13.2–13.8
  extended period simulation (EPS), 4.8
  friction, 4.7
  Hazen-Williams equation, 4.7
  Hazen-Williams roughness coefficient, 4.7
  head driven analysis, 13.6–13.8

hydraulic network solver formulation, 15.4–15.5
manning's equation, 4.8
manning's roughness factor, 4.8
methods of analysis, 4.8
  dynamic analysis, 4.8
  Hardy Cross, 4.8
  Newton-Rhapson method, 4.8
  steady-state analysis, 4.8
minor losses, 4.9
model application, 4.9–4.10
model calibration, 4.11
modeling temporal variations, 4.10–4.11
network characterization, 4.9
Newton's second law, 4.6
nodal demand estimation, 15.5–15.6
node flow analysis, 13.6
pressure-driven analysis, 13.2–13.8
Reynold's number, 4.7
role for contamination response, 15.3–15.4
semi-pressure driven analysis, 13.8–13.25
transient flows, 4.6
Network (hydraulic) redundancy, 15.1–15.2
Newton's second law, 4.6
N. fowleri, 2.11
Norovirus, 2.9
NRDAM, 11.22*t*
NRWA/ASDWA Guide for Security Decisions, 3.15–3.27
  general questions:
    distribution, 3.22
    entire water system, 3.15–3.18
    information storage, computers, controls, and maps, 3.24–3.25
    personnel, 3.23–3.24
    public relations, 3.26–3.27
    treatment plant and suppliers, 3.20–3.21
    water sources, 3.19
Nuclear contingencies, contamination of water, 17.6
Nuclear risks, 17.5
  China, 17.5
  Iran, 17.5
  North Korea, 17.5
  Pakistan, 17.5
  Russia, 17.5
Nucleic isolation, 18.10
Nukers, 5.13

optiMQ, 16.4–16.12
  definitions, 16.4–16.5
  domain of detection of a node, 16.5
  domain of pollution event, 16.4
  level of service, 16.4
  pollution event,
  redundancy, 16.5
  example application, 16.7–16.12
  methodology, 16.5–16.6
    detection likelihood, 16.6
    genetic algorithm (GA) parameters, 16.6*t*
    randomized pollution matrix construction, 16.5
    randomized row matching using genetic algorithm, 16.5
  optiMQ program, 16.6–16.7
    main screen, 16.7*f*

Passive infrared sensors, 12.6–12.7
Password crackers, 5.13
Path enumeration method, 7.11–7.15
  cut set, 7.11
  cut-set analysis, 7.11–7.13
  minimum cut sets, 7.12
  tie-set analysis, 7.13–7.15
Parasitic pathogens, 2.9–2.12
  acanthamoeba, 2.10
  cryptosporidium parvum, 2.10, 18.3
  entamoeba histolytica, 2.11
  microsporidia, 2.11
  naegleria, 2.11–2.12
Pathogens, 2.2*t*
  also see microbiological pathogens
Perimeter detection, 12.9
  elevated tanks, 12.9
Pump stations, 12.10
  raw water intake station, 12.11
  water treatment station, 12.10
Pesticides, 4.5
Photoelectric beam, 12.8
Photoelectric eye, 12.8
Physical threats, 1.6, 4.4, 8.3
Physiological effects, whole-body radiation, 17.4*t*
PL 107–188 (*see* Bioterrorism Act of 2002)
Plug flow, 4.18
Poisson rectangular pulse (PRP) model, 4.22–4.35
  premise, 4.22–4.25
  ported-coaxial buried cable sensor, 12.5
  definition sketch, 4.23*f*
  hourly patterns, 4.24*f*

Poisson rectangular pulse (PRP) model (*Conti.*):
  parameter estimation, 4.26–4.28
  parameters for, 4.23, 4.24*t*
  PRP example application, 4.28–4.35
  PRP model for demand simulation, 4.25–4.26
  PRPsym code, 4.25
Powdered activated carbon, 11.6
Population exposure, 11.29
Presidential Decision Directive (PDD) 61, 5.6
Presidential Decision Directive (PDD) 63, 3.7, 9.1
Presidents Commission on Critical Infrastructure Protection (PCCIP), 1.8
PRPsym code, 4.25–4.28
Public Health, Security, and Bioterrorism preparedness and Response Act (*see* Bioterrorism Act of 2002)

QUALNET, 4.13, 4.17–4.18

Rad, 17.3
Radiation, physiological effects, 17.4*t*
Radiation sickness, 17.4*t*
Radioactive contamination, 17.2–17.5
  Chernobyl, 17.3–17.5
  dangers of, 17.2–17.5
  Hiroshima, 17.2–17.3
  Nagasaki, 17.2–17.3
  Three Mile Island, 17.3–17.4
Radioactive fallout, 17.6
Rainwater contamination from radioactive material, 17.6
RAM-W^{SM}, 3.2, 5.9
Real-time monitoring, 17.7
  contingency planning, 17.7
Reliability assessment, 7.1–7.23
  logic trees, 7.1
  methods for, 7.9–7.22
    conditional probability approach, 7.15–7.17
    fault tree analysis, 7.17–7.22
    path enumeration method, 7.11–7.15
    state enumeration method, 7.9–7.11
  system reliability, 7.3–7.22
    bounds for, 7.6–7.9
    probability rules for, 7.3–7.6
  procedure, 7.2*f*
Reliability model, semi-pressure driven method, 13.25

Remote monitoring and control technology (RMCT), 14.1–14.20
  advanced remote monitoring devices, 14.11–14.12
    advanced spectrophotometers, 14.12
    biosentinels, 14.12
    multiparameter microchip sensors, 14.12
  engineering and designing, 14.3–14.4
  implementation, case studies, 14.5–14.11
    battelle project, 14.5–14.6
    T & E RMCT network, 14.11
    Washington, D.C. remote monitoring network, 14.6–14.7
    West Virginia RMCT network, 14.7–14.11
  lessons learned, effective implementation 14.12–14.13
    small-to-medium RMCT implementations, 14.12
    medium-to-large RMCT implementations, 14.13
    sensor device selection and implementation, 14.13
  SCADA system selection, 14.4–14.5
    advanced features, 14.4
    basic features, 14.4
  system components, 14.2–14.3
    communication media, 14.3
    control devices, 14.2
    online sampling, 14.2
    SCADA systems, 14.3
Remote monitoring devices (advanced), 14.11–14.12
  advanced spectrophotometers, 14.12
  biosentinels, 14.12
  multiparameter microchip sensors, 14.12
Residential indoor water use, 4.21*t*
Response planning, AMSA approach, 3.5*f*
Response, recovery, and remediation guidelines for, 3.28–3.37
  contamination events:
    articulated threat with unspecified material, 3.28–3.29
    major event, 3.29–3.30
    intrusion through SCADA, 3.32–3.33
    potential water contamination, 3.31–3.32
    structural damage, 3.33–3.34
Restricted-access material (RAM), 18.14
Reverse osmosis (RO), 17.9–17.10
  removal rates, 17.10*t*
Rings of defense, 5.10

Risk assessment methodology, 6.1–6.23
  agents, 6.2–6.7
    biological agents, 6.2–6.6
    chemical agents, 6.6–6.7
    threats posed by genetic engineering, 6.6
  efficient hydrologic tracer-test design (EHTD), 6.10–6.20
Risk assessment methodology
  basic design of, 6.10–6.11
  experimental example, 6.13–6.19
  range of capabilities, 6.11
  toxic release outcome prediction, 6.11–6.13
  experimental example, 6.19
  source-water protection, 6.7, 6.10
  toxic release outcome prediction, 6.11–6.13
Risk-based analysis for early warning systems 11.27–11.29
  risk-based decision making process for EWS, 11.30f
  spill risk model, 11.28–11.29
    application to Ohio River, 11.29
    model formulation, 11.28
    Monte Carlo simulation, 11.28
Risk-based decision making process for EWS, 11.30f
River spill models, see spill models
River oil spill models, summary of, 11.22t
  NRDAM, 11.22t
  RiverSpill, 11.22t
  ROSS, 11.22t
  ROSS2, 11.22t
  ROSS3, 11.22t
  WPMB, 11.22t
RMCT system components, 14.2–14.3
  communication media, 14.3
  control devices, 14.2
  online sampling, 14.2
  SCADA systems, 14.3
ROSS, 11.22t
  ROSS2, 11.22t
  ROSS3, 11.22t
Roraviruses, 2.9

Safe Water Drinking Act (SWDA) of 1974, 4.1, 4.20, 8.2, 14.16–14.17, 18.2
  amendments of 1986 (SWDAA), 14.16
  lead and copper rule (LCR), 14.16
  routine monitoring, 4.20
  trihalomethane rule, 14.16

Salmonella, 2.2t, 2.3, 8.4, 9.11, 9.13–9.23, 10.3–10.9, 18.3, 18.9, 18.11
  outbreak of in Gideon, MO, 9.13–9.23, 10.3–10.9
SANS Institute, 5.2, 5.15
  computer security incident handling guide, 5.15
SCADA encryption initiative, 5.6–5.8
SCADA fault trees, 5.9
SCADA system selection for RMCT, 14.4–14.5
  advanced features, 14.4
  basic features, 14.4
Scanners, 5.12
Security hardware and surveillance systems, 12.1–12.37
  access control, 12.16–12.26
    access control attributes, 12.17
    building access control concepts, 12.25
    card acess technologies, 12.21
    card reader systems, 12.19–12.21
    clear-zone areas, 12.23
    delivery access control, 12.24
    effective site access control, 12.23
    guard tour system, 12.21
    interior access control concepts, 12.25
    layered access control example, 12.26t
    layered security systems, 12.16–12.17
    locking systems, 12.17–12.18
    site perimeter access control, 12.21–12.23
    standoff distance, 12.23
    vehicle access control barriers, 12.24
    vehicle checkpoints, 12.24
  closed circuit television (CCTV), 12.11–12.16
    system components, 12.12–12.16
    special considerations, 12.16
  design and implementation considerations, 12.34–12.37
    access control system (ACS), 12.35
    closed circuit television (CCTV) system, 12.34
    central security areas, considerations, 12.36
  intrusion detection, 12.1–12.11
    alarm activation time, 12.2
    definition, 12.1
    design concepts and goals, 12.3–12.4
    general recommendations, 12.9–12.11
    intrusion detection sensor categories, 12.4, 12.4t

Security hardware and surveillance systems (*Conti.*):
  intrusion detection technologies, 12.4*t*
  issues to consider, 12.1–12.2
  performance characteristics, 12.2–12.3
  video motion detection, 12.9
  wireless sensors, 12.9
  monitor/control systems, 12.27–12.31
    access control applications, 12.29–12.31
    digital video management, 12.27–12.28
    intelligent video applications, 12.28–12.29
  special application, security system components, 12.31–12.33
    camera limitations, 12.32–12.33
    temporary devices, 12.31
Security self assessment guide, 3.2, 3.15–3.37
Security tools for vulnerability assessment, 3.2
  AMSA, 3.5*f*
  ASDWA, 3.2
  ASSET, 3.2
  NRWA, 3.2
  RAM-W^{SM}, 3.2
  VSAT, 3.2–3.3, 3.4*f*, 3.5
Security upgrades, 3.14
Sedimentation, 11.6
Seismic sensors, 12.5
Semi-pressure driven analysis, 13.8–13.25
  application of, 13.10–13.11
  available demand fraction (ADF), 13.10
    system wide ADF, 13.10
  EPS application, 13.15–13.25
  flowchart of algorithm, 13.9
  steady-state application, 13.11–13.13
  what-if scenarios, 13.13–13.15
Sensors, see intrusion detection
Sensor network protocol, 5.6
S. flexneri, 18.9
Shigella, 2.3–2.4, 18.11
Site perimeter access control, 12.21–12.23
  clear-zone areas, 12.23
  delivery access points, 12.24
  standoff distance, 12.23
  vehicle access control barriers, 12.24
  vehicle checkpoints, 12.24
Sniffer programs, 5.2
Social engineering, 5.2
Source-water protection, 6.7
Spill models, 11.15–11.22
  components of, 11.15
  fate module, 11.15
  flow module, 11.15
  river oil spill models, summary of, 11.22*t*
    NRDAM, 11.22*t*
    RiverSpill, 11.22*t*
    ROSS, 11.22*t*
    ROSS2, 11.22*t*
    ROSS3, 11.22*t*
    WPMB, 11.22*t*
  selected models, 11.16–11.22
    oil spill models, 11.21
    REMM model, 11.20
    Rhine Alarm-model, 11.18
    riverine spill modeling (RSMS), 11.16
    river oil spill models, summary of, 11.22*t*
    RiverSpill, 11.21
    R-TOT model, 11.10
  spill risk model, 11.28–11.29
    application to Ohio River, 11.29
    model formulation, 11.28
    Monte Carlo simulation, 11.28
  water quality transport module, 11.15
Spill risk model, 11.28–11.29
  application to Ohio River, 11.29
  model formulation, 11.28
  Monte Carlo simulation, 11.28
Spirilla, 2.3
Spread potential, 15.8
State enumeration method, 7.9–7.11
Supervisory control and data acquisition (SCADA), 1.6, 3.7, 3.12, 3.13, 3.34–3.35, 3.41, 5.2–5.3, 5.6–5.12, 14.2–14.13, 15.2
  communication protocols, 5.6
  encryption initiative, 5.6–5.8
  intrusion, 3.34–3.35, 5.2–5.3
  remote monitoring and control technology (RMCT), 14.2–14.13
    SCADA component of, 14.2–14.3
  security of, 5.9–5.10
    RAM-W, 5.9
    rings of defense, 5.10
    twenty-one step guide, 5.9–5.10
  SCADA system selection for RMCT, 14.4–14.5
    advanced features, 14.4
    basic features, 14.4
Surface Water Treatment Rule (1989), 4.1

Surveillance systems, see security hardware, 12.1–12.37
Synthetic organic chemicals, 11.6
System Administration, Networking and Security (SANS) Institute, 5.2, 5.15
  computer security incident handling guide, 5.15
System reliability methods, 7.9–7.22
  conditional probability approach, 7.15–7.17
  event-tree analysis, 7.9
  event trees, 7.9, 7.10f
  failure-mode approach, 7.3
  fault tree analysis, 7.17–7.22
    advantages of, 7.17
    disadvantages, 7.17
    example fault tree, 7.18f
    objective of, 7.17
    tree symbols, 7.19t
  path enumeration method, 7.11–7.15
    cut set, 7.11
    cut-set analysis, 7.11–7.13
    minimum cut sets, 7.12
    tie-set analysis, 7.13–7.15
  state enumeration method, 7.9–7.11
  survival-mode approach, 7.3
System unreliability, 7.13
System upgrades, 3.14

Taut-wire fence detection system, 12.6
Threat level adjustments, 12.30
Threats, 1.6–1.7
Three Mile Island, 17.3
Tools for cyber attacks, 5.12–5.14
  BIOS crackers, 5.13
  Hackers' Swiss knife, 5.13
  ICH hacking tools, 5.13
  key loggers, 5.13
  mail bombs, 5.13
  nukers, 5.13
  password crackers, 5.13
  scanners, 5.12
  Trojan horse, 5.12
  viruses, 5.12
  Windows NT hacking tools, 5.13
  worms, 5.12
Total Coliform Rule (1989), 4.1
Toxicity (relative) of poisons, 8.5t
Total trihalomethane formation potential (TTHMFP), 9.11

Toxic release outcome prediction, 6.11–6.13
Trihalomethane formation, 9.11
Tracers, 4.36–4.37, 4.38t
  calcium chloride, 4.36
  flouride, 4.36
  lithium chloride, 4.37
  sodium chloride, 4.37
Tracer study, 4.35–4.45
  definition, 4.35
  example studies, 4.39–4.45
    distribution system tracer study, 4.41–4.45
    tank tracer study, 4.39–4.41
  procedures for, 4.36–4.39
    injection procedures, 4.37
    monitoring program, 4.37, 4.39
    system operation, 4.39
    tracer selection, 4.36–4.37
Tracer testing (quantitative), 6.7, 6.10
Transient flows, 4.6
Trojan horse, 5.12
Typhoid fever, 2.3

Ultra filtration package plant, 14.6
Ultrasonic sensors, 12.7
U.S. EPA method 1623, 18.3
U.S. EPA's protocol, 1.10
U.S. EPA Water Protection Task Force (WPTF), 14.1
Utility composite network infrastructure, 5.5t
Utility information flow requirements, 5.4t
Utility information technology infrastructure, 5.2, 5.4t
  computerized maintenance management system, 5.2
  human resources, 5.2
  laboratory information management system, 5.2
  supervisory control and data acquisition system, 5.2

Valve location, optimization problem, 13.27–13.29
Valving, effect on reliability, 13.26–13.29
Vehicle access control barriers, 12.24
Vehicle checkpoints, 12.24
Vibrio, 2.6
Vibrio cholerae, 2.6, 6.6, 6.14–6.19
  application of risk methodology, 6.14–6.19
Video badging, 12.20

Video motion detection, 12.9
Viral pathogens, 2.7–2.9
  adenovirus, 2.7–2.8
  astroviruses, 2.8
  hepatitis A, 2.8
  hepatitis E, 2.8–2.9
  Norovirus, 2.9
  roraviruses, 2.9
Viruses (computer), 5.12
Volumetric detection, pump stations, 12.10
Raw water intake stations, 12.11
Water treatment stations, 12.10
VSAT, 3.2–3.3, 3.4f, 3.5
Vulnerabilities:
  of computer system infrastructure, 5.8–5.10
  of water supply systems, 4.3–4.5, 8.3
Vulnerability assessment, 3.1–3.2, 5.5–5.10
  AMSA asset-based approach, 3.5f
  common elements, 3.1–3.2, 3.11–3.14
  computer system infrastructure, 5.8–5.10
  facility vulnerability assessment and improvement identification, 3.4t
  points to consider, 3.11–3.14
  response, recovery, and remediation guidelines. 3.28–3.37
  security tools for, 3.2
    AMSA, 3.5f
    ASDWA, 3.2
    ASSET, 3.2
    NRWA, 3.2
    RAM-W^{SM}, 3.2
    VSAT, 3.2–3.3, 3.4f, 3.5
  self-assessment guide, 3.6
  small water systems, 3.15–3.27
  utility guide for security decision making, 3.3f

Water age, 4.19
WaterCAD model, 10.10
Water demands, 4.5–4.6
  baseline demands, 4.5
  demand multipliers, 4.5
  fire demands, 4.5
  projecting future demands, 4.5
  time varying demands, 4.5
Water demand simulation, 4.21–4.35
  assigning water demands, 4.21–4.22
  poisson rectangular pulse (PRP) premise, 4.22–4.25

definition sketch, 4.23f
hourly patterns, 4.24f
parameter estimation, 4.26–4.28
parameters for, 4.23, 4.24t
PRP example application, 4.28–4.35
PRP model for demand simulation, 4.25–4.26
PRPsym code, 4.25–4.28
residential indoor water use, 4.21t
uniform leakage, 4.22
Water distribution system (*see* Water supply systems)
Water ISAC, 3.7, 3.9–3.10t, 3.13, 3.40
  website locations, 3.9–3.10t
Water quality modeling, 4.11–4.18, 9.5–9.8, 9.11–9.23, 10.48–10.52, 14.17–14.19, 15.3–15.6
  applications of, 10.48–10.52, 14.19
  case studies, 9.11–9.23
    Gideon waterborne outbreak, 9.13–9.23
    North Marin Water Authority, 9.11–9.13
  connectivity and contamination spread potential, 15.6
  conservation of mass, 4.12, 9.5–9.8
    storage tanks, 4.12, 9.6–9.8
    substance concentrations, 4.12, 9.5–9.8
  current trends, 9.23–9.24
  dynamic water quality models, 14.19
  EPANET, 4.12–4.17, 9.7–9.10
    convective transport model, 4.17
    hydraulic simulation equations, 4.13–4.14
    other features, 4.16, 9.10
      travel time to a node, 4.16
      source tracing, 4.16
      water age, 4.16
    reaction rate model, 4.16, 9.9–9.10
      bulk phase reaction, 4.16, 9.9
      wall reaction, 4.16, 9.9–9.10
    water quality simulation equations, 4.15, 9.8–9.9
  network water quality, 14.17–14.18
  QUALNET, 4.13, 4.17–4.18
  role of, 15.3–15.4
  solution methods, 4.13
    Eulerian discrete-volume method (DVM), 4.13
    Eulerian finite-difference method (FDM), 4.13
    Lagrangian event-driven (EDM), 4.13

Lagrangian time-driven (TDM), 4.13
steady state water quality models, 14.18
storage effects, 4.18, 9.23–9.24
  complete mixing, 4.18
  mixing in storage tanks, 9.23–9.24
  plug flow, 4.18
tank effects, 4.18
  compartment modeling, 4.18
  nonuniform mixing, 4.16
water demand simulation, 4.21–4.35
Water quality simulation, 9.8–9.10
Water Sector Critical Infrastructure Advisory Group, 1.8
Water Supply Simulation Model (WSSM), 14.18
Water supply systems:
  building blocks of, 1.4*f*
  common elements of, 4.3, 9.2
  components of, 1.4*f*
  demands, 4.5–4.6
  diagram of, 1.3*f*
  features of, 4.2
  functions of, 4.2
  points of vulnerability, 4.3–4.5
Water use, 4.20–4.21
  residential indoor, 4.21*t*
Water use model, 4.6
Whole-body radiation, physiological effects, 17.4*t*
Windows NT hacking tools, 5.13
Wireless sensors, 12.9
World War II, 17.2
Worms (computer), 5.12
WPMB, 11.22*t*

Yersinia, 2.5
Yersinia enterocolitica, 18.9
Yersinia pseudotuberculosisis, 2.5

ABOUT THE EDITOR

Larry. W. Mays, Ph.D., P.E., P.H., is Professor of Civil and Environmental Engineering at Arizona State University and former chair of the department. He is the former Director of the Center for Research in Water Resources at the University of Texas at Austin, where he also held an Engineering Foundation Endowed Professorship. A registered professional engineer in seven states and a registered professional hydrologist, he has served as a consultant to many organizations. Professor Mays is the author of *Water Resources Engineering* (John Wiley & Sons, Inc.) and *Optimal Control of Hydrosystems* (Marcel Dekker, Inc.), coauthor of *Applied Hydrology* and *Hydrosystems Engineering and Management* (both from McGraw-Hill), coauthor of Groundwater Hydrology (John Wiley & Sons, Inc.), and editor-in-chief of the *Water Resources Handbook, Hydraulic Design Handbook, Water Distribution Systems Handbook, Stormwater Collection System Design Handbook, Urban Water Supply Handbook, Urban Water Supply Management Tools*, and *Urban Stormwater Management Tools* (all from McGraw-Hill). He was also the editor-in-chief of the *Reliability Analysis of Water Distribution Systems* (ASCE) and coeditor of *Computer Modeling of Free Surface and Pressurized Flow* (Kluwer Academic Publishers). The *Urban Water Supply Handbook* received the 2002 Honorable Mention in Engineering Award given by the Association of American Publishers. Among his honors include a distinguished alumnus award from the University of Illinois at Urbana-Champaign in 1999.